Fundamentals of EMI Phenomena in Radio Communication Systems

Miroslav Pajovic

Fundamentals of EMI Phenomena in Radio Communication Systems

Theories and Solution Methods

Springer

Miroslav Pajovic
San Jose, CA, USA

ISBN 978-3-031-88747-5 ISBN 978-3-031-88748-2 (eBook)
https://doi.org/10.1007/978-3-031-88748-2

This Springer imprint is published by the registered company Springer Nature Switzerland AG
The registered company address is: Gewerbestrasse 11, 6330 Cham, Switzerland

Preface

This book provides an overview of the most EMI-related topics with electromagnetic principles and solutions for better understanding and effectively controlling the adverse effects of electromagnetic interference (EMI) in modern radio communication systems. To be more expected in the future, EMI has become an alarming problem. The rapid growth of the number and complexity of radio wireless and IT systems in our environment has caused increasing EMI emissions, which usually results in performance degradation and operation limitations of radio communication and IT systems. This is especially critical in fast-growing 4G-LTE and 5G-NR mobile cellular networks, including WLAN, WIFI, machine-to-machine and Internet of Things (IoT) networks, among other wireless radio networks.

This engineering book can be helpful to technical personnel, who are involved in planning, designing, or maintaining radio equipment and systems and whose expertise is not in the EMI engineering area. The book can also benefit university students, practicing engineers, and others interested in discovering more about the origins and effects of EMI phenomena in today's radio communication systems.

Historically, EMI or radio frequency interference (RFI) has been an old EMI phenomenon that has existed for over a century, beginning with the use of radio electromagnetic waves as a communication medium. Later, with the invention and massive use of digital and computer equipment, power electronic circuits, switching power supplies, extended power and signal transmission lines, and various noisy electrical machinery and tools, so-called unintentional (or accidental) man-made EMIs were created. In addition to the manmade EMI, other unintentional interferences exist, such as *natural electrostatic* and *EMI* caused by *lightning*, *solar*, and other *cosmic* radiation.

Regarding where EMI exists and acts, we can consider two levels of EMI emissions: the circuit level (or the print circuit board level) and the system-level EMI. The primary scope and focus of this book are on the system-level EMI, whereas the circuit EMI level is considered well in more published professional and textbooks and articles.

This book is organized into four parts and 13 chapters to provide more profound insight into the mechanism of EMI phenomena in radio wireless systems.

Appendixes A, B, and C have been added at the book end, which reviews fundamental electromagnetic principles and laws relevant to understanding the origins and effects of EMI phenomena in radio systems. Appendix A provides the physics of constitutional electromagnetic laws (Faraday's, Ampere's, Gauss's, Ohm's, and others), governed by Maxwell's equations. Appendix B provides an overview of electromagnetic waves and radiators in open space, whereas Appendix C describes the fundamental physics of guided waves by transmission lines and waveguides.

Part I, titled the *Introduction to EMI Phenomena,* contains Chapters "General Concept and Types of EMI Emissions", "Digital and Pulse Signals, Random Variables, and Stochastic Processes", and "Nonideal Passive Circuit Components". Chapter "General Concept and Types of EMI Emissions" presents and describes the general concepts and types of EMI emissions, while Chapter "Digital and Pulse Signals, Random Variables, and Stochastic Processes" presents the fundamentals of *digital signals and spectra, random variables, and stochastic processes,* which are essential for studying and predicting EMI phenomena in radio communication systems. Chapter "Nonideal Passive Circuit Components" describes the nature of *non-perfect (realistic)* conductors and circuit components, which are directly linked to EMI emissions.

Part II, titled *Radio Communication Systems Concerning EMI Phenomena*, contains Chapters "Introduction to Digital Communication Systems", "Performance of Radio Digital Systems in Presence of EMI", "Overview of Broadband Radio Systems with High EMI Immunity", "Overview of 4G-LTE and 5G-NR Mobile Radio Systems from EMI Prospect", and "Performance of RF Cables and IT Signals Lines in Radio Systems with EMI". Chapter "Introduction to Digital Communication Systems" is an introduction to *digital communications,* while Chapter "Performance of Radio Digital Systems in Presence of EMI" describes the performance of radio digital systems in the presence of fadings and EMI. Chapter "Overview of Broadband Radio Systems with High EMI Immunity" overviews the concepts of broadband, multicarrier, and multiuser systems with high immunity to EMI, such as OFDM, Spread Spectrum techniques (CDMA, DDSS, FHSS), MIMO systems, and Adaptive Software Defined Radio (ASDR). Chapter "Overview of 4G-LTE and 5G-NR Mobile Radio Systems from EMI Prospect" provides an overview of the 4G-LTE and 5G-NR mobile cellular networks from the EMI perspective. At the end of Part II, Chapter "Performance of RF Cables and IT Signals Lines in Radio Systems with EMI" presents the performance of RF signal cables and lines as significant sources and victims of the EMI in radio communication systems.

Part III, titled *Conventional Methods for EMI Control in Radio Systems,* contains Chapters "EMI Filters: Theory and Design", "EMI Shielding for Radio Equipment", and "Grounding for Radio Systems and Facilities". Chapter "EMI Filters: Theory and Design" considers the concepts and applications of EMI filters, Chapter "EMI Shielding for Radio Equipment" describes *EMI shielding* for radio equipment, and Chapter "Grounding for Radio Systems and Facilities" presents *grounding systems* for radio equipment and facilities.

Part IV, titled *Natural Sources of EMI and Electromagnetic Disturbances,* contains Chapters "Physics of Lightning and Its Impact on Radio Systems on Earth" and "Solar and Other Cosmic Radiations with Effects on HF Radio and Satellite Data Communications", which, respectively, describe the electrical nature of lightning and *solar* and other cosmic EMI radiations that can severely affect the operation of terrestrial and satellite radio systems.

Appendixes—*Electromagnetic Fundamentals Essential to EMI Phenomena,* which include *Appendixes A, B, and C* titled, respectively: *Origins of Electric and Magnetic Fields Governed by Maxwell's Equations; Electromagnetic Radiators and Waves in Free and Open Space; and Electromagnetic Waves Guided by TLs and Waveguides.*

At the end, after each chapter, the related bibliographies (references), which are motivated and influence this book writing, are listed and recommended to readers, who are interested in future studies of EMI phenomena in modern radio communication systems.

San Jose, CA, USA

Miroslav Pajovic
pmiroslav@ieee.org

Acknowledgment

I am deeply grateful to my colleagues: *Prof. Dr. Dragan Olcan*, Faculty of Electrical Engineering, University of Belgrade, Serbia; *Pelin Salem, PhD*, Scientist, Wireless Compliance Department, Cisco Systems, Inc., USA; *Karen Burnham*, President and Chief Engineer, EMC United, Inc., USA; *Nicole Pothier McGillivray*, Owner and Chief Engineer, Carrington Shield, LLC, USA; *Prof. Dr. Mohamed Salem*, Electrical Engineering School, Sonoma State University, USA; and *John Savic*, Director, Wireless Compliance Department, Cisco Systems, Inc., USA; for their great contributions to the quality of this book.

I would also like to thank my colleague and friend *Zlatko Potocnik* from Belgrade, Serbia, who was enormously helping with the preparation of the book's graphical illustrations, which are among the substantial parts of this book.

Finally, I wish to thank my professors, colleagues, and friends from *Faculty of Electrical Engineering* and *Academic Radio Club* (YU1EXY), University of Belgrade; *SMIP Communication Department* (SV), Belgrade; and *Cisco Systems, Inc.*, Silicon Valley, CA, USA; where I spent many years of my time. The collaboration with these distinguished institutions and individuals has provide me valuable knowledge and practical experience in the areas of Radio communications, Antenna designs, and EMI/EMC engineering.

San Jose, CA, USA
December 2025

Miroslav Pajovic
pmiroslav@ieee.org

Competing Interests The author has no competing interests to declare that are relevant to the content of this manuscript.

Contents

Natural Sources of EMI and EM Disturbances

Introduction to Origins and Theory of EMI Phenomena

General Concept and Types of EMI Emissions

1 Introduction

Electromagnetic interferences (EMIs) are interfering electromagnetic fields and electric currents that interfere with the operation of electric and electronic equipment, degrading their performance. The nature and origins of EMI are described by the electromagnetic theories discussed in Appendices A and B, given at the end of this book. EMI emissions can be presented with a model consisting of an EMI *source* (or an EMI *emitter*), EMI *coupling path*, and EMI *receptor*, as illustrated by the block diagram in Fig. 1.

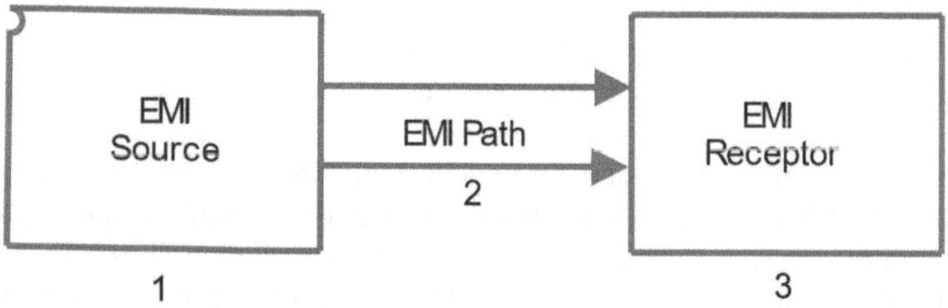

Fig. 1 General EMI emission concept model

Sources of EMI emissions can include *intentional* or *operating* radio and radar system emissions (also called radio frequency interference, RFI) and *unintentional EMI emissions for* various digital electronic and power electric equipment, power and signal lines, and machinery, as well as natural EMI sources such as lightning, electrostatic discharge (ESD), solar, and other cosmic emissions. Often, *unintentional* electromagnetic and radio frequency interference generally refers to

M. Pajovic, *Fundamentals of EMI Phenomena in Radio Communication Systems*, https://doi.org/10.1007/978-3-031-88748-2_1

electromagnetic interference (EMI), and this term is used in this book as in [1–4]. In some cases, the EMI from radio emission systems is called radio-frequency interference (RFI), as in [5].

The EMI may propagate and couple via various *intentional* and *unintentional* paths to a potential EMI *receptor.* When it has enough energy and certain *spectral content,* it is capable of causing undesired behavior or even malfunction of the receptor containing EMI-sensitive circuits, devices, and instruments. On the basis of the above-presented EMI model concept, we can build a global strategy to prevent, or at least minimize, the negative effects of EMI emissions on radio communication systems by (1) suppressing EMI emissions directly at the EMI *source*, (2) increasing the *resistance* of the EMI path, and (3) increasing the receptor's *immunity* to EMI emissions.

Hence, the first step is to avoid or minimize EMI effects at the EMI source. For example, a digital clock signal with fast *rise/fall* times is a significant EMI source because it has high-frequency and broadband harmonic content. With slower clock *rise/fall* times, the EMI frequency spectra narrow, leading to lower EMI emissions. However, the slower *rise/fall* times are limited because of the opposite system requirements regarding digital *signal quality* and *signal integrity,* which are primarily dictated by the speed of data transmission. Thus, system and circuit designers must make trade-offs regarding these opposite requirements. The effects of the digital signal *rise/fall* times on EMI emissions will be discussed later in chapter "Digital and Pulse Signals, Random Variables, and stochastic processes". Another common and traditional method of EMI control of EMI source emissions is electromagnetic shielding of the EMI source, which will be discussed in more detail in chapter "EMI Shielding for Radio Equipment".

In addition to the employment of EMI filters (discussed in chapter "EMI Filters: Theory and Design"), the placement and routing of signal lines parallel to a large conductive plane or ground is an effective method for reducing EMI-radiated emission from noisy signal lines and conductors as discussed in this chapter and chapter "Performance of RF Cables and IT Signals Lines in Radio systems with EMI". However, shielding of signal lines is the most effective method for signal lines, which are primary sources and "victims" of EMI. This topic will be considered in more detail in chapter "Performance of RF Cables and IT Signals Lines in Radio systems with EMI".

The writing of this chapter is motivated and influenced by references [1–14], which we recommend to readers interested in these topics.

2 Radiated and Conducted EMI Emissions

EMI coupling can occur via electromagnetic *radiation* and this type of EMI is called the *radiated* EMI. The radiated EMIs are categorized into far-field EMIs and near-field EMI (or cross-talk). Another form of EMIs coupling involves EMI *conductive* and *displacement* current flows via conductive or dielectric materials. This form of EMI is called *conducted* EMI. Both types of EMI are also considered well in [1–4]. Additionally, practical solutions for radiated and conducted EMI are presented well in [5].

Figures (2a, b) show the two EMI concepts, *radiated* and *conducted,* which include electronic device A, which contains an EMI source and unit B, which contains EMI-sensitive circuits.

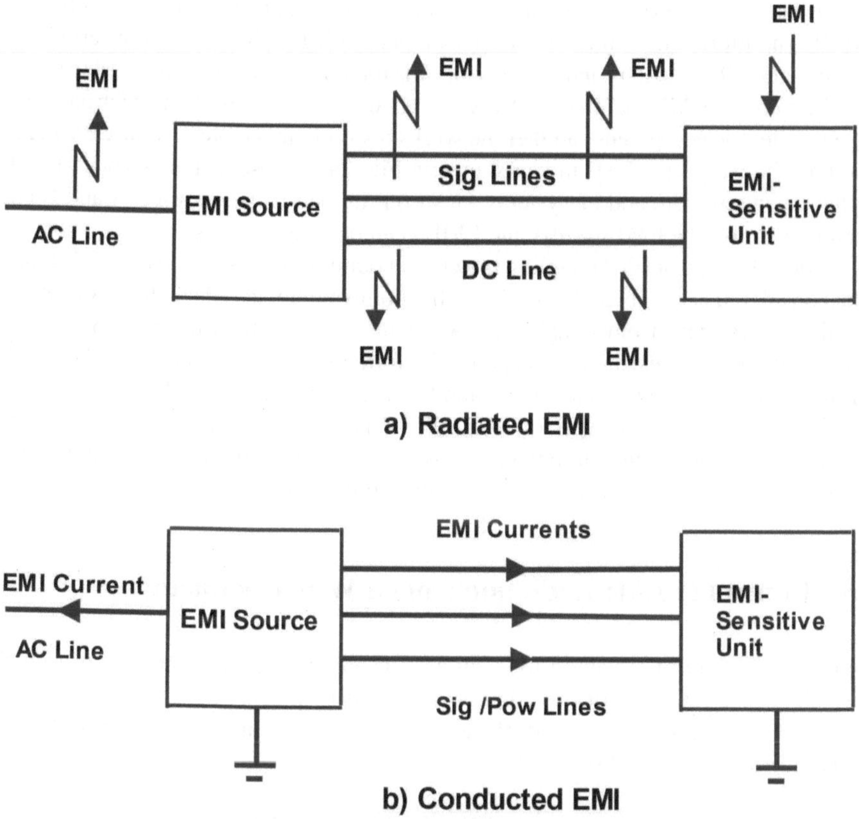

Fig. 2 EMI models: (**a**) radiated EMI; (**b**) conducted EMI

In Fig. 2a, the EMI source may radiate through various openings and holes on the device enclosure or over signal lines and AC/DC power lines, which interconnect devices A and B. In Fig. 2b, devices A and B are connected via power and signal lines that act as EMI paths between devices A and B, and over AC power lines into the AC power distribution network.

Hence, power cables and signal lines can be efficient EMI radiators, such as the long wire antennas discussed in Appendix B, whose radiation efficiency depends on the length and magnitude of the EMI current they carry. Therefore, the system cables (*signal* and *power*) should not be unnecessarily long because they can act as EMI wire radiators or as receptors for outside EMI. Accordingly, in preliminary system and equipment designs, system and equipment designers should indicate all system frequencies that may be critical in EMI terms, including the frequencies of potential EMI sources close to the system equipment.

Notably, AC cables can carry and *reradiate* higher-frequency EMI than can AC operating frequencies of 50 Hz or 60 Hz because high-frequency EMI can be coupled into cables from an EMI-contaminated environment. Additionally, induced high-frequency EMI currents at AC power lines and cords can flow into the power distribution net system containing many wires that may further re*radiate* EMI. Therefore, this is the case when *conducted* EMI emissions are converted into *radiated* EMI emissions and vice versa. This topic is also considered in [4, 6].

Therefore, EMI *radiated* and *conducted* emissions coexist in electronic and power electric equipment and their networks resulting in a more complex EMI situation. The simplified EMI model shown in Fig. 2a, b presents the essence of EMI phenomena in radio and IT systems. However, in the real world, units A and B can simultaneously be EMI *sources* and EMI *receptors* (i.e., victims).

Thus, the type of EMI coupling is a critical point of the system design and should be considered carefully and systematically. For example, to reduce the risk of eventual EMI problems from signal and power cables, various filtering and cable shielding methods are presented in chapters "Performance of RF Cables and IT Signals Lines in Radio systems with EMI" and "EMI Filters: Theory and Design". These mainly address the concepts of EMI filters and shielded cables, respectively. In terms of radiation from equipment enclosures, EMI shielding is presented and discussed in chapter "EMI Shielding for Radio Equipment".

3 Control of EMI Radiation from a Wire Conductor

One way to control radiated EMI emissions from a linear (wire) conductor (which usually acts as the EMI linear radiation source) is by placing the conductor parallel and next to a larger conducted plane or conducted ground. We demonstrate this method via the model shown in Fig. 3.

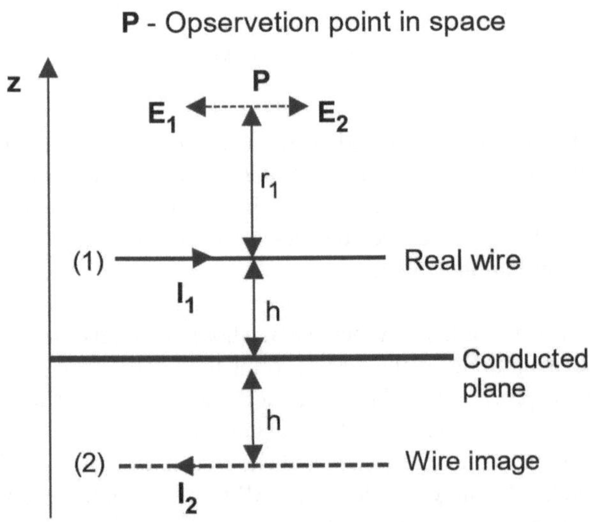

Fig. 3 Wire radiator (1) and its image (2) above and below a conducted plane

For the present model, the total maximal field $\vec{E}_t$ at point P in the +z-direction (normal to the conducted plane) is the vector sum of the *direct* field $\vec{E}_1$ from the real wire and the *reflected* field $\vec{E}_2$ from the conductive plane, or hypothetically from the wire *image* located below the plane at distance *h*. Hence, the sum of the field intensities E_1 and E_2, which propagate in the z-directon, is expressed in the complex-exponential form as

$$\tilde{E}_t = E_1\, e^{-i\beta(z-h)} + R{\cdot}E_2\, e^{-i\beta(z+h)} \tag{1a}$$

where *i* is the imaginary symbol, β is the propagation constant equal to $2\pi/\lambda_0$, and λ_0 is the wavelength in air. Additionally, *R* is the *reflection coefficient* close to -1 when the plane is a good conductor. This means that a phase of the field E_1 is changed by 180° when reflected from the conductive plane. Considering the direct and reflected waves at point P above the conducted plane, their magnitudes are practically equal, $E_1 \approx E_2$, and accordingly, Eq. (1a) becomes

$$\tilde{E}_t \approx E_1\left(e^{i\beta h} - e^{-i\beta h}\right)e^{-i\beta z} \tag{1b}$$

Owing to Euler's identity:

$$\frac{e^{i\beta h} - e^{-i\beta h}}{2} = i\sin(\beta h) \tag{2a}$$

Equation (1b) becomes

$$\tilde{E}_t \approx E_1\left[i2\sin(\beta h)\right]e^{-i\beta z} = E_1\tilde{F}(h)e^{-i\beta z} \tag{2b}$$

where *F(h)* is the *radiating factor* whose magnitude is given as

$$|F(h)| = 2\sin(\beta h) = 2\sin\left(\frac{2\pi}{\lambda_0}h\right) \tag{2c}$$

Additionally, the maximum $_d$ E1 in the far-field region is expressed via Eq. (B.15a) in Appendix B as

$$E^f{}_{1,\max} = \eta_0\frac{Il}{2\lambda_0 r_1} \tag{3a}$$

and substituting Eqs. (2c) and (3a) into Eq. (2b), $\left|\tilde{E}_{t,\max}\right|$ is expressed as

$$\left|\tilde{E}_{t,\max}\right| = \frac{\eta_0 Il}{2\lambda_0 r_1}\cdot 2\sin\left(\frac{2\pi}{\lambda_0}h\right) \tag{4}$$

For heights h in the range from 0 to $\lambda_0/4$, the $F(h)$ diagram (normalized by λ) is made via Eq. (2c) and shown in Fig. 4.

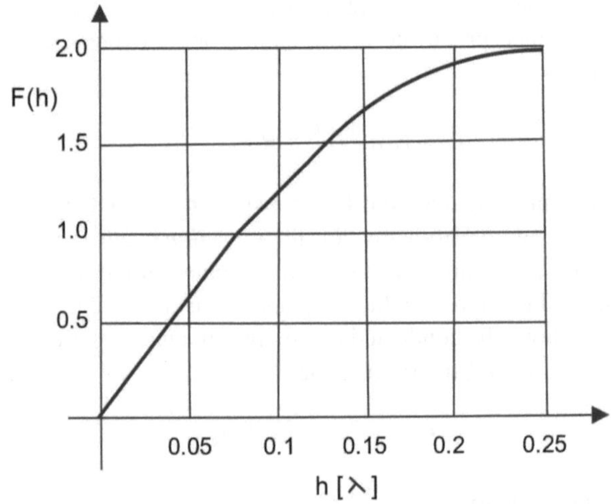

Fig. 4 The far-field radiating factor for a horizontal radiator above a conductive plane

Notably, the conducted plane acts as a radiation *suppressor* when $F(h) < 1$; accordingly, this factor is called a suppressing *factor,* $F_s(h)$. In contrast, when $F(h) > 1$, the plane enhances the conductor's radiation by acting as an antenna reflector. Therefore, this factor is called the *gain* factor, $F_g(h)$.

In the following example, we numerically illustrate the effect of a conducted plane on linear conductor radiation placed parallel to the plane at a different height h.

Example 1 Let us consider the radiation of the wire radiator at a frequency of *300 MHz*. The wire is long 10 cm and carries the average current $I = 1$ mA. Herein, we need to find the maximal *magnitude* of the radiated *electric* field at distance $r = 3\,\lambda_0$ in the far-field region when:

(a) The wire is in free space, far away from metallic planes and objects.
(b) The wire is above a conducting plane at the following distances: $0.01\,\lambda_0$, $0.1\,\lambda_0$, and $0.25\,\lambda_0$.

Solution

(a) The *electrical* length of the 10-cm wire in free space is $l = \dfrac{\lambda_0}{10}$ at 300 MHz ($f \approx 300$ MHz $\rightarrow \lambda_0 \approx 1$ m). Therefore, the wire belongs to the category of an electrically *small* radiator at the given frequency.
For an *electrically small* radiator, the maximum magnitude of the *electric field* in the *far-field* region can be determined via Eq. (B.15a) in Appendix B:

$$E^f_{1,\max} = \eta_0 \frac{I\,l}{2\lambda_0\, r} \approx 6.3\left(\frac{\text{mV}}{\text{m}}\right)$$

or in dBmV/m:

$$E^{f}_{1,\max} = 20\log\left(6.310^{-3}\cdot 10^{3}\right) \approx 16\,\mathrm{dBmV/m}$$

(b) When the wire is above the large conducting plane at distances $h = 0.01\ \lambda_0$, $0.1\ \lambda_0$, and $0.25\ \lambda_0$, the *radiation factors* $F(h)$ and $E^{f}_{1,\max}$ for $\lambda_0 = 1$ m are listed in Table 1.

Table 1 Factor $F(h)$ and $E^{f}_{1,\max}$ values for different h in λ_0

h	$F(h)$	$E^{f}_{d,\max}$
$0.01\ \lambda_0$	−18 dB	−2 dBmV/m
$0.1\ \lambda_0$	+1.4 *dB*	+17 dBmV/m
$0.25\ \lambda_0$	+6 dB	+22 dBmV/m

Furthermore, Fig. 5 shows 2D plots of the *radiation factor* in the *elevation plane* made by the expression for the *radiation factor* in polar coordinates θ:

$$F(\theta,h) = 2\cdot\sin\left[\frac{2\pi}{\lambda}h\right]\cos(\theta) \tag{5}$$

for $\lambda_0 = 1$ and $h = 0.01\ \lambda_0$, $0.1\ \lambda_0$, and $0.25\ \lambda_0$.

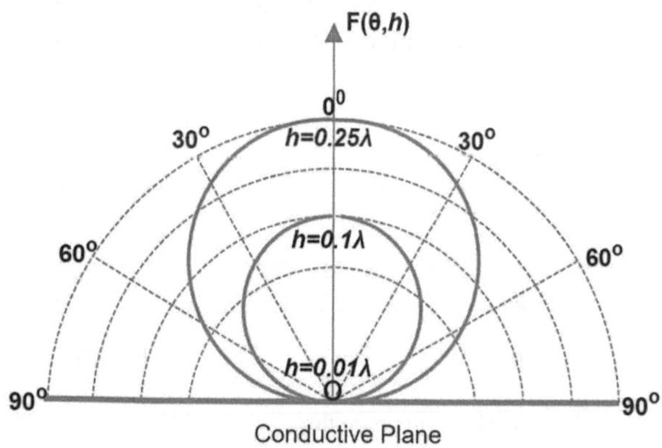

Fig. 5 Elevation radiation pattern of a horizontal linear radiator above conductive plane at heights **$h = 0.01\ \lambda_0$, $0.1\ \lambda_0$, $0.25\ \lambda_0$**

Note that Eq. (5) becomes Eq. (2c) when $\theta = 0°$:

$$\left|\tilde{F}(h)\right| = 2\sin\left(\frac{2\pi}{\lambda_0}h\right) \tag{5a}$$

4 Conclusion

By placing a parallel wire radiator very close to a conducted plane, for instance, $h = 0.01\ \lambda_0$ or lower, the wire radiation falls far below the radiation when the wire is in free space (or far from the conducted plane). In the present case, the radiation falls at -18 dB. However, when the wire is placed higher from the conducted plane, for example, at $h = 0.1\ \lambda_0$ or even $h = 0.25\ \lambda_0$, the radiation is increased by +1.4 dB or even +6 dB beyond the radiation when the wire is in free space. Herein, we can observe the suppression and gain effects of the conducted plane on the wire radiation.

Thus, the wire radiation increases when the wire height h above the conducting plane increases to $h = 0.25\ \lambda_0$. Here, the conducting plane starts to act as an antenna reflector. This is because fields from the *actual* wire and its *image* are in the phase in the far-field region above the plane. This is the best-case scenario for the antenna design, but from the EMI perspective, this is the worst-case scenario. Hence, in terms of EMI, a parallel linear radiator should be as close as possible to a conducted plane.

5 Differential-Mode Emission Vs. Common-Mode Emission

In the previous section we consider EMI emission related to a single conductor line. Here, we discuss the types of EMI emissions from two-coductor signal lines. First, we analyze the simplified current models shown in Fig. 6a, b, where the two radiation modes coexist, the differential mode (DM) and the common mode (CM), which both contribute to EMI emissions.

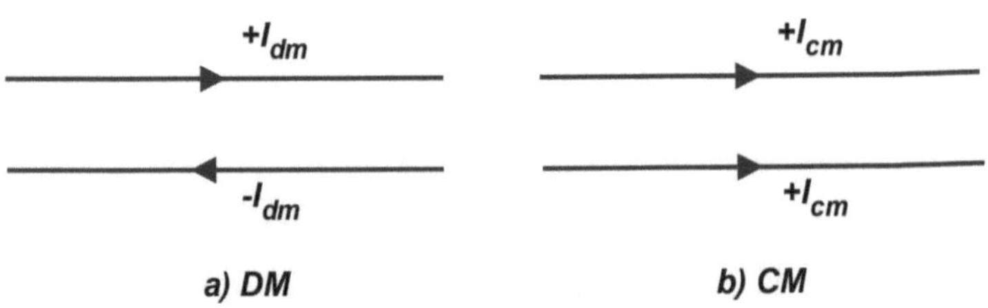

Fig. 6 (**a**) DM current flow and (**b**) CM current flow

Figure 6a shows two parallel transmission line (TL) wires 1 and 2 with currents I_{dm} that have equal magnitudes and opposite directions. This mode is called the *differential current* (DM), which is mainly an *intentional* (*operating*) mode in a differential transmission line. Here, the DM current flows down to the TL termination over wire 1 and returns to a TL source over wire 2, forming a *current* loop. In principle, this loop resembles the loop radiator discussed in Appendix B.

Figure 6b shows the same wire pairs but with currents I_{cm} that have the same magnitude and same directions. This current mode is called the *common mode* (CM), which may exist on two-conductor TLs or TLs with multiple conductors. In principle, this case is related to *linear* radiators, which are considered in Appendix B. This mode is typically *undesirable* from EMI standpoints. However, this mode is the operating mode for some applications, such as for regular antennas that are sources of *intentional* radiation. For the CM, the *return* current must flow back some way to its source (usually over distant conductors and/or through dielectrics), forming large current loops that may act as efficient EMI radiators.

Hence, in practice, the return CM currents are often "invisible" and may flow back to the source via various system wires and metallic structures, closing a CM current loop. Additionally, return CM currents may flow as displacement currents through the air and dielectrics. However, return CM currents tend to follow paths of least resistance.

In general, any electromagnetic radiation (DM or CM) is created by a current loop (smaller or larger one). A smaller current loop causes lower EMI emissions as noted in Appendix B.

5.1 Differential-Mode Emissions

Figure 7a, b shows the *differential* TLs with parallel wires 1 and 2 in the z–y plane, where the fields $\vec{E}_{1dm}$ and $\vec{E}_{2dm}$ are produced by the currents I_{dm} and $-I_{dm}$ outside the TL.

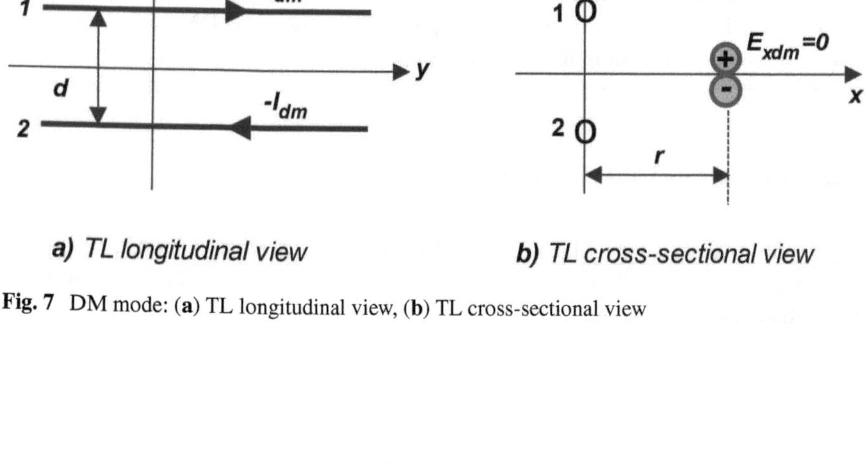

Fig. 7 DM mode: (a) TL longitudinal view, (b) TL cross-sectional view

According to the present model, the resulting field E_{dm} can be expressed as the sum of field components $\vec{E}_{1dm}$ and $\vec{E}_{2dm}$ as

$$\vec{E}_{dm} = \vec{E}_{1dm} + \vec{E}_{2dm} \tag{6}$$

where $\vec{E}_{1dm}$ is the field from the real radiator, wire 1, located above the x-y plane, and where $\vec{E}_{2dm}$ is the field from wire 2 located below the x-y plane.

Because the field components $\vec{E}_{1dm}$ and E_{2dm} are produced by the currents I_{dm} with opposite directions, the magnitude of the resulting field $\vec{E}_{dm}$ in the ±z-direction is expressed as

$$\left|\vec{E}_{dm}\right| = \left|\vec{E}_{1dm} - \vec{E}_{2dm}\right| > 0 \tag{6a}$$

Hence, the distance d between wires 1 and 2 is crucial for reducing or amplifying the intensity of the DM field around the wires. Additionally, the resulting field is zero in the x–y plane because the field components $\vec{E}_{1dm}$ and $\vec{E}_{2dm}$ cancel each other. This case is shown in Fig. 9b and is analytically expressed as

$$\left|\vec{E}_{1dm}\right| - \left|\vec{E}_{2dm}\right| = 0 \tag{6b}$$

In principle, the DM model from Fig. 7 is a *loop* (magnetic) *radiator* described in Appendix B. We can express the maximum magnitude of the *electric* field in the z-direction for the *near-field* and *far-field* regions via Eqs. (B.9) and (B.16a), which are given in Appendix B as

$$E^{n}_{dm,\max} = \eta_0 \frac{SI_{dm}}{2\lambda_0 r^2} = \eta_0 \frac{(l\,d)I_{dm}}{2\lambda_0 r^2}\left(\frac{V}{m}\right) \tag{7a}$$

$$E^{f}_{dm,\max} = \eta_0 \frac{\pi SI_{dm}}{\lambda_0^{\,2} r} = \eta_0 \frac{\pi l d I_{dm}}{\lambda_0^{\,2} r}\left(\frac{V}{m}\right) \tag{7b}$$

where S is the loop area given as $S = l \cdot d$. Additionally, l is the length of the electrically short TL since d is the separation between *TL* conductors 1 and 2. We use these expressions later in the numerical examples to illustrate the DM radiation of the differential TL as a rectangular current loop.

5.2 Common-Mode Emission

Figure 8a, b shows the CM current model with parallel wires 1 and 2, where the resulting CM electric field E_{cm} is produced by the two currents I_{cm} with the same magnitude and same direction.

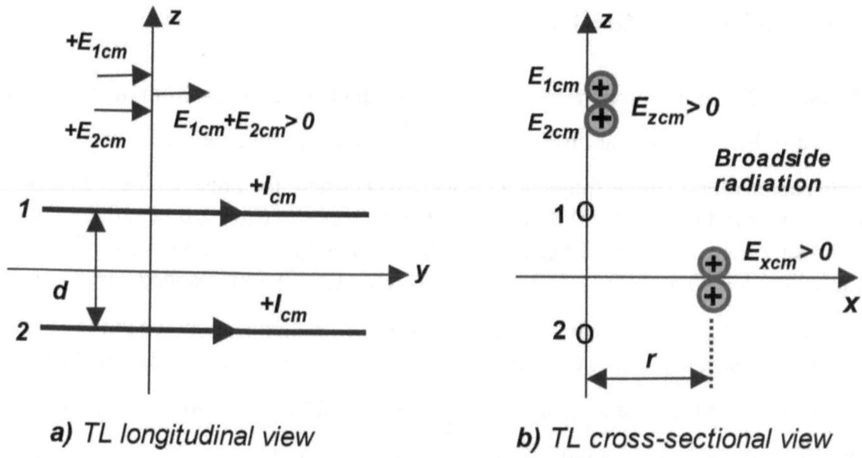

Fig. 8 CM mode: (**a**) TL longitudinal view, (**b**) TL cross-sectional view

The resulting CM field E_{cm} can be expressed as the sum of the field components $\vec{E}_{1cm}$ and $\vec{E}_{2cm}$:

$$\vec{E}_{zcm} \cong \vec{E}_{xcm} = \vec{E}_{1cm} + \vec{E}_{2cm} \tag{8a}$$

The fields $\vec{E}_{1cm}$ and $\vec{E}_{2cm}$ are produced by the two currents $+I_{cm}$ on conductors 1 and 2, respectively, in the same direction. Consequently, the CM resulting fields achieve a maximum in both planes, (z–y) and (z–x), i.e., this TL has broadside radiation. Thus, we can express the magnitude of the CM resulting fields in the (z–y) and (z–x) planes as

$$\left|E_{c,\max}\right| = \left|\vec{E}_{1cm}\right| + \left|\vec{E}_{2cm}\right| = 2\left|\vec{E}_{cm}\right| \tag{8b}$$

Furthermore, we can use the field expressions (B.7) and (B.15a) from Appendix B for an electrically short TL to calculate the maximal magnitude of the TL common-mode radiation in the *near-field* and *far-field* regions, respectively, as

$$E^{n}{}_{cm,\max} = \eta_0 \frac{(2I_{cm})\, l\lambda_0}{8\pi^2 r^3} \tag{9a}$$

$$E^{f}{}_{cm,\max} = \eta_0 \frac{(2I_{cm})\, l}{2\lambda_0\, r} \tag{9b}$$

Because there are two conductors of the same length and each with current I_{cm}, we consider the doubled current of I_{cm} ($2\, I_{cm}$) in Eqs. (9a), (9b).

5.3 DM vs. CM Emission

Example 2 Let us compare the levels of DM and CM radiation from the wire pair shown in Figs. 7 and 8 at distances of $r_1 = \frac{\lambda_0}{6}$ (*near-field* region) and $r_2 = 3\,\lambda_0$ (far-*field* region) for a signal frequency of 300 MHz, where the present TLs are electrically small radiators. The *DM* current is 1 mA since the CM current is 0.5 mA per conductor, i.e., the total CM current of the two conductors is 1 mA. Additionally, the wires' *length* and *separation* are $l = 10$ cm and $d = 3$ mm, respectively.

We consider the wire pair as (a) a small *loop* DM radiator, as shown in Fig. 7, and (b) two *linear* and parallel CM radiators with the same currents, as shown in Fig. 8.

(a) For the DM field at distances $r_1 = \frac{\lambda_0}{6}$ (*near-field* region) and $r_2 = 3\lambda_0$ (*far-field* region), the maximal field magnitudes are given by Eqs. (7a, 7b):

The solution for the *near* field at a distance $r_1 = \frac{\lambda_0}{6}$ is as follows:

$\eta_0 = 377\ \Omega$, $I_{dm} = 1$ mA, $l = 10$ cm, $d = 3$ mm, $\lambda_0 = 1$ m, $A = l\,d$.

$$E^n{}_{dm,\max} = \eta_0 \frac{I_{dm}\, l\, d}{2\lambda_0\, r_1^2} \approx 2 \cdot 10^{-3} \left(\frac{V}{m}\right)$$

or in dB: $E^n{}_{dm,\max} = 20\log\left(2 \cdot 10^{-3} \cdot 10^6\right) = 66\,\text{dB}\mu\text{V/m}$

The solution for the *far field* at $r_2 = 3\,\lambda_0$ is as follows:

$\eta_0 = 377\ \Omega$, $I_{dm} = 1$ mA, $l = 10$ cm, $d = 3$ mm, $\lambda_0 = 1$ m

$$E^f{}_{dm,\max} = \eta_0 \frac{\pi\, I_{dm}\, l d}{\lambda_0^2\, r_2}$$

$$E^f{}_{dm,\max} = 377 \frac{\pi\, I_{dm}\, l\, d}{\lambda_0 \left(3\lambda_0\right)^2} = 1.18 \cdot 10^{-4} \left(\frac{V}{m}\right)$$

$$E^f{}_{dm,\max} = 20\log\left(1.18 \cdot 10^{-4} \cdot 10^6\right) \approx 41.4\,\text{dB}\mu\text{V/m}$$

(b) Furthermore, for the CM field in the *near-field* region (at $r_1 = \frac{\lambda_0}{6}$) and in the *far-field* region (at $r_2 = 3\,\lambda_0$), the maximal field magnitudes are given by Eq. (9a), (9b):

Solution for the *near* field at $r_1 = \frac{\lambda_0}{6}$:

$\eta_0 = 377\ \Omega$, $2I_{cm} = 1$ mA, $l = 10$ cm, $\lambda_0 = 1$ m

The result of perfectly balanced TL is that the common currents I_{cm1} and I_{cm2} in the upper and lower halves of the differential TL are equal: $I_{cm1} = I_{cm2} = I_{cm}$. Thus, the DM and CM modes independently coexist at the perfectly balanced differential *TL* without being converted to each other. However, if at least one of the above-balanced conditions (equalities) is unsatisfactory, mode conversion exists in the TL.

When circuit theory is applied to the circuit from Fig. 10 at the TL load, it follows that:

$$V_{dm} = V_{L1} - V_{L2} = I_{cm1} Z_{L1} - I_{cm2} Z_{L2} + I_{dm} \left(Z_{L1} + Z_{L2} \right) \tag{10}$$

In the total balance condition where $I_{cm1} = I_{cm2} = I_{cm}$ and $Z_{L1} = Z_{L2} = Z_L$, the DM and CM voltages V_{dm} and V_{cm} become:

$$V_{dm} = V_{L1} - V_{L2} = 2 Z_L I_{dm} \tag{10a}$$

$$V_{cm} = V_L = Z_L I_{cm} \tag{10b}$$

In addition to the above balanced conditions, a differential line should be sufficiently distant from other TL lines or large conductive and dielectric objects to avoid coupling with the environment, which may also cause a TL imbalance that leads to mode conversion.

Figure 11 shows several examples in which the balance of differential lines can be disrupted.

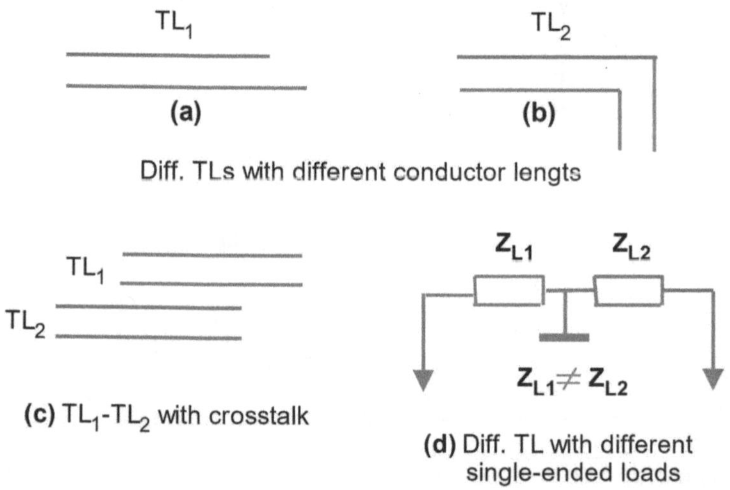

Fig. 11 Examples of unbalanced differential lines

Thus, imbalance of a TL differential line may exist if at least one of the following cases exists: (a, b) when the lengths of the TL conductors are not equal; (c) when excessive coupling (crosstalk) exists between too close TLs; and (e) when the TL single-ended load impedances are not equal. More information about the mode conversion is given in chapter "Performance of RF Cables and IT Signals Lines in Radio systems with EMI". Additionally, this topic is considered in detail in [7].

6.1 Examples of DM and CM Radiation Modes

Figure 12a, b shows the cases of the DM and CM radiation modes. Figure 12a shows the coupling of an external electromagnetic field into a nonperfectly balanced differential TL that produces CM currents I_{cm1} and I_{cm2} in the TL conductors. Furthermore, I_{cm1} and I_{cm2} are converted into the DM current I_{dm} in the TL, which creates DM radiation.

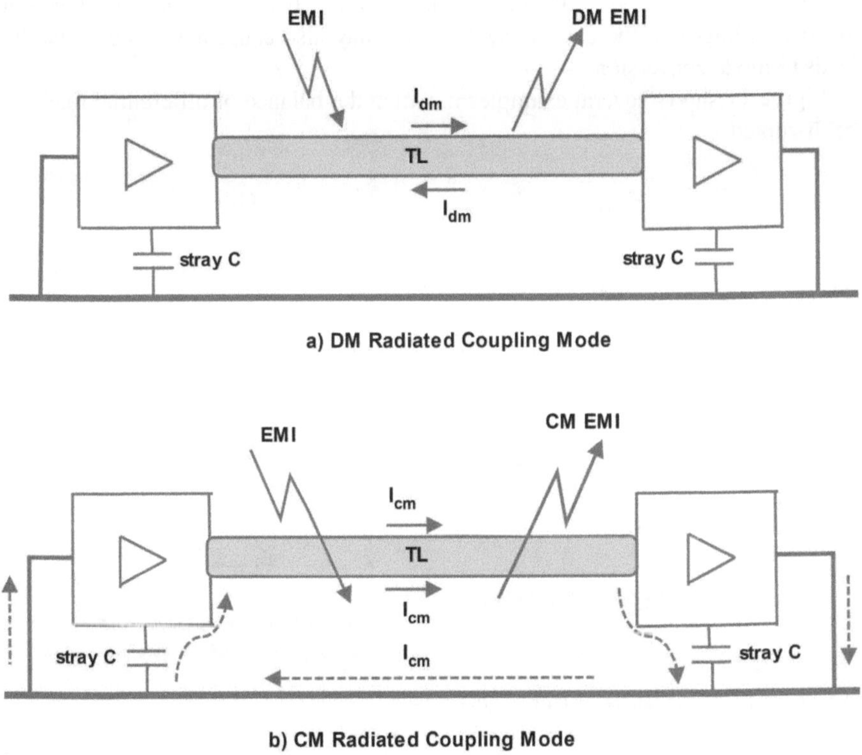

Fig. 12 DM and CM radiated coupling modes

Figure 12b shows the coupling of an external electromagnetic field into a TL-inducing current I_{cm} that causes CM radiation regardless of whether the TL is balanced.

More practical details about the coupling of waves into cable and signal lines are given in [8, 9].

In addition to the DM and CM, the so-called *antenna current mode* is also considered in the literature. Sometimes even the antenna mode and CM are considered the same mode. Figure 13 shows the concept of the antenna mode, which is similar to CM. In the antenna mode model, a close radiated current loop is the combination of displacement currents I_{dis}, which flow through the air from the antenna and ground, and conducted currents I_{con}, which flow through ground to the antenna. The antenna mode is similarly considered in [10].

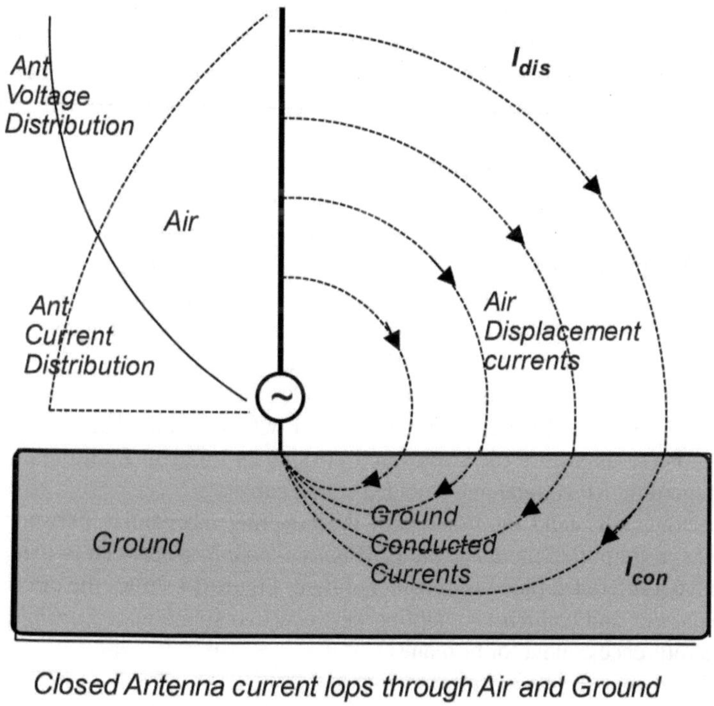

Fig. 13 Antenna current mode that includes air displacement currents and ground-conducted currents

However, a displacement current (or an equivalent current) does not physically exist. They are introduced in electromagnetic models and circuits (by Maxwell) analogous to the currents in conductive media or conductors. The real energy flow through air and vacuum can be explained by the Point vector, rather than currents, which is associated with the propagation of electromagnetic fields in air and vacuum, as described in Appendix A.

7 Near-Field EMI Coupling (Crosstalk)

The coupling of electromagnetic energy inside the *near-field* region and between two or more neighboring *TLs* is an effective mechanism of EMI transfer (see Appendix B). With respect to radiation, *TLs* act as loop radiators as noted in *Appendix B* and earlier in this chapter. The excessive coupling between *TLs* leads to a change in *TL* properties such as the *characteristic impedance, mode conversion,* and *signal propagation,* which may significantly affect or even disrupt the TL operating signals.

There are three types of *near-field* EMI *coupling*: (a) *inductive* or *magnetic* coupling, (b) *capacitive* or *electric field* coupling, and (c) circuit *common-impedance* coupling. The *inductive* and *capacitive* couplings are created by a *magnetic* and *electric* flux passing through the common area of two or more adjacent *TLs* or circuits. Additionally, when *TLs* or circuits share a reference (common) conductor with considerable impedance, the voltage developed on the common conductor contributes to *voltage* coupling between the TLs and circuits.

We discuss *inductive* and *capacitive* coupling between two *unshielded TLs* in this section, whereas EMI coupling associated with shielding TLs is discussed in chapter "Performance of RF Cables and IT Signals Lines in Radio systems with EMI" regarding RF coaxial and shielded signal cables.

To better understand the essence of the coupling mechanism between two or more *TLs*, a simplified model of the EMI *mutual coupling* between two unshielded *TLs* above a reference plane is considered here. Figure 14 shows the circuit model with *inductive* and *capacitive* coupling between two single-ended unshielded *TLs* above a reference conductor or plane.

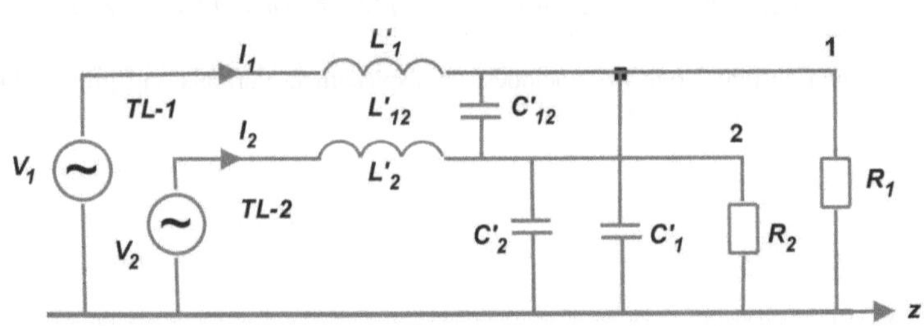

Fig. 14 Lumped circuit for inductive and capacitive coupling between two single-ended unshielded TLs

The per-unit-length elements of the lumped circuit are L'_1, L'_2, C'_1, and C'_2 as the *self-inductance* and *self-capacitance* of TL conductors 1 and 2, respectively, and L'_{12} and C'_{12} as the *mutual inductance* and *capacitance* between conductors 1 and 2, respectively.

For a *lossless TL*, the *TL equations* given in Appendix C are as follows:

$$\frac{\partial V(z)}{\partial z} = L'_1 \frac{\partial I}{\partial t} \equiv -i\omega L'_1 I(z) \tag{11a}$$

$$\frac{\partial I(z)}{\partial z} = C'_1 \frac{\partial V}{\partial t} \equiv -i\omega C'_1 V(z) \tag{11b}$$

Furthermore, on the basis of the *TL equations* and circuit theory, the *coupled TL equations* for two lossless coupled *TLs* (TL-1 and TL-2, shown in Fig. 14) are expressed as [7].

$$\frac{\Delta V_1(z)}{\Delta z} = -i\omega L'_1 I_1 - i\omega L'_{12} I_2 \tag{12a}$$

$$\frac{\Delta V_2(z)}{\Delta z} = -i\omega L'_{12} I_1 - i\omega L'_2 I_2 \tag{12b}$$

$$\frac{\Delta I_1(z)}{\Delta z} = -i\omega (C'_1 + C'_{12}) V_1 + i\omega (C'_{12}) V_2 \tag{12c}$$

$$\frac{\Delta I_2(z)}{\Delta z} = i\omega (C'_{12}) V_1 - i\omega (C'_2 + C'_{12}) V_2 \tag{12d}$$

where $\Delta V_1(z)$, $\Delta V_2(z)$, $\Delta I_1(z)$, and $\Delta I_2(z)$ are the voltage and current increments corresponding to TLs with electrically short segments long Δz along the z-axis.

To simplify the consideration of the coupling mechanism, let us assume that *TL-1* is an EMI noise source since *TL-2* is an EMI noise receptor. In this case, we can find the EMI coupled voltage and coupled current in *TL-2* by setting $I_2 = 0$, $V_2 = 0$, $\Delta V_2(z) = V^L_{2n}$ and $\Delta I_2(z) = I^C_{2n}$ in Eqs. (12b) and (12d). As a result,

$$\frac{\Delta V_2(z)}{\Delta z} = -i\omega L'_{12} I_1 \rightarrow V^L{}_{2n} = -i\omega L_{12} I_1 \tag{13a}$$

$$\frac{\Delta I_2(z)}{\Delta z} = i\omega C'_{12} V_1 \rightarrow I^C{}_{2n} = i\omega C_{12} V_1 \tag{13b}$$

where L_{12} and C_{12} are the *total mutual* inductance and capacitance, respectively, of a TL segment of length Δz. Therefore, $L_{12} = \Delta z L'_{12}$ and $C_{12} = \Delta z C'_{12}$, while L'_{12} and C'_{12} are the per-unit-length mutual inductance and capacitance, respectively, as mentioned before.

Note that the coupling between two *TLs* is a *mutual* process that simultaneously occurs in both directions, from *TL-1* to *TL-2* and vice versa. This general case is presented mathematically by the *coupled TL equations*, Eqs. (12a), (12b), (12c), (12d). To simplify the coupling analysis, we consider one-directional inductive and capacitive coupling in the following subsections. A similar approach is considered in [2].

7.1 Inductive Coupling

Figure 15 shows the equivalent circuit model for a *TL* segment as a part of the *TL* inductive coupling.

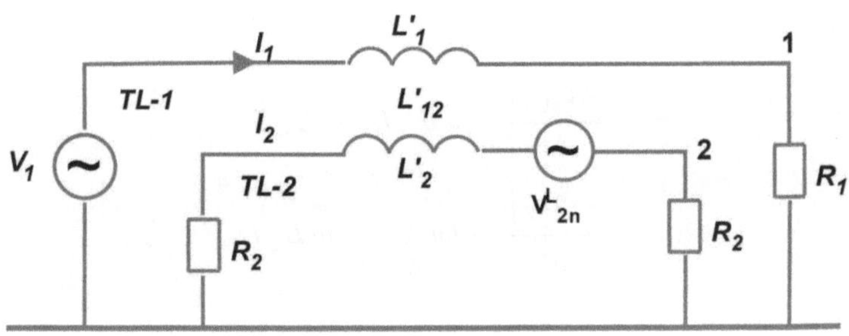

Fig. 15 Equivalent lumped circuit for TL inductive coupling

Per Eq. (13a), the *inductively* coupled voltage $V^L{}_{2n}$ in TL-2 directly depends on the *mutual inductance* L_{12}, which acts as an EMI path. Since the current I_1 in *TL-1* is a source of coupling, the induced voltage $V^L{}_{2n}$ can be expressed as

$$V^L{}_{2n} = -i\omega L_{12}\, I_1 \tag{14a}$$

The total *mutual inductance* L_{12} (in H) is defined on the basis of the per-unit-length mutual inductance L'_{12} given in Appendix C, Eq. (C.12c):

$$L_{12} = l \cdot L'_{12} = l\frac{\mu_0}{\pi}\ln\left(\frac{s_{12}}{r}\right)(H)(s_{12} > 6r), \tag{14b}$$

where l is the length of a *TL* and where s_{12} is the center-to-center spacing between *TL* wires *1* and *2*.

7.1.1 Capacitive Coupling

The *lumped* circuit of a *TL* segment for *capacitive coupling* is shown in Fig. 16.

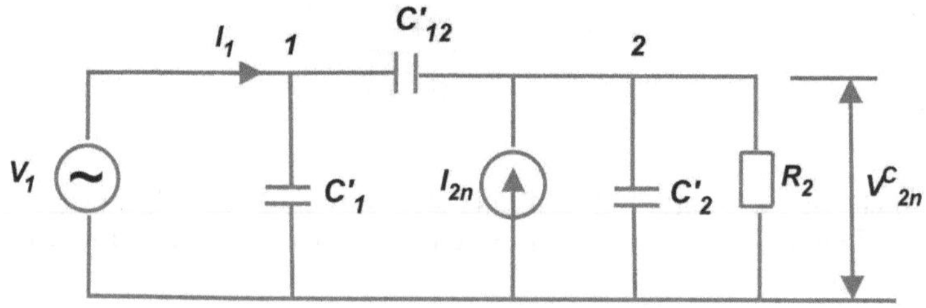

Fig. 16 Equivalent lumped circuit for TL capacitive coupling

Here, $V^C{}_{2n}$ is the *capacitively coupled* EMI voltage developed at the *TL-2* load R_2, caused by the source voltage V_{1s} (on the schematic V_1) of the *TL-1*. $V^C{}_{2n}$ is expressed on the basis of the equivalent circuit in Fig. 16 as

$$V^C{}_{2n} = \frac{Z_{C2}\,//R_2}{\left(Z_{C12} + Z_{C2}\,//R_2\right)} V_{1s} \tag{15}$$

where the symbol "//" represents the parallel connection between the impendences Z_{C2} and R_2. Hence, upon R_2, $Z_{C2} = \dfrac{1}{i\omega C_2}$, and $Z_{C12} = \dfrac{1}{i\omega C_{12}}$, Eq. (15) has the form

$$V^C{}_{2n} = \frac{C_{12}/\left(C_{12} + C_2\right)}{1 + 1/i\omega R_2\left(C_{12} + C_2\right)} V_{1s} \tag{16a}$$

Furthermore, when $1/i\omega R_2(C_{12} + C_2) \gg 1$, Eq. (16a) can be reduced to

$$V^C{}_{2n} = i\omega R_2 C_{12} V_{1s}\left(V\right) \tag{16b}$$

However, for $1/i\omega R_2(C_{12} + C_{20}) \ll 1$, Eq. (16a) is reduced to

$$V^C{}_{2n} = \frac{C_{12}}{C_{12} + C_2} V_{1s}\left(V\right) \tag{16c}$$

where C_{12} is the *TL* mutual capacitance, and where C_{10} and C_{20} are the self-capacitances of *TL-1* and *TL-2*, respectively, given in Appendix C as

$$C_{12} = l\frac{\pi\varepsilon}{\ln\left(\dfrac{s_{12}}{r}\right)} \tag{17a}$$

$$C_1 \cong l\frac{2\pi\varepsilon}{\ln\left(\dfrac{h_1}{r}\right)} \tag{17b}$$

$$C_2 \cong l \frac{2\pi\varepsilon}{\ln\left(\frac{h_2}{r}\right)} \tag{17c}$$

where h_1 and h_2 are the spacings of the *TL* conductors from the reference plane and l is the TL length. Notably, Eq. (17a) is valid when $s_{12} > 6\,r$ (r is the wire radius).

7.1.2 Numerical Illustration of Inductive and Capacitive Coupling

Here, we first find the voltage $V^L{}_{2n}$ *inductively* coupled into *TL-2* produced by the current $I_1 = 10$ mA in *TL-1*. The two single-ended and *coupled TLs* are shown in Fig. 14 with a spacing of $s_{12} = 10$ mm, both above the reference plane by $h_1 = h_2 = h = 10$ cm. Additionally, the radii of both TL conductors are the same: $r = 1$ mm, and the signal frequency in *TL-1* is $f = 100$ MHz.

Solution for Inductive Coupling

The *mutual* inductance between the *TLs* can be found via Eq. (14b) as

$$L_{12} = l\frac{\mu_0}{\pi}\ln\left(\frac{s_{12}}{r}\right) \approx 0.46\,\mu H$$

and further using Eq. (14a), the *inductively* coupled voltage is determined as

$$V^L{}_{2n} = (2\pi f)L_{12}I_1 \approx 2.8\,\mathrm{V}$$

Additionally, from the last equation, it can be seen that the *inductively coupled* voltage increases linearly with frequency.

Solution for Capacitive Coupling

For the *TL* circuit model shown in Fig. 16, we find the *capacitive-coupled voltage* $V^C{}_{2n}$ in *TL-2* generated by the source voltage $V_{1s} = 1$ V in *TL-1*. Additionally, as an example, we consider two cases in which *TL-2* is loaded by a low and larger resistor: 1 Ω and 500 Ω.

1. *The solution* for $R_2 = 1\ \Omega$ is

   For the parameters $s_1 = s_2 = s_{12} = 10$ mm, $l = 1$ m, and $\varepsilon_0 \approx (1/36\pi)\cdot 10^{-9}$ (F/m), the *TL mutual* and *self*-capacitance are, respectively.

$$C_{12} = l\frac{\pi\varepsilon_0}{\ln\left(\frac{s_{12}}{r}\right)} \approx 12\,pF, \quad C_2 \cong l\frac{2\pi\varepsilon_0}{\ln\left(\frac{h}{r}\right)} \approx 24\,pF$$

Additionally, for the above values C_{12} and C_2, $f = 100$ MHz, and $R_2 = 1\ \Omega$, the term in Eq. (16a) is $1/i\omega R_2(C_{12} + C_2) \gg 1$:

$$\frac{1}{(2\pi f)R_2(C_{12}+C_2)} \approx 44\Omega \gg 1$$

and accordingly, we will use Eq. (16b) to determine $V^C{}_{2n}$:

$$V^C{}_{2n} = 2\pi f R_2 C_{12} V_{1s} \approx 7.5\,\text{mV}$$

2. *The solution* for $R_2 = 500\ \Omega$ is $1/i\omega R_2(C_{12} + C_2) \ll 1$.

$$\frac{1}{(2\pi f)R_2(C_{12}+C_2)} \approx 0.09 \ll 1,$$

and we use Eq. (16c) to determine the coupled voltage $V^C{}_{2n}$:

$$V^C{}_{2n} = \frac{C_{12}}{C_{12}+C_2} V_{1s} \approx 330\,\text{mV}$$

Note that the capacitive-coupled voltage in the receptor *TL* is greater when the *TL* load resistance is greater.

7.2 Summary of Coupling Between Two Close and Parallel Unshielded TLs

Because *mutual coupling* between *TLs* can negatively affect the performance of both *TLs*, we need to keep the coupling to a low level as much as possible. There are several methods for this as mentioned before. The first is that parallel neighboring *TLs* should be as distant as possible. Second, the routing of *TLs* should be orthogonal to each other when possible because the *inductive coupling* can be significantly reduced, theoretically to zero.

If the solution methods above do not succeed enough or cannot be applied for the same reasons, *shielded TLs* or *twisted-differential pairs* can be used to overcome *EMI coupling* problems. These methods will be considered more in chapter "Performance of RF Cables and IT Signals Lines in Radio systems with EMI", where EMI coupling between regular signal lines and cables are discussed in detail. Additionally, readers interested in this topic are referred to references [11–14] for further reading.

References

1. C.R. Paul, R.C. Scully, *Introduction to Electromagnetic Compatibility* (Wiley, Hoboken, 2022)
2. H.W. Ott, *Electromagnetic Compatibility Engineering* (Wiley, Hoboken, 2009)
3. K.L. Kaiser, *Electromagnetic Compatibility Handbook* (CRC Press, Boca Raton, 2005)
4. T. Williams, *EMC for Product Designers* (Elsevier, Newnes, 2007)
5. M. Gruber (ed.), *The ARRL RFI Book, Practical Course for RFI*, 3rd edn. (2010)
6. F. Lattarulo, in *Electromagnetic Compatibility in Power Systems* (Elsevier, 2006)
7. S.H. Hall, H.L. Heck, *Advanced Signal Integrity for High-Speed Digital Designs* (Wiley, Hoboken, 2009)
8. F.M. Tesche, in *Plane Wave Coupling to Cables*, ed. by R. Perez. Chapter 4, Handbook of EMC (Academic Press, 1995)
9. E.F. Vance, in *Coupling to a Long Transmission Line*, ed. by R. Perez, Appendix C, Handbook of EMC (Academic Press, 1995)
10. E.D. Sude, *Earth Conduction Effects in Transmission Systems* (Dover Publications, Mineola, 1968)
11. R.B. Keller, in *Design for Electromagnetic Compatibility* (Springer, 2023)
12. C. Christopoulos, *Principles and Techniques of Electromagnetic Compatibility* (CRC Press, Boca Raton, 2023)
13. Y. Zhao, W. Yan, J. Sun, M. Zhou, Z. Meng, in *Electromagnetic Compatibility: Principles and Applications* (Springer, 2021)
14. C.R. Paul, *Electromagnetics for Engineers: Applications to Digital Systems and Electromagnetic Interference* (Wiley, Hoboken, 2003)

Digital and Pulse Signals, Random Variables, and Stochastic Processes

1 Introduction

This chapter considers digital signals in the time and frequency domain. The frequency domain or the spectrum of any signal is essential for the accurate prediction of the spectra of electromagnetic interference (EMI) created by those signals. Hence, to efficiently predict and control EMI emission, we need to acknowledge the spectra of the signals that produce EMI emission. The signals that may cause EMI emission can be *periodic*, *nonperiodic*, *pulse*, or *random*. For example, *periodic signals* are data clocks that are the common contributors to EMI emissions in digital equipment.

Nonperiodic and *random* signals usually present digital data streams and control signals in *communication* and *electronic* systems, and they may not be as EMI-critical as periodic signals. However, some nonperiodic pulses created by high-power natural EMI sources, such as *lightning* and *electrostatic discharge* events (ESDs), can disrupt operation and even harm electronic equipment. Additionally, *pulse* signals produced by abrupt connection and disconnection of various reactive loads in high-power distribution systems can cause EMI problems in radio and IT systems. We analyze nonperiodic and pulse signals in the following section of this chapter. The writing of this chapter is motivated and influenced by references [1–14].

M. Pajovic, *Fundamentals of EMI Phenomena in Radio Communication Systems*, https://doi.org/10.1007/978-3-031-88748-2_2

2 Periodic Signals

2.1 Time-Domain Periodic Signals

A waveform of a signal that periodically repeats itself in time can be simply determined in the future, and this signal belongs to the category of *periodic deterministic* signals. All these signals (deterministic and nondeterministic) are also considered well in the literature [1–4], which influenced the writing of this chapter.

As an example, a *periodic deterministic* signal in the time domain is shown in Fig. 1.

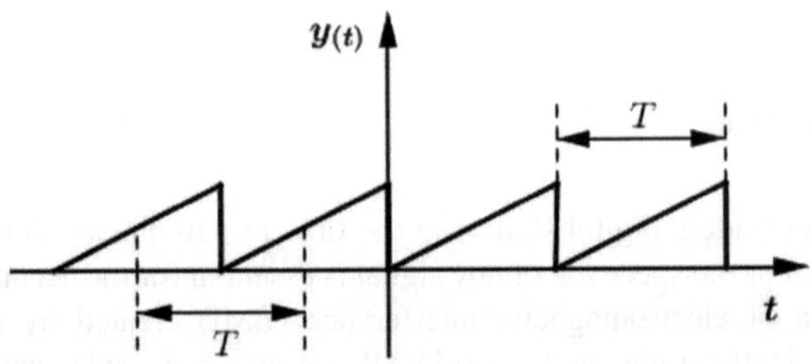

Fig. 1 Waveform of a periodic signal

This periodic signal function $y(t)$ is generally expressed as

$$y(t) = y(t \pm nT), \; n = 1,2,3,\ldots \tag{1}$$

where $y(t)$ repeats itself over each period T. The relationship between the signal period T and the signal *frequency* f is defined as

$$f = \frac{1}{T}(\text{Hz}) \tag{2}$$

where *Hz* is a frequency unit that refers to the number of signal periods (cycles) in a second. Additionally, a *radian* (*angular*) frequency is defined as

$$\omega = 2\pi f = \frac{2\pi}{T}(\text{rad}/\text{s}) \tag{3}$$

The simplest form of a *periodic* signal in the time domain is the *harmonic sinusoidal* (*sine*) signal expressed as

$$y_0(t) = \sin(t \pm nT) \quad n = 1,2,3,\ldots \tag{4}$$

This signal is shown in Fig. 2.

The presented $y(t)$ and $y_0(t)$ signals, like other *periodic* signals, commonly represent the time-domain forms of voltages, currents, and electric or magnetic fields.

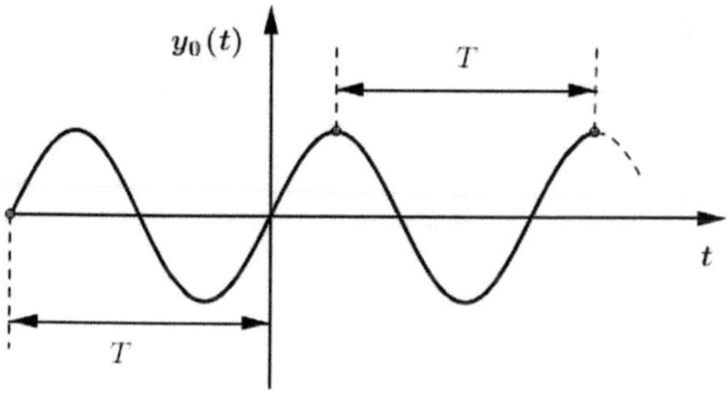

Fig. 2 Sinusoidal signal waveform

2.2 Fourier Series and Frequency Domain of Digital Signals

Any *periodic* signal can be presented in the time domain via the *Fourier series* of orthogonal trigonometric functions cos(x) and sin(x) as [5, 6]

$$y(t) = a_0 + \sum_{n=1}^{\infty} a_n \cos\left(2\pi n \frac{t}{T}\right) + \sum_{n=1}^{\infty} b_n \sin\left(2\pi n \frac{t}{T}\right) \tag{5}$$

where a_0, a_n, and b_n are the *Fourier* coefficients defined by the set of expressions

$$a_0 = \frac{1}{T}\int_{t'}^{t'+T} y(t)\mathrm{d}t; a_n = \frac{2}{T}\int_{t'}^{t'+T} y(t)\cos\left(2\pi n \frac{t}{T}\right)\mathrm{d}t;$$

$$b_n = \frac{2}{T}\int_{t'}^{t'+T} y(t)\sin\left(2\pi n \frac{t}{T}\right)\mathrm{d}t \tag{6}$$

where a_0 is an average *DC* signal value. Additionally, n denotes the signal harmonics. For instance, $n = 1$ denotes the *fundamental* frequency $f_0 = 1/T$ since $\omega_0 = 2\pi/T$ denotes the fundamental *radian* (angular) frequency. Thus, the series $n = 1, 2, 3\ldots$ refers to the fundamental (first), second, third, etc., signal harmonics.

Additionally, because the angular frequency of the first harmonic is $\omega_0 = 2\pi f_0$, where $f_0 = 1/T$, Eq. (5) can be expressed in terms of the harmonic frequency ($n\ \omega_0$) and time t as

$$y(\omega,t) = a_0 + \sum_{n=1}^{\infty} a_n \cos(n\omega_0 t) + \sum_{n=1}^{\infty} b_n \sin(n\omega_0 t) \tag{7}$$

Figure 3 shows a *periodic rectangular waveform* of pulses with amplitude A and duration τ.

The *Fourier series* coefficients a_0, a_n, and b_n for this signal can be expressed via Eq. (6) as

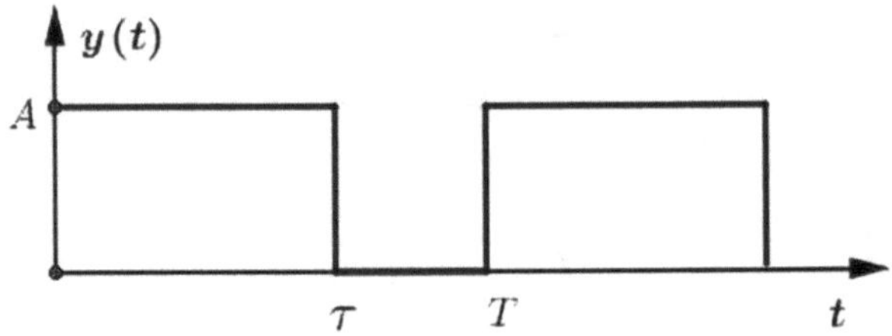

Fig. 3 Time-domain rectangular waveform

$$a_0 = \frac{1}{T}\left(\int_0^{\tau} A\,\mathrm{dt} + \int_{\tau}^{T} 0\cdot\mathrm{dt}\right) = \frac{A\tau}{T} \tag{8a}$$

$$a_n = \frac{2}{T}\int_0^{\tau} A\cos\left(2\pi n\frac{t}{T}\right)\mathrm{dt} = \frac{A}{n\pi}\sin\left(2\pi n\frac{\tau}{T}\right) \tag{8b}$$

$$b_n = \frac{2}{T}\int_0^{\tau} A\sin\left(2\pi n\frac{t}{T}\right)\mathrm{dt} = \frac{A}{n\pi}\left[1-\cos\left(2\pi n\frac{\tau}{T}\right)\right] \tag{8c}$$

The particular case of a rectangular waveform is the waveform with a 50%-duty cycle when $\tau = T/2$. For this case, the coefficients for *odd* harmonics are $a_n = 0$ and $b_n = 2A/n\pi$, whereas for *even* harmonics, both coefficients are zero (i.e., $a_n = 0$ and $b_n = 0$). Therefore, the digital signals with a 50%-duty cycle possess only *odd* harmonics. The *Fourier* series of these signal waveforms has the form of the modified Eq. (7) as

$$y(\omega,t) = \frac{A}{2} + \frac{2A}{\pi}\sin(\omega_0 t) + \frac{2A}{3\pi}\sin(3\omega_0 t) + \ldots \tag{9}$$

For example, Fig. 4 shows the time-domain appearance of this signal created by the sum of the first five harmonics.

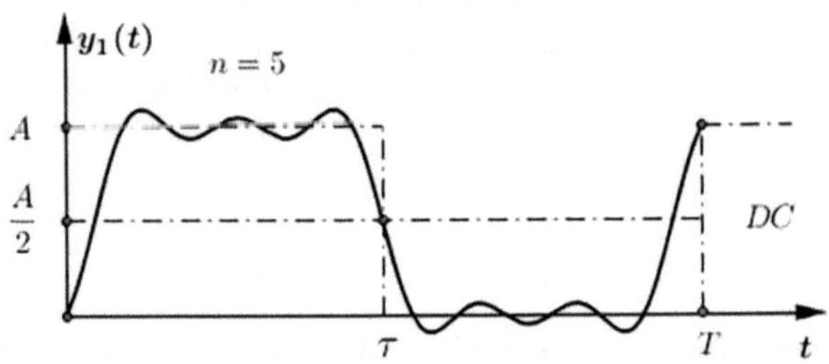

Fig. 4 Sum of the first five harmonics

Notably, by increasing the number of harmonics (i.e., for $n \to \infty$), the signal from Fig. 4 approaches the original rectangular signal shown in Fig. 3.

2.3 Fourier Series in Complex Exponential Form

Substituting the Euler identities

$$\cos(\omega t) = \frac{e^{i\omega t} + e^{-i\omega t}}{2} \text{ and } \sin(\omega t) = \frac{e^{i\omega t} - e^{-i\omega t}}{2i}$$

into Eq. (7), the *complex exponential* form of the *Fourier series* in general form (for $-\infty < n < +\infty$) is derived as

$$y(\omega,t) = c_0 + \sum_{n=-\infty}^{\infty} c_n\, e^{in\omega_0 t} \tag{10a}$$

or for the positive *n*-domain ($1 < n < +\infty$) as

$$y(\omega,t) = c_0 + \sum_{n=1}^{\infty} c_n e^{in\omega_0 t} + \sum_{n=1}^{\infty} c^*{}_n e^{-in\omega_0 t} \tag{10b}$$

where c_0 is an average constant, c_n is a complex coefficient, and $c^*{}_n$ is the conjugate value of c_n, both of which are expressed as follows:

$$c_n = \frac{1}{T} \int_{t'}^{t'+T} y(t) e^{in\omega_0 t} \mathrm{dt} \tag{11a}$$

$$c^*{}_n = \frac{1}{T} \int_{t'}^{t'+T} y(t) e^{-in\omega_0 t} \mathrm{dt} \tag{11b}$$

Using Eq. (11a), the coefficient c_n of the signal waveform with $y(t) = A$ for $0 \le t \le \tau$ and $y(t) = 0$ for $\tau < t \le T$) is derived as

$$c_n = \frac{1}{T} \int_0^{\tau} y(t) e^{in\omega_0 t}\, \mathrm{dt} = \frac{A}{in\omega_0 T}\left(1 - e^{-in\omega_0 \tau}\right)$$

$$= \frac{A}{in\omega_0 T}\left(e^{\frac{in\omega_0 \tau}{2}} - e^{-\frac{in\omega_0 \tau}{2}}\right) e^{-\frac{in\omega_0 \tau}{2}} \tag{12a}$$

Furthermore, considering Euler's identity: $\sin(\omega t) = \frac{e^{i\omega t} - e^{-i\omega t}}{2i}$, Eq. (12a) becomes

$$c_n = \frac{A}{in\omega_0 T} \cdot 2i \sin\left(\frac{n\omega_0 \tau}{2}\right) \cdot e^{-i\left(\frac{n\omega_0 \tau}{2}\right)} \tag{12b}$$

and rearranging this expression results in

$$c_n = \frac{A\tau}{T} \cdot \left| \frac{\sin\left(\frac{n\omega_0 \tau}{2}\right)}{\left(\frac{n\omega_0 \tau}{2}\right)} \right| e^{i\left(\frac{n\omega_0 \tau}{2}\right)} \tag{12c}$$

Also, substituting $\omega = 2\pi f_0$ into Eq. (12c), the magnitude $|c_n|$ of *n-th* harmonic is expressed as

$$|c_n| = \frac{A\tau}{T}\left|\frac{\sin(n\pi f_0\tau)}{(n\pi f_0\tau)}\right| \tag{12d}$$

Equation (12d) includes both negative and positive sides of the spectrum $(-\infty < f < +\infty)$ but considering only the positive (real) side of the spectrum $(0 < f < +\infty)$, the value of $|c_{n+}|$ should be doubled:

$$|c_{n+}| = 2\cdot\frac{A\tau}{T}\left|\frac{\sin(n\pi f_0\tau)}{(n\pi f_0\tau)}\right| \tag{13a}$$

Finally, based on Eq. (13a), the envelope (where all harmonic peaks lie) can be expressed as

$$|c_{n+}| = 2\cdot\frac{A\tau}{T}\left|\frac{\sin(\pi f\tau)}{(\pi f\tau)}\right| \tag{13b}$$

Hence, the expressions of $|c_n|$ and $|c_{n+}|$ presented by Eqs. (12a), (12b), (12c), (12d), (13a), and (13b) have the form of the function $\frac{\sin(x)}{x}$. Because negative frequencies do not exist in the physical world, the spectrum of a digital signal with a 50%-duty cycle is shown in Fig. 5 for positive frequencies only.

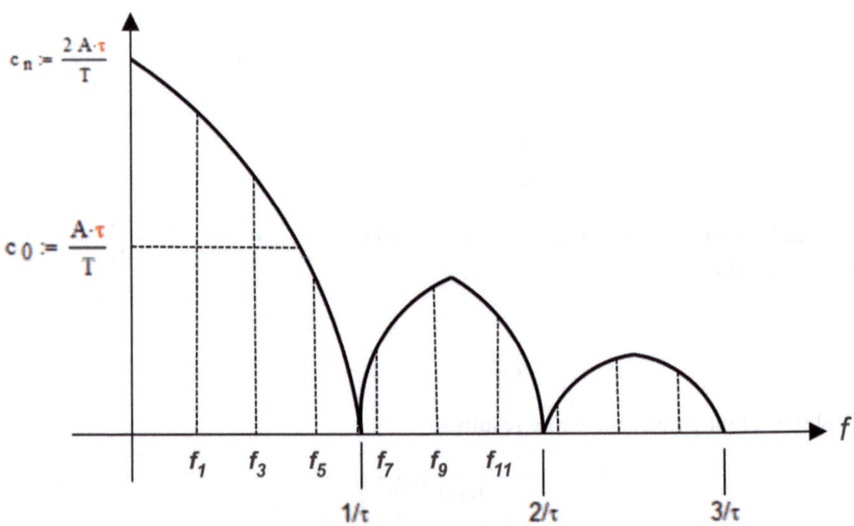

Fig. 5 The spectrum of a square waveform

As noted before, odd harmonics exist only for this specific signal.

Furthermore, Fig. 6 shows the bounds or envelopes of the harmonic peaks at the dB scale.

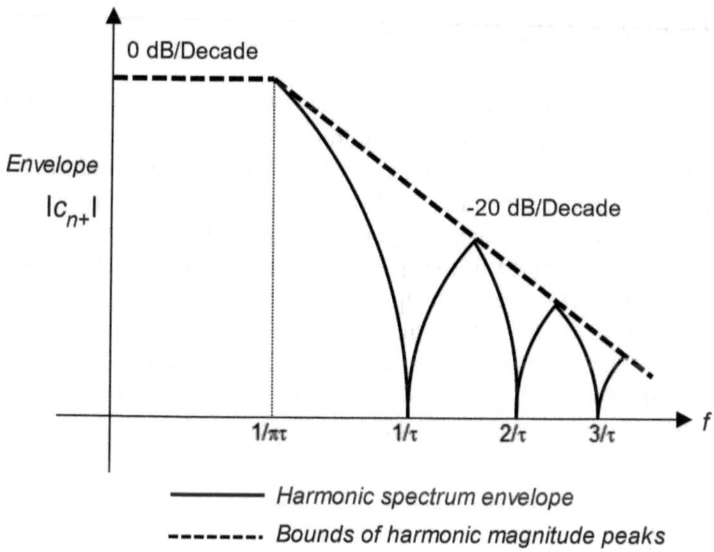

Fig. 6 The bound spectrum of a square waveform

This bound curve is analytically described by the expression

$$20\log\left(|c_n|\right) = 20\log\left(\frac{2A\tau}{T}\right) + 20\log\left(\left|\frac{\sin(\pi f\tau)}{(\pi f\tau)}\right|\right) \tag{14}$$

where the first bound for $0 \le f \le \frac{1}{\pi\tau}$ is normalized to 0 dB and for frequencies $\frac{1}{\pi\tau} < f \le +\infty$, the second bound slopes by −20 dB/decade. These bound curves actually show the frequency bandwidth of the considered digital signal.

3 Nonideal (Realistic) Digital Signals and Spectra

The shapes of the *periodic* digital pulses, which we considered in the preceding sections, are more idealistic and theoretical than real. However, considering ideal signals is useful as a first step in studying *real* digital signals that do not have a perfect rectangular waveform with sharp rise/fall edges. Therefore, we consider digital signals with real shapes in the following subsection.

3.1 Trapezoidal Digital Signals

We previously reported that rectangular signal pulses in the time domain have *rise/fall* edges with instant jumps and falls, resulting in infinitely wide spectra. Therefore, a wide spectrum is not desirable from an EMI standpoint, and its reduction directly leads to a reduction in EMI emission. However, other system requirements (such as signal integrity and signal quality requirements) cannot tolerate the slow *rise–fall* times of signals; consequently, the spectrum width reduction by this approach has certain limitations. However, a compromise between these two opposite requirements should be reached for the optimal system design.

We analyze the effects of real pulse *rise* and *fall* times on the signal frequency spectrum in this section. Figure 7 shows the *trapezoidal* pulse waveform in the time domain, whose edges are not as sharp as the edges of the previously considered rectangular signal waveform [7, 8].

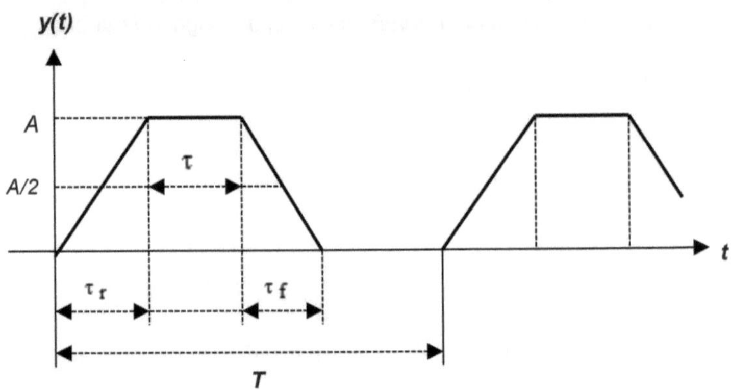

Fig. 7 Trapezoidal pulse waveform

This *trapezoidal* waveform can effectively represent real clocks and data streams in digital systems. Each pulse has an *amplitude* A, *width* τ, and *rise-* and *fall-times* τ_r and τ_f, as shown in Fig. 7. Here, we will see that the width of the signal spectrum is mostly dependent on the *rise* and *fall* times. To simplify the analytical analysis, we assume that the rise and fall times are equal ($\tau_r = \tau_f$), which is a fair approximation in practice.

We can present the *trapezoidal* waveform for positive frequencies via the Furrier series, Eq. (10b), where the coefficients $|c_n|$ are expressed by the product of two $\frac{\sin(x)}{x}$ functions of τ and τ_r and τ_f. For the simplified case when $\tau_r = \tau_f$, the coefficient $|c_n|$ is derived as

$$\left|c_{n+}\right| = 2\frac{A\tau}{T}\cdot\left|\frac{\sin\left(n\pi\tau/T\right)}{\left(n\pi\tau/T\right)}\right|\cdot\left|\frac{\sin\left(n\pi\tau_r/T\right)}{\left(n\pi\tau_r/T\right)}\right| \tag{15}$$

Additionally, we can present this discrete spectrum with a continuous envelope, substituting the term $\left(\frac{n}{T}\right)$ of Eq. (15) with frequency f:

$$\text{Envelope} = 2\cdot\frac{A\tau}{T}\left|\frac{\sin\left(\pi\tau f\right)}{\left(\pi\tau f\right)}\right|\left|\frac{\sin\left(\pi\tau_r f\right)}{\left(\pi\tau_r f\right)}\right| \tag{16}$$

The effect of the slower *rise/fall* times on the reduction in the signal spectrum is shown by the bound plot in Fig. 8.

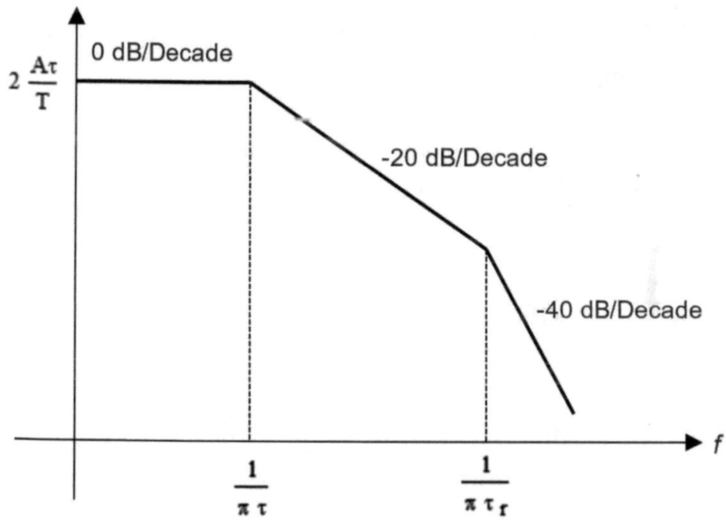

Fig. 8 Effect of the rise/fall time on the signal spectrum reduction

Here, the spectrum bound is presented by three lines:

$$\text{Line 1} = 20\log\left(\frac{2A\tau}{T}\right) \qquad (0\,\text{dB/decade})$$

$$\text{Line 2} = 20\log\left(\left|\frac{\sin\left(\pi f\tau\right)}{\left(\pi f\tau\right)}\right|\right) \qquad (-20\,\text{dB/decade})$$

$$\text{Line 3} = 20\log\left(\left|\frac{\sin\left(\pi f\tau_{r,f}\right)}{\left(\pi f\tau_{r,f}\right)}\right|\right) \qquad (-40\,\text{dB/decade})$$

where Line 1 has a slope 0 dB/dec with an amplitude level of $\frac{2A\tau}{T}$ for frequencies from $f = 0$ *to* $1/\pi\tau$. Line 2 has a slope of −20 dB/dec for frequencies from $f = 1/\pi\tau_r$ to $1/\pi\tau_r$, and Line 3 has a slope of −40 dB/dec for frequencies above $f = 1/\pi\tau_r$.

Thus, the slower signal rise and fall times contribute to the additional reduction in the frequency band by −20 dB/decade resulting in the total EMI reduction of −40 dB/decade. Additionally, Fig. 8 shows that the first spectral breakpoint occurs at $f_1 = 1/\pi\tau$ since the second spectral breakpoint occurs at $f_2 = 1/\pi\tau_{r,f}$.

Finally, note that the condition $\tau > \tau_{r,f}$ should always be satisfied. Additionally, shorter *rise/fall* times cause a wider EMI spectral content and consequently, the *rise/fall* times should be as slow as possible to reduce EMI emissions.

3.2 *Effects of the Signal Rate and Duty Cycle on the Signal Spectrum*

The *duty cycle D* of a pulse waveform is defined as the ratio of the pulse duration τ and the signal period *T:*

$$D = \tau/T \tag{17}$$

Accordingly, the coefficients $|c_{n+}|$ for the positive side of the frequency spectrum are expressed in terms of the *duty cycle D* via Eqs. (15) and (17) as

$$c_0 = AD; \quad |c_{n+}| = 2AD\left|\frac{\sin(n\pi D)}{(n\pi D)}\right| \cdot \left|\frac{\sin(n\pi\tau_r f_0)}{(n\pi\tau_r f_0)}\right| \tag{18}$$

where $f_0 = 1/T = D/\tau$.

Figure 9 shows the spectral bounds in terms of the *fundamental* frequency f_0 and the *duty cycle D*.

The first spectral breakpoint occurs at $f_1 = 1/\pi\tau = f_0/\pi D$, and the second spectral point occurs at $f_1 = 1/\pi\tau_r$. Additionally, a decrease in the duty cycle *D* leads to a reduction in the pulse amplitude from 2AD to $2AD_1$.

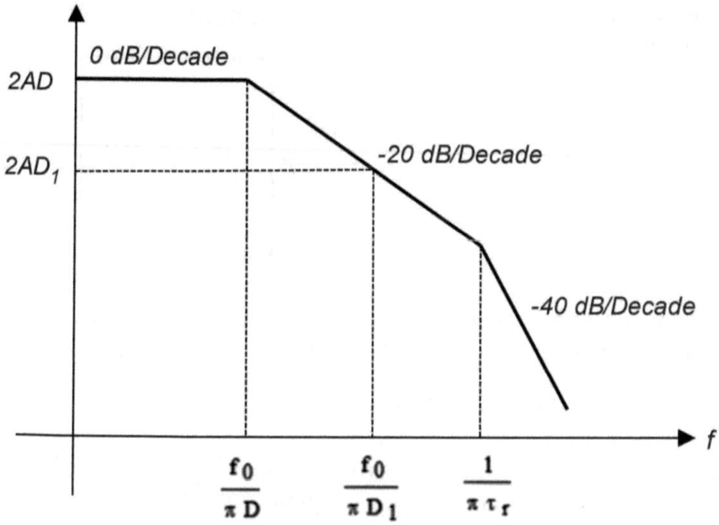

Fig. 9 Spectral bounds corresponding to the duty cycle D of a digital signal

4 Response of Linear Systems to Periodic Signals

We can consider systems to be linear when they satisfy the following conditions:

1. When the signals $y_1(t)$ and $y_2(t)$ are the system responses to the input signals $x_1(t)$ and $x_2(t)$, respectively, and the signal sum $y_1(t) + y_2(t)$ is the system response to the signal sum $x_1(t) + x_2(t)$. This condition is illustrated by the block diagram in Fig. 10.

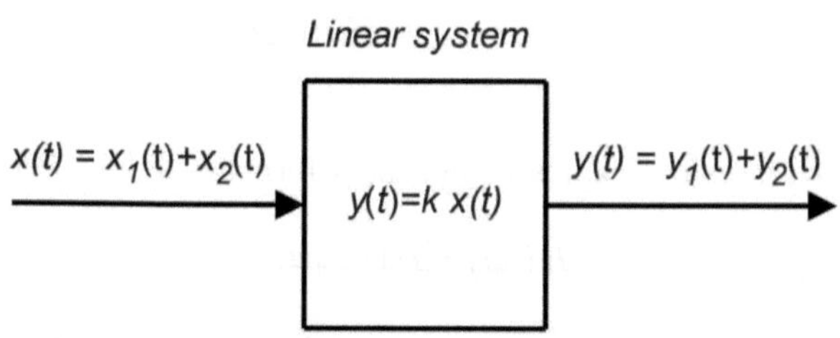

Fig. 10 First condition of system linearity

2. If the output signal $y(t)$ is a system response to the input signal $x(t)$, then the signal $a \cdot y(t)$ is the system response to the signal $a \cdot x(t)$ where a is an arbitrary constant. This condition is illustrated by the block diagram in Fig. 11.

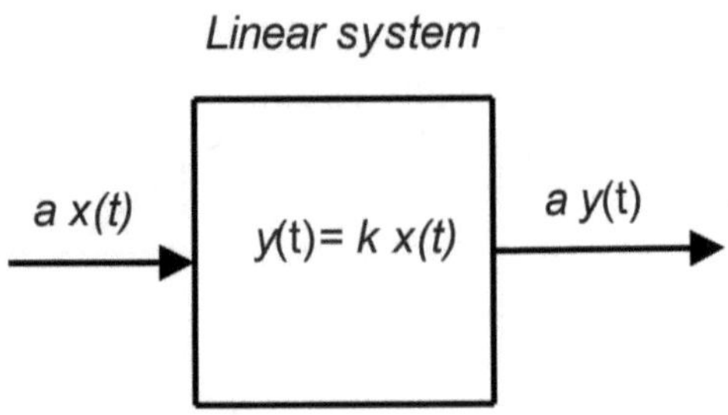

Fig. 11 The second condition of system linearity

Let us now consider the *unit impulse function* $\delta(t)$ also known as *Dirac's* impulse, as the *input* signal of a *linear* system. This theoretical unit function is defined as

$$\delta(t) = 1 \text{ for } t = 0 \text{ and } \delta(t) = 0 \text{ for } t \neq 0 \tag{19}$$

The $\delta(t)$ impulse has an extremely narrow duration (theoretically zero), resulting in its extremely wide spectrum (theoretical infinity). The impulse $\delta(t)$ is a differential of a step function $f(t)$ (i.e., $\delta(t) = df(t)/dt$) where $f(t) = 1$ for $t \geq 0$ and 0 for $t < 0$).

Figure 12 shows a linear system where the input signal is the $\delta(t)$ impulse and its system response is the *unit function* $h(t)$.

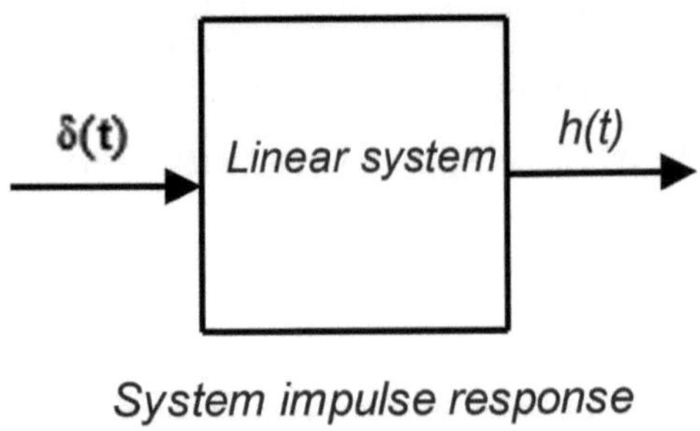

Fig. 12 $h(t)$ is a system response to the impulse function $\delta(t)$

The function $h(t)$ can be expressed in *phasor* form as

$$H(i\omega) = |H(i\omega)|\psi_h \tag{20}$$

where $H(i\omega)$ is the extremely wide spectrum of the *function* $\delta(t)$ that represents the *transfer function* of a linear system. Therefore, the system *transfer function*, which has an infinitely wide spectrum is useful for linking the phasor forms of the *output* and *input* signals $|Y(i\omega)|\lfloor\varphi_Y$ and $|X(i\omega)|\lfloor\varphi_X$, respectively, via the system transfer function $H(i\omega)$:

$$|Y(i\omega)|\varphi_Y = |H(i\omega)|\psi_h \cdot |X(i\omega)|\varphi_X \tag{21}$$

or in the complex form as

$$\tilde{Y}(i\omega) = \tilde{H}(i\omega)\cdot\tilde{X}(i\omega) \tag{22}$$

where the output signal magnitude and phase are expressed as

$$|Y(i\omega)| = |H(i\omega)|\cdot|X(i\omega)| \text{ and } \varphi_Y = \psi_h + \varphi_X \tag{23a}$$

Figure 13 shows the block diagram for this case.

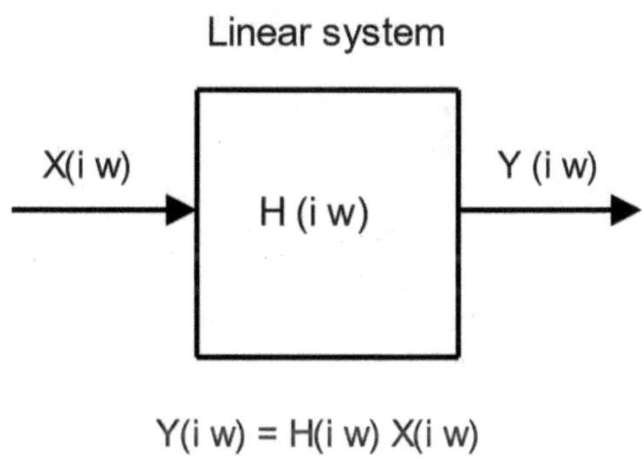

Fig. 13 System transfer function$H(i\omega)$

Regarding a linear system, if the frequency-domain impulse response is $H(i\omega)$ and the corresponding time-domain response is $h(t)$, then the time-domain output signal $y(t)$ can be obtained via the *convolution integral* as

$$y(t) = \int_{-\infty}^{+\infty} h(t-\tau)\cdot x(\tau)\,d\tau \tag{23b}$$

Hence, knowing $H(i\omega)$, the corresponding time-domain response $h(t)$ can be determined via Eq. (23b) by applying to a narrow pulse at the system input, whose duration is much smaller than the smallest system time constant.

Note that if the input signal $x(t)$ is a *periodic digital* signal, then its *Fourier series* is the infinity sum of *cosine* harmonics:

$$x(t) = c_0 + \sum_{n=1}^{\infty} |C_n| \cos(n\omega_0 t + \alpha) \tag{24}$$

Owing to the infinitely wide *transfer function* $H(i\omega)$, each $x(t)$-harmonic passes through the system creating the corresponding cosine or sine response because $y(t) = k\,x(t)$, where k is a system constant independent of frequency.

5 Nonperiodic and Pulse Signals

Pulses without a periodical repetition in time (or that occur once in time) are nonperiodic or single-pulse signals. The regular nonperiodic or pulse signals from digital circuits and systems are not so critical from an EMI standpoint. Nevertheless, if they originate from strong pulses and interference such as lightning, ESD, or power transients, we need to have knowledge of the *spectral content* of these signals. However, the behavior of nonperiodic and pulse signals is not easy to predict because they usually behave randomly in time. In the preceding sections, we considered *deterministic* signals whose behavior can be predicted in the future, but the future behavior of *random* signals in time can be approximately predicted and described by *probabilistic* means only.

Thus, a *nonperiodic* signal can be considered a single pulse or a train of pulses that do not repeat themselves in time. We can consider a single pulse as a case of periodic train pulses, whose period T expands to infinity keeping the pulse amplitude A and width τ unchanged, and then consider the spectral content of the remaining pulse. Therefore, let us consider a single pulse from the square wave in Fig. 14a, whose spectral content is shown in Fig. 14b.

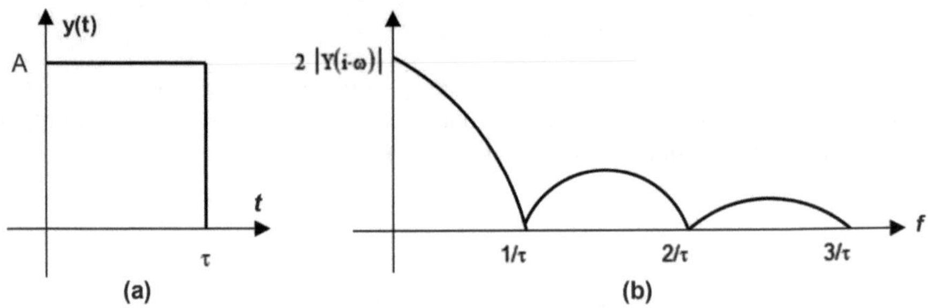

Fig. 14 (**a**) A single pulse in the time domain and (**b**) its spectrum

When we keep the pulse amplitude A and width τ unchanged while increasing the period T, the *envelope* of the spectral (harmonic) components will obtain the form via Eq. (13b) as

$$\text{Envelope} = \frac{2A\tau}{T}\left|\frac{\sin(\pi f \tau)}{(\pi f \tau)}\right| \tag{25}$$

By increasing the period T of a periodic signal, its fundamental frequency $\left(f_0 = \frac{1}{T}\right)$ decreases, causing the spectral components to shift closer to each other. The shape of the envelope per Eq. (25) remains the same because it depends on the unchanged pulse width τ. Additionally, Eq. (25) shows that the harmonic amplitudes decrease from $\frac{2A\tau}{T}$ to 0 when T increases. Furthermore, when T continues to increase to infinity, the discrete harmonic components approach the continuous spectrum.

To analyze the spectral content of a single pulse mathematically, we need to employ the *Fourier transform* of the time-domain signal $y(t)$ as

$$Y(j\omega) = \int_{-\infty}^{\infty} y(t) e^{-i\omega t}\, \mathrm{dt} \tag{26}$$

where the *inverse Fourier transform* has a time-domain form as

$$y(t) = \frac{1}{2\pi} \int_{-\infty}^{\infty} Y(j\omega) e^{i\omega t}\, d\omega \tag{27}$$

Hence, the *Fourier transform* of the single pulse $y(t)$ shown in Fig. 14a can be obtained as follows:

$$Y(j\omega) = \int_{0}^{\tau} A e^{-i\omega t}\, \mathrm{dt} = -\frac{A}{j\omega}\left(e^{-i\omega\tau} - 1\right)$$

$$= -\frac{A}{j\omega} e^{-\frac{i\omega\tau}{2}} \left(e^{-\frac{i\omega\tau}{2}} - e^{\frac{i\omega\tau}{2}}\right) = -\frac{2A}{j\omega} e^{-\frac{i\omega\tau}{2}} \left[2i\sin\left(\frac{1}{2}\omega\tau\right)\right]$$

$$\left|Y(j\omega)_{+}\right| = 2A\tau \left| \frac{\sin\left(\frac{1}{2}\omega\tau\right)}{\frac{1}{2}\omega\tau} \right| \tag{28}$$

The spectrum magnitude $|Y(j\omega)|$ is doubled because we consider it only in the positive part of the spectrum, as shown in Fig. 14b.

Note that there is an analogy between the *Fourier series* of *periodic* signals and the *Fourier transform* for *nonperiodic* signals. However, in terms of quantity, there is a major difference between the two because the amplitude of individual harmonics has infinitely small values as noted previously.

The relationship between the properties of the *Fourier transform* and the *Fourier series* is highly useful for signal analysis. Knowing the *Fourier transform* of a single pulse, it is possible to determine the *complex-exponential* coefficients for the *Fourier series* of a *periodic* waveform of such pulses by replacing ω with $n\omega_0$ into $Y(j\omega)$ and dividing this result by the signal period T:

$$c_n = \frac{1}{T} \cdot Y(jn\omega_0) \tag{29}$$

In the literature, such as in [9], there are tables of *Fourier transforms* for variously shaped pulses based on Eqs. (8a–c) that we can use to determine the *complex exponential* coefficients for the *Fourier series*.

6 Digital Random Signals

We considered *deterministic periodic* signals in the preceding sections, whose time domain is known a priori. The future behavior of *random* signals in the time domain can be more or less predicted *probabilistically* only. These signals are also considered in detail in [10, 11].

Random signals can be any data that carry arbitrary information. One of the *random* signals is the PCM-NRZ digital waveform. PCM-NRZ refers to "*pulse-code modulation—nonreturn to zero,*" which means that the signal does not return to zero after changing its logic state.

The *random* waveform $y(t)$, with the related *supporting random functions* $n(t)$ and $n(t - t_0)$, is shown in Fig. 15a–c, respectively.

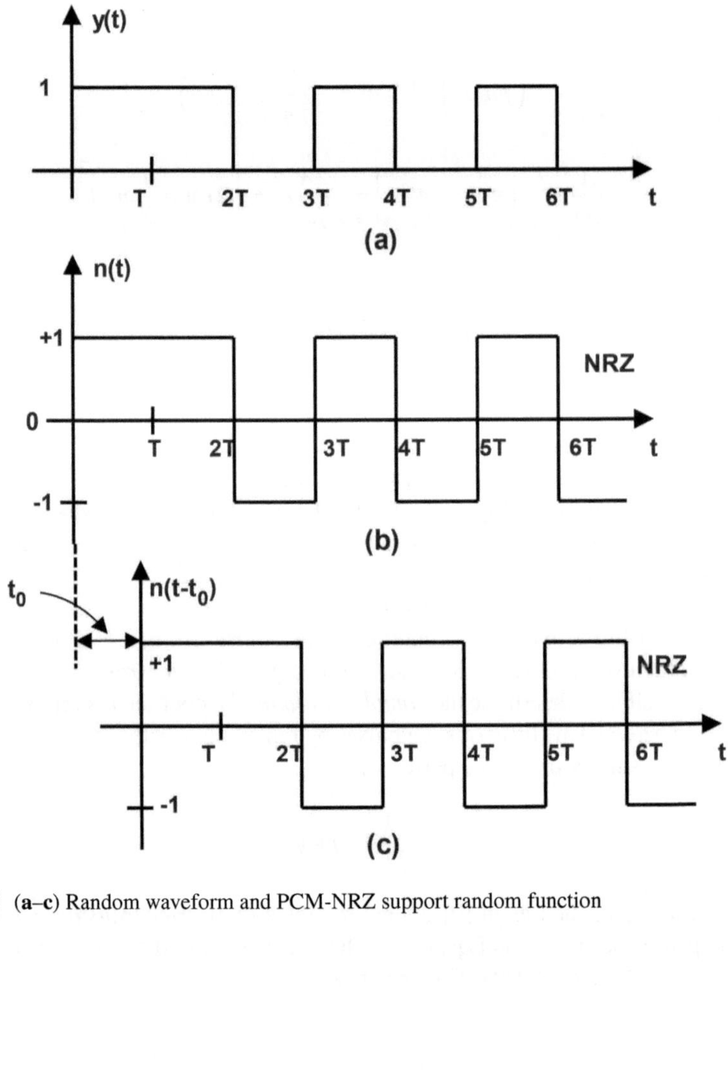

Fig. 15 **(a–c)** Random waveform and PCM-NRZ support random function

The function $y(t)$ can be described by the expression

$$y(t) = \frac{1}{2}\left[1 + n(t)\right] \tag{30}$$

where $n(t)$ and its shift $n(t - t_0)$ by t_0 time is the supporting function. The function $y(t)$ has two logic states (±1) and *equal probability* in a bit interval $n\,T < t < (n + 1)\,T$. Notably, Eq. (30) gives a fair approximation of the data waveform $y(t)$.

Furthermore, the *autocorrelation* function $R_y(t_0)$ of the *random* signal $y(t)$ is defined as the product of that signal and its shift in time by t_0:

$$R_y(t_0) = \overline{y(t)\cdot y(t - t_0)} = \lim_{t\to\infty} \frac{1}{T} \int_{-T/2}^{T/2} y(t)\cdot y(t - t_0)\,\mathrm{dt} \tag{31}$$

where $R_y(t_0)$ represents the *statistical average* of all possibilities. Furthermore, substituting (30) into (31) leads to

$$R_y(t_0) = \frac{A^2}{4}\left[1 + \overline{n(t)\cdot n(t - t_0)}\right] \tag{32}$$

The calculation of $R_y(t_0)$ provides the following results:

$$R_y(t_0) = 1 - \frac{|t_0|}{T} \quad \text{for } |t_0| < T \tag{33}$$

$$R_y(t_0) = 0 \qquad \text{for} |t_0| > \mathrm{T} \tag{34}$$

The shape of the $R_y(t)$ function is shown in the plot in Fig. 16.

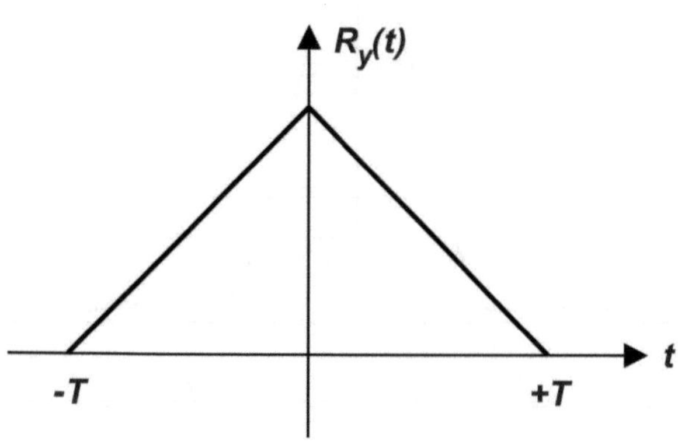

Fig. 16 Autocorrelation function of the random signal $y(t)$

Furthermore, the frequency spectrum of the *random* signal $y(t)$ can be given in terms of the *power spectral density* $G_y(\omega)$. Per the Wiener–Khinchine theorem, the power spectral density of a random signal is defined as the *Fourier transform* of its *autocorrelation function*.

$$G_y(\omega) = \int_{-\infty}^{\infty} R_y(t_0) \cdot e^{-i\omega t_0} \mathrm{d}t_0 \tag{35}$$

The *average* power of the random signal $y(t)$ in the frequency domain can be expressed as

$$P_{AV} = \int_{-\infty}^{\infty} G_y(f) \mathrm{d}f \tag{36}$$

Here, the *average* power with a statistically larger number of samples should be taken into the calculation. Therefore, for the PCM-NRZ waveform, the *power spectral density* is given as

$$G_y(f) = \frac{1}{4}\left(A^2 \delta(f) + \mathrm{T}\left| \frac{\mathrm{A}\sin(\pi fT)}{\pi fT} \right|^2 \right) \tag{37}$$

and plotted in Fig. 17.

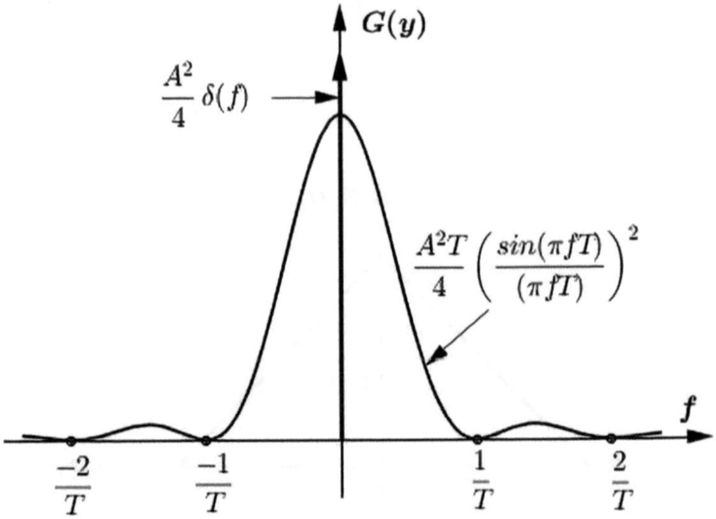

Fig. 17 Power spectral density of PCM-NRZ waveform

7 Prediction of EMI Spectra from Trapezoidal Signal Spectra

7.1 Common-Mode Emission Spectrum

Here, we consider the spectrum of the EMI radiation from a linear conductor (as a CM-mode radiator) with the current I_{cm} of a trapezoidal periodic signal. This waveform is shown in Fig. 7.

For the case of an electrically *short electric* radiator, the maximum field in the *far-field* region can be expressed by Eq. (B.15a) from Appendix B as

$$E^{f}{}_{cm,\max} = \eta_0 \frac{I_{cm}\, l}{2\lambda_0\, r} \tag{38}$$

and substituting $\lambda_0 = \dfrac{c}{f}$ into the expression, the intensity of the *CM* field is linearly proportional to the signal frequency *f* as

$$E^{f}{}_{cm,\max} = \eta_0 \frac{I_{cm} l}{2rc} \cdot f \tag{38a}$$

where $\eta_0 = 377\ \Omega$, l is the conductor length, I_{cm} is the CM current of the trapezoidal waveform, f is *the* spectral frequency, c represents the speed of the light $\left(\approx 3 \cdot 10^8\ {}^{m}\!/\!_{s}\right)$, and r represents the radial distance from the radiator surface.

Figure 18 shows the radiation of the spectral bounds of the trapezoidal waveform.

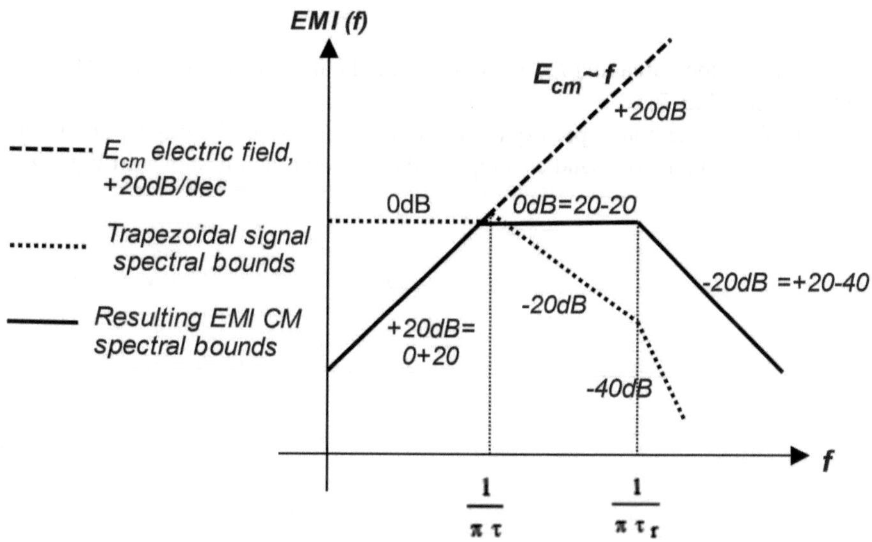

Fig. 18 Resulting EMI CM spectral envelope of a trapezoidal signal

According to Eq. (38a) of the field intensity E_{cm} and the trapezoidal common-mode current I_{cm}, the resulting CM EMI spectrum is plotted. First, the E_{cm} intensity increases at a rate of +20 dB/decade up to the first frequency bound $f_1 = 1/_{\pi\tau}$ (due to $+20\ dB + 0\ dB = +20\ dB$). Beyond f_1, the E_{cm} intensity remains constant (0 dB/decade) up to the second frequency bound $f_2 = 1/_{\pi\tau_r}$ (due to $+20\ dB - 20\ dB = 0\ dB$). After the f_2 bound frequency, the spectral bound E_{cm} envelope decreases to 0 at a rate of −20 dB/decade (due to $+20\ dB - 40\ dB = -20\ dB$).

Hence, on the basis of the known radiator parameters and the spectrum of a digital waveform, we can fairly predict the spectrum of the CM radiation from a linear conductor as an EMI source.

7.2 *Differential-Mode Emission Spectrum*

Now we consider the DM radiation spectrum of a current loop with the current I_{dm} that also carries the trapezoidal signal waveform. This radiator belongs to the category of small current loop or differential-mode radiators. In this case, the maximum DM field in the *far-field* region is given in Appendix B by Eq. (B.16a) as

$$E^f{}_{dm,\max} = \eta_0 \frac{\pi S I_{dm}}{\lambda_0^2 r}\left(\frac{V}{m}\right) \tag{39}$$

Substituting $\lambda_0 = \frac{c}{f}$ into Eq. (39), we obtain the DM field intensity in terms of frequency as

$$E^f{}_{dm,\max} = \eta_0 \frac{\pi S I_{dm}}{r c^2} \cdot f^2 \tag{39a}$$

where S is the loop area and I_{dm} is the differential current of the trapezoidal waveform shown in Fig. 7.

According to the bound per Eq. (39a) and the bound plot of a trapezoidal waveform, the resulting EMI bound envelope of the DM emission is shown in Fig. 19.

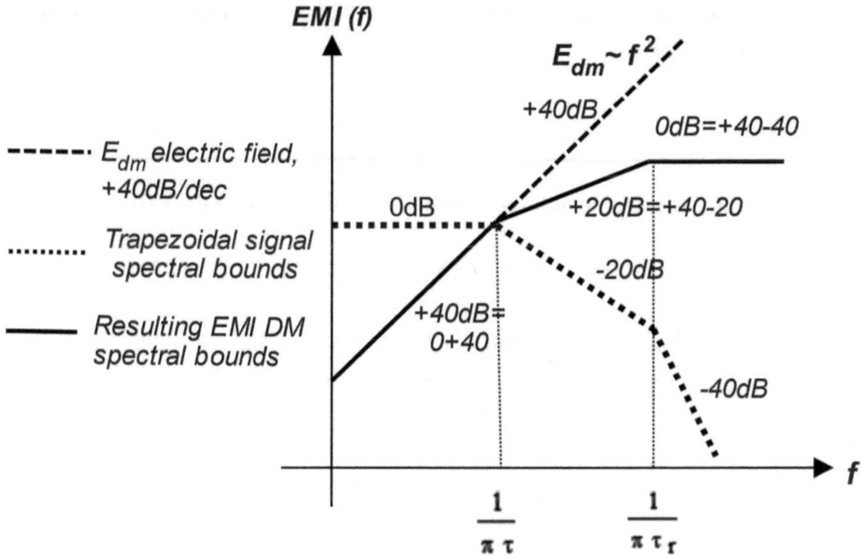

Fig. 19 Resulting DM spectral envelope of a trapezoidal signal

Therefore, according to Eq. (39a), the intensity of the E_{dm} field is proportional to f^2 and the resulting field increases by +40 dB/decade from *0* Hz to the first frequency bound $f_1 = 1/\pi\tau$ (due to *+40 dB – 0 dB = +40 dB*). Beyond f_1, the DM field intensity increases at a rate of +20 dB/decade up to the second spectral bound $f_2 = 1/\pi\tau_r$ (due to *+40 dB – 20 dB = +20 dB*), and next, the resulting bound envelope becomes constant, 0 dB/decade (due to *+40 dB – 40 dB = 0 dB*).

Therefore, on the basis of the known radatior parameters and the spectrum of the trapezoidal waveform, we can predict the spectrum of the radiated field created by a loop radiator as a source of DM radiation.

8 Definitions of Random Variable, Cumulative Distribution Function (CDF), and Probability Density Function (PDF)

A function $\boldsymbol{x}(n)$ is called a random variable $\boldsymbol{x}$, whose variable n is a set of probability results. For a given number x, the function set $\langle \boldsymbol{x}(n) \le x \rangle$ is a possible event for any real number x. As a result, the probability of this event, $P(\boldsymbol{x} \le x)$, is a function of the number x. This function is called the *cumulative distribution function* (CDF) or simply the *distribution function F(x)*, expressed as [12–14]

$$\mathrm{CDF} = F(x) = P\{\boldsymbol{x} \le x\} \quad \text{for } -\infty < x < \infty \tag{40}$$

Furthermore, the derivative of this function is called the *probability density function* $f(x)$ (PDF):

$$\text{PDF} = f(x) = \frac{dF(x)}{dx} \tag{41}$$

Additionally, the CDF can be determined by the integral of the PDF as

$$\text{CDF} = F(x) = \int_0^1 f(x)\,\text{dx} \tag{41a}$$

For example, Fig. 20a, b graphically illustrates the CDF and PDF for an event with 8 outcomes, each with the same probability P_k, where $k = 1, 2, 3, \ldots, 8$.

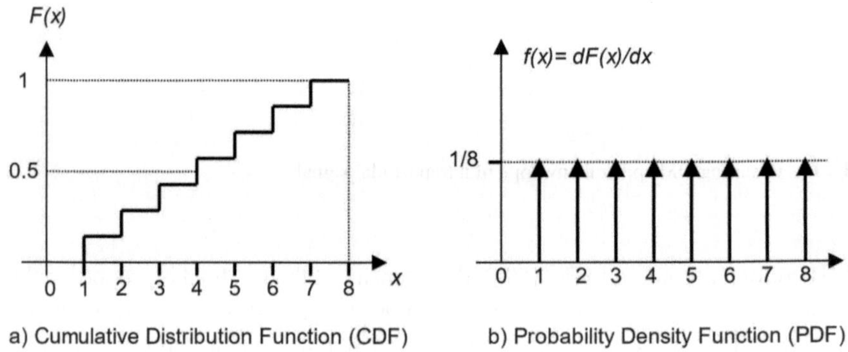

Fig. 20 (**a**) CDF and (**b**) PDF illustration

The CDF is not a decreasing function and exists in the range $0 < F(x) < 1$, since a normalized $f(x)$ varies between 0 and 1.

Furthermore, the expected value of a real random variable $\boldsymbol{x}$ can be determined by the integral

$$E\{\boldsymbol{x}\} = \int_{-\infty}^{\infty} x\, f(x)\,\text{dx} \tag{42a}$$

where $f(x)$ is the PDF of $\boldsymbol{x}$. Generally, the expected value of a function $g(x)$ is given as

$$E\{\boldsymbol{g}(\boldsymbol{x})\} = \int_{-\infty}^{\infty} g(x)\, f(x)\,\text{dx} \tag{42b}$$

When $\boldsymbol{x}$ contains discrete values x_n with probability p_n, $E\{\boldsymbol{x}\}$ is expressed as

$$E\{\boldsymbol{x}\} = \sum_n x_n\, P\{\boldsymbol{x} = x_n\} = \sum_n x_n\, p_n \tag{43}$$

The mean value m of the expected random variable determines the center gravity of the PDF function. Another important parameter of a PDF is its variance σ^2, which is defined as

$$\sigma^2 = \int_{-\infty}^{\infty} (x-m)^2 f(x)\mathrm{d}x \tag{44}$$

For $\boldsymbol{x}$ with discrete values x_n, the deviance σ^2 is expressed as

$$\sigma^2 = \sum_n (x_n - m)^2 P\{\boldsymbol{x} = x_n\} = E\left\{(\boldsymbol{x}-m)^2\right\} \tag{45}$$

Using the useful relationship $E\{(\boldsymbol{x} - m)^2\} = E\{(\boldsymbol{x})^2\} - E^2\{\boldsymbol{x}\}$, where $E\{\boldsymbol{x}\} = m$, Eq. (45) becomes

$$\sigma^2 = E\left\{(\boldsymbol{x})^2\right\} - m^2 \tag{46}$$

The functions defined above are used as important parameters in chapter "Performance of Radio Digital Systems in Presence of EMI" when we consider radio signals in randomly changed transmission channels with fading and EMI.

9 Stochastic Signal Processes

We consider a carrier signal presented by the function $x(t) = A\cos(\omega_c t + \phi)$, whose parameters are the amplitude A, frequency $\omega_c = 2\pi f_c$, and phase ϕ. When these signal parameters are constant in time, the signal $x(t)$ is a deterministic signal. However, when these parameters randomly vary, this is the case for stochastic (random) signals whose parameters are random variables of a stochastic process. The mean of a random process, $m(t)$, is the expected value $E\{\boldsymbol{x}(t)\}$ of the random variable $\boldsymbol{x}(t)$ at the fixed time τ and is given as [12–14]

$$m(\tau) = E\{\boldsymbol{x}(\tau)\} = \int_{-\infty}^{\infty} x f(x,\tau)\mathrm{d}x \tag{47}$$

The *autocorrelation function* of this random process is presented by two random variables $\boldsymbol{x}(t_1)$ and $\boldsymbol{x}(t_2)$ at times τ_1 and τ_2 as

$$R(\tau_1,\tau_2) = E\{\boldsymbol{x}(\tau_1)\boldsymbol{x}(\tau_2)\} = \int_{-\infty}^{\infty} x_1 x_2 f\left[(x_1,\tau_1);(x_2,\tau_2)\right] dx_1 dx_2 \tag{47a}$$

For instance, the *autocorrelation* of the random variable $\boldsymbol{x}(t)$ with two random processes is given by Eq. (48).

$$\boldsymbol{x}(t) = \boldsymbol{a}_1 \cos(\omega_c t) + \boldsymbol{a}_2 \sin(\omega_c t) \tag{48}$$

where both variables $\boldsymbol{a}_1$ and $\boldsymbol{a}_2$ have $m(t) = 0$ and the same variance σ^2 with the ω_c= constant.

Because $m(t) = E\{\boldsymbol{x}(t)\} = 0$ and $E\{\boldsymbol{a}_1\boldsymbol{a}_2\} = 0$, and because random processes a_1 and a_2 are independent of each other, the autocorrelation of $\boldsymbol{x}(t)$ expressed by Eq. (47a) becomes

$$R(\tau_1,\tau_2) = \sigma^2 \cos\left[\omega_c(\tau_1 - \tau_2)\right] \tag{48a}$$

where $\sigma^2 = E\{\boldsymbol{a}_1^2\} = E\{\boldsymbol{a}_2^2\}$.

9.1 Stationary Stochastic Signals

A stochastic signal (or process) is stationary if its PDF is unchanged in time. In particular, $f(x,t) = f(x)$, which results in

$$m = E\{\boldsymbol{x}(t)\} = \text{constant}$$

$$E\{\boldsymbol{x}(\tau_1)\boldsymbol{x}(\tau_2)\} = E\{\boldsymbol{x}(t)\boldsymbol{x}(t+\tau)\} \quad \text{for} \quad \tau = \tau_1 - \tau_2$$

$$R(\tau) = E\{\boldsymbol{x}(t)\boldsymbol{x}(t+\tau)\}$$

Hence, the autocorrelation of a stationary random signal depends only on the time difference τ. Thus, a stationary random signal can be uniquely defined in terms of its *mean* and *autocorrelation*. Additionally, the power spectrum density $S(\omega)$ of a stationary signal $\boldsymbol{x}(t)$ is represented by the Fourier transform (FT) of its *autocorrelation* $R(\tau)$:

$$S(\omega) = \int_{-\infty}^{\infty} R(\tau)e^{-i\omega\tau}d\tau \tag{49a}$$

since the inverse FT gives $R(\tau)$:

$$R(\tau) = \int_{-\infty}^{\infty} S(\omega)e^{i\omega\tau}d\omega \tag{49b}$$

where for $\tau = 0$ there is

$$R(0) = \int_{-\infty}^{\infty} S(\omega)d\omega \tag{49c}$$

Now, $S(\omega)$ becomes the *average (mean) signal power* in a given frequency band $\Delta\omega$.

9.2 A Response of a Linear System to Stochastic Signals

Previously we considered the response of a linear system to a deterministic signal. Now we consider the response of the linear system to *stochastic* (*random*) signals. Fig. 21 shows a block diagram for this case.

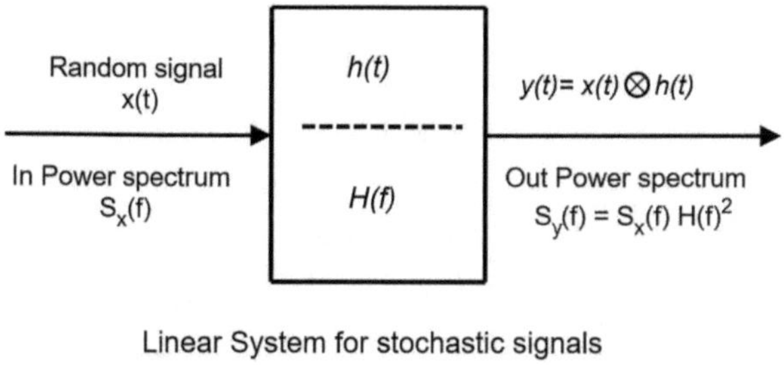

Fig. 21 A stochastic signal over a linear system

Here, the output signal *y(t)* is represented by the convolution of the input signal *x(t)* and system impulse response *h(t)*. Additionally, the output power spectrum $S_y(f)$ of the output signal *y(t)* is expressed as $S_y(f) = S_x(f)\ |H(f)|^2$, where $S_x(f)$ is the power spectrum of the input signal $x(t)$.

9.3 Q(x) Gaussian Distribution Function

The $Q(x)$ function (or complementary cumulative distribution function (CCDF)) is commonly used for the characterization of the *symbol error rate* and many other stochastic processes that follow the normal distribution or Gaussian distribution. Q(x) is expressed as

$$Q(x) = 1 - \int_x^{+\infty} f_G(z)\,\mathrm{dz} = 1 - \int_x^{+\infty} \frac{1}{\sigma\sqrt{2\pi}} e^{-\frac{1}{2}\left(\frac{z-m}{\sigma}\right)^2} \mathrm{dz} \tag{50a}$$

where $f_G(z)$ is a Gaussian PDF expressed as

$$f_G(z) \frac{1}{\sigma\sqrt{2\pi}} e^{-\frac{1}{2}\left(\frac{z-m}{\sigma}\right)^2} \tag{50b}$$

For example, Fig. 22 shows the three zero-mean Gaussian PDFs for $\sigma = 0.5$, 1, and 2.

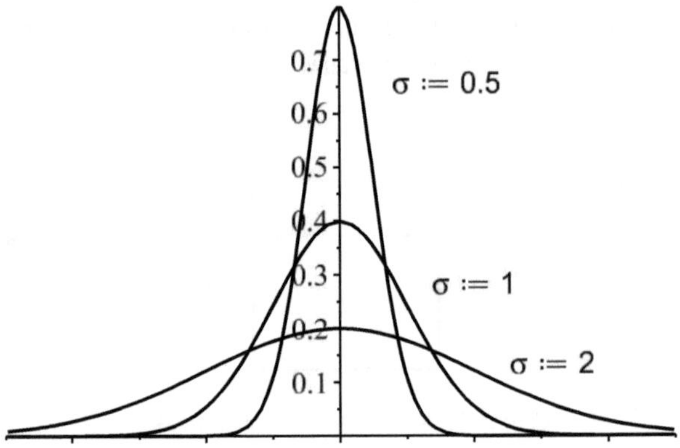

Fig. 22 Gaussian PDFs with different variances σ (0.5, 1, 2) and $m = 0$

Therefore, the Gaussian probability distribution is represented by the area below the Gaussian PDF function $f_G(z)$. Additionally, for unity variance ($\sigma^2 = 1$) and zero mean ($m = 0$), the function $Q(x)$ can be expressed as

$$Q(x) = 1 - \frac{1}{\sqrt{2\pi}} \int_x^{+\infty} e^{-\frac{1}{2}\left(\frac{z}{\sigma}\right)^2} \mathrm{dz} = \frac{1}{2} erfc\left(\frac{x}{\sqrt{2}}\right) \tag{50c}$$

where $erfc(x) = \frac{2}{\sqrt{\pi}} \int_0^x e^{-z^2} \mathrm{dz}$.

The Gaussian probability distribution, or normal distribution, is further discussed in chapter "Introduction to Digital Communication Systems" regarding additive white Gaussian noise (AWGN) in digital transmission channels. Additionally, several other probability distributions are discussed in chapter "Performance of Radio Digital Systems in Presence of EMI", such as the Rayleigh, Ricean, and Nakagami gamma distributions, which consider signal fadings (slow and fast fading).

The readers interested in further reading the stochastic signals and stochastic processes are referred to [15].

References

1. B.P. Lathi, R. Green, *Linear Systems and Signals*, 3rd edn. (Oxford University Press, Oxford, 2017)
2. C.R. Paul, R.C. Scully, *Introduction to Electromagnetic Compatibility* (Wiley, Hoboken, 2022)
3. C.R. Paul, *Analysis of Linear Circuits* (McGraw Hill, New York, 1989)
4. C.D. Mc Gillem, G.R. Cooper, *Continues and Discrete Signals and System Analysis* (Holt, Rinehart, and Winston, New York, 1974)

5. E.M. Stein, R. Shakarchi, *Fourier Analysis: An Introduction* (World Book Publishing Co., Chicago, 2011)
6. *A Student's Guide to Fourier Transforms: With Applications in Physics and Engineering*, 3rd edn. (Cambridge University Press, Cambridge, 2011)
7. E.C. Jordan, in *Reference Data for Engineers* (Indianapolis, 1985)
8. S.J. Mason, H.J. Zimmerman, *Electronic Circuits, Signals, and Systems* (Holt, Rinehart, and Winston, New York, 1974)
9. E. Kreyszig, *Advanced Engineering Mathematics*, 8th edn. (Wiley, New York, 1999)
10. A. Papoulis, *Probability, Random Variables, and Stochastic Processes* (McGraw-Hill, New York, 2002)
11. A.B. Carlson, *Communication Systems* (McGraw-Hill, New York, 1968)
12. W. Stark, *Introduction to Digital Communications* (Cambridge University Press, Cambridge, 2023)
13. J. Speidel, in *Introduction to Digital Communications*, 2nd edn. (Springer, 2021)
14. S-M M. Yang, in *Modern Digital Radio Communication Signals and Systems*, 2nd edn. (Springer, 2020)
15. M. Kovacevic, Z. Djurovic, in *Fundamentals of Stochastic Signals, Systems, and Estimation Theory*, 2nd edn. (Springer and Academic Mind, 2008)

Nonideal Passive Circuit Components

1 Introduction

Electrical circuits typically contain different passive components, such as *resistors*, *capacitors*, *inductors*, and pieces of *conductors*, that connect the circuit components to each other. This chapter analyzes the properties of nonideal or realistic passive circuit components and conductors. This topic is vital regarding the properties of the lossy signal lines, EMI filters, and EMI shielding, which are discussed in chapters "Performance of RF Cables and IT Signals Lines in Radio systems with EMI", "EMI Filters: Theory and Design", and "EMI Shielding for Radio Equipment", respectively.

Real circuit components and conductors behave differently at DC and frequency. They act as pure (ideal) R, L, and C components at DCs that have *parasites* at all frequencies, introducing negative effects in electronic circuits and signal lines. In the following sections, we will see that nonideal (realistic) conductors and passive circuit components have *parasites* such as

(a) Nonideal *conductors* with *parasitic resistance* and *inductance*.
(b) Nonideal *resistors* with *parasitic inductance* and *capacitance*.
(c) Nonideal *capacitors* with *parasitic resistance* and *inductance*.
(d) Nonideal *inductors* with *parasitic resistance* and *capacitance*.

As part of regular electronic circuits, filters, and TLs, all these components and conductor parasites enhance and stimulate EMI coupling and emissions, being parts of high-frequency circuits, filters, and *TLs*. In the following sections, we consider the *low-frequency* and *high-frequency* properties of nonideal conductors and circuit components influenced by their parasites.

The writing of this chapter is motivated and influenced by the references [1–6].

M. Pajovic, *Fundamentals of EMI Phenomena in Radio Communication Systems*, https://doi.org/10.1007/978-3-031-88748-2_3

2 Dielectrics versus Conductors

As discussed in Appendix A, there are two types of currents in media materials: the *conducted* currents I_c, which are dominant in good conductors, and the *displacement* currents I_d, which are dominant in dielectrics. Additionally, a good conductor is associated with very high conductivity σ since, in contrast, a good dielectric is associated with very low conductivity.

Generally, materials can act as dielectrics and conductors, called *lossy conductors* or *lossy dielectrics*. According to Maxwell–Ampere's equation and Ohm's law ($\vec{J}_c = \sigma \vec{E}$), both of which are discussed in Appendix A,

$$\vec{\nabla} \times \vec{H} = \vec{J}_c + \vec{J}_d = \sigma \vec{E} + i\omega\varepsilon \vec{E} \tag{1}$$

or

$$\vec{\nabla} \times \vec{H} = (\sigma + i\omega\varepsilon) \cdot \vec{E} \tag{1a}$$

Equation (1a) can also be expressed in another form for lossy dielectric media as

$$\vec{\nabla} \times \vec{H} = i\,\omega\tilde{\varepsilon} \cdot \vec{E} \tag{1b}$$

where $\tilde{\varepsilon}$ is the *complex permeability* expressed as

$$\tilde{\varepsilon} = \varepsilon' - i\varepsilon'' = \varepsilon'\left(1 - i\frac{\sigma}{\omega\varepsilon'}\right) = \varepsilon'\left(1 - i\frac{\sigma}{s}\right) \tag{1c}$$

where $s = \varepsilon'\omega$ is the dielectric *susceptance* and where ω is the angular frequency ($\omega = 2\,\pi f$).

Therefore, the *displacement* current is the dominant current in a good dielectric ($I_d \gg I_c$) because it possesses much greater *susceptance* than *conductance* ($s \gg \sigma$). However, for good conductors, the opposite is true: the *conducted* current is dominant in a conductor ($I_c \gg I_d$) because a good conductor possesses much greater *conductance* σ than *susceptance* s ($\sigma \gg s$).

Equation (1) indicates that the conducting current $\vec{J}_c$ is proportional to the electric *conductance* σ, whereas the *displacement* current $\vec{J}_d$ is proportional to the dielectric *susceptance* $s = \varepsilon'\omega$. Therefore, for good conductors and good dielectrics, the Maxwell–Ampere equations are as follows:

$$\vec{\nabla} \times \vec{H} = \sigma \vec{E}; \quad \vec{\nabla} \times \vec{H} = i(\varepsilon'\omega)\vec{E} \tag{2}$$

As noted in Appendix A, both currents (I_c and I_d) produce a circulated magnetic field $\vec{H}$ around a conductor and dielectric that carry those currents.

Another important parameter for *lossy* dielectrics is the *loss tangent* (or just the *tangent* δ), which is defined as

$$\tan(\delta) = \frac{\varepsilon''}{\varepsilon'} = \frac{\sigma}{\varepsilon'\omega} = \frac{\sigma}{s} \tag{2a}$$

Additionally, substituting Eq. (2a) into Eq. (1c), the complex propagation constant $\tilde{\varepsilon}$ can be expressed in terms of the *tangent* δ as

$$\tilde{\varepsilon} = \varepsilon'\left(1 - i\frac{\sigma}{s}\right) = \varepsilon'(1 - i\tan\delta) \tag{2b}$$

Let us now consider a *lossy dielectric* conductor and lossy *conductor* separately. In Appendix C, the *propagation constant* is introduced in *complex* form for *lossy* dielectric media as

$$\gamma = \sqrt{Z'Y'} = \sqrt{i\,\omega\mu(\sigma + i\omega\varepsilon)} = \alpha + i\,\beta \tag{3a}$$

where $Z' = i\,\omega\mu$ and $Y' = (\sigma + i\omega\varepsilon)$ are the per-unit-length *impedance* and *admittance* of the material, respectively. However, for *good dielectrics* where $s = |\varepsilon\,\omega| \gg \sigma$, Eq. (3a) becomes

$$\gamma = \alpha + i\,\beta = i\,\omega\sqrt{\mu\,\varepsilon} = i\,\beta \tag{3b}$$

where from Eq. (3b), the propagation constant is

$$\beta = \omega\sqrt{\mu\,\varepsilon} \tag{3c}$$

Furthermore, for a *lossy conductor*, where $\sigma \gg |\varepsilon\,\omega|$, Eq. (3a) leads to the simplified forms

$$\gamma \approx \sqrt{i\,\omega\mu\sigma} = \sqrt{\omega\mu\sigma}\,(1+i) \tag{4}$$

Hence, according to Eqs. (3a) and (4), the *attenuation constant* α and the *propagation constant* β are equal in the *lossy* conductor

$$\alpha = \beta \cong \sqrt{\frac{\omega\mu\sigma}{2}} = \sqrt{\pi f \mu\sigma} \tag{5}$$

where $\omega = 2\pi f$.

Another important media parameter is the *intrinsic* impedance of a lossy conductor ($\sigma \gg \omega\varepsilon$), which is expressed as

$$\tilde{\eta} = \sqrt{\frac{X'}{Y'}} = \sqrt{\frac{i\,\omega\mu}{(\sigma + i\omega\varepsilon)}} \cong \sqrt{\frac{i\,\omega\mu}{\sigma}} = \sqrt{\frac{\omega\mu}{\sigma}}\,(1+i) \tag{6a}$$

In *lossless dielectric media*, where $\sigma = 0$, the impedance η has a simple form:

$$\eta = \sqrt{\frac{i\omega\mu}{(\sigma + i\omega\varepsilon)}} \cong \sqrt{\frac{\mu}{\varepsilon}} \tag{6b}$$

The parameters of arbitrary media (which can be treated both as lossy conductors and lossy dielectrics) are summarized in Table 1.

Table 1 Conductive media versus dielectric media

	Ampere's law	Propagation constant	Intrinsic impedance
Arbitrary media	$\vec{\nabla} \times \vec{H} = (\sigma + i\omega\varepsilon)\vec{E}$	$\gamma = \alpha + i\beta$ $= \sqrt{i\omega\mu(\sigma + i\omega\varepsilon)}$	$\tilde{\eta} = \sqrt{\frac{i\omega\mu}{(\sigma + i\omega\varepsilon)}}$
Lossy conductive media	$\vec{\nabla} \times \vec{H} = \sigma\vec{E}$	$\gamma = \alpha + i\beta$ $= \sqrt{\frac{\omega\mu\sigma}{2}}(1+i)$	$\tilde{\eta} = \sqrt{\frac{\omega\mu}{\sigma}}(1+i)$
Lossy dielectric media	$\vec{\nabla} \times \vec{H} = i\omega\tilde{\varepsilon}\,\vec{E}$ $\tilde{\varepsilon} = \varepsilon' - i\varepsilon''$	$\gamma = i\omega\sqrt{\mu\varepsilon} = i\beta$ $\beta = \omega\sqrt{\mu\varepsilon}$	$\eta \cong \sqrt{\frac{\mu}{\varepsilon}}$

3 Skin Depth at Conductors

When electromagnetic waves penetrate a conductive material, the amplitude of the field inside the conductor decreases exponentially with increasing penetration depth, expressed by $E_{\max}\, e^{-\alpha\delta}$. The $E_{\max}$ represents the field amplitude at the conductor surface, α represents the *attenuation* constant, and δ represents the depth of wave penetration into the conductor. This phenomenon is called the *skin effect* [1, 2]. In particular, the skin depth δ is associated with the depth in the conductor where the wave amplitude is reduced to the level of $1/e$ (~37%):

$$E_{\max} e^{-\alpha\delta} = 1/e \tag{7a}$$

and normalizing Eq. (7a) with E_{max} results in

$$e^{-\alpha\delta} = 1/e \tag{7b}$$

Furthermore, substituting α from Eq. (5) into Eq. (7b) yields:

$$\delta = \frac{1}{\alpha} = \frac{1}{\sqrt{\pi f \mu_0 \sigma}} \tag{8}$$

Thus, the parameter δ numerically determines the *skin depth,* which is dependent on the frequency f, magnetic permeability μ, and material conductivity.

4 Wire Conductors

Low-impedance wires, traces on printed circuit boards, or other linear conductors, and high-frequency TLs are essential for electronic systems because they connect circuit components and system devices to each other. When the conductor impedance is not low, *signal loss* occurs at operating frequencies because the conductor impedance may significantly vary with frequency.

4.1 Wire Resistance and Inductance

Notably, *DC* and low-frequency currents penetrate deeper into a conducting material, whereas higher-frequency currents tend to flow closer to the conductor surface because of the *skin effect* (discussed in Sect. 2). Owing to this effect, the conductor impedance (resistance and inductance) depends on the frequency [1, 4].

Figure 1a–d shows (a) a wire long l with radius r, (b) the equivalent *RL* circuit that electrically represents the wire conductor, (c) the current distribution at the wire cross-sectional area for dc/low-frequency signals, and (d) for high-frequency currents when the skin depths are $\delta = r$ and $\delta < r$.

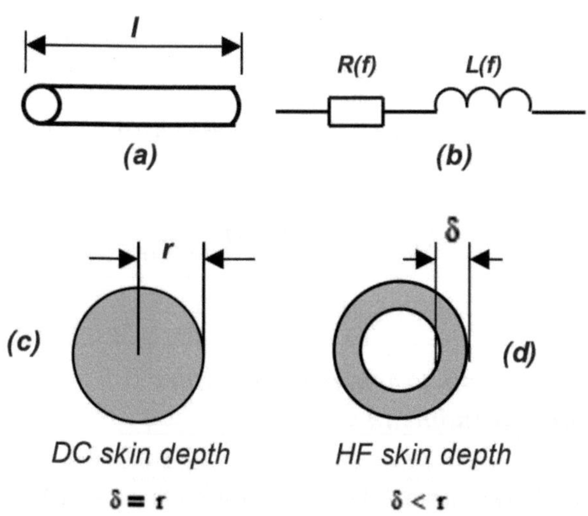

Fig. 1 (**a**) Finite-length wire, (**b**) Wire RL equivalent circuit, (**c**) DC skin depth, (**d**) HF skin depth

Here, we consider the wire *DC/low-frequency* and *high-frequency* resistance, as well as the *internal inductance*. In Appendix C, we consider the wire's *internal inductance* related to the *magnetic flux* inside the wire conductor and the wire's *external inductance* associated with the *magnetic flux* around the conductor. We note in Appendix C that the *external conductor inductance* is significantly greater than the *internal inductance*. However, for some specific conductor applications, it is helpful to consider the *internal inductance* of conductors, as we will do in the next section.

4.2 Conductor DC/Low-Frequency Resistance and Internal Inductance

Figure 1c shows the cross-sectional area ($S = \pi r^2$) entirely occupied by the current where the skin depth is $\delta = r$, where the per-unit-length *dc/low-frequency resistance* R'_{lf} is defined as

$$R'_{lf} = \rho \frac{1}{S} = \frac{1}{\sigma S} = \frac{1}{\sigma} \frac{1}{\pi r^2} \left(\frac{\Omega}{m} \right) \quad (9)$$

where $\rho = 1/\sigma$ is the per-unit-length wire *resistance* in $\left(\frac{\Omega}{m} \right)$.

Additionally, the wire's per-unit-length *dc/low-frequency internal inductance* has a constant value independent of frequency [4]:

$$L'_{lfi} = \frac{\mu_0}{8\pi} = 50 \left(\frac{nH}{m} \right) \quad \text{for} \quad \delta = r \quad (10)$$

Thus, at DC and very low frequencies, the conductor resistance and *internal* inductance are independent of frequency.

Additionally, the per-unit-length *impedance* associated with DC/low-frequency resistance and *internal inductance* is given as

$$Z'_{lfi} = R'_{lf} + i(2\pi f) L'_{lfi} \left(\frac{\Omega}{m} \right) \quad (11)$$

4.3 Conductor High-Frequency Resistance and Internal Inductance

Figure 1d shows the skin depth and the wire cross-sectional area ΔS occupied by the high-frequency current. The following expressions show the high-frequency dependency of the per-unit-length *resistance* R'_{hf} and wire *internal inductance* L'_{hf}, where the *skin depth* is $\sigma \ll r$:

$$R'_{hf} = \frac{1}{\sigma \Delta S} = \frac{1}{\sigma \left[\pi r^2 - \pi (r - \delta)^2 \right]} \approx \frac{1}{\sigma [2\pi r \delta]} = \frac{r}{2\delta} R'_{lf} \quad (12)$$

and substituting Eqs. (8) and (9) into Eq. (12), we obtain

$$R'_{hf} \approx \frac{1}{2r}\sqrt{\frac{\mu_0}{\pi\sigma}}\sqrt{f}\left(\frac{\Omega}{m}\right) \tag{13}$$

Furthermore, the per-unit-length *internal* inductance for *high* frequencies where $\delta \ll r$ is given by [4]

$$L'_{hfi} = \frac{1}{4\pi r}\sqrt{\frac{\mu_0}{\pi\sigma}}\frac{1}{\sqrt{f}} = \frac{2\delta}{r}L'_{lfi}\left(\frac{H}{m}\right) \tag{14}$$

Additionally, the internal high-frequency wire impedance is

$$Z'_{hfi} = R'_{hfi} + i(2\pi f)L'_{hfi}\left(\frac{\Omega}{m}\right) \tag{15}$$

Figure 2a, b shows (a) the plots of R'_{lf} and R'_{hf} and (b) the plots of L'_{lfi} and L'_{hfi} with frequency.

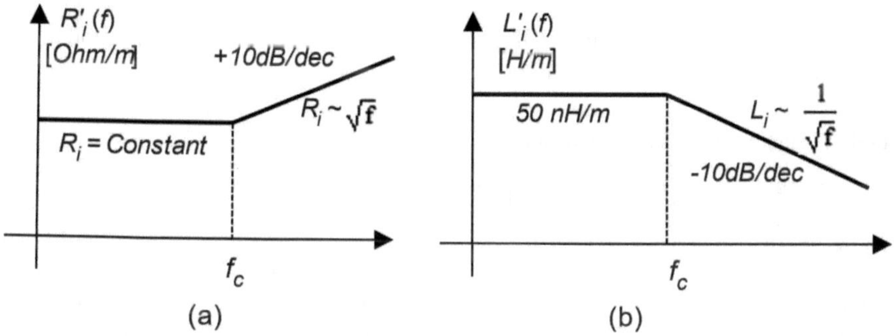

Fig. 2 Plots of (**a**) internal resistance and (**b**) internal inductance

Plot (a) shows that the resistance R'_i is constant up to the cutoff frequency f_c, where the *skin effect* starts to exist; furthermore, R_i increases by a rate of +10 dB/decade because R_i is proportional to the square root of the frequency f, $R'_{hf} \sim \sqrt{f}$, according to Eq. (13). Additionally, it can be seen in plot (b) that the *internal inductance* L'_{lfi} is constant up to the cutoff frequency f_c, where the *skin effect* starts to exist. Furthermore, this inductance decreases by a rate of 10 dB/decade because one is inversely proportional to the frequency square root, $L'_{hfi} \sim \frac{1}{\sqrt{f}}$, according to Eq. (14).

The cutoff frequency f_c is a boundary between the *low-frequency* and *high-frequency* regions of the *intrinsic* parameters of the conductor. Thus, f_c can be derived via Eqs. (9) and (13) for $R'_{lf} - R'_{hf}$ or for $L'_{lfi} - L'_{hfi}$ via Eqs. (10) and (14):

$$f_c = \frac{4}{\sigma_c \mu_0 \pi r^2} \tag{16}$$

In the following example, we calculate the impedance of the wire conductor via the above expressions (4)–(12).

Example 1 Let us find the per-unit-length impedance of a copper wire with a diameter d = 2 mm (radius r = 1 mm) long *1 m* at frequencies of 10 kHz, 100 kHz, 1 MHz, 10 MHz, 100 MHz, and 1 GHz via Eqs. (4)–(12).

Solution We first find the cutoff frequency f_c via Eq. (16) for the copper conductivity $\sigma_c = 5.8 \cdot 10^7 \frac{S}{m}$, *magnetic permeability* in free space $\mu_0 = 4\pi\, 10^{-7}\left(\frac{H}{m}\right)$, and wire radius r_w = 1 mm. According to (16) and the present parameters, we obtain f_c:

$$f_c = \frac{4}{\sigma_c \mu_0 \pi r_w^{\,2}} = 17.47\,\text{kHz}$$

The solutions for the per-unit-length wire *resistance*, *internal inductance*, and *internal impedance* are listed in Table 2.

Table 2 Summarized results of Example 1

f	$R'_{lf}\left(m\Omega/m\right)$	$L'_{lfi}\left(nH/m\right)$	$R'_{hf}\left(\Omega/m\right)$	$L'_{hfi}\left(nH/m\right)$	$\lvert Z'_i\rvert\left(\Omega/m\right)$
DC & 10 kHz	~5.5	50	–	–	$\sim 5.5 \cdot 10^{-3}$
100 kHz	–	–	0.013	21	0.013
1 MHz	–	–	0.042	6.6	0.042
10 MHz	–	–	0.13	2.1	0.13
100 MHz	–	–	0.41	0.66	0.41
1 GHz	–	–	1.3	0.21	1.36

Note that f = 10 kHz belongs to the lower-frequency region because f_c = 17.47 kHz, whereas at frequencies above f_c, it belongs to the *high-frequency* region. Additionally, the values of the wire *resistance*, *internal* inductance and, consequently, *internal* (*intrinsic*) *impedance* significantly increase with frequency.

4.4 *Conductor External Inductance*

Hence, in Appendix C, we note two types of conductor inductances: *internal* and *external* (or *loop*) inductances, which are associated with the *magnetic flux* flow within and outside the conductor. Additionally, we noted that the *external* (*loop*) inductance was significantly greater than the *internal* inductance.

Figure 3 shows an isolated two-conductor transmission line with conductor 1 that carries the current I_d in the direction from left to right and returns in the opposite direction over conductor 2.

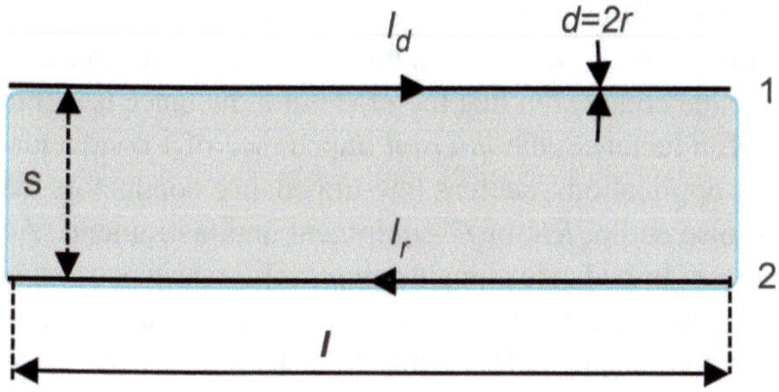

Fig. 3 Two-conductor TL and loop inductance

For the two-conductor line, with a conductor radius r and conductor separation s, the per-unit-length *external* inductance and impedance are expressed in Appendix C:

$$L'_{e2} \approx \frac{\mu_0}{\pi} \ln\left(\frac{s}{r}\right)\left(\frac{H}{m}\right) \tag{17a}$$

$$Z'_{e2} = X'_{e2} = (2\pi f) L'_{e2} \tag{17b}$$

In the next example, we numerically illustrate Eqs. (17a), (17b) by comparing a loop inductance with an internal inductance of the same conductor from Example 1.

Example 2 For the cases in Fig. 3, with conductors 1 and 2 having long $l = 1$ m, separated by $s = 4$ mm, and with the same radius $r = 1$ mm, the per-unit-length *external* inductance at $f = 100$ MHz is determined via Eq. (17a):

$$L'_{e2} \approx \frac{\mu_0}{\pi} \ln\left(\frac{s}{r}\right) \approx 0.55\left(\frac{\mu H}{\mathrm{m}}\right)$$

Additionally, the per-unit-length *inductive reactance*, at the same frequency, is determined via Eq. (15b):

$$Z'_{e2} = (2\pi f) L'_{e2} \approx 345\left(\frac{\Omega}{\mathrm{m}}\right).$$

Let us now compare the *internal* impedance from Example 1 and this *external* impedance at the same frequency of 100 MHz. Thus, in both examples, the copper conductors have identical dimensions (r and l). Thus, per Table 1, the calculated per-unit-length *internal impedance* of the conductor in Example 1 is as follows:

$$\left|Z'_{hfi}\right| \approx 0.42\left(\frac{\Omega}{\mathrm{m}}\right),$$

For two conductors in series, the *internal impedance* is $2 \cdot 0.42 \approx 0.84\ \Omega/\text{m}$. When the *internal* (or *intrinsic*) impedance is compared with the loop impedance $Z'_{e2} \approx 345\ (\Omega/\text{m})$, the difference between these impedances is huge. With a further increase in the conductor length and/or frequency, the difference between the two impedances increases further. By this, we confirmed that the *external* impedance of a conductor is much greater than its *intrinsic (internal)* impedance.

Regardless of the conclusion that the *external* inductance is significantly greater than the *internal* inductance, the *internal* impedance of a conductor can be important for sensitive applications, such as low-impedance conductors that connect sensitive and low-noise audio, *RF*, or *IT* equipment and instruments. Additionally, the conductor's *internal* impedance contains an *internal* resistance that is dependent on the frequency, which also contributes to a loss in long transmission lines. We further consider this topic in chapters "Performance of RF Cables and IT Signals Lines in Radio systems with EMI", "EMI Shielding for Radio Equipment", and "Grounding for Radio Systems and Facilities", which are related to lossy signal lines, EMI shielding, and system grounding.

5 Overview of Conductor and Dielectric Parameters

Here, we present some valuable tables below for typical conductors and dielectrics data. Table 3 shows the *relative* conductivities σ_r of typical conductors (relative to the copper conductivity, $\sigma_c = 5.8 \cdot 10^7 \frac{S}{m}$) and the *relative* permittivity ε_r of typical dielectrics (relative to the permittivity of air, $\varepsilon_0 \approx \frac{1}{36\pi} 10^{-9} \frac{F}{m}$).

Furthermore, we present Table 4 for the skin depth of the copper with frequency, which is obtained via Eq. (8).

Table 3 Conductor and dielectric properties

Conductors	σ_r	Dielectrics	ε_r
Silver	1.05	Air	1
		Styrofoam	1.05
Copper	1.00	Polyethylene foam	1.6
		Wood dry	1.5–4
Gold	0.70	Teflon	2.1
Aluminum	0.61	Polyethylene	2.3
Zinc	0.29	Polystyrene	2.5
Brass	0.26	Rubber	2.5–3
Nickle	0.20	Paper	3
Bronze	0.18	Nylon	3.0
Tin	0.15	Silicone rubber	3.1
Steel	0.10	Plexiglas	3.45

Table 3 (continued)

Conductors	σ_r	Dielectrics	ε_r
Lead	0.08	PVC	3.5
Monel	0.04	Epoxy resin	3.6
Stainless steel	0.02	Quartz	3.8
Iron	0.17	Epoxy glass	2.7
Beryllium	0.10	Bakelite	2.9
Mu-metal@1 kHz	0.03	Glass (Pyrex)	5.0
Permalloy@1 kHz	0.03	Mylar	5.0
–		Mica	5.4
–		Porcelain	6.0
–		Neoprene	6.5
–		Polyurethane	6.8
–		Silicon	11.8
Fresh/saltwater	$\sigma = 0.001/5$ S/m	Fresh/saltwater	80/81

Table 4 δ of Cu with frequency

f	δ
50/60 Hz	9.3/8.5 mm
1 kHz	2.1 mm
10 kHz	0.66 mm
100 kHz	0.21 mm
1 MHz	0.066 mm
10 MHz	0.02 mm
100 MHz	$6.6 \cdot 10^{-3}$ mm
1 GHz	$2.1 \cdot 10^{-3}$ mm

6 Nonideal Resistors

Resistors are the most common components in electronic circuits whose purpose is to obtain the desired resistance. In terms of technology, there are different types of resistors, such as (a) wire-wound resistors, (b) carbon-composition resistors, (c) metal-film resistors, and (d) thin-film (SMD) resistors. With respect to packaging and PCB mounting, low-frequency or "through-hole" resistors and high-frequency or surface mounting device (SMD) resistors exist. Nonideal (real) resistors are also considered well in [3–6].

The *ideal* resistor has magnitude *R* and a phase of zero degrees (i.e., its *reactance* is zero) in the broadband frequency range:

$$Z_R = R\angle 0^\circ = R + i0 \tag{18}$$

Furthermore, Fig. 4 shows the equivalent circuit of the *nonideal (real)* resistor in the broadband frequency spectrum.

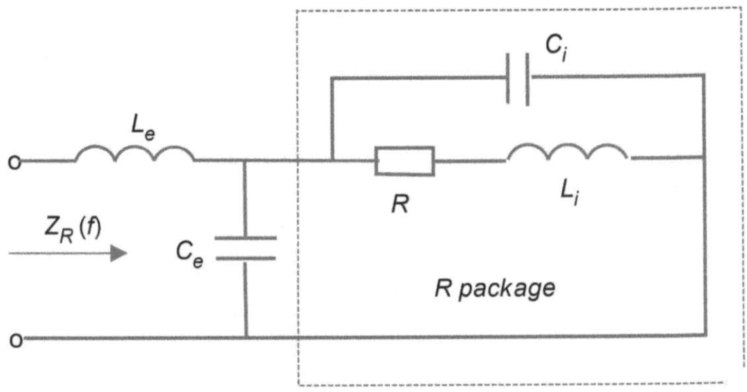

Fig. 4 Equivalent circuit of a nonideal resistor at high frequencies

This high-frequency *equivalent* circuit contains the *external*, leads *inductance* L_e and the *external*, leads *capacitance* C_e to the reference conductor or ground, as well as the *internal* (package) *inductance* L_i and *internal* (package) capacitance C_i. These inductances and capacitances represent undesired *parasites* of the actual or nonideal resistor.

On the basis of the equivalent circuit, the expression for the impedance of the *actual* resistor is

$$\tilde{Z}_R = iX_{Le} - iX_{Ce} \, / / \tilde{Z}_{R\_pack} \tag{19}$$

or

$$\tilde{Z}_R = iX_{Le} + \frac{-iX_{Ce} \cdot \tilde{Z}_{R\_pack}}{-iX_{Ce} + \tilde{Z}_{R\_pack}} \tag{19a}$$

where X_{Le} and X_{Ce} are, respectively:

$$X_{Le} = 2\pi f L_e ; X_{Ce} = \frac{1}{2\pi f C_e} \tag{19b}$$

Furthermore, $\tilde{Z}_{R\_pack}$ is part of the resistor impedance that includes the resistor package:

$$\tilde{Z}_{R\_pack} = -iX_{Ci} \, / / \left(R + iX_{Li}\right) \tag{20}$$

or

$$\tilde{Z}_{R\_pack} = \frac{-iX_{Ci} \cdot \left(R + iX_{Li}\right)}{-iX_{Ci} + \left(R + iX_{Li}\right)} \tag{20a}$$

where

$$X_{Ci} = \frac{1}{2\pi fC_i}; X_{Li} = 2\pi fL_i \tag{21}$$

The *external* parasites (L_e and C_e) depend on the length and proximity to the ground of the resistor leads, whereas the *internal* (package) parasites L_i and C_i depend on the package size and packaging technology. The next example illustrates the use of Eqs. (18)–(21).

Example 3 For example, Fig. 5 shows the frequency response of the nonideal carbon resistors, whose magnitudes $|\tilde{Z}_R|$ are normalized with the corresponding ideal resistances R at frequencies up to 10 GHz. The carbon resistors have four different resistances: $R = 1\ \Omega$, $10\ \Omega$, $100\ \Omega$, and $10\ \text{k}\Omega$ with the same inductive and capacitive *parasites*: $L_e = 0.5$ nH, $L_i = 0.3$ nH, $C_e = 0.2$ pF, and $C_i = 0.2$ pF.

The present plot shows that the impedance of the low-resistivity resistors exceeds the ideal (declared) values of 1 Ω and 10 Ω at frequencies above 100 MHz and 1 GHz. This is because the resistors' *inductive parasites* dominate over the small resistor values (1 Ω and 10 Ω) above the noted frequencies, i.e., $\omega\ (L_e + L_i) > R$. Additionally, the plot shows that the impedance of the resistors with higher resistance is below the ideal (declared) resistance of 100 Ω and 10 kΩ at frequencies above 20 MHz and 1 GHz, respectively. This is because the *capacitive parasites* dominate over the decleared resistor value R above the noted frequencies, i.e., $1/2\pi f(C_e + C_i) < R$.

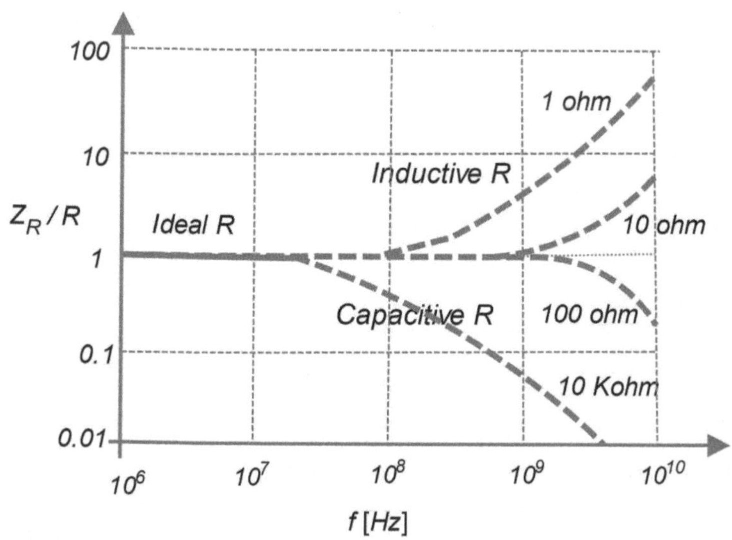

Fig. 5 Frequency response of nonideal resistors with different resistances R

Therefore, with respect to resistor package technologies, a *wire-wound* resistor (called a "low-frequency" resistor) behaves as an inductor or capacitor at higher frequencies depending on its resistance. *Carbon-composite* resistors (called "medium-frequency" resistors) have fewer *parasites* than *wire-wound* resistors do. However, we need to use larger-package resistors, such as wire-wound resistors or carbon resistors, for power electronic applications.

Another resistor technology is thin-film technology or surface-mounted device technology (SMD). Resistors of this type belong to the category of high-frequency resistors, which have the most miniature *parasites* because they possess small package dimensions and very short connecting leads. The standard SMD sizes are 0603, 0402, 0201, and 0101. Thus, SMD thin-film resistors are intended for low-power and high-frequency applications, where carbon and wire-wound resistors are not applicable.

7 Nonideal Capacitors

An ideal capacitor has impedance expressed as

$$\tilde{Z}_C = -i\frac{1}{2\pi fC} = \frac{1}{2\pi fC}\lfloor -90^\circ \qquad (22)$$

where the magnitude of the impedance of the perfect capacitor decreases linearly (on a log scale of −20 dB/decade) with frequency since its phase stays constant by −90° with frequency. Nonideal (realistic) capacitors are considered well in [3–6].

Figure 6 shows the *equivalent* circuit of a *nonideal* or realistic capacitor.

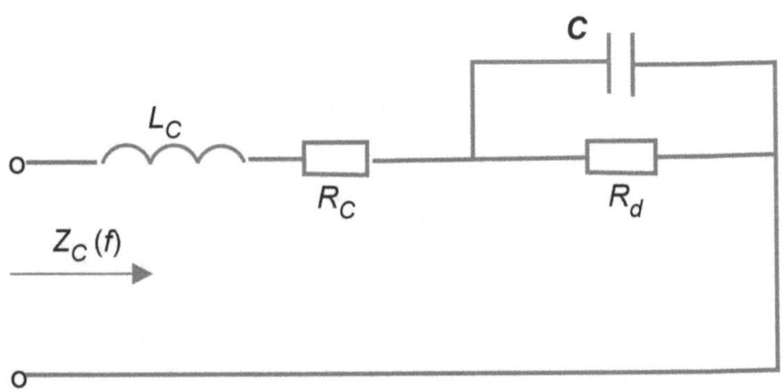

Fig. 6 Equivalent circuit of a nonideal capacitor at high frequencies

In addition to the original capacitance C, parasites such as R_C, L_C, and R_d exist in this circuit, where R_C and L_C are the resistance and inductance of the capacitor connecting leads, respectively, and where R_d is the dielectric resistance resulting in dielectric loss. The longer leads and larger capacitor package introduces more *parasites* that consequently have a greater impact on the frequency response of the capacitor. Additionally, associated with dielectric loss, the parameter called the *loss tangent* is defined as

$$\tan(\delta) = \frac{R_C}{X_C} \tag{23}$$

where X_C is the declared *capacitive reactance* of the capacitor equal to $1/(2\pi fC)$.

The following expression shows the impedance $Z_C(f)$ of the imperfect capacitor per the circuit in Fig. 6:

$$Z_C(f) = R_C + i(2\pi f)L_C + R_d // \left(\frac{1}{i2\pi fC}\right) \tag{24}$$

or

$$Z_C(f) = R_C + i(2\pi f)L_C + \frac{R_d \cdot \left(\frac{1}{i2\pi fC}\right)}{R_d + \left(\frac{1}{i2\pi fC}\right)} \tag{24a}$$

Additionally, the series parasite resonance f_r of the present capacitor circuit is expressed as

$$f_r = \frac{1}{2\pi\sqrt{L_C C}} \tag{25}$$

Hence, the parasitic inductance of the nonperfect capacitor may produce undesired resonance effects within the capacitor's package.

In terms of packaging technology, several types of capacitors exist: *electrolytic, tantalum, ceramic,* and *thin film (SMD)* types. The *electrolytic* capacitor has a larger capacitance, on the order of *1–1000 μF*, whereas *tantalum* technology allows a larger capacitance to be made in a smaller package. The *ceramic* capacitors have a smaller capacitance (on the order of *pF* to *μF*) but have a wide frequency response. Additionally, some *high-frequency* capacitors contain low-loss NPO and X7R dielectrics, as given in manufacturing catalogs.

As an example of using Eq. (24a), Fig. 7 shows the frequency response of a typical capacitor $C = 10$ nF with the parasites $R_C = 0.1\ \Omega$, $L_C = 0.5$ nH, and $R_d = 20$ MΩ.

Furthermore, according to Eq. (25), the *series* resonance of the present capacitor is $f_r \approx 70$ MHz.

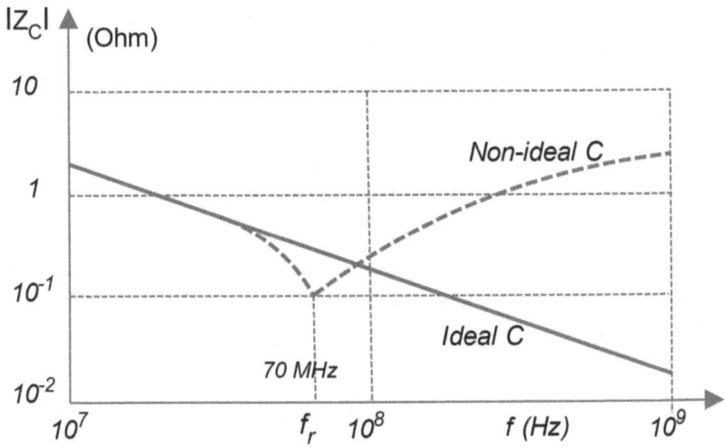

Fig. 7 Frequency response for a realistic capacitor

Usually, *ceramic* capacitors are used for diverting high-frequency EMI-conducted currents from power supply sources and power IC pins. As an example, Fig. 8 shows the decoupling model for diverting high-frequency EMI currents from an ASIC DC power pin to the ground.

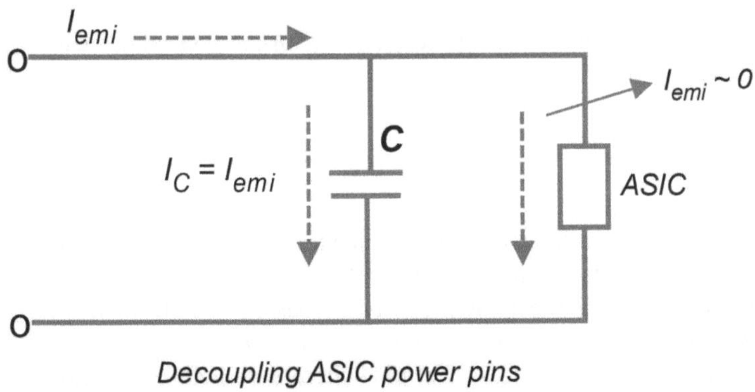

Fig. 8 Decoupling of EMI currents from ASIC power pins

Here, Z_{ps} is the impedance of a DC power ASIC input impedance that we need to protect from EMI current flow into the PS. To divert the EMI currents to the ground, we need to place the decoupling capacitor *C* with a much lower impedance than the PS impedance. The *C* placement should be too close and parallel to that *PS* output to reduce the C parasite inductance at the minimum. For a given circuit model to divert most of the current I_{noise}, Z_C should be much smaller than the impedance Z_{ps} ($Z_C \ll Z_{ps}$) at the frequencies of interest.

As an example, this decoupling method is effective up to 100 MHz when the capacitor C is 10 nF, whose frequency impedance profile is shown in Fig. 7.

In the end, the decoupling method is efficient only when the circuit impedance magnitude is much greater than the impedance magnitude of the decoupling capacitance at frequencies of interest, and when the decoupling capacitor is placed as close as possible to the circuit, we need to protect it from EMI currents. A decoupling capacitor acts as a simple low-pass filter that diverts high-frequency EMI currents from the protected circuit to the ground. We consider the low-pass filters in more detail later in the section on EMI Filter Theory and Applications, chapter "EMI Filters: Theory and Design".

8 Nonideal Inductors

The perfect or *ideal* inductor with the inductance L has an inductive reactance only, which is expressed in the imaginary or phasor forms as

$$\tilde{Z}_L = i2\pi fL = |2\pi fL| + 90° \tag{26}$$

Thus, the magnitude of the impedance of the perfect inductor linearly increases with frequency (on a log scale of +20 dB/decade), whereas its phase remains constant at +90° with frequency.

However, in reality, an inductor is imperfect at higher frequencies because its parasites are dependent on the inductor's technology. Figure 9 shows the equivalent circuit for the *nonideal* inductor at high frequencies [3–6].

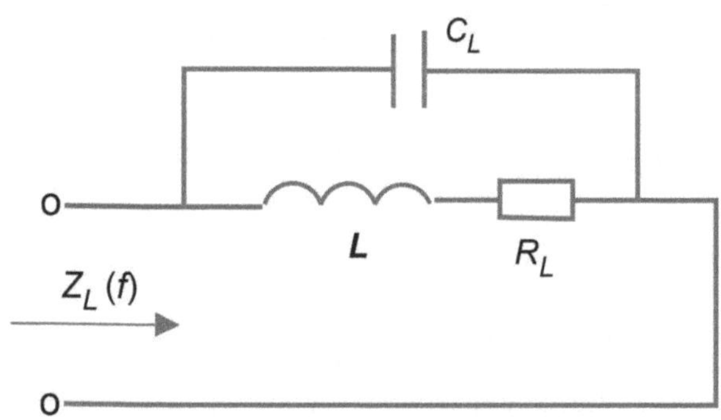

Fig. 9 Equivalent circuit model for a nonideal inductor at high frequencies

The equivalent circuit model contains the parasitic series resistance R_L and parasitic parallel capacitance C_L, which leads to the nonideal behavior of the inductor at high frequencies. Additionally, this model includes the *resistance*, *inductance,* and *capacitance* of the inductor connecting leads, but these parasites are significantly smaller than the internal package parasites and can be neglected. Per the given circuit model, the impedance of the nonperfect (realistic) inductor is expressed as

$$\tilde{Z}_L(f) = (R_L + i2\pi fL)//\left(\frac{1}{i2\pi fC_L}\right) \tag{27}$$

or

$$\tilde{Z}_L(f) = \frac{(R_L + i2\pi fL)\left(\frac{1}{i2\pi fC_L}\right)}{(R_L + i2\pi fL) + \left(\frac{1}{i2\pi fC_L}\right)} \tag{27a}$$

Because this inductor circuit is a parallel *LC* circuit, its *antiresonance* frequency is

$$f_{ar} = \frac{1}{2\pi\sqrt{LC_L}} \tag{28}$$

Per Eq. (27), Fig. 10 shows the frequency response for the ideal 1 μH inductor with an impedance magnitude equal to $Z_L = 2\pi fL$ (a solid curve) and the inductor impedance for the same inductor but with a *parasitic resistance* of $R_L = 0.3\ \Omega$ and a parasitic *capacitance* of $C_L = 1$ pF (a dashed curve per Eq. 27). The antiresonance frequency of this nonideal inductor is $f_{ar} \approx 160$ MHz.

The magnitude of the inductor impedance increases progressively with frequency up to the antiresonance frequency $f_{ar} \approx 160$ MHz (per Eq. 28), where the impedance is purely *inductive*. After the frequency f_{ar}, the magnitude $|Z_L(f)|$ decreases with increasing frequency, where the inductor impedance becomes purely *capacitive*.

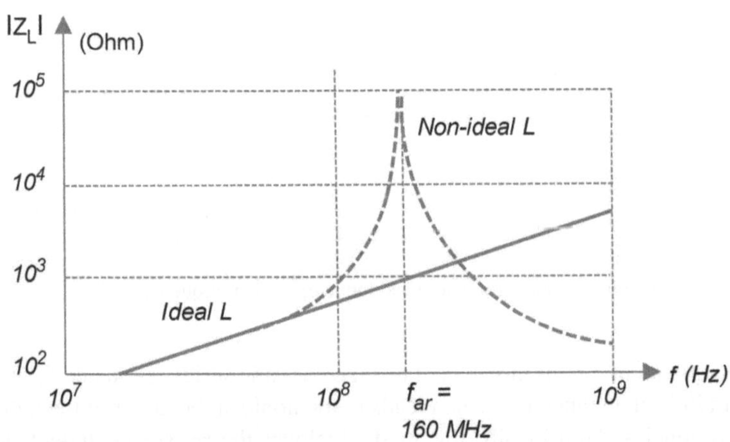

Fig. 10 Frequency response of a nonideal inductor—the solid trace refers to an ideal inductor and the dashed trace refers to a nonideal inductor

For example, we will show below how the present inductor of 1 μH can block *high-frequency* currents from passing into a power supply. By placing the inductor in series with the + rail of the power supply, we can effectively block EMI and RF currents from flowing into the power supply. This case is shown in Fig. 11.

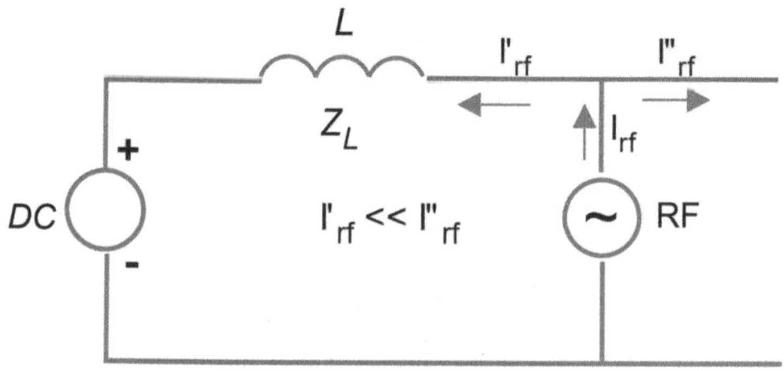

Fig. 11 Inductive protection of DC PS from RF currents

The inductor has an impedance magnitude of $|\tilde{Z}_L| = 1.25$ kΩ at 200 *MHz* according to Eq. (27), and accordingly, the inductor may effectively block *RF* and *EMI* currents from flowing into the power supply up to 200 MHz.

8.1 *Inductors with Ferromagnetic and Ferrite Cores*

As noted in Appendix A, the inductance of an inductor can be efficiently increased by winding the inductor on a ferromagnetic or ferrite core with high relative permeability μ_r. However, the critical point of this method is the *saturation* of the inductor's core. The saturation directly depends on the current intensity in the inductor's windings (i.e., the intensity of the magnetic field inside the core), which may significantly cause the loss of the declared inductance. Therefore, these inductors are specified for the maximum operating RMS current or power.

The core saturation is caused when the magnetic flux B in the core starts to saturate, i.e., no future follows the increase in the magnetic field H. In addition to the current intensity in the inductor's wirings, the core saturation mainly depends on the type and volume of the core magnetic material. For example, Fig. 12 shows the approximate $B - H$ and μ_r curves for pure iron as a function of the magnetic field intensity.

The plot μ_r shows the increase and decrease in the core relative permeability μ_r with magnetic intensity H. As shown, μ_r sharply jumps to a maximum and progressively decreases when the core reaches the saturation region. Because the inductance of any coil inductor is proportional to the core relative permeability μ_r (as discussed *in Appendix A*), the choke inductance follows the μ_r increase and decrease.

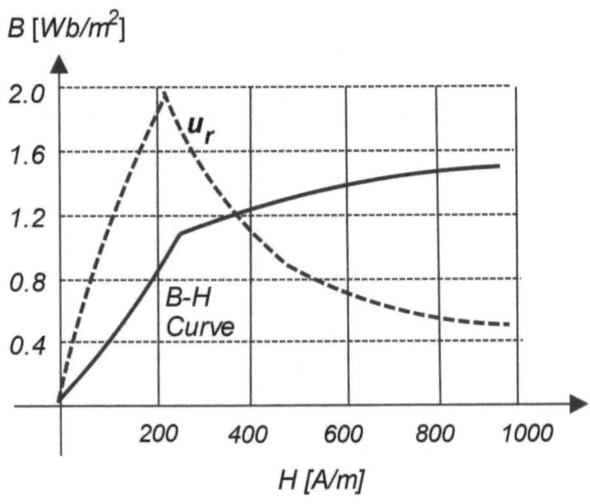

Fig. 12 *B–H* and μ_r curves for pure iron

The inductor's core usually saturates during high-current transients (surges), which commonly exist at the power mains. Notably, the basic main filters are commonly intended to suppress EMI time-steady currents and they are not as efficient against larger and faster current transients. The main filters are discussed in more detail in chapter "EMI Filters: Theory and Design", including equipment protection against high-current transients.

References

1. D.K. Cheng, *Field and Wave Electromagnetics*, 2nd edn. (Pearson Education, London, 2013)
2. S.H. Hall, H.L. Heck, *Advanced Signal Integrity for High-Speed Digital Designs* (Wiley, Hoboken, 2009)
3. L. Tihanyi, *Electromagnetic Compatibility in Power Electronics* (IEEE Press, Piscataway, 1995)
4. C.R. Paul, R.C. Scully, *Introduction to Electromagnetic Compatibility* (Wiley, Hoboken, 2022)
5. H.W. Ott, *Electromagnetic Compatibility Engineering* (Wiley, Hoboken, 2009)
6. C.W. Sayre, *Complete Wireless Design*, 2nd edn. (McGraw-Hill, New York, 2008)

Radio Communication Systems Performance Concerning EMI

Introduction to Digital Communication Systems

1 Introduction

The fundamentals of digital communication systems with additive white Gaussian noise (AWGN) are discussed in this chapter. Radio systems with AWGN, fading, and EMI are presented and discussed in chapter "Performance of Radio Digital Systems in Presence of EMI". The writing of this chapter is motivated and influenced by the references [1–11].

The purpose of any communication system is to transfer information data or messages from one location to another, i.e., from an information source to an information destination. Media for transporting information can include signal cables (copper or optical) and radio-wireless channels. The types of signals that carry information can be used to categorize communication systems into two main groups: analog and digital communications.

Analog communication systems are traditional systems in which a high-frequency analog signal (the carrier) is modulated by a baseband analog information waveform or by baseband digital information waveforms (sequences of binary digits). These digital waveforms are converted into an analog baseband waveform by modulating analog subcarriers in a modem. The subcarriers are low-frequency (audio) sinusoidal signals whose parameters (amplitude, frequency, or phase) vary according to the information baseband waveform. At the receiver side, the baseband waveform is extracted from a carrier through the demodulation process, which is the reverse of the modulation process.

Digital communication systems are modern and highly attractive in today's communications landscape because of the growing demand for faster and more reliable data transfers. These systems have extensive flexibility in terms of signal processing options and technological integration with various digital and information technology (IT) equipment. Most importantly, digital communication systems provide much higher reliability and transfer capacity than analog systems do.

M. Pajovic, *Fundamentals of EMI Phenomena in Radio Communication Systems*, https://doi.org/10.1007/978-3-031-88748-2_4

In a digital communication system, digital information signals modulate a high-frequency analog carrier by keying its amplitude, frequency, phase, or combination of these parameters. When the input information signal is an analog, it is first converted into discrete signal sequences by sampling. These sequences are then quantized and formatted into symbol sequences that are compatible with the digital communication system and a corresponding propagation channel.

Figure 1 shows the block diagram of a data communication system, including the transmitter (TX) and receiver (RX) blocks. The transmitter contains the source and channel encoders, modulator, and power amplifier (PA), whereas the receiver includes a low-noise amplifier (LNA), demodulator, channel decoder, and source decoder. In general, the TX and RX are connected via a communication channel (CH), which is affected by AWG noise and external electromagnetic interference (EMI).

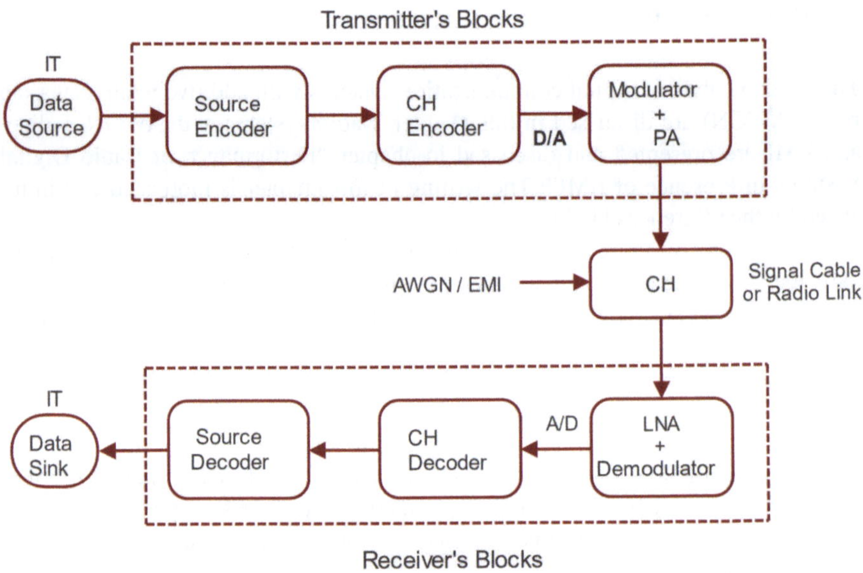

Fig. 1 Block diagram of a digital communication system

The diagram shows that an information bit stream enters a source encoder, whose functions are to encode and format the input bit streams into symbol streams suitable for processing by the channel encoder.

Next, in the channel encoder, redundant error control bits are added to symbol sequences, and new digital waveforms are used to modulate high-frequency carriers in the TX modulator. The modulated carriers are then amplified by the output power amplifier (PA) and transmitted through the communication channel.

In the RX system section, the received high-frequency analog carriers are amplified by the input low-noise amplifier (LNA) and then demodulated into a digital waveform suitable for processing by the channel decoder, where the redundant error control bits are removed. Finally, the digital waveform is further decoded and reformatted in the source decoder into the original information bit streams sent from the TX section.

One more substantial process critical to the operation of digital communication systems is the synchronization between the functions of the TX and RX communication blocks. This process is crucial for ensuring asynchronous operation both within each block and between the TX and RX blocks. Synchronization is provided in both the time and frequency domains. For example, *coherent* systems require synchronization of their frequency reference with the channel carrier frequency and phase, whereas *noncoherent* systems only require synchronization with the carrier frequency.

2 Transmitter Block

The transmitter is a part of the communication system that provides *frequency upconversion* for low-frequency baseband (information) digital waveforms. In this process, the baseband waveform is "translated" into high frequencies where it can be transmitted over longer distances. The following text describes the basic functional blocks of the transmitter.

2.1 Source Encoding

Usually, computers, analog-to-digital (A-to-D) conversion modules, and other digital IT and instrumentation equipment serve as sources and sinks (destinations) for data. The data are input signals for communication systems, particularly for *source encoders*. As previously noted, the task of a *source encoder*, as illustrated in Fig. 1, is to convert and format the original information contained in the input signal into a sequence of bits suitable for signal processing in the channel encoder.

The bit sequence may consist of computer-generated data, digitized voice, video, images, etc. The process in the source encoder involves formatting bits into a stream of information symbols (characters), where the information symbols are blocks containing a specific number of bits. For example, information symbols (or bit blocks) could be 5-bit symbols representing 32 (2^5) distinct symbols (e.g., letters and numbers) or 8-bit symbols for the ASCII code representing 128 (2^8) characters (letters, numbers, and punctuation).

Regarding digital symbols, the fundamental information unit in digital signals is a binary digit or bit since a digital baseband waveform is represented by a symbol M, where M refers to the number of symbols, each containing b bits:

$$M = 2^b \tag{1}$$

Thus, the information symbol M corresponds to a specific combination of bits, as expressed by Eq. (1).

Therefore, the process of formatting an input bit stream into sequences of M binary digits is called source encoding. In addition to formatting, source encoding commonly includes digital data compression to increase the system's spectral efficiency. Spectral efficiency is the measure of how effectively a communication system transmits data over a given bandwidth, typically expressed in bits per second per Hertz (bps/Hz).

2.2 Channel Encoder

To reduce the probability of errors in the received data streams, specific control redundant bits (or blocks) are added to the source binary sequences at the channel encoder. This process is called channel encoding. The added redundant bits are used at the receiver for *error bit detection* and *correction,* minimizing the effects of system and channel noise and interference that may seriously affect system signals. Generally, a channel coding process effectively increases the reliability and quality of the communication system.

Thus, the task of the channel encoder is to produce a digital waveform suitable for the applied signal modulations that are well suited for random-changing channels owing to multipath propagation and interference. In this context, the primary role of an encoder is to participate in the *error detection* and *correction* processes to reduce the bit error probability (BER) in the received information bits.

There are more methods of channel encoding (i.e., error detection and correction methods), which will be discussed further in this chapter.

2.3 Digital Modulator

The task of the digital modulator is to convert digital sequences produced by the channel encoder into an analog bandpass waveform. The waveform is optimized for transmission through a specific propagation channel. There are several levels of bandpass modulation techniques. The simplest method is binary digital modulation, which includes binary pulse amplitude modulation (BPAM), binary phase shift keying (BPSK), and binary frequency shift keying (BFSK).

Generally, a digitally modulated bandpass waveform can be considered (or mapped) as the sum of signals $x_i(t)$, where $i = 1, 2, \ldots, M$ where $M = 2^k$, with k representing the number of transmitted bits per carrier state. For binary modulation, the $x_i(t)$ signal is mapped into two waveforms $x_1(t)$ and $x_2(t)$, each associated with a specific amplitude, phase, or frequency carrier state. Hence, for the case of binary modulations, $M = 2$ ($M = 2^k = 2^1 = 2$).

Additionally, higher-level or M-level modulations correspond to $M > 2$, allowing a modulator to transmit $k > 1$ bits per carrier state at a time.

Note that baseband modulation is not required before bandpass modulation in digital communication systems, as baseband binary sequences can directly modulate the amplitude, phase, or frequency of the channel carrier.

After the modulation, the modulated signal is amplified to the required level by the output power amplifier (PA) and filtered to remove the noise and distortions generated within the transmitter due to thermal noise, nonlinear TX stages, and signal processes.

3 Receiver Block

The receiver is a part of the communication system responsible for frequency-down conversion of received bandpass (channel) waveforms. This section describes the receiver's essential functions, where signal processing occurs in the opposite direction to the processing in the transmitter.

3.1 *Bandpass Demodulator*

The task of the bandpass demodulator is to restore the analog bandpass channel waveform received by the receiver's antenna and shape it to resemble the waveform sent by the TX modulator. It must also achieve effective synchronization with the TX blocks to overcome undesired variations in the amplitude, phase, and frequency of the received signal. The importance of maintaining synchronization between base stations and end devices in mobile networks is to ensure that data packets arrive at the correct time and frequency, preventing interference.

Additionally, the critical function of the channel demodulator is channel equalization, which minimizes the intersymbol interference (ISI) caused by the dispersion of the received signal due to multipath propagation in the communication channel.

The receiver's demodulator includes typical filtering processes such as

(a) Filtering of unwanted high-frequency components during the frequency-down conversion of the channel bandpass waveform.
(b) Filtering to achieve suitable pulse shaping.
(c) Filtering in conjunction with signal equalization compensates for signal degradation due to poor system impulse response and the communication channel.

After filtering, the demodulated signal is ready for further processing in the channel decoder.

3.2 Channel Decoder

The demodulated waveform from the demodulator flows into the channel decoder. Its main task is to recover the original information digital sequence via the code employed by the TX encoder and the redundant bits received with the original data. The efficiency of the decoder is measured by the number of bit errors in the decoded bit sequence. A critical system parameter is known as the probability of bit error (P_e) or the bit error rate (BER), which will be discussed later in this chapter.

3.3 Source Decoder, Reformatter, and Data Storage

The source decoder converts the digital sequences from the channel decoder into a sequence on the basis of the source encoding algorithm implemented in the TX source encoder. However, owing to undetected errors in the communication channel caused by signal distortions in the TX stages and external EMI in the channel, the received signal at the RX decoder is an approximation of the signal sent from the TX.

When the original information is not digital data but analog signals, the receiver must convert the digital data back into original analog waveforms, such as voice, music sounds, images, or videos.

The storage of the received information typically involves computers, servers, data storage, terminal equipment, and similar devices.

4 Types of Communication Channels

The signal propagation medium or communication channel can be a wireline or wireless link. Wireline channels include transmission lines (coaxial or differential pairs) for slow and mid-speed data transfers and optical fiber cables for high-speed data transmissions. Wireless links may use various frequency spectra, from low frequencies (in sub-MHz ranges) to high frequencies such as microwave spectra (>1 GHz) and further up to mm-Wave regions, i.e., up to 100 GHz or beyond.

The transmitted signal may be corrupted across the transmission channel by random phenomena such as additive thermal noise and spurious signals from the system's electronic devices, EMI/RFI on the transmission channel from other communication systems, and ambient noise and EMI from manmade and natural sources (atmospheric noise, lightning discharges, solar activities, and cosmic radiation). Therefore, appropriate and accurate channel models are based on the knowledge of the random behavior of a communication channel, using stochastic analytical analysis in conjunction with empirical studies from measurements on realistic communication channels.

The model is more predictable for a wireline channel, as the dominant noise in the channel is additive Gaussian noise, which is well studied in Refs. [1–4]. Assuming that the channel is linear (when the components of the received waveform y_i(t) are linearly proportional to the signaling components of the transmitted waveform x_i(t)), it is referred to as the *time-invariant linear Gaussian channel* model shown in Fig. 2.

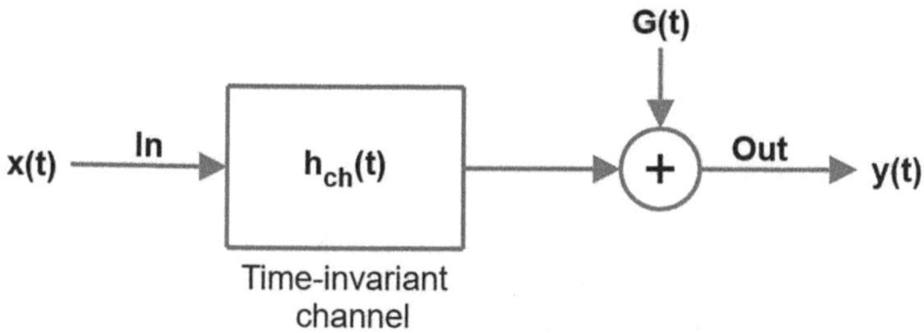

Fig. 2 Time-invariant Gaussian linear channel

The block $h_{ch}(t)$ represents the channel as a *linear time-invariant* filter with an impulse response. After the filter, Gaussian noise $G(t)$ is added to the channel, and the output signal $y(t)$ is expressed as

$$y_i(t) = x_i(t) * h_{ch}(t) + G(t) \tag{2a}$$

where the operator "$*$" denotes the *convolution* of the input signal $x_i(t)$ and the filter's linear transfer function $h_{ch}(t)$ in the time domain. The function $h_{ch}(t)$ is expressed as a function of Dirac's impulse δ(t) as

$$h_{ch}(t) = k\delta(\mathrm{t}), \tag{2b}$$

where k is a channel constant.

Equation (2a) can also be expressed in integral form as

$$y_i(t) = \int_{-\infty}^{\infty} x_i(t-\tau) h_{ch}(\tau) d\tau + G(t) \tag{2c}$$

Thus, the linear time-invariant Gaussian model is suitable for modeling wireline and line-of-sight (LOS) wireless communication channels, such as terrestrial and satellite links. However, this model is incorrect for radio communication systems with random channels, such as mobile network channels and ionospheric links, where multiple propagation paths exist between the radiation source (TX antenna) and the information destination (RX antenna) in addition to the external EMI.

In mobile radio communication channels with waves reflecting and refracting objects, the situation becomes more complex, as these radio channels change randomly over time. Therefore, a more accurate model for random radio channels uses the *random-time-changing* transfer function $H_{ch}(t,\tau)$ instead of the *time-invariant* filter $h_{ch}(t)$ (Fig. 3).

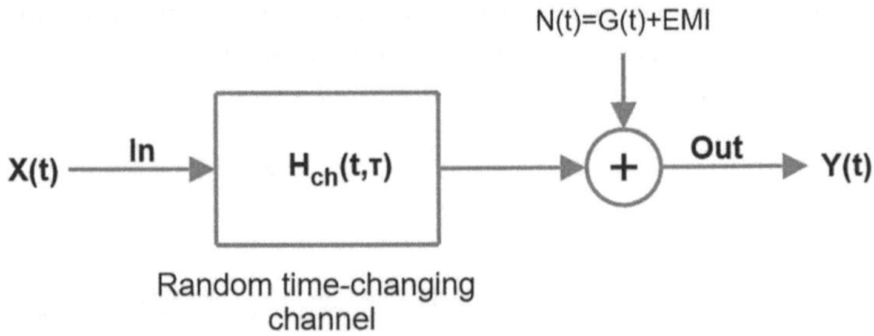

Fig. 3 Randomly time-changing channel (probabilistic model)

Here, the signal at the channel output is expressed as

$$Y_i(t) = x_i(t) * H(t,\tau) + N(t) \tag{3a}$$

or

$$Y_i(t) = \int_{-\infty}^{\infty} x_i(t-\tau) H(t,\tau) d\tau + N(t) \tag{3b}$$

where τ is a random signal delay and where $N(t)$ is the ambient Gaussian noise among other channel noise and EMI, expressed by the sum $N(t) = N(t)_{awgn} + N(t)_{emi}$.

Thus, the random channel transfer function $H(t,\tau)$ is a more suitable parameter for modeling random and nonlinear radio channels. The performance of random time-changing radio channels is discussed in more detail in chapter "Performance of Radio Digital Systems in Presence of EMI".

5 Channel Noise and Channel Capacity

5.1 *Additive Gaussian Noise*

Figure 1 illustrates the white Gaussian noise (or AWGN) introduced into a transmission channel that is commonly present in communication systems and electronic equipment. This noise uniformly covers the entire channel bandwidth with a constant amplitude and broadband (flat) spectrum.

White Gaussian noise originates from random electron fluctuations in resistive components of electronic equipment and primarily affects the RX performance. Because the power and velocity of these fluctuations are proportional to the absolute temperature in Kelvins (K), this noise is also called thermal noise or

Johnson–Nyquist noise. It can be described by a Gaussian random process, hence the name Gaussian noise. The amplitudes of the thermal noise N can be defined at any time t by the standard Gaussian probability density function (PDF) or $P(x)$:

$$P(x) = \frac{1}{\sigma\sqrt{2\pi}} e^{-\frac{1}{2}\left(\frac{x-\mu}{\sigma}\right)^2} \tag{4}$$

where σ is the RMS voltage of the noise, σ^2 is the probability variance related to the RMS noise power, and μ is the mean. The plots for the zero-mean Gaussian distributions for several σ values are provided in chapter "Digital and Pulse Signals, Random Variables, and Stochastic Processes", Fig. 22.

The ions and electrons in materials move because the kinetic energy increases with temperature. The movement of electrons and other charged particles in a circuit is observed as electric currents, whereas the net difference in electric potentials over some distance in the circuit is characterized as a voltage.

According to Nyquist, each resistive element (resistors or conductors) can be represented by a circuit of a noise generator with voltage v_n and noiseless resistance R. Nyquist experimentally reported that the average noise power N can be expressed as

$$N = \frac{v_n^{\ 2}}{4R} = KTB(\mathrm{W}) \tag{5a}$$

where the mean voltage value is zero ($\overline{v}_n^{\ 2} = 0$). K is the Boltzmann constant ($=1.38 \cdot 10^{-23} J/K$), T is the noise temperature in kelvin (K), and B is the noise bandwidth in Hz. Accordingly, system-generated noise can be modeled having zero mean power and the power spectral density $N_0/2$.

The power noise in a circuit with two resistances R (source and load resistance) can be expressed as

$$P = \int_t^{\infty} v_n(t) \cdot i_n(t)\,dt = \frac{1}{4R}\int_t^{\infty} v_n(t)^2\,dt = \frac{\sigma_n^{\ 2}}{4R} \tag{5b}$$

where σ_n^2 is the RMS power of the noise or the probability variance.

Comparing Eqs. (5a), (5b), we can see the equivalency: $P = N = \frac{v_n^{\ 2}}{4R} = \frac{\sigma_n^{\ 2}}{4R} = KTB$.

Owing to the similarity between the power noise distributions in communication channels and electrical circuits, this channel noise is called ambient white noise and is usually treated as AWGN.

Other noises generated in equipment with active components are low-frequency *flicker* noise, *Schottky* noise, and high-frequency *transition* noise, which are *nonthermal noises*. Figure 4 shows the total power frequency spectrum of noise (thermal and nonthermal) generated in typical solid-state amplifiers or buffers consisting of passive and active components.

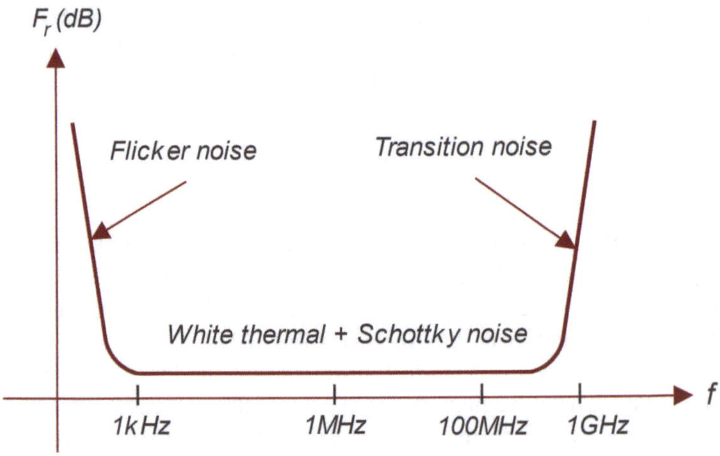

Fig. 4 Spectral power density of flat noise in an electronic circuit

Flicker noise originates from time-varying voltage–current relationships in solid-state devices. In passive circuits, this relationship (resistance) is constant and not associated with semiconductor elements. The cause of flicker noise is the non-linear operation of semiconductor elements (diodes and transistors). This noise is "colored" with a power spectral density near DC, decreasing with frequency *f* proportionally to *1/f*, as shown in Fig. 4.

Schottky noise primarily exists in active devices, causing random fluctuations in the average current flow in these devices. In particular, this noise exists in diodes, bipolar transistors, and FET transistors. In bipolar transistors, Schottky noise follows the flat and broad spectrum of thermal noise, as shown in Fig. 4. Hence, it behaves as white noise characterized by a Gaussian distribution.

Transition noise is high-frequency noise that occurs when active devices operate above their transition frequencies and increase rapidly with frequency, as shown in Fig. 4. This phenomenon occurs when the vibration period of a current carrier decreases and approaches transit time delays during recombination at the junctions of active electronic elements or FET channels.

In addition to internal AWGN generated in electronic devices, *external* noise and interference, such as *atmospheric*, *lightning*, *solar-cosmic*, and *EMI* noise (narrow-band and broadband) from other systems, exist in open communication channels. EMI noise is discussed in more detail in chapter "Performance of Radio Digital Systems in Presence of EMI", whereas *lightning* and *solar-cosmic noise* are covered in chapters "Physics of Lightning and Its Impact on Radio Systems on Earth" and "Solar and Other Cosmic Radiations with Effects on HF Radio and Satellite Data Communications", respectively.

5.2 SNR and Channel Spectral and Power Efficiency

A critical parameter of digital transmission channel performance is the signal-to-noise ratio per bit (or SNR per bit), expressed as

$$\mathrm{SNR}_b = \frac{E_b}{N} = \frac{E_b}{N_0 B} \tag{6a}$$

where E_b is the *signal power density per bit* and where N_0 is the *spectral density of thermal or white noise* (in Joules) expressed as $N_0 = N/B$ where N is the noise RMS power and where B is the noise bandwidth, or in a radio system, an RX IF bandwidth. Additionally, N was expressed by Eq. (5a) as $N = KTB$.

The *signal-to-noise ratio per symbol* (or SNR per symbol) is used mainly for multilevel modulated signals that carry an M-number of symbols and is expressed as

$$\mathrm{SNR}_s = \frac{E_s}{N_0 B} \tag{6b}$$

where E_s is the energy *per symbol* related to the energy per bit E_b as

$$E_s = kE_b \tag{7}$$

Here, k is the number of bits in a symbol expressed as $k = \log_2(M)$, where M is the number of symbols.

The bit error rate (BER) is one of the main performance parameters for digital communication systems and is a function of the SNR or E_b/N_0, expressed by Eq. (6b). This probabilistic quantity is defined at the RX decoder output as the ratio between the number of error bits and the total number of received bits in an observing time interval. The most reliable method for determining the BER is empirical measurement, especially in cases of nonlinear radio channels with impulsive noise and interference.

5.3 Channel Capacity

The channel data capacity represents *the maximum rate at which data can be transmitted across a channel without error*. The Shannon–Hartley theorem defines the maximum data rate, stating that the data channel capacity is limited by the channel bandwidth and the SNR. Therefore, the theorem for the channel data capacity C for the AWGN channel is expressed as

$$C_b = B\log_2\left(1 + E_b/\left(N_0\, B\right)\right) = B\log_2\left(1 + \mathrm{SNR}_b\right)\text{bits/s} \tag{8a}$$

The important conclusions from the *S-H* theorem are as follows:

1. When the *data channel rate R* is less than or equal to the channel capacity ($R \le C$), error-free data transmission across the channel is possible through appropriate coding.
2. When the *data rate from the source* exceeds the channel capacity ($R > C$), no code can achieve error-free data transmission through the channel.

Conclusions (1) and (2), with the following Eqs. (9a), (9b) highlight Shannon's limit on the bandwidth or *spectral efficiency* of the communication channel. The *spectral channel efficiency* is denoted as η_b, expressed as

$$\eta_b = \frac{R}{B}(\text{bits/s/Hz}) \tag{9a}$$

Considering (1) and Eqs. (8a) and (9a),

$$\eta_b = \frac{R}{B} \le \frac{C}{B} = \log_2\left(1 + \frac{E_b}{N_0 B}\right)(\text{bits/s}), \tag{9b}$$

Shannon's limit on power efficiency can be expressed as

$$\frac{E_b}{N_0} > \frac{2^{\eta_b} - 1}{\eta_b} \tag{9c}$$

From Eq. (9b), the ideal case follows: for $\eta_b = 1$, $R = C$.

Thus, in addition to *spectral efficiency* η_b, Shannon's limit on *power efficiency* (SNR) is a vital channel parameter expressed by Eq. (9c). This limit refers to the ability of the communication system to reliably transfer information at the lowest possible SNR or achieve an acceptable BER (or symbol-error rate, SER) of received data.

The channel capacity C_b is measured in bits/s, which can also be considered in symbols/seconds as C_s. Substituting Eq. (7) into Eq. (8a), the channel capacitance is expressed as

$$C_s = B\log_2(1 + kE_b/N) = B\log_2(1 + \text{SNR}_s)\,sym/\text{s} \tag{9d}$$

where k is the number of bits in a binary symbol.

The capacitance C_S is more suitable for channels with multilevel (or M-level) signals modulations, which will be discussed in Sect. 6 regarding *modulation methods*.

5.4 Capacity of Discrete Memoryless Channels (DMCs)

In discrete memoryless channels, the input and output signals are considered discrete random variables. The memoryless nature of the channel indicates that an output discrete signaling component depends statistically only on the present input discrete signaling component. There are two variations of DMC channels: (a) with *continuous input* and *continuous output* (CCMC) and (b) with *discrete input* and *continuous output* (DCMC). These channels are modeled as vector channels, where the input and output signals are considered n-dimensional vectors. These channel models are suitable for calculating the capacity of specific M-level signal modulations.

Figure 5 shows the generic DMC channel model. The input and output discrete random variables (X and Y) contain discrete components x_i and y_j, where indices i and j are integers $i = 1,2, \dots M$ and $j = 1,2, \dots N$.

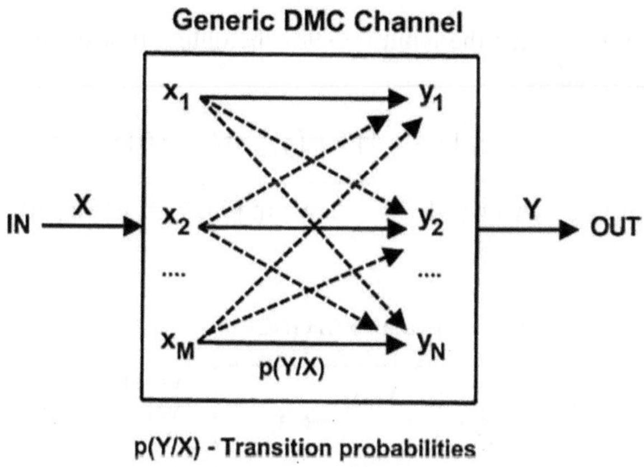

Fig. 5 DMC channel model

The transition probabilities *p(Y/X)* depend on the type of channel. For the AWGN channel, *p(Y/X)* follows the Gaussian probability density function (PDF) and defines a memoryless channel. The following expression, Eq. (10a), determines the DMC channel's crossover probability *P*(*Y*/*X*) when the numbers of the input and output samples are equal, i.e., $M = N$.

$$P(Y/X) = P(y_1,y_2,\dots,y_N / x_1,x_2,\dots,x_N) = \prod_{i}^{N} P(y_i / x_i) \tag{10a}$$

Furthermore, Fig. 6 shows a binary symmetric channel (BSC), a particular case of the DMC channel, where the channel input and output have two possible bit values [0, 1].

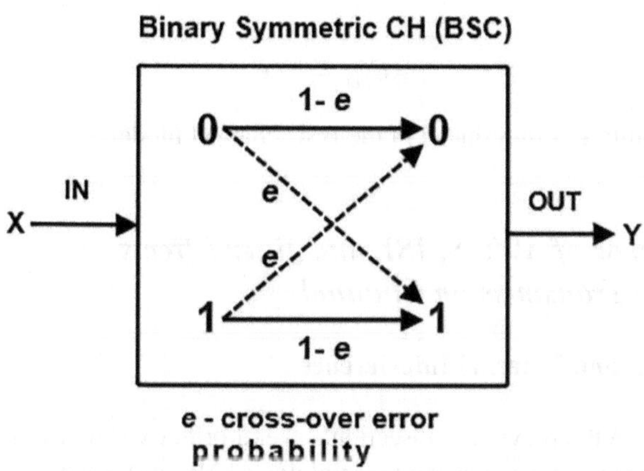

Fig. 6 BSC channel model

The probability of receiving a wrong bit is called the *crossover probability* or just the *error probability* e expressed as

$$e = P(y_i = 1/x_1 = 0) = P(y_i = 0/x_1 = 1) \tag{10b}$$

The probability p when the input 0 yields the output 0 or the input 1 yields the output 1 is expressed as

$$p = P(y_i = 1/x_i = 1) = P(y_i = 0/x_i = 0) = 1 - e \tag{10c}$$

The binary erasure channel (BEC), shown in Fig. 7, is also a particular case of a DMC channel.

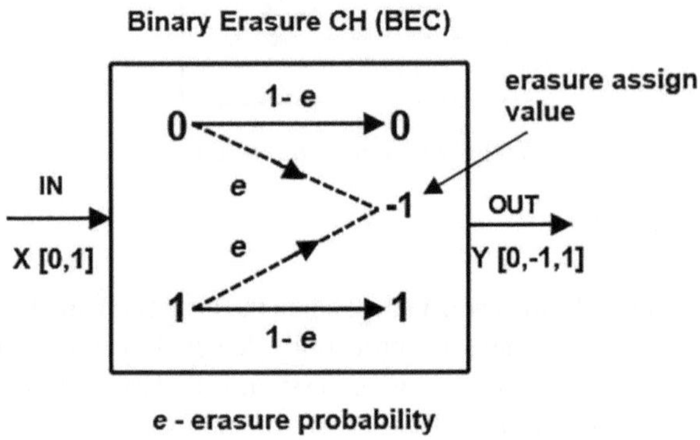

Fig. 7 BEC channel model

The input of the BEC channel has two random values, [0, 1], whereas the channel output has three values, [1, −1, 1], where the bit −1 is the erasure bit that is lost during transmission. If erasure bits do not appear in a symbol at the receiver, then no bit errors have occurred in the transmission. The capacity of the BEC channel is given by

$$C_{BEC} = 1 - e \tag{10d}$$

where probability e is introduced in the BSC channel model.

5.5 *Overview of AWGN, ISI, and Jitter Effects on the Transmission Channel*

5.5.1 AWGN and External Interference

As mentioned in the previous subsection, several primary sources of channel noise and interference exist, such as the *thermal noise* (AWGN) generated in the *TX* and *RX* equipment, *RFI* from other users of the communication channel, and EMI from various transients (manmade and natural). They are classified as system-generated noise and signal distortions, as well as channel noise and interference.

The effects of all this noise, signal distortions, and EMI increase the system error probability, leading to degradation of the received signal when the SNIR (Signal-to-Noise/Interference Ratio) at the receiver is reduced below the acceptable level. The acceptable SNIR level is a threshold quality level.

5.5.2 Intersymbol Interference

All the channels have a limited bandwidth, which causes the spread of a pulse waveform passing through the channel. When the channel bandwidth exceeds the pulse bandwidth, pulse spreading is mild, but the system *spectral efficiency* is low. When the channel bandwidth is smaller than the pulse bandwidth, the pulse duration is greater, resulting in overlap between adjacent symbol pulses. This overlap is referred to as intersymbol interference (ISI).

The ISI degrades system performance by increasing the system bit error rate (BER) or symbol error rate (SER). This happens when the pulses have sharp edges, such as those near a rectangular shape (broadband digital signals). In the case of *sinc* or *trapezoidal* pulses with slower rise/fall times, the pulse bandwidth is narrow and more suitable for transfer through narrowband channels.

5.5.3 Timing Jitter

For uniformly time-spaced pulses, such as analog signal samples or system digital clocks, the jitter is zero, meaning that the pulses' time positions are stable over the time of interest. However, when jitter exists, the pulses are no longer uniformly spaced. Jitter is commonly a random process, causing the pulse position in a sequence to be imprecisely known. In this case, a low-level broadband spectral content is induced, similar to quantization noise. A stable clock reference and good power supply isolation adequately control the jitter.

5.6 *Types of Data Errors Introduced in Data Transmissions with AWGN and Impulse EMI*

The two main types of data transmission errors are as follows:

1. *Random errors*: These errors are caused by AWGN in the channel, affecting the transmitted symbols independently.
2. *Burst errors*: These *errors* are caused by *impulse* EMI noise in the channel, affecting several consecutive bits and resulting in errors occurring in bursts.

When both AWGN and impulse EMI noise affect the channel, which is common, the resulting errors are called *compound errors*.

The writing of this section is influenced by [1–6].

6 Digital Modulation Methods

Digital modulations offer significant advantages over traditional analog modulation methods, such as AM, FM, and PM, providing much higher data rates, greater channel capacity with limited bandwidth and greater immunity to channel noise and the EMI. These topics are covered well by the reference [1, 3, 4, 7, 8], which influenced the writing of this section.

Digital modulation methods are classified into two distinct methods in terms of coherence: *coherent* and *noncoherent* modulations.

Coherent modulation involves *cohesive detection* in the receiver, where the receiver's carrier is phase-locked with the transmitter's carrier. Hence, *coherent detection* involves firm synchronization between the receiver's local carrier and the transmitter's carrier. This enables reliable detection on the basis of the correlation between the received data with the EMI and the receiver's carrier.

The *noncoherent* digital modulation does not require the receiver's carrier to be phase-locked with the transmitter's carrier, resulting in less complexity but a higher probability of data errors compared with *coherent* modulation.

6.1 General Representations of Digitally Modulated Signals in Complex Form

Digital modulations can be performed on source information signals or signals in the channel encoder. The first modulation is called *baseband* modulation, which produces a baseband (source) waveform for further processing. The second modulation is called *bandpass* (channel) modulation, which creates a bandpass waveform optimized for transfer over the communication channel. A bandpass waveform is optimized for transfer over the communication channel. The baseband modulation process can also be performed directly on radio-frequency (RF) channel carriers, where a baseband waveform is directly mapped to the channel carriers and subcarriers, which further simplifies and improves the modulation.

A modulated signal (baseband or passband) is a signal whose energy spectrum is concentrated around the carrier frequency ω_c (=$2\pi f_c$). Any digitally modulated carrier can be represented analytically in quadrature form as

$$\tilde{s}(t) = A(t)\cos\left[\omega_c t + \phi(t)\right] = s_I(t)\cos(\omega_c t) - s_Q(t)\sin(\omega_c t) \tag{11a}$$

where $A(t)$ is the magnitude of the modulated signal, expressed as

$$A(t) = \sqrt{s_I(t)^2 + s_Q(t)^2} \text{ and } \phi(t) = \tan^{-1}\left[\frac{s_Q(t)}{s_I(t)}\right] \tag{11b}$$

Owing to the *sine* and *cosine* terms in Eq. (11a) being orthogonal components, the signal $\tilde{s}(t)$ can be represented in IQ complex form as

$$\tilde{s}(t) = s_I(t) + is_Q(t) \tag{11c}$$

This modulation is known as IQ baseband modulation, where I denotes an in-phase component and Q denotes the quadrature component of the modulated signal. The block diagram of IQ modulation is shown in Fig. 8.

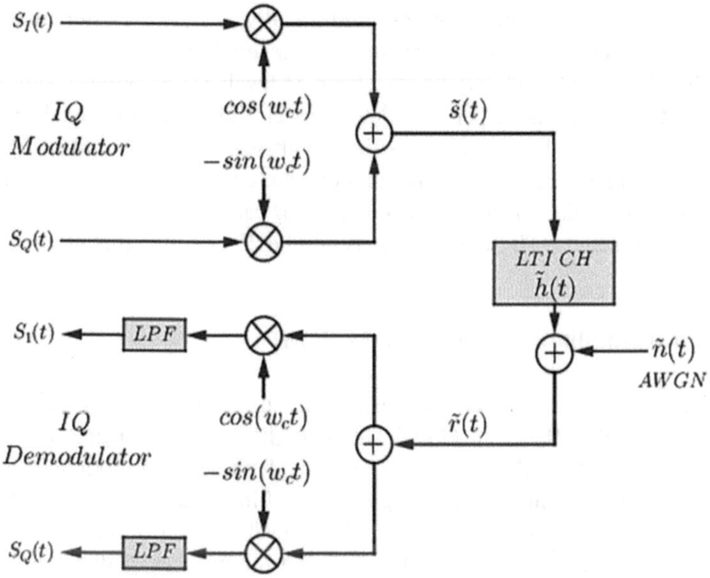

Fig. 8 IQ baseband form of a passband modulated signal $\tilde{r}(t)$

LPF blocks in the IQ demodulator are low-pass filters.

Consider a *linear time-invariant* (LTI) channel with a *transfer function* $\tilde{h}(t)$ as the channel response to the impulse $\delta(t)$. This transfer function is also presented in complex IQ form as

$$\tilde{h}(t) = h_I(t) + ih_Q(t) \tag{12a}$$

The effect of the channel on the modulated signal is represented by the *linear convolution* of the modulated signal $\tilde{s}(t)$ and the channel transfer function $\tilde{h}(t)$, $\tilde{h}(t) = \tilde{s}(t) * \tilde{h}(t)$.

The received signal at the receiver's front end is represented by the sum of the convolution $\tilde{s}(t) * \tilde{h}(t)$ and AWG noise $\tilde{n}(t)$:

$$\tilde{r}(t) = r_I(t) + ir_Q(t) = \tilde{s}(t) * \tilde{h}(t) + \tilde{n}(t) \tag{12b}$$

Finally, the demodulated received signal is filtered by low-pass filters (LPFs), as shown in Fig. 8.

6.2 Types of Digital Modulations: M-PAM, M-PSK, M-QAM

As discussed earlier in this chapter, three parameters (amplitude, phase, and frequency) of a signal carrier can carry data information in the form of digital symbols. The corresponding modulations are *pulse amplitude* modulation (PAM), *phase shift keying* modulation (PSK), and *quadrature amplitude modulation* (QAM), which combine PAM and PSK modulations. Additionally, if only a carrier frequency is modulated (shifted), it is called *frequency shift keying* (FSK).

All the mentioned modulations may have *M* levels (or M orders), significantly increasing the data capacity per modulated carrier. Thus, the modulated carrier can carry up to *M* digital symbols at a time with each symbol containing *k* bits. Accordingly, higher-order digital modulations include M-PAM, M-PSK, M-QAM, and M-OFSK modulations where the common *M* can reach 256, or even higher for massive data transfers. These modulations will be described in this section.

6.2.1 Pulse Amplitude Modulation (M-PAM)

M-PAM modulation is a one-dimensional modulation with only an in-phase component *I*, with $s_Q(t) = 0$ in Eq. (11c). Thus, all the information data are encoded (mapped) in the amplitude of a single carrier with an M-PAM modulator transmitting a series of data symbols consisting of *k* bits of information. The number of data symbols can be $m = 1, 2, \ldots, M$, where *M* is a level or order of the modulation. As previously mentioned, the relationship between the number of data symbols (M) and the number of bits (*k*) of each symbol is $M = 2^k$.

The expression for graphically creating the M-PAM signal space or constellation is given as

$$s(t) \rightarrow A_m = 2m - 1 - M \quad \text{for } m = 1, 2, \ldots, M \tag{13}$$

On the basis of Eq. (13), Fig. 9 shows the M-PAM constellation in an IQ plane.

Figure 9 shows that for M-PAM signaling, the constellation points are located only on the *I* axes at points ±1, ± 3, …, ± $(M - 1)$. For example, if 2 bits are placed in one transmitted symbol, it results in 4-PAM modulation ($M = 2^k = 2^2 = 4$).

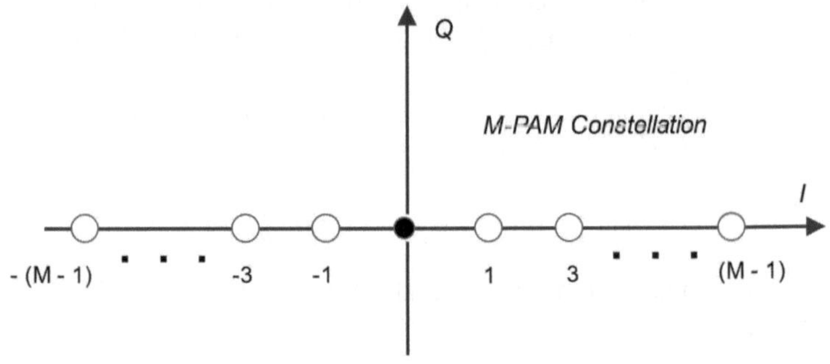

Fig. 9 M-PAM Constellation

6.2.2 Phase Shift Keying Modulation of M-order (M-PSK)

In M-PSK modulation, all the information symbols are encoded in the carrier phase. Like the M-PAM, M-PSK transmits a series of data symbols to M, where each symbol has k bits ($M = 2^k$). The data symbols are modulated upon M-PSK mapping, and the expression for a created M-PSK signal is given as

$$\tilde{s}_m(t) = A\cos\left[\omega_c t + \phi_m(t)\right] \tag{14a}$$

where the phase term $\phi_m(t)$ is given as

$$\phi_m(t) = -\frac{2\pi(m-1)}{M} \quad \text{for } m = 1,2,\ldots,M \tag{14b}$$

Equations (14a) and (14b) can also be expressed as a combination of in-phase I and quadrature-phase Q components in the I-Q plane as

$$\tilde{s}_m(t) = A\left\{\cos\left[\frac{2\pi(m-1)}{M}\right] + i\sin\left[\frac{2\pi(m-1)}{M}\right]\right\} \tag{14c}$$

for $m = 1, 2, \ldots, M$.

Figure 10, which is based on Eq. (14c), shows the constellations for the 8-PSK and 16-PSK modulated signals.

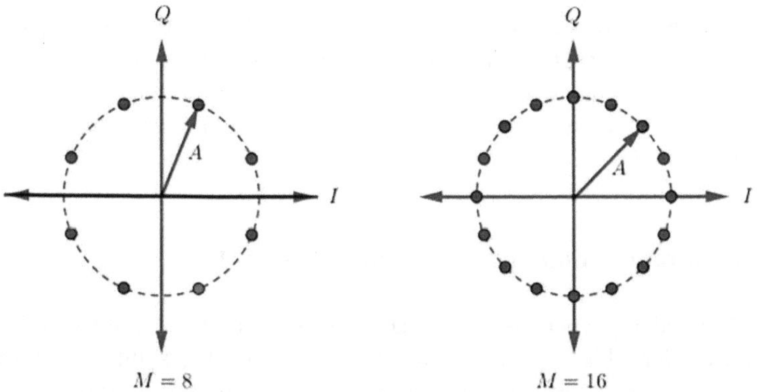

Fig. 10 8-PSK and 16-PSK constellations

A special case of M-PSK modulation is the traditional 4-PSK or QPSK modulation, whose vector constellation is shown in Fig. 11.

Fig. 11 QPSK Modulation in the IQ-plane

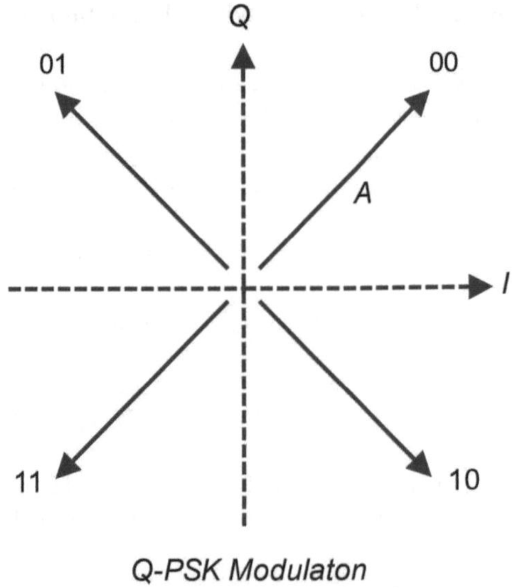

Figure 11 shows a combined phasor and signal space diagram for the QPSK signal represented by four discrete IQ-phasors and four signal points. The QPSK modulation allows the transfer of two bits per phase change (or two-phase state) since the carrier amplitude A remains unchanged. Thus, this modulation enables a double data transfer compared with BPSK in the same channel bandwidth and time interval.

A comparison of the transfer capacity of the 8-PSK and 16-PSK modulations with that of the QPSK shows that the 8-PSK and 16-PSK modulations have, respectively, two times and four times greater transfer bit capacity than the QPSK modulation. However, increasing the M-order for PSK modulation has technical limits. To overcome this, other modulations, such as *M-quadrature amplitude modulation* (M-QAM), have been developed, which is described in the next subsection.

6.2.3 Quadrature Amplitude Modulation (M-QAM)

M-QAM modulation encodes (maps) data bits into both the amplitude and phase of the signal carrier. This modulation transmits a series of data symbols, each consisting of k bits in the information symbol ($M = 2^k$). To aid the receiver's decoder in making correct decisions when bit errors are received, data symbols are mapped via Gray coding. This is achieved by converting the input data symbols to gray-coded symbols before mapping them into the desired QAM constellation. Gray coding restricts incorrect symbol decisions to single-bit errors by ensuring that adjacent symbols in the TX signal constellation differ by no more than one bit.

A QAM constellation is represented by two orthogonal PAM signals. For example, a 16-QAM quadrature signal is created by two quadrature 4-PAM constellations ($4 \times 4 = 16$), or a 64-QAM quadrature signal is created by two quadrature 8-PAM constellations ($8 \times 8 = 64$). The generic expression used to generate quadrature PAM signals of arbitrary dimension D is given as

$$s(t) \rightarrow A_d = 2d - 1 - D \text{ for } d = 1, 2, \ldots, D \tag{15}$$

For example, to generate 16-QAM, the PAM constellation dimension is $D = \sqrt{16} = 4$, whereas for 64-QAM, we need two PAM signals, each with dimensions of $D = \sqrt{M} = \sqrt{64} = 8$.

As an example, Fig. 12 shows a 16-QAM constellation with 16 signal points (tips of signal vectors), where each point carries a 4-bit symbol mapped according to the Gray code.

To implement Gray code, a Karnaugh lookup table is used to map the points in the 16-QAM constellation, and the mapping progress through this constellation is shown in Fig. 12.

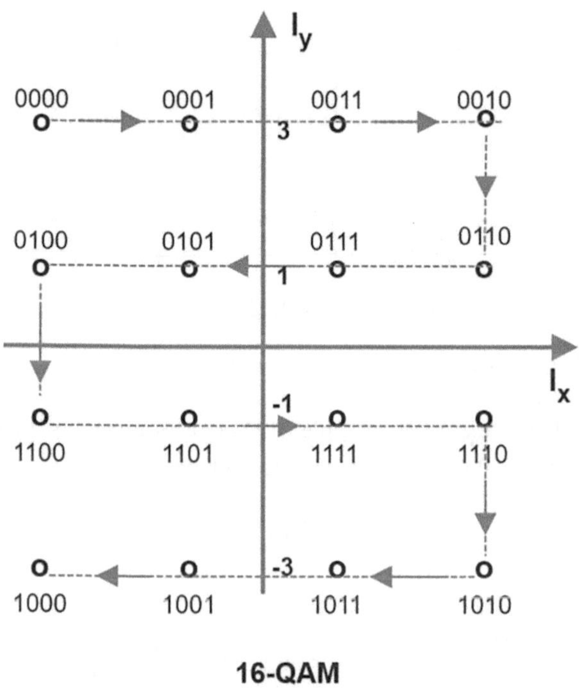

Fig. 12 16-QAM Constellation mapped per Gray code

QAM, as a combination of both *phase* and *amplitude* modulation methods, exploits various carrier phase shifts and two or more discretely modulated carrier amplitudes. All amplitude/phase states or shifts are assigned to different and distinct bit combinations. For example, the 16-QAM shown in Fig. 12 provides the transfer of 4 bits per carrier AM and PM states. The total number of AM/PM states is 16, including 4 amplitude and 12 phase states.

Furthermore, Fig. 13 shows the concept of higher-order QAM (64-QAM), where 64 6-bit symbols are mapped according to Gray code.

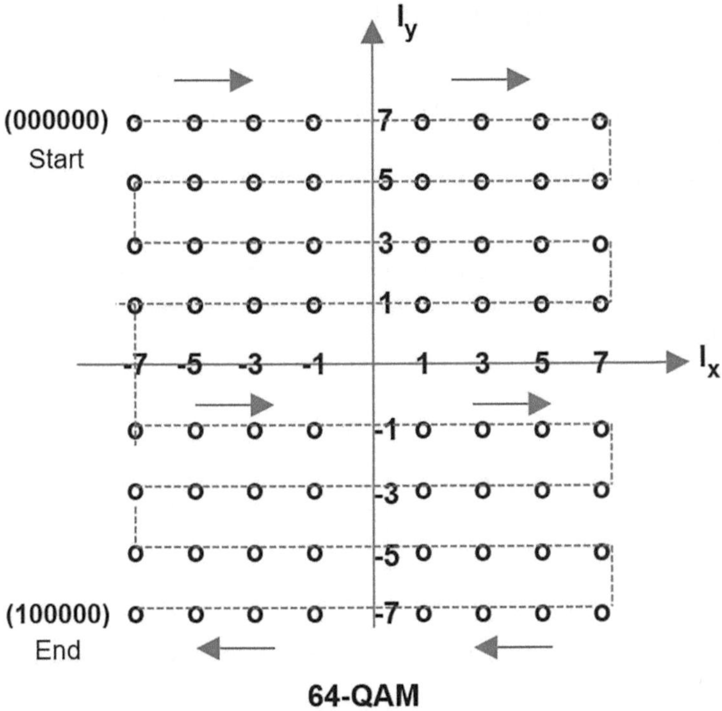

Fig. 13 64-QAM Constellation mapped per Gray code

Figure 14 shows the look-up table (LUT) of Karnaugh's mapping for the 64-QAM signals to achieve the Gray coding principle.

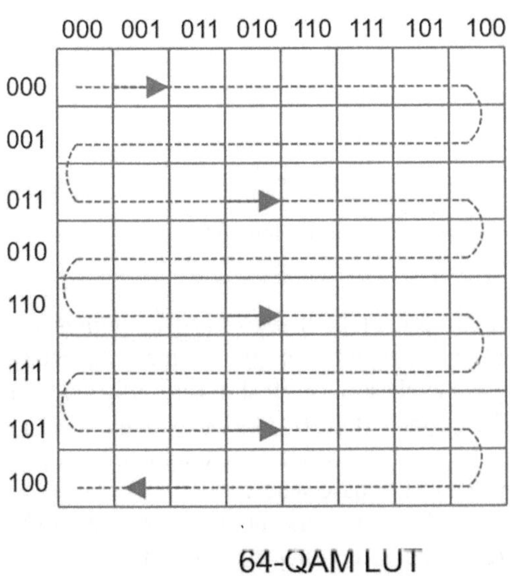

Fig. 14 LUT for 64-QAM signal mapping

It is evident that more data can be transferred within an allocated bandwidth or time interval when the number of QAM levels increases. However, increasing the modulation level comes at a cost: the carrier's AM/PM states become closer to each other, leading to higher noise and adjacent signal interference, which increases the probability of the BER in communication transmission. To use a higher modulation level, efficient error-correction techniques, as discussed earlier in this chapter, can be employed to increase the signal-to-noise/interference ratio (SNIR).

For example, 64-QAM is extensively used in 5G networks to achieve high data rates within limited spectral bandwidths. By combining amplitude and phase variations, the 64-QAM allows for efficient use of resources, making it ideal for high-speed mobile communication systems.

6.3 *Digital Modulations in Terms of Symbol Error Rate (SER) vs. SNR*

The expressions for the symbol error rate (SER) for different modulations are listed in Table 1 [9–11].

Table 1 Probability of SER for M-PAM, M-PSK, and M-QAM

Modulation	Probability of symbol error rate ($P_{e,s}$)
M-PAM	$P_{e,s} = 2\left(\frac{M-1}{M}\right)Q\left(\sqrt{\frac{6\log_2(M)}{(M^2-1)}\cdot\frac{E_b}{N_0}}\right)$
M-PSK	$P_{e,s} = 2Q\left(\sin\left(\frac{\pi}{M}\right)\sqrt{2\log_2(M)\frac{E_b}{N_0}}\right)$
M-QAM	$P_{e,s} = 1-\left[1-\left(2-\frac{2}{\sqrt{M}}\right)Q\left(\sqrt{\frac{3\log_2(M)}{(M-1)}\cdot\frac{E_b}{N_0}}\right)\right]^2$

The above relationships assume that the errors in the in-phase (I) and quadrature-phase (Q) signaling components are independent.

In Table 1, Q(x) is the *Gaussian complementary distribution function*, which can be expressed by the erfc function as given in chapter "Digital and Pulse Signals, Random Variables, and Stochastic Processes", Eq. (52c):

$$Q(x) = \frac{1}{2}erfc\left(\frac{x}{\sqrt{2}}\right) \tag{16}$$

The probability of the bit error rate γ_b for M-level modulations is expressed by the ratio $\frac{E_b}{N_0}$, whereas the probability of the symbol error rate γ_s is expressed as $\gamma_s = \frac{E_s}{N_0} = \frac{kE_b}{N_0} = k\gamma_b = \gamma_b \log_2(M)$. Additionally, $\sigma^2 = N_0/2$, where σ^2 is the variance of the PDF.

For a comparison of commonly used digital modulations in terms of SER, calculated per the equations in Table 1, Fig. 15 shows the plots of the probabilities of SER for 16-PAM, 16-PSK, and 16-QAM signals over the AWGN channel with the same *spectral efficiency* but with different *power efficiencies* per symbol.

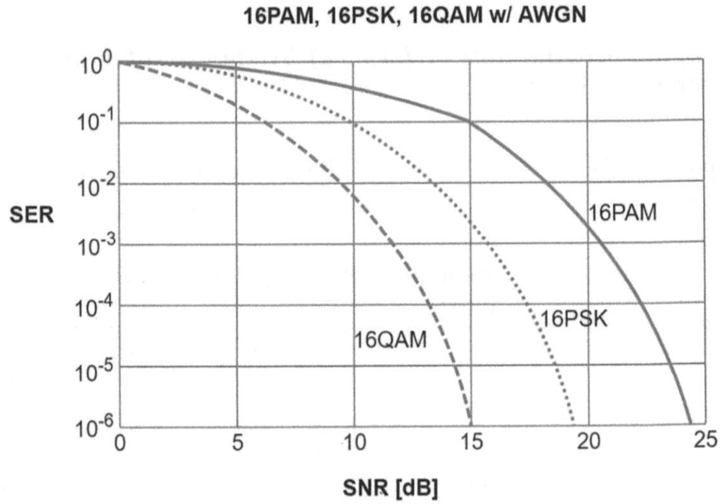

Fig. 15 SER of 16-PAM, 16-PSK, and 16-QAM vs. SNR (E_b/N_0)

This example shows that the higher-level M-QAM modulation has better *power efficiency* per symbol than the M-PSK and M-PAM modulations do. The plots show that to obtain $SER = 10^{-6}$, for 16-QAM, the *signal-to-noise per bit* ($SNR_b = E_b/N_0$) should be at least ~15 dB, whereas for the same SER, M-PSK and PAM require ~19 dB and ~24 dB, respectively. The *power efficiency* difference approaches null for lower modulation levels (M < 4).

6.4 Orthogonal Frequency Shift Keying (OFSK)

M-OFSK modulation belongs to the category of multicarrier modulations, where the frequencies of *M*-transmitted carriers vary according to source data symbols, and adjacent carriers from different signal sets in the radio bandwidth are orthogonal to each other. In M-OFSK modulation, M-carrier waveforms are generated with each waveform carrying symbols with k bits. As previously noted, the relationship between the number of bit symbols, M, and the number of bits, k, inside a symbol is expressed as $M = 2^k$ ($k = \log_2 M$). For $M = 2$, we have traditional binary 2-FSK (BFSK) modulation.

A set of analytical expressions for M-OFSK waveforms $\langle a_1, a_2, a_3, \ldots, a_M \rangle$ are given as

$$s_{MFSK}(t) = \begin{cases} A\cos\left[\omega_{c1}t + \phi_1(t)\right] \\ A\cos\left[\omega_{c2}t + \phi_2(t)\right] \\ \quad \cdots\cdots \\ \quad \cdots\cdots \\ A\cos\left[\omega_{cM}t + \phi_M(t)\right] \end{cases} \tag{17}$$

for $A = \sqrt{2E_b/T_b}$ where E_b is the energy per symbol bit and where T_b is the bit time. Here, for *coherent* M-OFSK, $\phi_1(t) = \phi_2(t) = \ldots = \phi_M(t) = 0$, whereas for *noncoherent* M-OFSK, $\phi_1(t) \neq \phi_2(t) \neq \ldots \neq \phi_M(t)$.

Figure 16 shows two signal sets, i and j, whose adjacent signals are orthogonal.

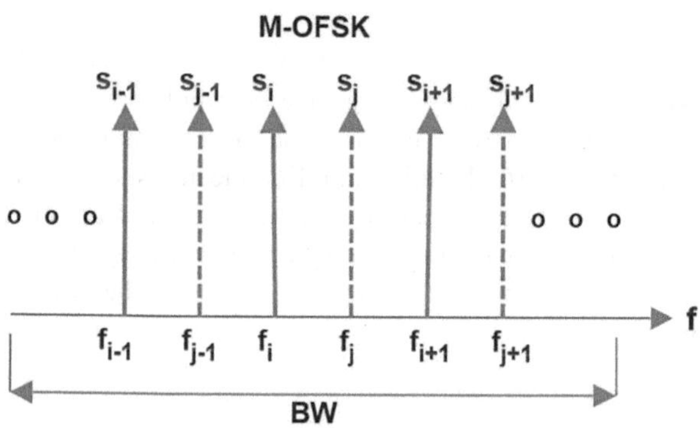

Fig. 16 Two M-OFSK signal sets, i and j, orthogonal to each other

The orthogonality in a time period T is analytically presented as

$$\int_0^T s_i(t)s_j(t)dt = \begin{cases} 1, & i = j \\ 0, & i \neq j \end{cases} \tag{18}$$

where $0 \leq t \leq T$, $1 \leq i \leq M$, and $1 \leq j \leq N$, where usually $M = N$.

Next, the orthogonality of the given waveforms expressed by Eq. (17) can be simply represented in a signal space in a diagonal matrix form as

$$\vec{s} = \begin{bmatrix} \vec{s}_1 \\ \dots \\ \vec{s}_M \end{bmatrix} = A_b \begin{bmatrix} A_b e^{i\phi_1} & \cdots & 0 \\ \vdots & \ddots & \vdots \\ 0 & \cdots & A_b e^{i\phi_M} \end{bmatrix} \tag{19}$$

where $A_b = \sqrt{E_b}$ is the signal voltage in volts.

The M-OFSK communication model for the AWGN channel is shown in Fig. 17.

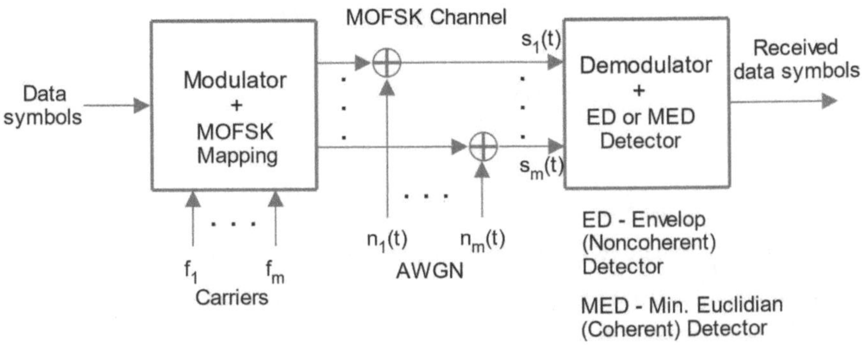

Fig. 17 M-OFSK communication model with AWGN

The present modulation scheme involves an envelope detector (ED) for noncoherent FSK modulation, which selects a received symbol corresponding to the maximum envelope value. However, in the case of coherent modulations (when $\phi_1(t) = \phi_2(t) = \ldots = \phi_M(t) = 0$), the minimum Euclidean distance (MED) is needed.

With respect to *coherent* detection, the optimized receiver calculates the Euclidean distance between each received symbol and the symbols from the reference array. The symbols from the reference array with the MED are chosen as received symbols.

7 Overview and Applications of Error Control Codes

Error control codes are considered well in the references [1, 8–10], which influenced the writing of this section.

Bit or symbol errors primarily occur in communication systems with simpler signal modulation in the presence of extensive channel noise and EMI. To minimize the probability of data errors, more sophisticated *modulation* techniques and *error correction* coding methods should be employed. There are two main groups of error control coding methods: *block* coding and *convolution* coding. *Block* coding involves data blocks with relatively large lengths that include redundant bits for error control. This coding is suitable for systems where data delay (latency) is not critical. For systems where data delay is critical, *convolution error control* coding is the preferred solution.

Error control coding and modulation methods can be combined in a unique process known as *coded modulation,* which offers a more effective solution for balancing the maximum data transmission rate with a low error probability. An example of this is the use of trellis codes. Figure 18 shows a system block diagram with two functional layers: layer 1 (the physical layer), which handles the channel modulation/demodulation process, and layer 2, which handles the channel encoding/decoding and error detection/correction processes.

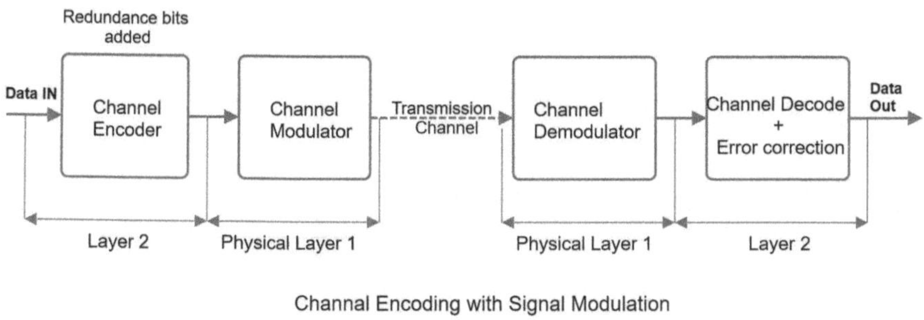

Fig. 18 Channel encoding with signal modulation

Thus, the coded modulation method unifies functional layers 1 and 2 into one layer.

Previously, we discussed wire-line, line-of-sight radio, and satellite links, which are typically modeled as linear Gaussian links with minimal channel noise and negligible external interference. Assuming that only Gaussian noise (AWGN) exists in these links (which is a fair assumption), less complex coding/decoding schemes can be used. However, in the case of nonlinear wireless links and networks with randomly changing parameters and unpredicable channel EMIs, more sophisticated error-correction methods and modulation techniques are needed.

7.1 *Classifications of Error Control Codes*

In summary, the error-control or forward error correction (FEC) codes are summarized in the block diagram in Fig. 19.

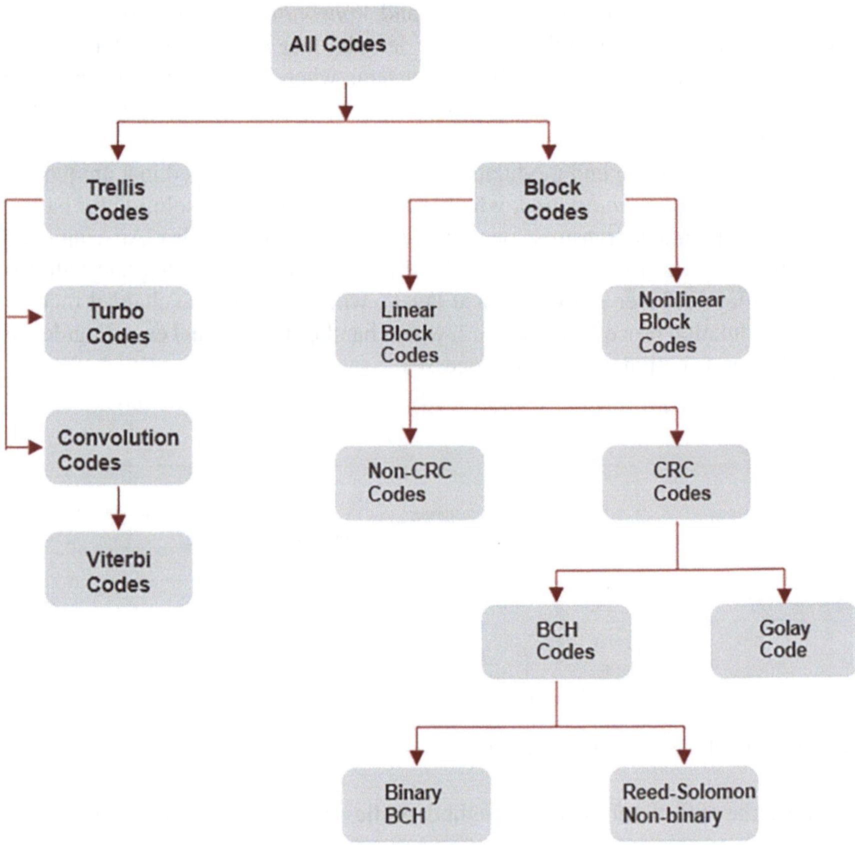

Fig. 19 Classification of error control codes or FEC

The error control codes are classified into two main groups: *convolution* codes and *block* codes. These are listed as follows:

- Standard *Convolution Codes:* Turbo codes, Trellis Modulation codes, Recursive Systematic codes, Self-Orthogonal codes, and Viterbi Convolutional codes.
- Standard *Linear Block Codes* include Hamming codes, Cyclic codes, Bose–Chaudhuri–Hocquenghem (BCH) codes, and Reed–Solomon (RS) codes.

Because detailed descriptions of convolution and block codes are out of the scope of this book, we refer readers interested more in this topic to references [1–3].

7.2 *Block Coding*

Block coding is widely used in massive data transfer to and from data storage, cellular mobile networks, Wi-Fi networks, and satellite communications, whereas *convolution* coding is employed in communication systems where low data delay (latency) is needed, i.e., in real-time systems, such as remote control and real-time measurement systems.

Generally, a block code contains vector M blocks, each with a fixed length of n bits. Vector M is called a *codeword*, and the corresponding coding maps information bits into codewords. A simple example of block coding is to add three identical bits as a control block before each information symbol. If all three bits are not identical at the receiver decoder, the decoder uses a majority rule to determine which received codeword is correct. However, when two or three bits are incorrectly received, automatic request control (ARQ) is employed to resend the information block along with the associated error control block. This is a simple and reliable error correction method for reducing the bit error probability, but the price is a lower transmission rate because of the resumption of control and information blocks during the ARQ process.

The block diagram in Fig. 20 illustrates the basic principle of block coding to minimize the BER in a channel under noise and interference.

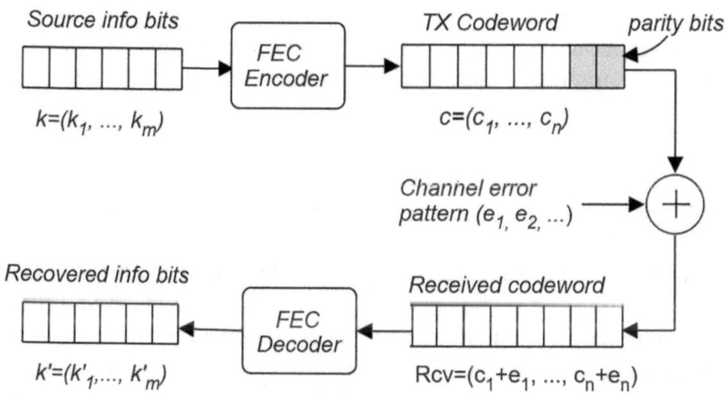

Fig. 20 *Block encoding process on AWGN channel*

Another standard block code is the CRC code. This code is highly efficient for error detection because it can be designed to detect many error combinations. Moreover, implementing this code in encoding and error detection circuits is highly practical. The CRC code, along with other codes, is one of the most important cyclic codes used for *error detection* in data networks and storage systems. The standard CRC codes include CRC-12, CRC-16, and CRC-CCITT.

For example, Reed–Salman (RS) codes are important in telecommunications for error correction in satellite and deep-space communications, ensuring data integrity in challenging environments.

7.3 Applications of Error Control Coding

The following error-control techniques are applied in specific cases:

1. For AWGN channels, forward error correction (FEC) block codes are usually employed. Typical applications include wire-line and line-of-sight links, such as satellite and space communication links.

2. The ARQ method is typically used for *compound-error* channels with low and medium data rates. Typical applications include telephone and radio channels.
3. *Block codes* are widely used to provide error control for massive data storage systems.
4. *The Trellis coding* method combines *convolution coding* and signal *modulation* into a single function. Trellis modulation coding (TMC) is used in the new generation of modems developed for the radiotelephony and WiFi channels. TMC is also employed in new 5G cellular networks, which will be discussed in chapter "Overview of 4G-LTE and 5G-NR Mobile Radio Systems from EMI Prospect".

7.4 Quality Parameters of Error Control Coding

Specific channel coding suitable for particular radio systems and channels is selected on the basis of parameters such as *coding fidelity*, *coding gain*, and *coding efficiency* [6].

7.4.1 Coding Fidelity

The quality of a digital communication system is defined by the bit error probability (BEP), which is estimated via bit error rate (BER) measurements. BER is the number of erroneous bits in a received bit stream of n bits. A lower BER results in better fidelity of the communication system. When the observation interval is sufficiently long (i.e., when many bits are transmitted), the BER approaches the bit error probability, which is a long-term statistical index of the communication system performance.

7.4.2 Gain of Coding

The gain of a coding method is the effective reduction in the SNIR in a communication system with respect to the system without coding when both channels have the same BER and data rate (R).

7.4.3 Coding Efficiency

The *coding efficiency* is defined in terms of the *coding rate*. If each information symbol at the encoder input has k bits and the corresponding codeword at the encoder output has c bits, then its length is $c = k + r$, where r is the number of redundant bits added by the encoder. In this case, the *coding rate* R_c is expressed as

$$R_c = \frac{k}{c} = \frac{k}{k+r} \tag{20}$$

Thus, a larger *coding rate* R_c results in greater *coding efficiency,* up to a maximum of 1.

References [12–16] are recommended for further reading to readers interested in the topics in this chapter.

References

1. W. Stark, *Introduction to Digital Communications* (Cambridge University Press, 2023)
2. J. Speidel, *Introduction to Digital Communications* (Springer, 2018)
3. R.G. Gallager, *Principles of Digital Communication* (MIT, 2008)
4. U. Madhow, *Fundamentals of Digital Communication* (Cambridge University Press, 2008)
5. T.M. Cover, J.A. Thomas, *Elements of Information Theory*, 2nd edn. (Wiley, 2006)
6. J. Gavan, *Principles of Radio Systems and Relevant EMI Environment Effects, Analysis, and Computation*, ed. by R. Perez. Chapter 19 of Handbook of EMC (Academic Press, 1995)
7. S.W. Ellingson, *Radio Systems Engineering* (Cambridge University Press, 2016)
8. S-M.M. Yang, *Modern Digital Radio iCommunication Signals and Systems*, 2nd edn. (Springer, 2020)
9. A. Goldsmith, Wireless Communications (Cambridge University Press, 2005)
10. M.K. Simon, M-S. Alouini, *Digital Communication over Fading Channels: A Unified Approach to Performance Analysis* (Wiley, 2000)
11. M. Viswanathan, *Wireless Communication Systems in MathLab*, 2nd edn. (2020)
12. S. Ling, C. Xing, *Coding Theory—A First Corse* (Cambridge University Press, 2004)
13. S. Lin, D.J. Costello, *Error Control Coding* (Pearson Prentice-Hall, 2004)
14. J.G. Proakis, *Digital Communications*, 5th edn. (McGraw-Hill, 2008)
15. B. Sklar, *Digital Communications—Fundamentals and Applications*, 2nd edn. (University of California, Los Angeles, 2010)
16. R.E. Ziemer, *Principles of Communications* (Wiley, 2009)

Performance of Radio Digital Systems in Presence of EMI

1 Introduction

Digital communication systems with AWG noise are considered in chapter "Introduction to Digital Communication Systems". Here, we consider radio digital systems with fading and EMI in the transmission channel. In these situations, the primary goal of radio systems is to operate properly in a real electromagnetic environment without causing undesired EMI effects on other radio systems and themselves [1]. To meet this goal, the radio system should be appropriately planned and designed by selecting an adequate system architecture with optimal signal processing (*coding* and *modulations*). In this way, the expected system performance can be reached, and the effects of the channel fadings and EMI are minimized. In addition to optimal signal processing, conventional EMI prevention methods, discussed in chapters "EMI Filters: Theory and Design", "EMI Shielding for Radio Equipment", and "Grounding for Radio Systems and Facilities", should be applied. These methods include EMI *filtering,* EMI *shielding, EMI* and *safety grounding*, including the signal and power cables discussed in chapter "Performance of RF Cables and IT Signals Lines in Radio Systems with EMI". Proper system planning also includes an effective trade-off between EMI system protection and system *power* and *spectral* efficiency. This may be achieved by *multiplex* (TDM and FDM), *multicarrier (MC), multiple access point*, and MIMO *transmissions,* as well as by adequate forward error correction (FEC) techniques. All these transmission methods and techniques are discussed in chapter "Overview of Broadband Radio Systems with High EMI Immunity".

Figure 1 shows a block scheme for a standard radio digital system with a random radio channel, including signal processing relevant to the effective EMI and fading controls in the radio system.

M. Pajovic, *Fundamentals of EMI Phenomena in Radio Communication Systems*, https://doi.org/10.1007/978-3-031-88748-2_5

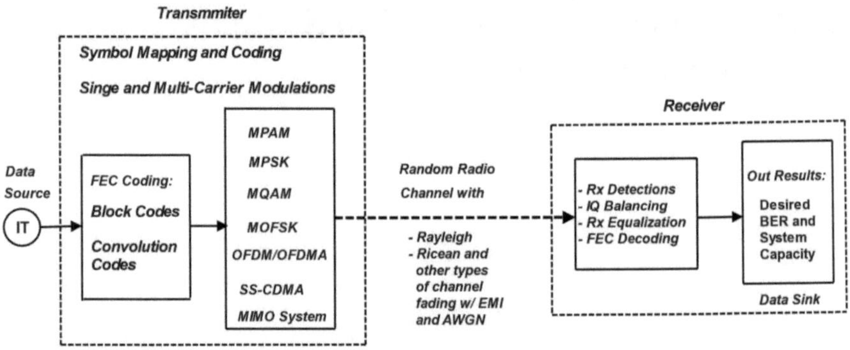

Fig. 1 Summary of signal processing techniques relevant for effective EMI control in random (nonstationary) radio channels

In particular, Fig. 1 shows the data symbol formatting and baseband coding functions, the baseband waveform mapping into a single carrier or multiple carriers, and the emission of the modulated signal over the radio channel. The actual radio channels include, for instance, Rayleigh, Ricean, or Nakagami fading, as well as EMI from other radio systems or itself. After being amplified by a low-noise amplifier (LNA) and filtered by the receiver's front-end filters, the received signal is subjected to several processing steps, such as RF detection, IQ balancing, FEC decoding, and linear equalization. An overview of FEC techniques is discussed in chapter "Introduction to Digital Communication Systems", where digital signal processing techniques, such as the digital modulations MPAM, MPSK, and MQAM, are considered.

In the following two sections, we discuss a radio transmitter as a main EMI or RFI source with undesired effects on radio receivers as victims of EMI or RFI. References [1–12] motivated and influenced the writing of this chapter and recommended for further study of the performance of radio digital systems.

2 Radio Transmitter as a Source of EMI and RFI

Radiofrequency interference (RFI) is a particular case of EMI that radio transmitters produce. A transmitter can interfere with other radio systems by its desired emissions and undesired spurious (signal distortions) produced inside the transmitter owing to the imperfections and nonlinearity of the signal processing TX circuitry and filters.

Figure 2 shows a basic block diagram of a radio system where the transmitter TX radiates RFI/EMI out in the environment over the TX antenna. The primary victims of these emissions are receivers of other radio systems and their own receivers of the considered radio system.

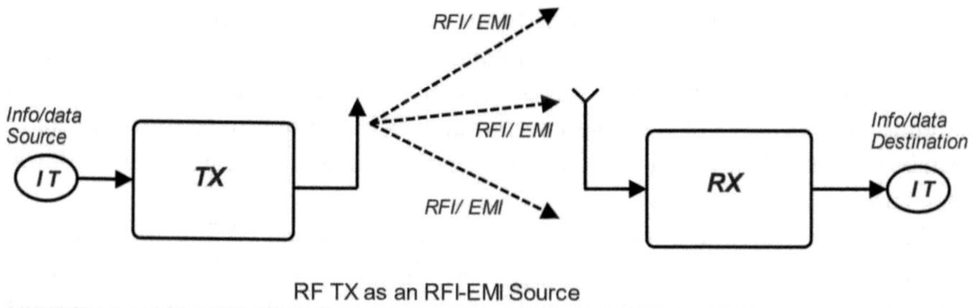

Fig. 2 A TX as an RFI/EMI source

The frequency spectra allocated to radio-wireless communication systems by the ITU is divided into bands that extend from extremely low to extremely high frequencies. These frequency spectra are listed in Table 1.

Table 1 The frequency spectra

Band designation	Frequency range	Meanings
VLF	3–30 kHz	*Very low frequencies*
LF	30–300 kHz	*Low frequencies*
MF	300–3000 kHz	*Medium frequencies*
HF	3–30 MHz	*High frequencies*
VHF	30–300 MHz	*Very high frequencies*
UHF	300–3000 MHz	*Ultra high frequencies*
SHF	3–30 GHz	*Super high frequencies*
EHF	30–300 GHz	*Extremely high frequencies*

The frequency spectrum allocated to radar systems and satellite communications are listed in Table 2.

Table 2 Radar and satellite bands

	Radar systems	Satellite systems
Letter symbols	Frequency bands [GHz]	Frequency bands [GHz]
L	1–2	1.525–1.710
S	2–4	2.5–2.7
C	4–8	3.4–4.2 4.5–4.8 5.85–7
X	8–12	–
Ku	12–18	10.7–13.25 14–14.5
K	18–27	17.7–20
Ka	27–40	27.5–30

As mentioned above, the present bands are allocated to different services and purposes by the International Telecommunication Union (ITU). The updated ITU's official websites provide more details about actual frequency-band allocations. Additionally, because radio system performance is vulnerable to *EMI/RFI*, the radio bands are controlled by domestic and international spectrum regulations and EMC standards.

Thus, transmitters are the main contributors to RFI and EMI, especially today when their number rapidly increases globally. The TX radiation spectrum includes the desired (operating) bandwidth and *out-of-band* spectrum associated with spurious signals. Owing to the desired TX signals being a significant part of RFI/EMI, the TX operating bandwidth (BW) and TX power level should be under constant control.

The TX bandwidth mainly depends on the baseband-information waveforms, passband (channel) waveforms, and type of applied signal modulations. The modulation method in broadband multiple-carrier radio systems and multilevel M-QAM modulation is often used. For example, MQAM modulation is widely used in today's radio digital systems because it is a highly spectrally efficient modulation method, as discussed in chapter "Introduction to Digital Communication Systems".

For example, a 16-QAM-modulated signal theoretically occupies an infinitely broad spectrum, but only 2.5 kHz of the spectrum around a 20-kHz subcarrier contains ~98% transmitted power, as shown in Fig. 3.

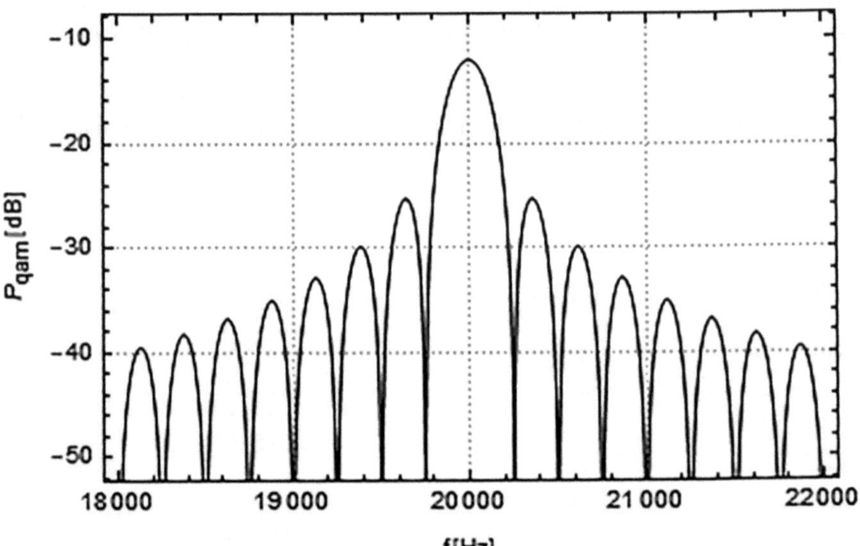

Fig. 3 16-QAM spectrum

The power spectral density (PSD) plot of the 16-QAM signal at a 1-Ω load is made according to the following QAM expression:

$$PSD_{QAM} = A_{QAM}\left\{\sin c^2\left[2\pi\left(\frac{T_{QAM}}{2}\right)(f - f_c)\right] + \sin c^2\left[2\pi\left(\frac{T_{QAM}}{2}\right)(f + f_c)\right]\right\} \quad (1)$$

where A_{QAM} is the amplitude of the signal envelope expressed as

$$A_{QAM} = \left(\frac{A_{avg}}{2}\right)^2 \cdot \frac{T_{QAM}}{2} \quad (2)$$

where A_{avg} is the average amplitude of the QAM signal and where T_{QAM} is the QAM symbol time associated with the symbol rate, $R = 1/T_{QAM}$. Additionally, f_c is a carrier for the QAM signal.

On the basis of the plots presented above, the implementation of guard bandwidths between single-carrier channels is required to achieve a trade-off between good *spectral efficiency* and low *adjacent channel interference* (*ACI*).

2.1 Classification of EMI in Radio Systems

The main types of EMI/RFI interference produced by a radio transmitter are as follows [2]:

(a) *In-band* interference is caused by normal TX operation at frequencies that fall into operating bands of receivers of other or own radio systems. The ways to control this EMI are to control the TX power level, operating frequencies and bandwidths, TX signal distortions, and radiation pattern of the TX antenna.

(b) *Cochannel interference* (CCI) is associated with undesired signals received in the RX IF bandwidth from the own TX and duc to multipropagation conditions on the channel. The CCI can be linear or nonlinear. The nonlinear CCI is produced by nonlinear effects in the TX stages, mainly in the TX final stage (PA and filters). The CCI may cause harmful effects even at a far-located RX.

(c) *Adjacent channel interference* (ACI) occurs when a transmitted signal of one channel passes through adjacent channels because of insufficient RX front-end selectivity. This usually occurs when the TX and/or RX lack efficient interchannel isolation, i.e., when guard bandwidths between adjacent channels are not sufficiently wide.

(d) *Out-of-band* interference exists when a TX transmits RFI/EMI out of its operating bands. This interference may be broadband or impulse EMI, as well as harmonic and nonharmonic. The cause of this EMI can be the lack of filtering sufficiency of the TX IF and TX output filters. Furthermore, high TX power levels of the EMI emissions received by a close-located RX may saturate the RX front–end stages and produce broadband and harmful distortions in the received signal [2]. This EMI is classified as *nonlinear* interference. The common way to minimize these EMIs is to control the TX power level and linearity of the TX final stages.

(e) Another important source of TX nonlinearity in digital radio systems is related to sampling and quantization processes in conjunction with the present fast Fourier transform (FFT) algorithm.

2.2 Types of EMI Noise in Radio Systems

The main sources that affect radio system performance are (a) *additive white (ambient and thermal) noise* known as *AWGN,* (b) *signal distortions (linear and nonlinear),* and (c) *mutual electromagnetic interference* [2].

(a) *Additive noise* in radio systems can be *internal* or *external,* as discussed in chapter "Introduction to Digital Communication Systems". *Internal* or equipment noise is generated in radio system components (transmitters, receivers, encoders/decoders, information transcoders, etc.). *External* additive noise originates in a propagation medium or elsewhere from the environment.
(b) *Signal distortions* can be harmonic and intermodulation products, primarily caused by the nonlinearity of the TX and RX functional blocks and circuits, such as power amplifiers, frequency converters, filters, and other nonlinear signal-processing circuits. The signal distortions may be *amplitude*, *frequency,* or *phase* distortions.
(c) *Self-mutual electromagnetic interference* is associated mainly with *fading* caused by *multipath* propagations of the operating signaling components in a dispersive medium. The multipath propagations are caused by wave reflection, diffraction, and refraction. In general, there is scattering due to various obstacles in a propagation medium. Therefore, the desired signaling components travel over different paths toward the receiver's antenna and interfere with each other at the receiver's front end. Various types of fading in radio channels are discussed in this chapter. Fading can generally be considered as self-EMI.
(d) Additionally, the EMI/RFI originating from other radio systems operating at the same or close frequencies are classified as *intersystem interference.* If EMI originates from equipment belonging to the same system, it is classified as *self-interference*.

2.3 Narrowband and Broadband EMI

From the time and frequency domain perspectives, EMI can be *nonimpulsive* and *impulsive*. The *impulsive* EMI is a broadband EMI that is especially critical for *digital radio systems* because it may significantly decrease the signal-to-interference/noise ratio (SINR) and, consequently, increase the bit error rate (BER) or symbol error rate (SER).

The impulsive EMI algorithm is a complex random phenomenon that requires complex statistical analysis. The National Telecommunication and Information Administration (NTIA) developed real-world physical and statistical EMI models in the time and space domains. The three types of EMI are included in this model:

(a) *Class A*—considers a *narrowband* EMI whose spectrum is narrower than an EMI receptor's bandwidth. EMI *Class A* represents EMI noise associated with ambient noise, which is due mainly to the operations of radio systems. This may be, for example, related to coherent pulse trains typical for digital radios, radars, etc.
(b) *Class B*—considers the cases in which the *broadband* EMI has a wider spectrum than the EMI receptor's bandwidth. *Class B EMI* considers highly impulsive EMI, which usually represents manmade or natural broadband EMI (i.e., nonmassage noise and interference) from manmade sources, atmospheric, and solar noise. This EMI class is often coherent impulsive noise that may cause an overlap of signal data symbols in the victim RX, resulting in intersymbol interference (ISI).
(c) *Class C—Hybrid* interference processes (more realistic) are combinations of EMI Classes A and B, which have a non-Gaussian stochastic distribution.

2.4 Nonlinearity of the Power Amplifier as a Main Source of EMI Signal Distortion

The nonlinearity of the TX power amplifier (PA) results from the spectral spread of the transmitted signal outside the given operating bandwidth. This spectral spread is commonly quantified by the adjacent channel power rate (ACPR), i.e., the ratio of power sent into an adjacent frequency channel relative to the power sent in the operating (desired) channel. Another consequence of signal distortion from PA nonlinearity is quantified by the error vector magnitude (EVM), which determines the level to which the carrier amplitude and phase are in error concerning the desired waveform amplitude/phase at the nominal sampling time.

Several methods might be used to improve the PA linearity, such as (a) the use of a *quasilinear* (QL) region of the transfer (input/output) characteristic of the PA, (b) *predistortion* linearization as an improvement for the QL in/out region, (c) *feedforward* linearization, and (d) *feedback* linearization [3].

The *predistortion* is the broadening PA quasilinear region, effectively modifying the input signal to compensate for the PA nonlinearity. The PA nonlinearity can be sufficiently represented by the third-order memoryless polynomial:

$$y(t) \cong k_0 + k_1 x(t) + k_2 x^2(t) + k_3 x^3(t) \tag{3}$$

where $y(t)$ and $x(t)$ are the PA output and input signals, respectively. Additionally, the parameter k_0 represents the DC signaling component that can be neglected without concern. Hence, the desired polynomial (full linearity) is obtained when $k_0 = k_2 = k_3 = 0$, where Eq. (3) is $y(t) = k_1\, x(t)$.

However, for some cases, we can ignore the nonlinearity caused by the parameter k_2 (i.e., the second-order nonlinearity) because it produces the output signal at frequencies far from a carrier frequency. Thus, we consider only polynomials

$$y(t) \cong k_1\, x(t) + k_3\, x^3(t) \tag{3a}$$

Hence, only the third-order nonlinearity is considered. This nonlinearity can be overcome simply by using a diode or FET transistor (at the PA input) with the *I-V* characteristic suitable for compensating for the third-order nonlinearity. In this case, Eq. (3a) becomes

$$y(t) \cong k_1\, x(t) + k_3\, x^3(t) \cong a_1\, x(t) \tag{3b}$$

A component (such as a diode or FET transistor) makes the input signal $x(t)$ linear at the PA output (after being affected by the PA nonlinearity). The case of the *predistortion* linearization is illustrated in Fig. 4.

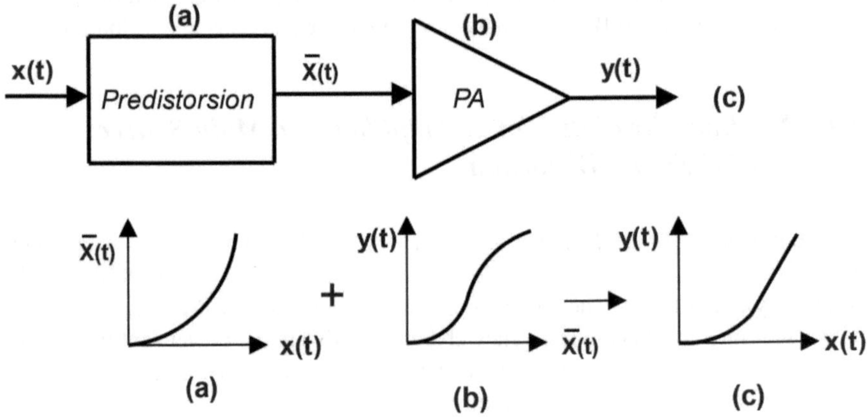

Fig. 4 Predistortion linearization by shaping the PA input signal

The plots show (a) a nonlinear I–V curve of a diode or FET at the PA input, (b) the PA nonlinear transfer function, and (c) the effective, quasilinear PA transfer function created by the signal predistortion process.

However, *predistortion* linearization is not always a perfect method because the PA characteristics of all PAs are not given by Eq. (3a). Sometimes, PA characteristics might have higher-order nonlinearity beyond the third order. The polynomial parameters k_i may vary with temperature and have an unstable bias. Additionally, the PA is not always memoryless, which is one of the conditions for modeling the PA characteristics via Eq. (3).

Another relatively simple and real-time effective method is *feed-forward linearization*. It is illustrated by the block scheme in Fig. 5.

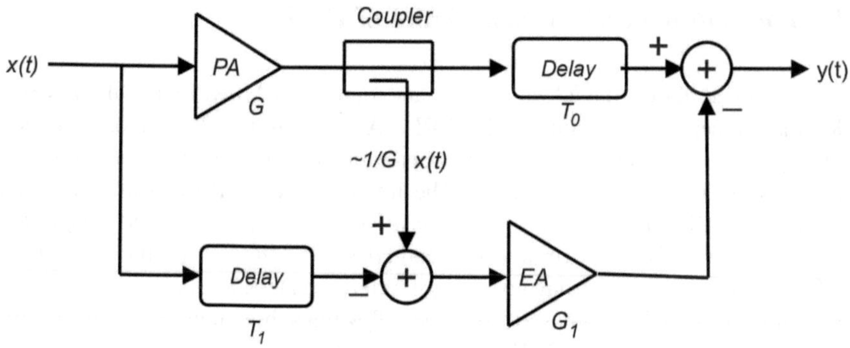

Fig. 5 PA feedforward linearization

The PA input signal x (t) is selected from the PA input and delayed by time T_1, resulting in the signal $x\ (t - T_1)$. This signal is used as a real-time reference for the linearization process. Additionally, a sample of the PA output signal is taken by a directional coupler when attenuated by ~1/G, resulting in a signal close to the input signal x (t). Furthermore, the subtraction of this signal from the delayed signal x (t − T_1) results in the difference signal x (t) − x (t − T_1), which is associated with the nonlinear distortion (or a signal error) produced by the PA. This signal difference is further amplified by an error amplifier EA with gain G_1, where $G_1 \ll G$. In some practical cases, the EA amplifier is not needed. Additionally, the delayed blocks T_0 and T_1 are used to prevent the signal delays (related to propagation times through the PA and the EA) from being treated as distortions.

The advantage of *feed-forward* linearization is that there is no need for prior knowledge (or assumption) of the PA nonlinear transfer function because the linearization process automatically starts and stops when the PA linearity reaches the desired degree. However, some disadvantages exist, such as that delays must be precisely determined over a larger bandwidth. Additionally, the error signal amplifier EA should have excellent linearity. Regardless of these requirements, this linearization method is quite effective and is often used in PAs of digital systems. The more effective but more complex linearization scheme is the *feed-forward* linearization described in Ref. [3].

Additionally, effective analog low-pass filtering at the PA end is requested regarding PA nonlinearity. Analog low-pass filtering and other analog filtering methods are discussed in detail in chapter "EMI Filters: Theory and Design".

In the end, proper antenna integration with the PA is crucial. This integration includes proper impedance and balance matching between the PA and the antenna. This is important because the antenna is an integral part of the PA. Therefore, better impedance and balance matching between the PA and the antenna results in lower PA nonlinearity. This topic is discussed in more detail in chapter "Performance of RF Cables and IT Signals Lines in Radio Systems with EMI".

3 The Radio Receiver as a Victim of EMI

As mentioned above, the TX is a dominant source of EMI/RFI since the RX is a dominant receptor or "victim" of EMI/RFI. As a result, the RX is more complex because it needs to neutralize all the imperfections of the TX emissions and EMI and noise in the propagation media. The important RX performance parameters include the RX *gain, dynamic range, sensitivity*, *frequency selectivity, amplitude* and *frequency stability*, *error detection* and *correction capability*, and *immunity* to *RX* saturation caused by strong nonlinear interference at the RX input. Some critical RX parameters are briefly described in the following subsections. This topic is also discussed in more detail in [2, 3].

3.1 Rx Sensitivity, Selectivity, and Gain

The RX *sensitivity* is defined as the ability of the RX to provide the desired signal level for successful detection at a minimal SNIR at the antenna input. The interfering and disturbing signals may be from other communication systems, ambient noise, radio channel *fadings*, external interference (intentional and unintentional), or signal distortions from system radio equipment. Notably, *external interference* (EMI/RFI) and equipment signal distortions have a greater impacts on the RX *sensitivity* than flat thermal and ambient noise does. Therefore, radio sensitivity depends fundamentally on external noise, interference, and internal RX noise. Additionally, the received external noise and interference (EMI) depend on the antenna radiation pattern and directivity.

The task of the receiver *gain* G is to increase the level of a received signal to be strong enough for A/D conversion or analog detection. For an A/D-based receiver, the gain should be large enough to ensure that the operating signals dominate ambient noise, EMI, quantized noise, and spurious noise generated by the A/D converter.

RX *selectivity* is the ability of the RX to separate desired signals from undesired interfering signals and signal distortions. RX *selectivity* is directly linked to the RX bandwidth (BW). Thus, the limit of the RX selectivity is related to its front-end filters, which should reduce the noise and interference received by the RX antenna, which may flow into the receiver's IF stage. However, some noise and nonlinear interference converted by the RX mixer may fall into the RX IF bandwidth and cannot be attenuated by the RX front-end filter.

Typically, before the RX front-end filters, *preselector* filters (or band filters) are employed, which exclude signals or entire frequency bands that are not of interest. Removing the bands that are not of interest reduces the need for wideband linearity and significantly simplifies the frequency downconversion process. Therefore, a preselector does not influence the selectivity of the RX bands of interest. Thus, most receivers need additional filters after the preselectors whose tasks are (a) filtering spurious products created by frequency conversion and intermodulation processes, (b) filtering anti-alias in A/D-based receivers, (c) isolating one operating signal from other operating signals inside the RX band, and (d) limiting the RX bandwidth to maximize the SINR at the RX detector input.

3.2 Rx Intermodulation Distortion and RX Dynamic Range

In general, intermodulation distortions (IMDs) can be produced in the RX when two or more signals with different but close frequencies excite nonlinear RX front–end circuits; as a result, the spectrum of the RX output signals contains harmonic and intermodulation products associated with the desired and interfering signals at the RX demodulator input. Thus, when the RX receives interfering signals from other radio systems, they are mixed with the system operating signals in nonlinear RX circuits, resulting in IMD products. These IMD interferences are different from the IMD interference generated by a radio transmitter.

In today's radio systems, the levels of the interfering signals from other radio systems exponentially increase, and they reduce the so-called *spurious-free dynamic range* (SFDR) of the RX. The broadband RX dynamic range is associated mainly with the second order of the intermodulation product (IMP). For moderate IMP, the SFDR can be expressed as [2]

$$\mathrm{SFDR}_2 \cong \frac{1}{2}\left[174 + P_{IP2} - F_r - 10\log(B)\right], \tag{4a}$$

For a *narrower-band* RX, the SFDR is related mainly to the third order of the IMP (P_{IP3}) and can be described as

$$\mathrm{SFDR}_3 \cong \frac{2}{3}\left[174 + P_{IP3} - F_r - 10\log(B)\right] \tag{4b}$$

where 174 dBm/Hz is the level of white-thermal noise, F_r is the RX overall noise figure or factor, and B is the RX IF bandwidth.

Furthermore, excessive interference at the RX input may further reduce the RX *dynamic* range and *sensitivity*. In this extreme case, the RX front-end preamplifier and first mixer stages can saturate (in a nonlinear regime), causing additional disturbing harmonic and IMD products in the RX. In this situation, RX automatic gain control (AGC) cannot compensate for the signal variation caused by high-level interference [2].

3.3 IQ Imbalance in the Receiver

Receivers with IQ signal processing (or quadrature downconversion) are widely used in radio digital communications. The main advantage of this downconversion method is that one has *direct frequency image rejection* without needing separate image filtering, such as conventional superheterodyne receivers need. The known quadrature conversion schemes are Hartley and Waver conversion [3]. If not suppressed enough, the image bands act as self-EMIs in the receiver.

Figure 6 shows a basic block scheme for an IQ receiver where image bands are directly rejected.

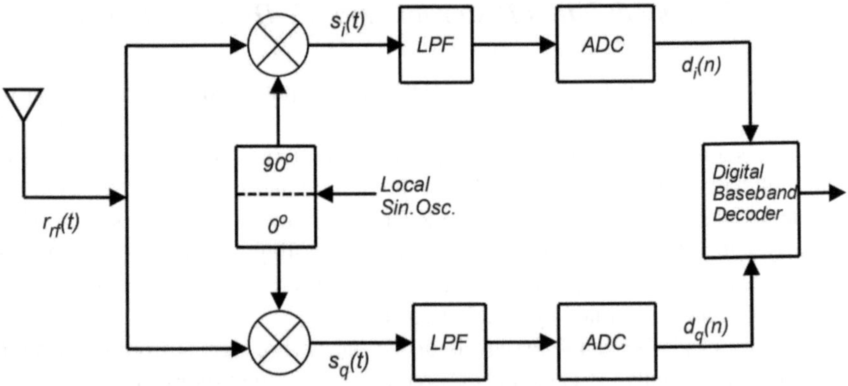

Fig. 6 Direct conversion (IQ) receiver block diagram

IQ conversion is an efficient process but is imperfect because of the present IQ imbalance between the I and Q channels. This imbalance can be in the amplitudes and phases of the signals $s_I(t)$ and $s_Q(t)$. Additionally, these signals exercise DC-offsets. The IQ imbalance is caused by a nonperfect balance of RF IQ front-end circuits and cross-talk between branches I and Q. Additionally, undesired nonorthogonality between I and Q signals is caused by the inaccurate and unstable phase of an output signal of the RX local oscillator. The realistic IQ signal s(t), before any compensation, is presented by the following expression [3]:

$$s(t) = \{DC_i + A\cos(\omega_c t)\} + \{DC_q + \Delta A \sin(\omega_c t + \Delta\phi)\} \tag{5}$$

where DC_i and DC_q are DC offsets introduced in the I and Q signals, respectively, while ΔA and $\Delta\phi$ are the differences in amplitude and phase between the I and Q signaling components, respectively.

IQ imbalance is more critical for higher-level modulations that are used in many of today's communication systems, such as 4G-LTE, WLAN, IEEE 802.11 (WiFi), and 5G-NR. To avoid complex and expensive RF front-end circuits, digital signal processing (DSP) algorithms for the compensation of the IQ imbalance and DC offsets of the digital signals $d_i(n)$ and $d_q(n)$ are developed, both of which are shown in Fig. 6.

3.4 Signal Pulse Shaping and RX Equalization

Owing to all practical communication systems having limited bandwidths, digital signals propagating through these systems result in intersymbol interference (ISI), which may significantly degrade the system's performance. Today's radio communication systems operate in a limited spectrum to increase the system's spectral efficiency and minimize EMI/RFI effects from other communication systems. Several approaches are applied to minimize the ISI level in a communication system, such as appropriate *pulse shaping* at the TX side and *linear equalization* at the RX side.

The pulse shape with the narrowest bandwidth that avoids ISI is a *sinc* pulse, although other pulses, such as *raised-cosine* and *square-root-cosine pulses,* are commonly used. For example, the rectangular-shaped pulse is the worst case regarding high ISI because of its broad spectrum caused by its extremely sharp edges, as discussed in chapter "Digital and Pulse Signals, Random Variables, and Stochastic Processes".

Because radio channels have limited bandwidths, they can be considered bandpass filters whose transfer function *H (f)* is expressed as

$$H(f) = |H(f)| e^{i\phi(f)} \tag{6}$$

where $|H(f)|$ and $\phi(f)$ are the magnitude and phase of the channel impulse response, respectively. The group delay of this filter model is given as

$$\tau_g(f) = -\frac{1}{2\pi}\frac{d\phi(f)}{df} \tag{7}$$

Note that a communication channel without ISI distortions has a constant amplitude response and linear phase response inside its entire operating bandwidth, i.e., it has a constant group delay $\tau_g(f)$. Thus, ISI occurs when the amplitude response and group delay are not constant. A channel with a nonconstant group delay is considered a *dispersive* channel. For example, when a pulse train propagates through a dispersive channel, the pulse's components arrive at a receiver at different times, which may interfere with each other at a destination point. This phenomenon is called intersymbol interference (ISI), and it is typical for multipath radio channels, as noted before in this chapter.

The ISI at dispersive channels affects mainly high-data-rate radio systems with a time-variant multipath nature. To reasonably control and minimize the ISI, three approaches should be considered: (a) satisfying the *first Nyquist criterion*: force the ISI to zero via signal–pulse shaping, such as *sinc*, *raised–cosine*, and *square–root raised–cosine shaping*; (b) *partial signaling*: introduce a controlled amount of ISI at the TX side that will be used for the ISI compensation process at the RX side; (c) employ algorithms and *linear equalizers* to counter the ISI at the RX. Here, the Viterbi algorithm is used with *maximum likelihood sequence estimation* (MLSE). The MLSE algorithm is the most efficient for long-channel responses, among other ISI compensation methods, but it is the most complex. One is used in the cases of severe ISI distortions in the channel. For moderate ISI cases, *linear channel equalization* simplifies and makes implementation easier.

Linear channel equalization is an efficient approach to compensate for the ISI because it is based on a model of linear time-invariant filter characteristics that can be considered by using standard signal processing theory. Three common types of linear channel equalizers are used: (a) the *zero-forcing* (ZF) equalizer, (b) the *minimum square error* (MSE) equalizer, and (c) the least mean square adaptive (LMSA) equalizer.

For example, Fig. 7 shows how a ZF equalizer works by compensating for the channel response.

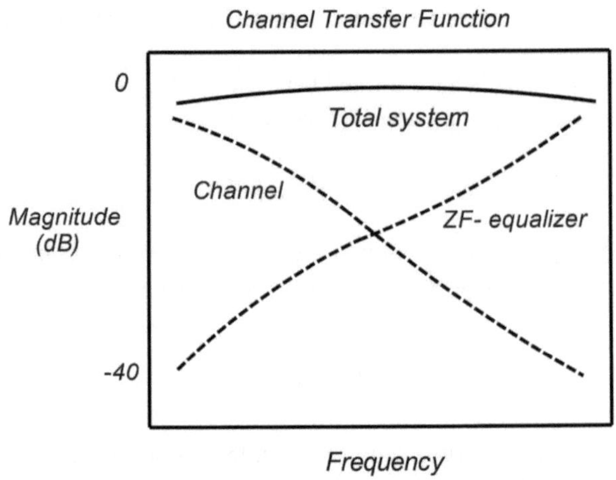

Fig. 7 Analytic principle of channel equalization

Here, the channel transfer function *H (f)* is approximated to the FIR filter response when the equalizing function ZF (f) is the transfer function of the infinity impulse (IIR) filter. Thus, the functions ZF (f) and H (f) are inversely proportional, resulting in linear data transmission.

$$\mathrm{ZF}(f) \cong \frac{1}{H(f)} \tag{8}$$

On the basis of adaptive capabilities, linear channel equalization can be categorized into two groups:

(a) *Preset* equalization when the channel is considered *linear time-invariant* (LTI) during the data transmission interval.
(b) *Adaptive* equalization when the channel is considered a time-variant channel, whose parameters are determined upon adaptive algorithms adopting the data transmission for the current channel conditions. There are more adopting algorithms, such as *least mean square* (LMS), recursive lattice square (RLS), conventional square root, and fast Kalman algorithms.

More about optimal pulse shaping and equalization methods to minimize the ISI and ICI in radio systems can be found in [2, 3].

4 Digital Transmission over Random Multipath Radio Channels

The parameters of transmission channels, such as transmission cables and radio line-of-sight (LOS) terrestrial and satellite links, are mainly constant over time and belong to the category of time-invariant channels discussed in chapter "Introduction to Digital Communication Systems". Generally, radio wireless channels are quite different because signal waves travel over multiple paths with different propagation times and attenuations due to reflections and scattering at various obstructions, such as buildings, trees, and vehicles. All these situations cause signal fading. This phenomenon is considered in detail in the literature [4–6]. Here, various radio communications and noncommunication electromagnetic interference (EMI) methods should be added. Moreover, when the transmitter and/or the receiver are in motion, the Doppler effect also affects the propagating signal.

Hence, radio wireless channels might be very cruel compared with AWGN channels, and they require different modeling approaches and more effective algorithms for channel coding to obtain enough quality signal reception. Therefore, optimal signal coding, modulations, and error-control methods should be employed, including spatial diversities or MIMO techniques.

4.1 Characteristics of Multipath Propagation

As mentioned earlier, the cause of multipath propagation is the fading manifested in the random changes in the received signal parameters such as the amplitude, phase, frequency, and angle of arrival at the receiver antenna. Hence, these changes occur in the time, frequency, and space domains. With respect to the time change, fading is classified as *slow* or *fast*. *Slow* fading is associated with slow spatial signal variations that tend to have a normal Gaussian power distribution, as discussed in chapter "Introduction to Digital Communication Systems".

Fast fading results from multiple reflections from obstructions in a propagation channel, where the radio signal propagates over different paths with different lengths, causing fast deviations in the received signal strength. There are two cases of multipath channels: *static* channels, when both the transmitter and receiver are stationary, and *dynamic* channels, when the transmitter and/or receiver or obstructions in the channel are in motion. Therefore, in the *static* multipath channel, a signal varies primarily in the time domain, whereas in a dynamic channel, the signal varies in the time, spatial, and frequency domains.

Hence, the signal sent by a mobile radio system may have many components (copies) with randomly distributed amplitudes, phases, frequencies, and angles of arrival at the receiver, causing the signal to fade in time and space. Additionally, in a dynamic multipath channel, signal fading occurs in the frequency domain due to the Doppler effect. Particularly in mobile systems, the Doppler effect occurs because of the relative movement of a mobile user to or away from a base station (BS). The phase change, $\Delta\phi$, of the received signal due to the difference in the propagation lengths, Δl, is given as

$$\Delta\phi = \beta\Delta l = \frac{2\pi}{\lambda}\Delta l\cos(\theta) = \frac{2\pi}{\lambda}(v\Delta t)\cos(\theta) \quad (9)$$

where λ is the wavelength of the signal, v is the relative speed of a receiver, Δt is the time needed for the receiver to travel the distance Δl, and θ is the angle of wave arrival to the receiver. On this basis, the Doppler frequency shift f_D is given as

$$f_D = \frac{1}{2\pi}\frac{\Delta\phi}{\Delta t} = \frac{v}{\lambda}\cos(\theta) = \left(\frac{v}{c}\right)f_s\cos(\theta) \quad (10)$$

where the relationship between λ and the signal frequency f is given as $\lambda = c/f_s$, where c is the speed of the light.

Note that when a mobile receiver is moving toward the arriving signal with frequency f_s, the received frequency increases with the frequency f_D, i.e., $(f_s + f_D)$, but when the receiver moves away from the arriving signal, the received frequency is decreased by the frequency f_D, i.e., $(f_s - f_D)$. Figure 8 shows a moving object and a stationary base station (BS). When the object approaches point 0 by speed v, the frequency Doppler shift f_D is positive relative to the signal frequency f_s.

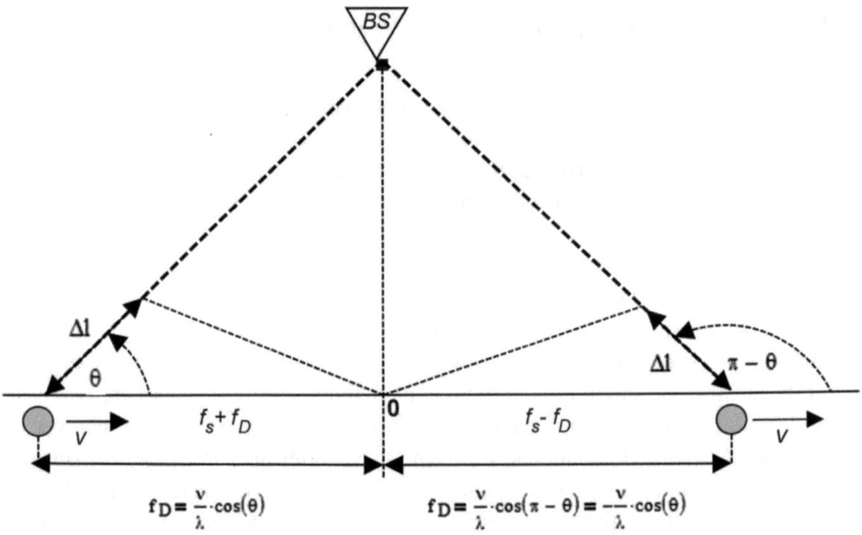

Fig. 8 Doppler effect

Conversely, when the object moves away from point 0, f_D is negative relative to f_s. The above statement can be confirmed via the following Eqs. (11a), (11b):

$$f_D = \left(\frac{v}{\lambda}\right)\cos(\theta) \tag{11a}$$

$$f_D = \left(\frac{v}{\lambda}\right)\cos(\pi - \theta) = -\left(\frac{v}{\lambda}\right)\cos(\theta) \tag{11b}$$

4.2 *Multipath Channel Parameters: Time Dispersion, Coherent Bandwidth, and Coherent Time*

4.2.1 Time Dispersion Parameters

The *time dispersion* parameters are important for frequency-wideband (pulse) signals passing through a radio channel with a finite bandwidth. These parameters are the *mean excess delay*, the *rms delay spread*, and the *excess delay spread.*

The *mean excess delay* $\langle\tau\rangle$ is given by the *first moment* of the power delay profile $P(\tau_i)$ of a signal pulse train expressed as

$$\tau = \frac{\sum_0^{N-1} \tau_i P(\tau_i)}{\sum_0^{N-1} P(\tau_i)} \tag{12a}$$

The *rms delay spread* is expressed as the square root of the second central moment of the PDP, expressed as

$$\sigma_\tau = \sqrt{\tau^2 - \tau_i^2} \tag{12b}$$

where $\langle\tau^2\rangle$ is

$$\tau^2 = \frac{\sum_0^{N-1} \tau_i^2 P(\tau_i)}{\sum_0^{N-1} P(\tau_i)} \tag{12c}$$

All delays are considered from the first signal arriving at the receiver at τ_0.

4.2.2 Coherent Bandwidth

In a multipath channel, the time-domain *power delay profile* and the frequency-domain *power spectral response* are mathematically connected via the Fourier transform. The power spectral response is called the channel *coherence bandwidth*, which refers to a statistical parameter of the frequency range where the channel is considered flat. In the coherence bandwidth, the amplitudes of two frequency signals are strongly correlated. To describe the dispersive nature of a multipath channel completely, both the *power delay profile* (PDP) and the *coherence bandwidth* domain are needed.

Depending on the degree of the amplitude correlation of two frequency-separated signals, there are different definitions of the coherence bandwidth in practice, such as when the signal correlation above 50% corresponds to $B_c \approx 0.02/\sigma_\tau$ and when the correlation above 90% corresponds to $B_c \approx 0.2/\sigma_\tau$ [5].

4.2.3 Coherence Time and Doppler Spread

The coherence time T_c and Doppler spread that was discussed earlier are in duality, and their relation is

$$T_c \approx \frac{1}{B_D} = \frac{\lambda}{v} \tag{13}$$

where λ is the wavelength of the received signal and v is the speed of the TX, RX, or surrounding scattering object motion.

More information about the multipath channel parameters mentioned here can be found in [2, 7].

4.3 Summary of Slow and Fast, Frequency-Selective Fading

In terms of frequency, radio communication channels are classified as *frequency-nonselective* (or *frequency-flat*) channels and frequency-*selective* channels. For example, Rayleigh fading and Rician fading are *flat* fading. The fading cases will be discussed in the next section. The frequency-selective fading acts as an FIR filter (finite impulse response filter) with complex coefficients that change over time. An FIR filter includes a response of two-dimensional, real-color random Gaussian processes.

During *nonselective frequency* fading, the spread of the pulse signal delay t_{spr} is smaller than the pulse duration T_s ($t_{spr} < T_s$). However, during *frequency-selective* fading characterized by rapid time and frequency fluctuations of the received signal, the spread of the signal delay τ_{spr} is significantly greater than that of T_s ($t_{spr} > T_s$).

As mentioned, the received signal is affected by the *Doppler effect* when the transmitter and/or receiver are in motion. This effect causes the undesired spread of the signal frequency spectrum. The total Dopler frequency spread, $f_m = f_2 - f_1$, is around the carrier frequency f_c, $(f_c - f_1) < f_c < (f_c + f_2)$. This spread also corresponds with the Doppler spread or bandwidth B_d with respect to the modulated signal bandwidth B_s, when fading is classified as *fast* and *time-selective* when $B_d > B_s$, or *slow* and *time-nonselective* when $B_d < B_s$.

In the diagram shown in Fig. 9, all types of fadings are systematized as follows: (a) in the time domain, in terms of the signal time spread t_{spr} and the signal pulse duration T_s; and (b) in the frequency domain, in terms of the coherent channel bandwidth B_c, Doppler spread B_d, and the signal bandwidth B_s.

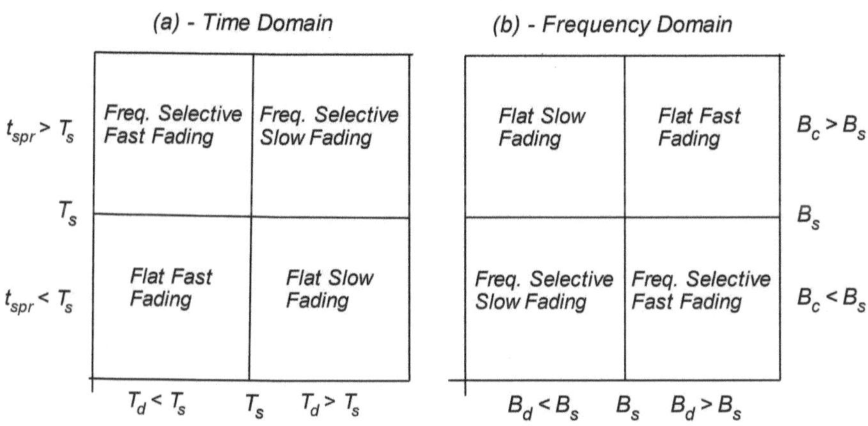

Fig. 9 Fadings in the (**a**) time domain and (**b**) frequency domain

4.4 *Concept of a Multipath Channel*

For example, Fig. 10a, b shows (a) the basic multipath-channel concept and (b) the corresponding Tx and Rx *discrete power delay profiles* (DPDPs).

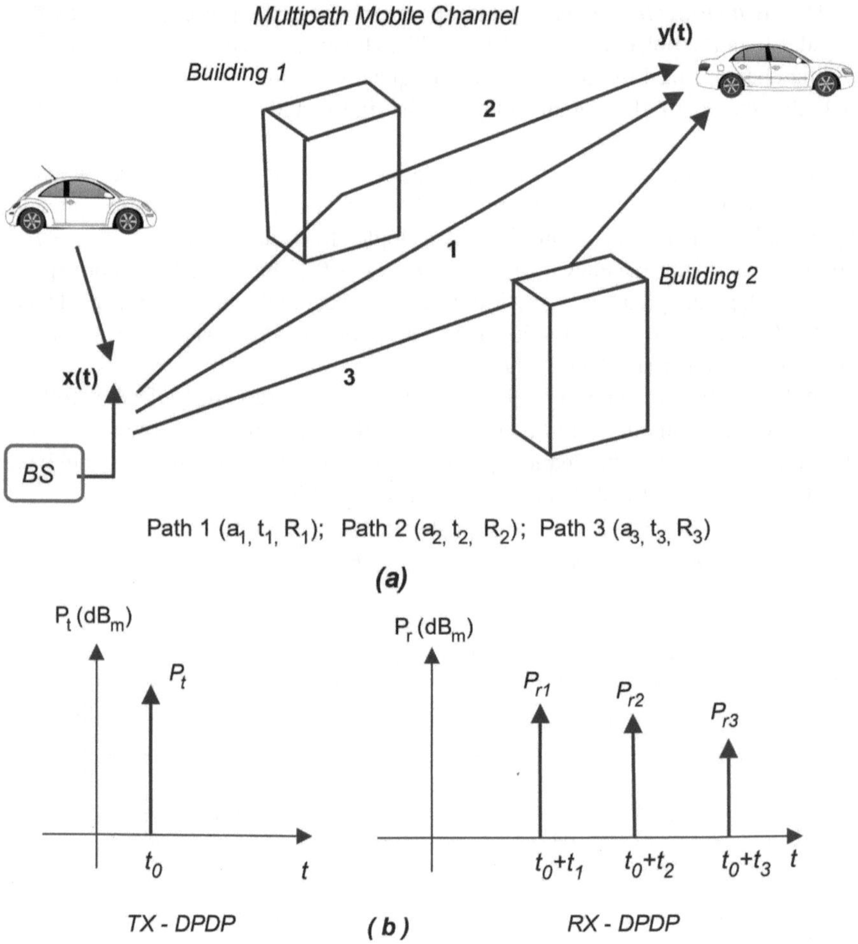

Fig. 10 (**a**) 3-path channel and (**b**) corresponding DPDP

Here, the channel is time-variant, and its parameters are (a_1, t_1, R_1), (a_2, t_2, R_2), and (a_3, t_3, R_3) characterize paths 1, 2, and 3, respectively, where a_i denotes the signal amplitude *attenuation*, t_i denotes the *signal delay*, and R_i denotes the *random channel variables* that represent random variations in the signal gain and phase. The random variables can be Ricean, Nakagami, Rayleigh, or others. For example, if the line-of-sight (LOS) signaling component exists in the multipath channel, then the Rician distribution is most suitable for the channel model. Moreover, in the absence of the LOS signal, the Rayleigh distribution is well suited for multipath channels.

The complex attenuation $a_i(t)$ of an *i-th* path in continuous time can be expressed as

$$\tilde{a}_i(t) = a_i(t)\, e^{-i\omega_c t_i(t)}, \tag{14a}$$

while the complex multipath channel impulse response is expressed as

$$h(t,t_i) = \sum_{i=0}^{N-1} a_i(t)\delta\left[t - t_i(t)\right] \tag{14b}$$

where $h(t, t_i)$ is the response of the *i-th* path to the unit (Dirac's) impulse $\delta[t - t_i(t)]$.

A channel can be *time invariant* even in multipath propagation when slow flat fading exists in the observing time interval. $h(t)$ for the time-invariant channel is expressed as

$$h(t) = \sum_{i=0}^{N-1} a_i(t)\delta(t - t_i) \tag{14c}$$

Additionally, the multipath channel *impulse response* can be a function $h(t, \tau)$ with two variables: the instantaneous time t and the propagation delay τ. Hence, a received signal $r(t)$, as the random complex signal at a specific time interval and random delays, can be expressed by the *convolution* of the complex channel impulse response $h(t, \tau)$ and the transmitted signal x (t):

$$r(t) = \int_0^{\infty} x(t-\tau)h(t-\tau,\tau)\cdot d\tau \tag{15}$$

5 Common Random Distributions in Multipath Radio Channels

These topics in this chapter are given in detail in the references [7–9], which influenced the writing of the chapter.

5.1 *Ricean and Rayleigh Distributions*

There are more PDFs (probability distribution functions) that can describe the fadings, such as Rayleigh, Ricean, and Nakagami, which are all considered well in mathematics. As noted earlier, the Ricean distribution is more general and well suited for estimating the contributions of LOS and NLOS (nonline-of-sight) signaling components to multipath fading. In this case, the random multipath signaling components (scatters) are added to the LOS component. Suppose a signal is sent across a *frequency-flat* channel. In that case, the received signal can be expressed as a sum of the line-of-sight signaling component and all scattered (reflected) components that converge to the destination point.

Hence, the line-of-sight (LOS) component of the received signal $m_{los}(t)$ can generally be described as a time-variant complex random variable:

$$m_{los}(t) = m_I(t) + im_Q(t) = re^{i(\omega_d t + \theta_\rho)} \tag{16}$$

where r, ω_d, and θ_ρ are the amplitude, Doppler frequency, and phase of the LOS signal, respectively.

Additionally, the sum of all scattered components can be described as

$$\mu_{sct}(t) = \mu_I(t) + i\,\mu_Q(t) \tag{17}$$

where $\mu_I(t)$ and $\mu_Q(t)$ are statistically uncorrelated real values of Gaussian random variables whose envelope $|\mu_{sct}(t)|$ is a Rayleigh amplitude.

Furthermore, the sum of the scattered components and the LOS component (the Rician case) can be expressed as

$$R_\rho(t) = \mu_{sct}(t) + im(t) \tag{18a}$$

where the envelope $|R_\rho(t)|$ of the sum is a Rician random amplitude given as

$$|R_\rho(t)| = |\mu_{sct}(t) + im(t)| \tag{18b}$$

Hence, the envelope of the received signal with both the LOS component and scattered components describes the Rician random process. The Rician distribution is typically characterized by the Rician factor K, which is defined as the ratio of the power of the LOS component and the power of all scattered components of the received signal:

$$K(\sigma) = \frac{|m(t)|^2}{|\mu_{sct}(t)|^2} = \frac{\rho^2}{2\sigma^2} \tag{19}$$

For the generic Rician process, the probability distribution function (PDF) of the received signal amplitude is given as

$$Ric(x) = \frac{x}{\sigma^2} e^{-\frac{1}{2}\left(\frac{x^2+\rho^2}{\sigma^2}\right)} I_0\left(\frac{\rho x}{\sigma^2}\right) \quad \text{for } x \geq 0 \tag{20}$$

where I_0 is a modified Bessel function of the first kind.

Furthermore, by substituting Eq. (19) into Eq. (20), the Rician PDF expression with the factor K becomes

$$Ric(x,K) = \frac{x}{\sigma^2} e^{-\left(\frac{x^2}{2\sigma^2}+K\right)} I_0\left(\frac{x}{\sigma}\sqrt{2K}\right) \quad \text{for } x \geq 0 \tag{21a}$$

Per Eq. (21a) and for K/σ pair parameters: $K/\sigma = 0.5/1$; $K/\sigma = 4/0.35$; $K/\sigma = 22/0.15$, Fig. 11 shows the probability distribution function (PDF) for Rician flat fading, which approaches the Rayleigh distribution for a low K (0.5) and the Gaussian distribution for a larger K (22).

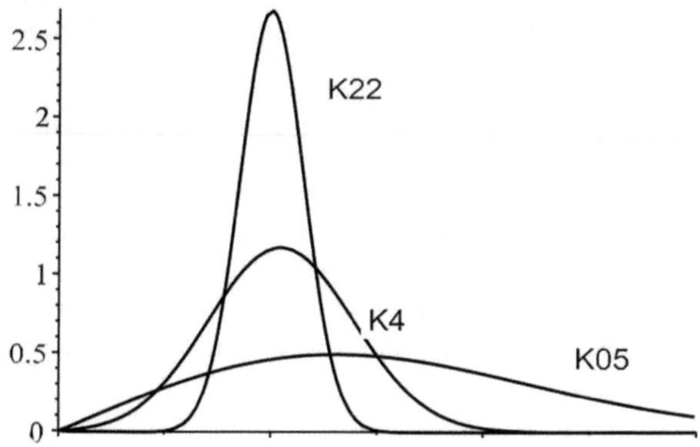

Fig. 11 PDF of Ricean flat fading

Therefore, the Rician random process, such as the Rayleigh process, is modeled by two uncorrelated real-valued *colored* Gaussian processes.

Note that when $K = 0$, Ricean becomes the Rayleigh distribution:

$$Ray(x,K) = \frac{x}{\sigma^2} e^{-\left(\frac{x^2}{2\sigma^2}\right)} \text{ for } x \geq 0 \tag{22}$$

The plots of the Gaussian PDF and Rayleigh PDF, which contain two *quadrature-colored* zero-mean Gaussian PDFs, are shown in Fig. 12.

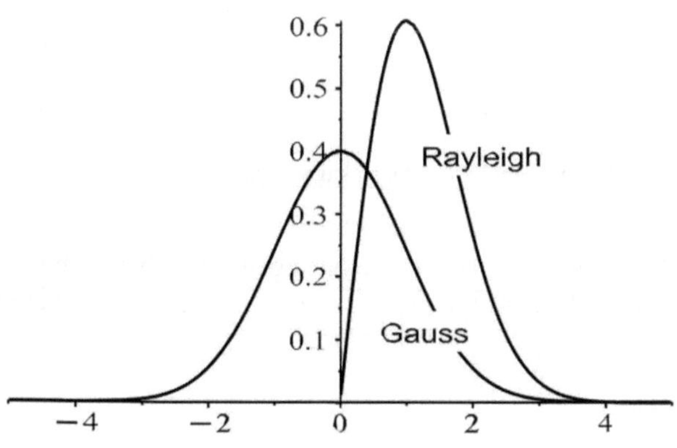

Fig. 12 Rayleigh PDF vs. Gaussian PDF

5.2 Nakagami Gamma Distribution of Fast Fading

The Nakagami distribution is also one of the generalized PDFs associated with the Gamma function. This distribution is a parametric distribution with two parameters: the shape distribution parameter m and the spread distribution parameter ω. The following expressions present the Nakagami probability density function (PDF).

$$f\left(x;,m;,\omega\right)=\frac{2}{\Gamma\left(m\right)}\left(\frac{m}{\omega}\right)^{m}x^{2m-1}e^{-\left(\frac{m}{\omega}\right)x^{2}};x\geq 0;m\geq 0.5 \qquad (23)$$

where $\Gamma(m)$ is the gamma function defined by Euler's integral $\Gamma\left(m\right)=\int_{0}^{\infty}t^{m-1}e^{-t}dt$.

Additionally, m is the Nakagami fading parameter, and ω defines the mean power of the distribution function $f(x;m,\omega)$. The effect of fading can be adjusted by the parameter m. For example, Fig. 13 shows Nakagami PDF fading per Eq. (23) for several parameters m and ω.

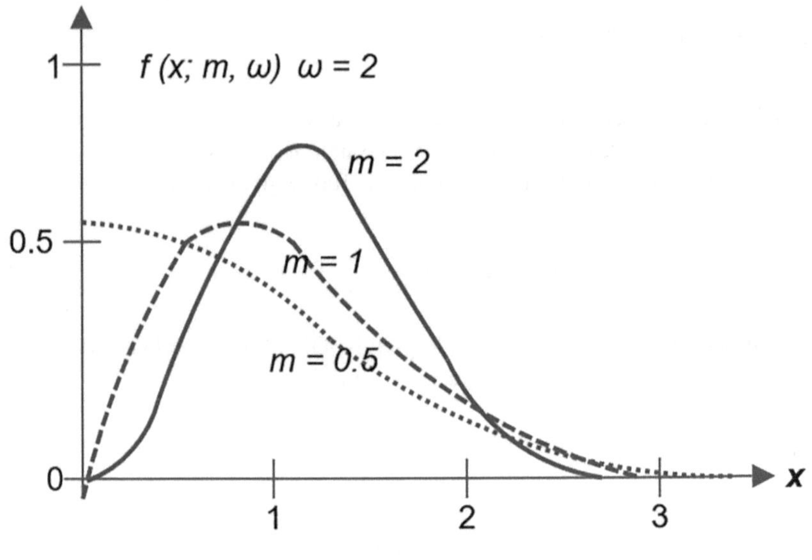

Fig. 13 Gauss (m = 0.5), Rayleigh (m = 1), Nakagami (m > 1)

For example, for m = *0.5* and 1, the Nakagami distribution becomes one-sided Gaussian and Rayleigh distributions, respectively, and for $m > 1$, there is a Nakagami distribution.

5.3 Symbol Error Rates (SER) of Rayleigh and Ricean Channels with AWGN

The symbol error rate or probability P_{es} of the symbol error is defined by the following expression:

$$\mathrm{SER} = \int_0^{\infty} p(x) P_{es}(x) dx \tag{23a}$$

where $p(x)$ is a probability density function (PDF) for radio channels with AWGN or with some fading (Rayleigh, Rician, Nakagami, or so on) and where $P_{es}(x)$ is the probability of symbol error for a digital modulation at a specific SNR. As noted earlier, the Rician stochastic process can represent multipath channels that include AWGN and Rayleigh fading as exceptional cases.

The probability of the symbol error rate (SER or P_{es}) in the Rician fading channel with AWGN can be determined via the *momentum generation function* (MGF) [7, 8].

$$M_s(\gamma_b, \phi) = \frac{(1+K)\sin^2(\phi)}{(1+K)\sin^2(\phi) + g\gamma_s} e^{-\frac{Kg\gamma_s}{(1+K)\sin^2(\phi)+g\gamma_s}} \tag{24}$$

where g is the parameter dependent on the modulation type and is listed in Table 3 for the MPSK and MQAM modulations. Additionally, γ_s is the ratio between the symbol energy E_s and the noise energy N_0.

$$\gamma_s = E_s / N_0 = (E_b / N_0)\log_2(M) = \gamma_b \log_2(M) \tag{24a}$$

where $\gamma_b = E_b/N_0$. Here E_b is a bit of energy.

The expressions of the theoretical average SER for the MPSK and MQAM modulations are listed in Table 3 [7–9].

Table 3 Rician SER for the MPSK and MQAM modulations

Modulations	g	Average P_{es} (SER)
MPSK	$\sin^2\left(\frac{\pi}{M}\right)$	$P_{es}(\gamma_b) = \frac{1}{\pi}\int_0^{\pi(M-1)/M} M_s(\gamma_b, \phi) d\phi$ (25a)
MQAM	$\frac{1.5}{M-1}$	$P_{es}(\gamma_b) = \frac{4}{\pi}\left(1 - \frac{1}{\sqrt{M}}\right)\left\{M_s(\gamma_b, \phi) d\phi - \left(1 - \frac{1}{\sqrt{M}}\right)\int_0^{\pi/4} M_s(\gamma_b, \phi) d\phi\right\}$ (25b)

The plots in Fig. 14 are made via Eqs. (25a, b), where the differences between the SERs in the Rayleigh fading channel with AWGN are shown for the 16-PSK and 16-QAM modulations.

Fig. 14 SERs in the Rayleigh fading channel for 16-PSK and 16-QAM modulations

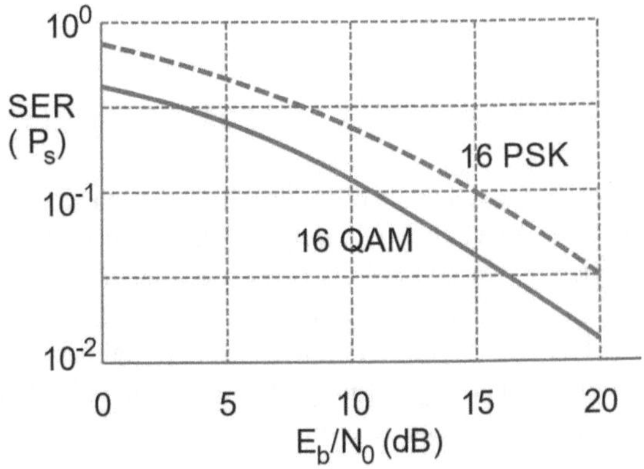

The advantage of the QAM modulation over the PSK modulation in terms of the power efficiency can be seen here.

Additionally, the plots in Fig. 15 are made via Eqs. (24) and (25b), where the differences between SERs in the Rician fading channel with different K factors are shown for the 16-QAM modulation.

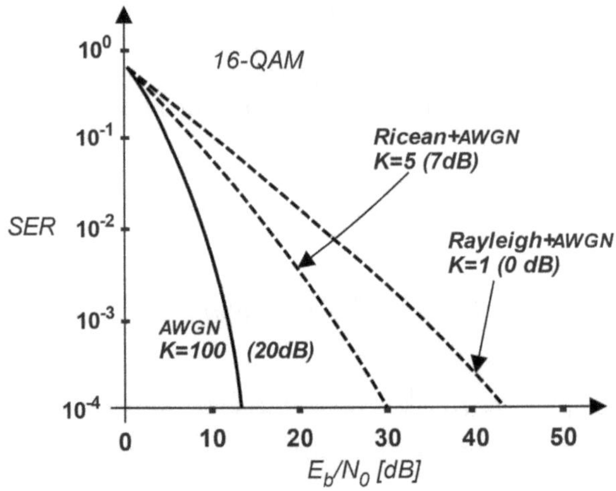

Fig. 15 16 QAM with AWGN and Ricean flat fading

For the QAM, the Rician fading channel is close to the AWGN channel when K increases to 100 (20 dB) and close to the Rayleigh fading channel when $K = 1$ (0 dB).

6 Impulse EMI in the Radio Channel

The presence of impulse EMI noise is often experienced in radio wireless communication channels. Therefore, different statistical models for impulse EMI noise have been considered and proposed in the literature. Two common models are considered and proposed for *narrowband* impulsive EMI: the Middleton Class A model and the Bernoulli-Gaussian (BG) model. Impulse EMI noise was first studied by Middleton [11] when he developed a model for impulse EMI in memoryless and narrowband communication channels. This model noise contains trains of EMI impulses with varying intensities and durations, where the individual pulses occur randomly without overlapping each other.

However, the above-proposed models do not sufficiently represent the burst-impulse EMI (where overlapped impulses propagate in groups), which is often the case in severe fading radio channels and channels impacted by natural impulse EMI, such as lightning and solar flares, which will be discussed in chapters "Physics of Lightning and Its Impact on Radio Systems on Earth" and "Solar and Other Cosmic Radiations with Effects on HF Radio and Satellite Data Communications", respectively. The models that better fit the burst-impulse EMI nature of radio channels are the Middleton class *B* model and the Markov BG model. These models are also discussed in Ref. [12].

Figure 16 shows nonoverlapped impulses and bursty impulses of EMI noise.

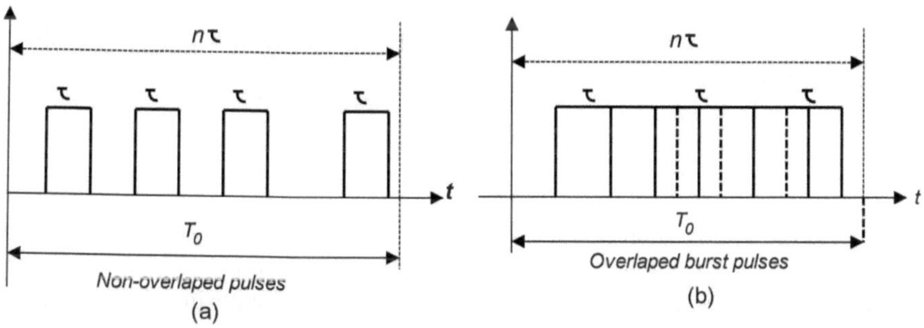

Fig. 16 EMI nonoverlapped and bursty impulses

6.1 *Class A Impulsive EMI*

The Class A EMI model includes the probability density function (PDF - $P_{e,b}$) of an EMI noise sample x_k, which is theoretically expressed as

$$P_{e,b}\left(x_k;\mu,\sigma_m^{\ 2}\right)=\sum_{m=0}^{K\to\infty}P_mN\left(x_k;\mu,\sigma_m^{\ 2}\right) \tag{26}$$

where P_m is the Poisson probability of the variance $\sigma^2{}_m$ of an impulse m, expressed as

$$P_m = \frac{A^m}{m!} e^{-A} \tag{26a}$$

The term $N(x_k; \mu, \sigma^2_m)$ in Eq. (26) is an equivalent Gaussian PDF (with zero mean and variance σ^2_m) expressed as

$$N\left(x_k; 0, \sigma_m^{\ 2}\right) = \frac{1}{\sigma_m \sqrt{2\pi}} \exp\left(-\frac{x_k^{\ 2}}{2\sigma_m^{\ 2}}\right) \tag{26b}$$

where the variance σ^2_m is expressed as

$$\sigma^2_{\ m} = \sigma^2_{\ g} + \left(\frac{m}{A}\right)\sigma^2_{\ i} = \sigma^2_{\ g}\left(1 + \frac{m}{A\Gamma}\right) \tag{27a}$$

Here σ^2_i is the variance (power) of the impulse EMI noise and σ^2_g is the variance (power) of the AWGN (without impulsive EMI noise). Additionally, the parameter Γ is the power ratio of the power of Gauss's noise and the power of the observed impulses, expressed as

$$\Gamma = \sigma^2_{\ g} / \sigma^2_{\ i} \tag{27b}$$

The parameter A in Eq. (27a), also called the *impulsive index*, is the impulse noise probability whose value is between 0 and 1 ($0 \le A \le 1$). In particular, this parameter represents the density (a number) of impulses with width τ in the observation interval T, expressed as

$$A = \frac{\eta \tau}{T} \tag{27c}$$

where η is the average number of impulses per second, τ is the average duration of each impulse, and T is the observation interval with unit time ($T = 1$). Hence, the parameter A describes the density of EMI impulse noise that occurs per the Poisson distribution expressed by Eq. (26a).

For example, we compare the BERs of MPSK signals with AWGN only and AWGN in addition to impulse EMI noise to determine the impact of impulse EMI on radio channel performance. It is more suitable if we use the following BER expression, which is equivalent to Eq. (25b) for MPSK signals [10]:

$$P_{e,mpsk} = \left(\frac{M-1}{M}\right)\left[(1-A)Q\left(\sqrt{\frac{E_b}{N_0}}\right) + AQ\left(\sqrt{\frac{1}{\left(\frac{1}{A\Gamma}+1\right)}} \bullet \sqrt{\frac{E_b}{N_0}}\right)\right] \tag{28}$$

where E_b is the energy of a signal bit. Additionally, we can use the erfc function here instead of $Q\ (x)$, expressed as $Q(x) = \frac{1}{2} erfc\left(x / \sqrt{2}\right)$.

Figure 17 shows the BER plots based on Eq. (28) for a 16-PSK signal with AWGN and with AWGN + impulse EMI, where the A/Γ pair parameters are A = 0.25/*Γ* = 0.1 and A = 0.01/*Γ* = *0.01*. The first A/*Γ* pairs are related to the medium impulse EMI, whereas the second pair is related to the severe impulse EMI.

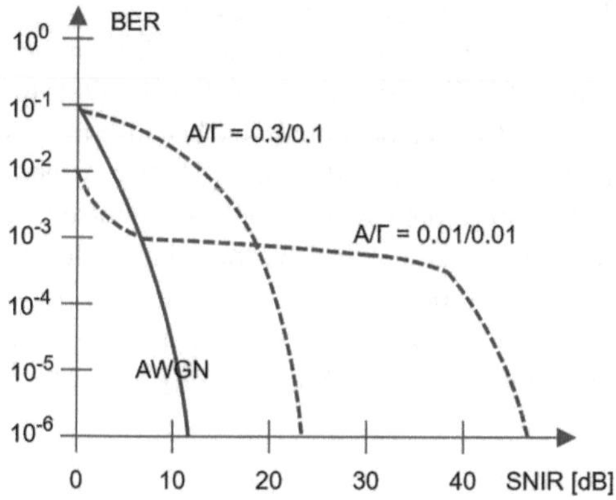

Fig. 17 Impulse EMI with AWG noise in a 16-PSK radio channel with (**a**) 16-PSK + AWGN, and (**b**) and (**c**) 16-PSK + AWGN+EMI class A: Γ/A = 0.2/0.1 and Γ/A = 0.01/0.01, respectively

We can see from the present plots that the impulsive EMIs have a significant effect on the 16-PSK signal transmission performance in terms of the BER and system power efficiency. For example, to provide BER = 10^{-6}, the SNR for the 16-PSK signal with AWGN should be only ~10 dB, whereas the SNRs for the PSK signal with AWGN and impulsive EMI should be ~22 dB or even ~44 dB, depending on the impulsive EMI intensity and duration of random parameters A and Γ.

In general, all signal modulations (regardless of their order) are more or less impacted by impulse EMI, which may have serious problems for radio digital systems.

6.2 Class B Burst-Impulsive EMI

Many empirical studies and measurements on practical communication channels have shown that impulse EMI also occurs in bursts of overlapped impulses with a memory nature. These EMIs are more difficult to predict than the narrowband impulse EMI.

However, to accurately predict these complex EMIs, several statistical models have been proposed, which are mainly based on the concept of a partitioned Markov chain. This approach takes into account the memory nature of impulse burst noise. To model impulse EMI noise with memory, two models are used in the literature, the Markov–Middleton and Markov–Gaussian models, which are modifications of the Class A and Bernoulli–Gaussian models, respectively, in which both include Markov chains. These models are well discussed in Ref. [12].

When the density A of EMI impulses is very low in single carrier (SC) modulation, the bust-impulse EMI has very few symbols. Therefore, the probability of error in SC modulation (for very low A) does not depend on the average power (variance) of the impulse EMI. However, when A increases, the average power of the impulse EMI becomes a significant matter of the error probability in the SC modulation.

However, in the case of MC modulations, such as OFDM, the average power of impulse EMI is a matter of the error probability because the EMI power is spread along all subcarriers, affecting every MC symbol. When A increases, MC modulation is more affected but still more resistant to burst-impulse EMI than SC modulation is. This is because parameter A in MC modulation does not affect the SNR requirement as it does in SC modulation.

MC modulations, such as OFDM, are discussed in chapter "Overview of Broadband Radio Systems with High EMI Immunity".

7 Capacity of a Radio Channel with EMI

In chapter "Introduction to Digital Communication Systems", the channel capacity C with AWGN (additive white Gaussian noise), or just white noise, is defined by the Shannon–Hartley theory as

$$C = B_w \log_2\left(1+\frac{S}{N_0 B_w}\right) = B_w \log_2\left(1+\frac{S}{N_w}\right) \text{bits/s} \tag{29a}$$

where N_0 is the spectral density of the white noise and where N_w is the power of the white noise in the bandwidth B_w, given as $N_w = N_0 B_w$.

Additionally, the *spectral efficiency* C_{se} of a channel with white noise N_w is defined as

$$C_{se} = \frac{C}{B_w} = \log_2\left(1+\frac{S}{N_w}\right) \text{bits/Hz/s} \tag{29b}$$

Furthermore, Eq. (29b) gives the capacity of the channel with AWGN and EMI interference:

$$C = B_w \log_2\left(1+\frac{S}{N_w + N_{emi}}\right) \text{bits/s} \tag{30a}$$

where N_{emi} is the power of EMI, which commonly exists in radio channels in addition to AWGN. Furthermore, Eq. (30a) can be rewritten as

$$C = B_w \log_2\left[1+\frac{1}{\frac{N_{emi}}{S}+\frac{N_w}{S}}\right] = B_w \log_2\left[1+\frac{1}{K^{-1}+\mathrm{SNR}_w^{-1}}\right] \tag{30b}$$

or

$$C(K,\mathrm{SNR}_w) = B_w \log_2\left[1+\frac{K\cdot\mathrm{SNR}_w}{K+\mathrm{SNR}_w}\right] \tag{30c}$$

where $K = S/N_{emi}$ (the signal-to-EMI power) and $\mathrm{SNR}_w = S/N_w$ (the signal-to-white-noise power).

Figure 18 shows the frequency-normalized capacity of the channel with AWGN and EMI for different K parameter values from 4 (6 dB) to 10^4 (40 dB).

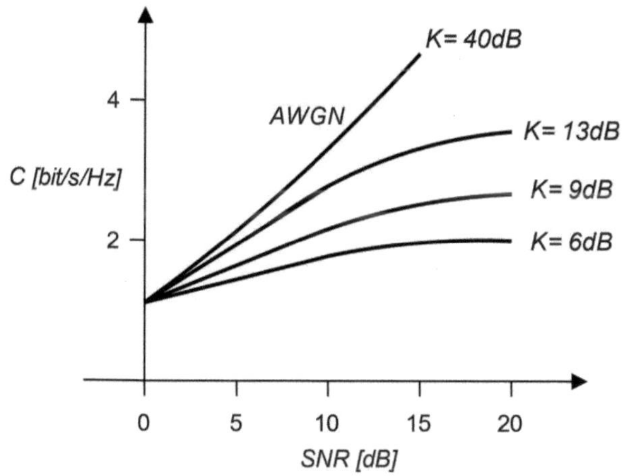

Fig. 18 Frequency-normalized capacity of channels with AWGN and EMI

As expected, it can be seen that the channel approaches the AWGN channel when the K increases (i.e., EMI decreases).

References

1. ITU CCIR, *Definition of Spectrum Use and Efficiency* (ITU, Geneva, 1990)
2. J. Gavan, in *Radio Systems Parameters and Performance Criteria Concerning Computation and Mitigation of EMI Effects*, ed. by R. Perez, Chapter 20 of Handbook of EMC (Academic Press, 1995)
3. S.W. Ellington, *Radio Systems Engineering*, 1st edn. (Cambridge University Press, 2016)

4. R.L. Hauput, *Wireless Communications Systems: An Introduction* (IEEE Press-Wiley, 2020)
5. K.Q.T. Zhang, *Wireless Communication Principles, Theory and Methodology* (Wiley, 2015)
6. D. Tse, P. Viswanath, *Fundamentals of Wireless Communications* (Cambridge University Press, 2005)
7. M.K. Simon, M-S. Alouini, *Digital Communication over Fading Channels: A Unified Approach to Performance Analysis* (Wiley, 2000)
8. A. Goldsmith, *Wireless Communications* (Cambridge University Press, 2005)
9. M. Viswanathan, *Wireless Communication Systems in MathLab*, 2nd edn. (2020)
10. A. Papoulis, *Probability, Random Variables, and Stochastic Processes* (McGraw-Hill, 2002)
11. D. Middleton, *An Introduction to Statistical Communication Theory*, vol. 960 (McGraw-Hill New York, 1960)
12. T. Shongwe, A.J. Han Vink, H.C. Ferreira, *A Study on Impulsive Nise and its Models* (South African Institute of Electrical Engineering, 2015)

Overview of Broadband Radio Systems with High EMI Immunity

1 Introduction

This chapter considers the concepts of broadband wireless communication systems with high capacity and spectral efficiency, and high immunity to EMI existing on radio communication channels. Broadband systems employ the following techniques and communication methods: time division multiplexing (TDM), frequency division multiplexing (FDM), orthogonal frequency division multiplexing (OFDM), spread spectrum (SS) and code division multiplexing access (CDMA), spatial diversity (SD), spatial multiplex MIMO (multiple-input-multiple-output), and adaptive software-defined radio (ASDR) systems. All these techniques and methods are well considered in the literature [1–20], which has influenced the writing of this chapter, and recommended to readers who are interested in further study of these topics.

As discussed in chapter "Introduction to Digital Communication Systems", for single carrier (SC) modulations, the carrier's amplitude, phase, and frequency are modulated by M data symbols where each symbol corresponds to one carrier amplitude, phase, or frequency state. The standard modulation techniques are MPAM, MPSK, MQAM, and MOFSK, as discussed in chapter "Introduction to Digital Communication Systems". For example, 4-PSK has 4 phase states where each state carries 2 data bits ($M = 2^2 = 4$). Additionally, 256-QAM has 256 amplitude–phase states where each carrier state carries 8 data bits ($M = 2^8 = 256$). Accordingly, 256-QAM modulation requires just one-fourth of 4-PSK bandwidth to transfer the same amount of data.

However, owing to the growing demand for more transfer data capacity, single-carrier systems have limited data capacity or data rates. Multiple-carrier modulation (MCM) with a larger number of parallel narrowband SCs is a solution for improving the system capacity, spectral efficiency, and immunity to selective fading and EMI on transmission channels.

M. Pajovic, *Fundamentals of EMI Phenomena in Radio Communication Systems*, https://doi.org/10.1007/978-3-031-88748-2_6

2 Multiplexing Systems

Multiplexing or *division multiple access* methods are used to further increase the data throughput, i.e., transfer the data capacity of a communication system. These methods also provide better spectral and time efficiency for a communication system.

Hence, multiplexing is the concept of simultaneously sending more individual signals across a communication channel. This method is commonly used for telephony, telemetry, and massive data transfers over terrestrial and satellite communications. One example of terrestrial communication is the current mobile cellular network. Two basic types of digital multiplexing are discussed in this section: *time division multiplexing* (TDM) and *frequency division multiplexing* (FDM) [1, 5].

2.1 Time-Diversity Multiplex (TDM)

In the TDM technique, a group of signals are sampled in time with a specific sampling rate and multiplexed for transfer across a single communication channel. This method enables the simultaneous transfer of computer data with other digital signals, such as digital voice, digital facsimile, and digital TV/video signals, into a single high-rate data stream.

Figure 1 shows the principle concept of digital TDM, where the digital data are multiplexed in a timely manner on bit-to-bit or byte-to-byte interval bases.

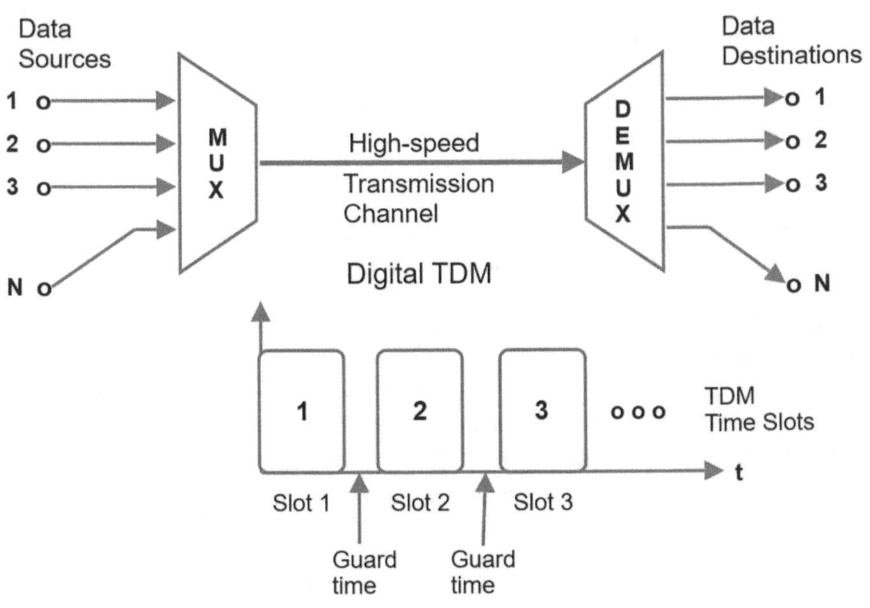

Fig. 1 TDM multiplex

The multiplex procedure is achieved via the multiplex block or switch (MUX), which sequentially selects bits or bytes from each input and forwards them in time slots across the high-speed transmission channel. At the receiver side, the demultiplexing block (DEMUX) separates the received bits or bytes in an original form and directs them to dedicated destinations. Thus, the entire communication system bandwidth is successively shared with N source signals sent in a short time slot.

There are two types of TDM: *synchronous* or *deterministic* TDM and *asynchronous* or *statistical* TDM. In synchronous TDM, time slots have constant lengths and the corresponding individual channels have a constant delay and bandwidth. For instance, the TDM is used in the ISDN, PDH, and SDH systems. In contrast, *asynchronous* TDM has a variable delay and bandwidth for a specific individual communication channel, where time slots have a variable length selected on demand. Asynchronous TDM is typically applied for X25, ATM, or IP systems.

Hence, in both synchronous and asynchronous TDM, data formatting and time sequencing are needed before the data are sent to specific destinations over the transmission channel. In this way, the system receiver can adequately identify and direct data to the allocated user. The time framing procedure involves dividing a large data stream into smaller time frames and sequencing them into single time slots separated by guard times. Figure 2a, b shows the concepts of *synchronous* and *asynchronous* TDM frames, respectively.

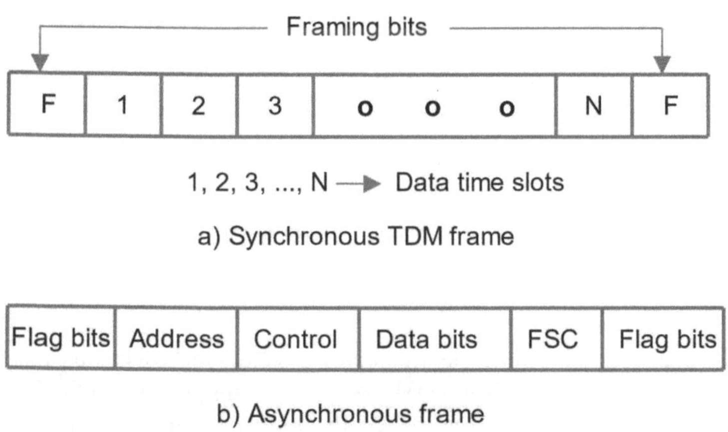

Fig. 2 TDM frames; (**a**) synchronous, (**b**) asynchronous

In both cases, the data frames are bounded by format bits (case *a*) and by bit blocks or flags (case *b*).

2.2 *Frequency-Division Multiplex (FDM)*

FDM multiplexing involves multicarrier data transfer, where each *M*-level modulated carrier with its bandwidth is allocated to one user. Therefore, the system bandwidth is divided into subbandwidths corresponding to each carrier. Additionally, subbandwidths are separated by specific guard bandwidths to minimize interchannel interference. Figure 3 shows the principle of FDM multiplexing, where a system bandwidth is divided into subbandwidths separated by guard bandwidths.

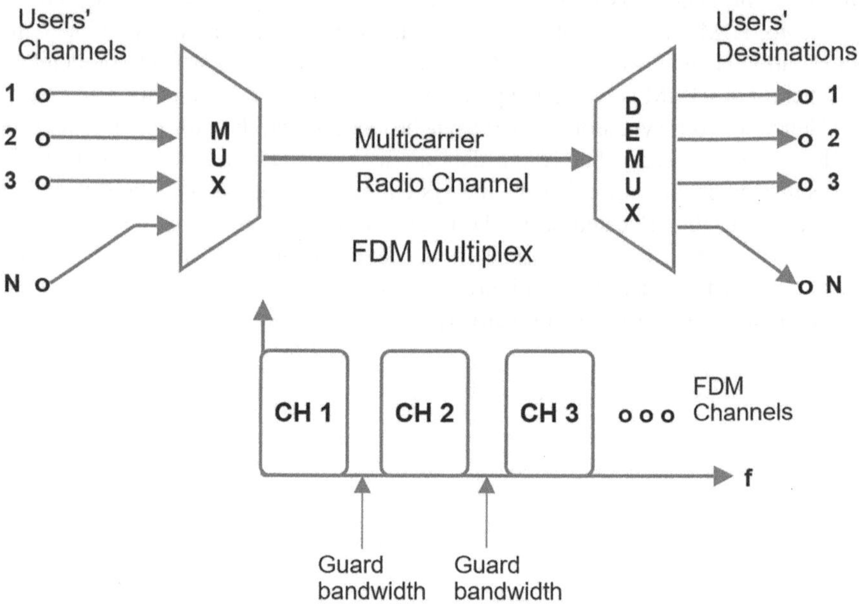

Fig. 3 FDM multiplex

FDM multiplexing requires less complex system synchronization to maintain efficient user connections than TDM does. This multiplexing method generally involves multicarrier transmissions, where the orthogonal *frequency* division multiplex (OFDM) is also a type of transmission. The OFDM will be discussed in the next section.

3 Multicarrier Signal (MC)

The idea of using a multicarrier (MC) is that a broadband channel with frequency-selective fading is divided into narrowband SC channels, each with flat fading, where complex equalization and decoding methods are no longer needed in the receiver. To obtain the maximum spectral efficiency, adjacent SC subchannels are set to be orthogonal. Therefore, the transmission is very efficient against frequency-selective fading and impulsive EMI, as discussed in chapter "Performance of Radio Digital Systems in Presence of EMI". More information about MC modulations is given in [2–6].

The practical release of multicarrier signals was proposed by Weinstein and Ebert [5], where multicarrier signals are generated via discrete Fourier transform (DFT). Because perfect orthogonality among adjacent SC channels is difficult to maintain in multipath and random radio channels, guard intervals are still added between the SC channels to avoid intersymbol interference (ISI). Figure 4 shows the bandwidth of a broadband communication system that may consist of many single-carrier bandwidths called subchannels [6].

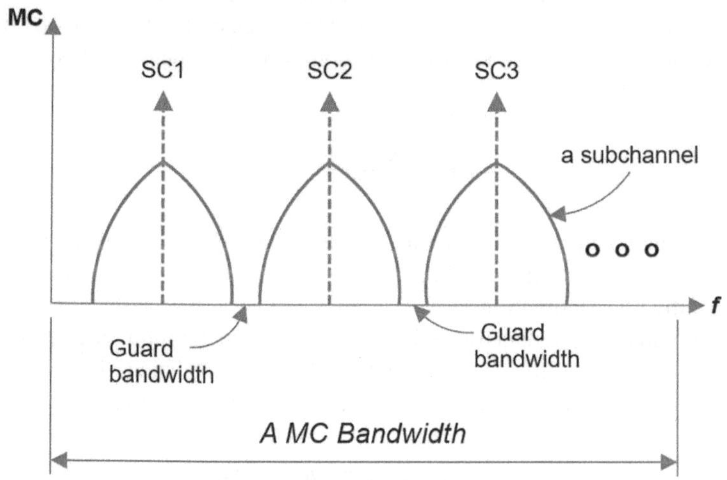

Fig. 4 BW of a multicarrier modulated signal with guard bandwidths between single-carrier bandwidths

The complex envelope of a multicarrier signal, as a sum of m bursts transmitted at rate $1/T_b$, is expressed as [6]

$$s_{mc}(t) = \sum_{m} s_m(t - mT_b) \tag{1}$$

where $s_m(t)$ is the *m-th* signal burst and where T_b is the burst duration. The complex envelope of $s_m(t)$ is expressed as

$$\tilde{s}_m(t) = \sum_{n=0}^{N-1} \tilde{S}_{n,m} e^{i(\omega_0 + n\Delta\omega)t} w(t) \tag{1a}$$

where N is the number of subcarriers, $\tilde{S}_{k,m}$ is the complex symbol transmitted by the n-th subcarrier of the m-th burst, and w(t) is the transmitted pulse. Additionally, ω_0 is the carrier frequency (i.e., commonly the central frequency of the MC spectrum), and $\Delta\omega$ is the separation between subcarriers.

Performing the Fourier transform $F\{s(t)\}$ to both sides of Eq. (1a) leads to

$$\tilde{S}(f)=\sum_{n=0}^{N-1}\tilde{S}_n W(\omega-n\Delta\omega) \tag{1b}$$

where $W(\omega) = F\{w(t)\}$.

In multicarrier transmission, the sequences of information symbols for a given burst are transmitted in the frequency domain. Each burst is transmitted by a different subcarrier in the time interval T. The simplest case of multicarrier data transfer is the conventional frequency-division multiplex discussed in the previous section.

For single carrier (SC) modulation, the complex envelope of an N-symbol burst can be expressed as

$$s_{sc}(t)=\sum_{n=0}^{N-1}\tilde{S}_n\cdot w(t-nT) \tag{2}$$

where $\tilde{S}_n$ denotes the n-th data symbol, w(t) is the transmitted pulse, and T is the symbol duration. According to Eq. (2) and Eq. (1b), the SC modulation is a dual version of the MC modulation.

To ensure that the SC data transmission is without intersymbol interference (ISI), the orthogonality condition between pulses should be satisfied. This condition is satisfied when the convolution between adjacent pulses $w(t - nT)$ is zero:

$$\int_{-\infty}^{+\infty} w(t-nT)\cdot w^*(t-n'T)\,dt=0 \quad \text{for}\, n\neq n' \tag{2a}$$

The operator '*' denotes a complex conjugate of the pulse function w(t).

Additionally, to ensure that the multicarrier transmission is without ICI, the orthogonality condition between subcarrier channels in the frequency domain is satisfied if the convolution between adjacent subcarrier bandwidths $W(f - n\Delta f)$ is zero:

$$\int_{-\infty}^{+\infty} W(f-n\Delta f)\cdot W^*(f\quad n'\Delta f)\,df-0 \qquad n\neq n' \tag{2b}$$

where Δf is the frequency separation between adjacent subcarrier bandwidths. Additionally, the operator '*' denotes a complex conjugate of the function W(f).

With SC-modulated signals in linear channels, pulses $w(t)$ can be orthogonal even if they overlap in time. Additionally, with MC-modulated signals in a linear channel, the functions $W(f)$ can be orthogonal even if they overlap in the frequency domain. For example, for SC modulations, a pulse $w(t) =$ sinc (t/T) satisfies the orthogonality condition (2a). Additionally, for MC modulations, the orthogonality condition (2b) is satisfied for the function $W(f) =$ sinc $(f/\Delta f)$.

However, in the case of a nonlinear MC channel with multipath propagations, the received symbols may significantly overlap in time when received, regardless of the orthogonality conditions that are satisfied. This problem is minimized by introducing a specific guard interval between adjacent data symbols with a specific length on the TX side. To avoid intersymbol and interchannel interference (ISI, ICI), in addition to the orthogonality conditions discussed above, the length of a guard interval should be longer than the overlap burst time, including the impulse responses of the TX/RX frequency-selective stages and the communication channel itself.

3.1 Orthogonal Frequency Division Multiplex (OFDM)

More information about OFDM modulation is given in references [3–7], which influenced the writing of this section.

The orthogonal frequency division multiplex (OFDM) is a version of orthogonal multicarrier transmission. The guarding between the OFDM symbols is obtained when the first part of a transmitted symbol with a certain length (called the *CP*) is retransmitted as a guard interval and added to the following symbol. This process is performed for each OFDM symbol sent from the TX section. In the RX section, the *cyclic prefix* (CP) is removed from the received OFDM symbols.

Figure 5 shows a sent data symbol with the added *cycle prefix CP* whose duration is T_G. Therefore, T_B is equal to the sum of the original data sequence duration T and the guard interval T_G.

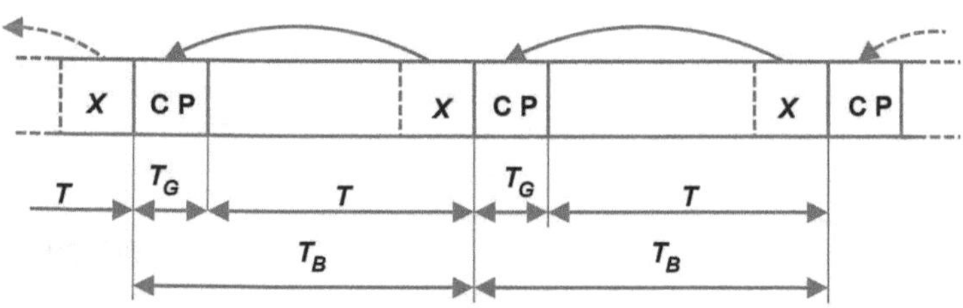

Fig. 5 CP-OFDM symbol structure

Hence, after adding the cyclic prefix CP, the OFDM evolves to its variant CP-OFDM.

The basic functional block diagram of the CP-OFDM system with M-PSK and M-QAM modulations in a communication channel with frequency-selective Rayleigh fading is shown in Fig. 6.

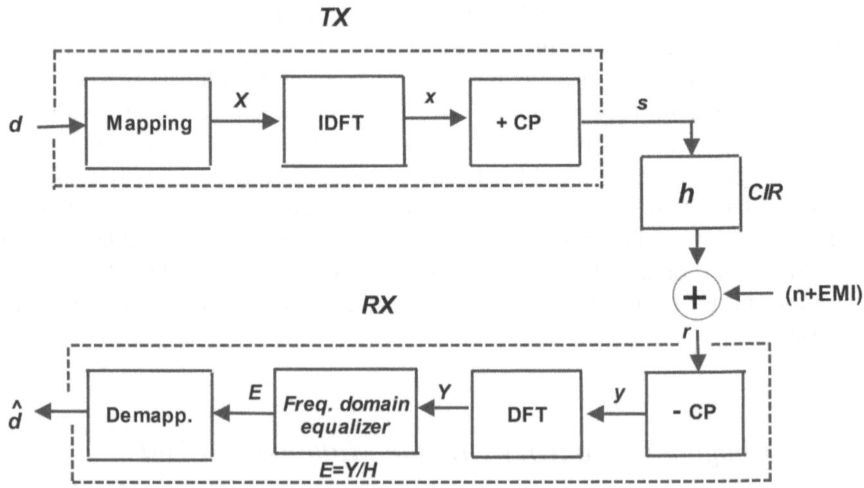

Fig. 6 Basic block diagram of the CP-OFDM system with a channel with Rayleigh's selective frequency fading

On the TX side, the first block is for the mapping data symbols, the second block involves the inverse discrete Fourier transform (IDFT), and the third block is a CP adder unit. On the RX side, a CP remover block, discrete Fourier transformer (DFT), a frequency equalizer, and a data symbol demapper follow. Regarding the channel, *h* (CIR) is associated with the channel impulse response, and accordingly, *H* is the corresponding channel's frequency response. Between the Tx and RX, we consider radio channels with Rayleigh fading and EMI.

As an example, the symbol error rate (SER) is plotted for the OFDM system with 64-PSK-CP-OFDM and 64-QAM-CP-OFDM modulations via Monte Carlo simulation. The resulting plots are shown in Fig. 7.

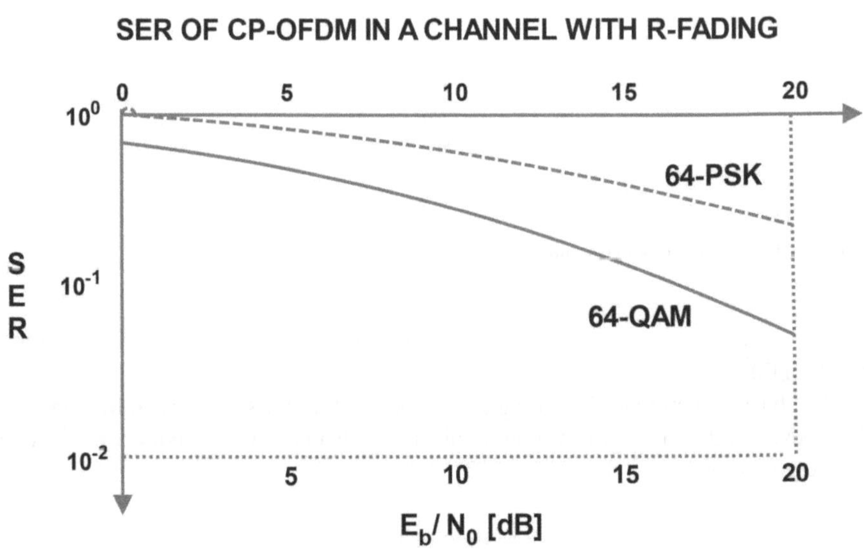

Fig. 7 SER of CP-OFDM in a channel with frequency-selective Rayleigh fading

CP-OFDM with 64-QAM modulation has significantly better power efficiency than 64-PSK-CP-OFDM modulation at an SNR_b (E_b/N_0) of 20 dB. This is generally true for higher modulation levels, such as for M > 16.

3.2 *Applications of Orthogonal Frequency Division Multiplex (OFDM)*

Therefore, the main reason why OFDM has become the dominant method for mobile and stationary radio systems is that frequency-selective fading channels can be treated as a sum of narrowband frequency flat fading channels. For example, multicarrier systems are widely employed in wireless local area networks (WLANs) via OFDM and its variant orthogonal frequency division multiple access (OFDMA). Some examples of multicarrier transmissions with the number of careers are wireless LAN 802.11a with approximately 48 carriers, 4G-LTE with up to 1200 carriers, digital audio broadcasting (DAB) with approximately 1575 carriers, and digital video broadcasting terrestrial (DVB-T) with approximately 8000 carriers. Additionally, some forms of OFDM are used in *wireless broadband* access systems, IEEE 802.16 *(a, d, e),* referred to as WiMAX. Furthermore, OFDMA can be found to be efficient in wireless regional area networks (WRANs) proposed by the IEEE. Related to the MCM, HSOPA (*High-Speed* OFDM Packet Access) [3GP06] is useful in the existing 4G-LTE and upcoming 5G-NR cellular networks, as will be discussed in chapter "Overview of 4G-LTE and 5G-NR Mobile Radio Systems from EMI Prospect".

3.3 *Orthogonal Frequency Division Multiple Access (OFDMA)*

The OFDMA is a multiuser variant of OFDM used mainly for downlinks, providing multiple accesses over the same channel at a time with a different set of subcarriers allocated to a dedicated user. Here, a regular OFDM scheme is used for the uplinks. Therefore, OFDMA works in conjunction with the OFDM.

The way subcarriers are given to a specific user can result in the best system performance in terms of fewer problems with channel fading and RFI/EMI from other users present on the same channel. The common way to assign carriers to users is that a user terminal selects the best subcarrier channel with the highest signal-to-noise ratio (SNR) available at the moment. This dynamic real-time carrier assignment is best obtained in conjunction with adaptive software-defined radio (ASDR), which is also discussed in this chapter.

As mentioned before, OFDMA is commonly used in wireless broadband access systems, such as IEEE WLAN, WiMAX, 4G-LTE, and 5G-NR technologies.

The writing of this section is influenced by [7].

4 Spatial Diversity (SD) and Spatial Multiplex (MIMO)

The writing of this section is influenced by Refs. [7–11], for which we refer interested readers to further study these topics.

4.1 Spatial Diversity (SD)

The simple and traditional efficient technique for reducing the effects of fast fading and impulsive EMI in radio channels are *spatial diversity* methods. In general, the *diversity* methods are associated with receiving the same desired signal over two or more propagation paths and selecting one that is less distorted at the receiver side. This technique improves radio system performance in terms of system immunity to EMI and channel fading. The standard diversity methods include the following [8]:

(a) The *spatial diversity* method is very efficient against channel fading and EMI. This diversity works when two or more effectively spaced antennas receive the same signal, resulting in noncorrelated signals at the RX input. Hence, to receive noncorrelated signals and achieve efficient diversity performance, the receiving antennas should be separated by several operating wavelengths λ. There are more criteria for choosing one of the received signals, such as the traditional method, which selects the strongest received signal for the demodulating process.

(b) *Polarization diversity* is based on zero autocorrelation of the desired signal received by two orthogonally polarized antennas, while the antennas' spacing should be relatively small (up to λ/16). This method is inefficient and not recommended for *HF* ionospheric communications because the polarization of HF waves randomly changes across ionospheric paths, as discussed in chapter "Solar and Other Cosmic Radiations with Effects on HF Radio and Satellite Data Communications". *Polarization diversity* is more suitable for *radio satellites* or *line-of-sight terrestrial* links where the polarization is relatively unchangeable.

(c) *Angular diversity* is based on receiving the desired signal by two or more RX antenna radiation lobes separated by certain angles. This method can also be used with a single TX antenna with two or more feeds when the antenna has multiple radiation lobes. This diversity is usually obtained by using phased antenna arrays, with multiple lobes controlled on basis of the real-time propagation channel conditions.

(d) *The hybrid* diversity method is a combination of the methods above.

For example, the concept of traditional *spatial diversity* is shown in Fig. 8. This is *single input multiple outputs* (*SIMO*), where the system contains only one transmitter and multiple (M) receivers.

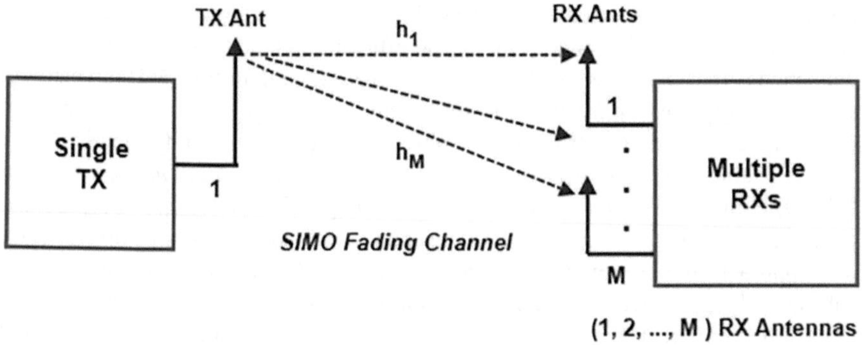

Fig. 8 SIMO spatial diversity concept

If M signals independently fade over individual transmitting paths, then we can assume that the signals propagate over independent channels. In this case, the probability of selecting a well-performing channel for RX processing is high. Therefore, if multiple copies of the data signal are received over independent fading channels, all the copies are affected differently by fading and the external EMI. On this basis, several selection methods, such as *maximum ratio selection* (MRS), *equivalent gain selection* (EGS), and *selection combining* (SC) methods [8, 20], are considered to detect fewer received signals in the literature and practice.

An example of the SC efficiency measured on the channel with a QPSK signal, AWG noise, and Rayleigh fading is shown in Table 1.

Table 1 N receivers vs. SNR to reach $SER = 10^{-4}$

QPSK; $SER = 10^{-4}$	
RCV number	SNR (dB)
1	36
2	20
3	15
4	12
10	7

The SER of 10^{-4} is reached with a nonspatial diversity ($N = 1$) when the SNR ≈36 dB, whereas with 10 receivers ($N = 10$), the same *SER* is reached when SNR ≈ 7 dB only. Therefore, Table 1 shows the significant improvement in a communication system's power efficiency and immunity to EMI with increasing number of receiving antennas.

4.2 *Spatial Multiplexing (MIMO)*

As discussed earlier in this chapter, radio wireless communication channels are exposed to hostile electromagnetic environments that cause rapid amplitude and phase fluctuations (fading) of signals, signal dispersion in the time and frequency domains, cochannel interference, EMI/RFI from other wireless communication systems and natural EMI sources. One of the efficient methods for overcoming these challenges in a radio system is the use of a *multi-input-multioutput* (MIMO) method, which is a type of *spatial multiplexing* technique. A significant improvement in communication system performance is observed when this method is applied.

The main difference between *spatial diversity* and *spatial multiplexing,* or MIMO, is that all signals received over MIMO spatial channels are accepted and processed in the RX, not only the strongest one, as in traditional spatial diversity. By this approach, the communication system capacity (or transmission data rate) is substantially increased by the MIMO, similar to orthogonal frequency-division multiplexing (OFDM), where more subchannels simultaneously carry different parts of the transmitted data. However, in a high-scattering environment (with fading and intense EMI), when more radio channels are allocated in a limited frequency band, spatial multiplexing has an advantage in terms of a higher multiplexing gain (data capacity) without increasing the signal power and bandwidth.

Hence, MIMO systems use *spatial multiplexing* techniques, which include multiple transmitters and multiple receivers. Figure 9a, b shows a general MIMO scheme concept containing N transmitters and M receivers.

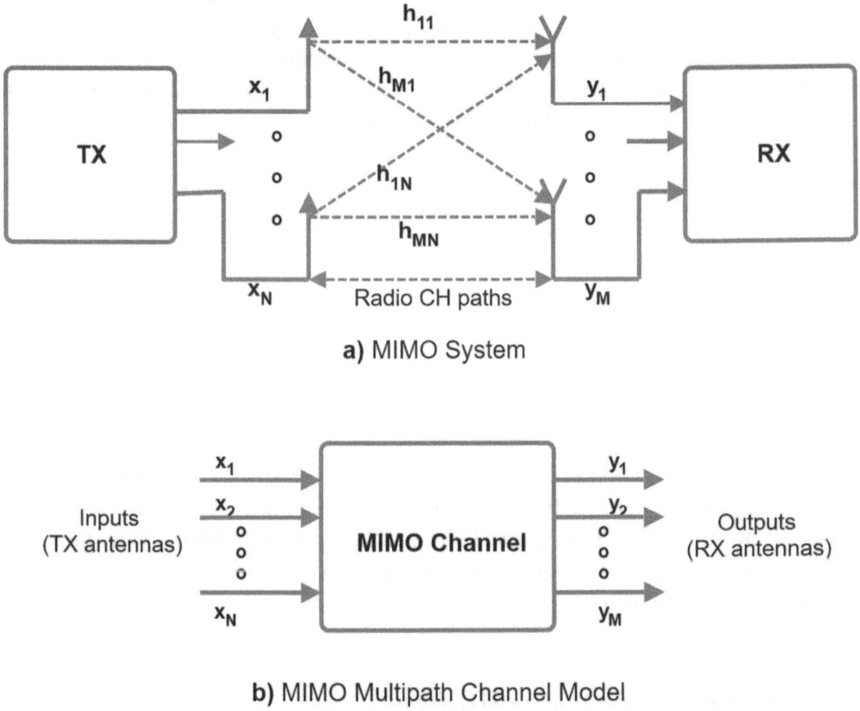

Fig. 9 (**a**) MIMO configuration; (**b**) MIMO model

Owing to the MIMO system containing total (N × M) paths, its transfer function can be accordingly expressed in the form of the (N × M) matrix as

$$\boldsymbol{y} = \boldsymbol{H} \cdot \boldsymbol{x} + \boldsymbol{n} \tag{3}$$

or

$$\boldsymbol{y} = \begin{bmatrix} y_1 \\ \vdots \\ y_M \end{bmatrix} = \begin{bmatrix} h_{11} & \cdots & h_{M1} \\ \vdots & \ddots & \vdots \\ h_{1N} & \cdots & h_{MN} \end{bmatrix} \cdot \begin{bmatrix} x_1 \\ \vdots \\ x_N \end{bmatrix} + \begin{bmatrix} n_1 \\ \vdots \\ n_M \end{bmatrix} \tag{4}$$

where $\boldsymbol{H}$ is the transfer function of the MIMIO system.

MIMO systems usually use multibeam antenna arrays at the TX and RX sides with M and N lobes, respectively instead of M-transmitters and N-receivers. These radiating lobes can be uncorrelated and correlated; accordingly, the MIMO systems are divided into *uncorrelated* and *correlated* systems. Here, we consider two uncorrelated and correlated MIMO channels: AWGN only and AWGN with fading and EMI.

4.3 Capacity of MIMO Multiplexing with AWGN

On the basis of Eq. (29a), chapter "Performance of Radio Digital Systems in Presence of EMI", the capacity of an *uncorrelated* MIMO system (with enough separated antenna array lobs) for $M \geq N$, is given as

$$C = NB_w \log_2\left(1 + \frac{M \cdot S}{N \cdot N_w}\right) \text{ bits/s} \tag{5}$$

where B_w and N_w are the AWGN bandwidth and power, respectively. Thus, we consider M transmitted and N received signals. Therefore, the total power of the entire MIMO system is divided equally among N receiving channels with the same gain.

Furthermore, for the *correlated* MIMO system with AWGN, the capacity is given as

$$C = B_w \log_2\left(1 + \frac{(M \cdot N) \cdot S}{N_w}\right) \text{ bits/s} \tag{6}$$

Here, only one channel whose gain is the product (M·N) is considered.

4.4 Capacity of MIMO Multiplexing with AWGN and EMI

In the cases of *uncorrelated* and *correlated* ($M \times N$) MIMO systems with AWGN and EMI, the equivalent SNR_e is given as

– Uncorrelated MIMO:

$$\text{SNR}_{e,unc} = \frac{M \cdot (K \cdot \text{SNR}_w)}{N \cdot (K + \text{SNR}_w)} \tag{7}$$

– Correlated MIMO:

$$\text{SNR}_{e,cor} = \frac{\text{MN} \cdot (K \cdot \text{SNR}_w)}{(K + \text{SNR}_w)} \tag{8}$$

– where $K = S/N_{\text{EMI}}$ and $\text{SNR}_w = S/N_w$.

Furthermore, on the basis of chapter "Performance of Radio Digital Systems in Presence of EMI", Eq. (30c), the capacity of an *uncorrelated* MIMO system is given as

$$C(\text{SNR}_w) = NB_w \log_2\left[1 + \text{SNR}_{e,unc}\right] = NB_w \log_2\left[1 + \frac{M}{N} \cdot \frac{K{\cdot}\text{SNR}_w}{K + \text{SNR}_w}\right] \tag{9}$$

For a *correlated* MIMO system, the capacitance is given as

$$C(\text{SNR}_w) = B_w \log_2\left[1 + \text{SNR}_{e,cor}\right] B_w = \log_2\left[1 + MN \cdot \frac{K \cdot \text{SNR}_w}{K + \text{SNR}_w}\right] \tag{10}$$

For example, Fig. 10a, b shows the capacity versus K (S/N_{EMI}) for (1 × 1), (2 × 2), and (5 × 5) MIMO systems, which are *uncorrelated* and *correlated,* on the basis of Eqs. (13a) and (13b) for the parameter SWR_w (S/N_w) = 8 dB.

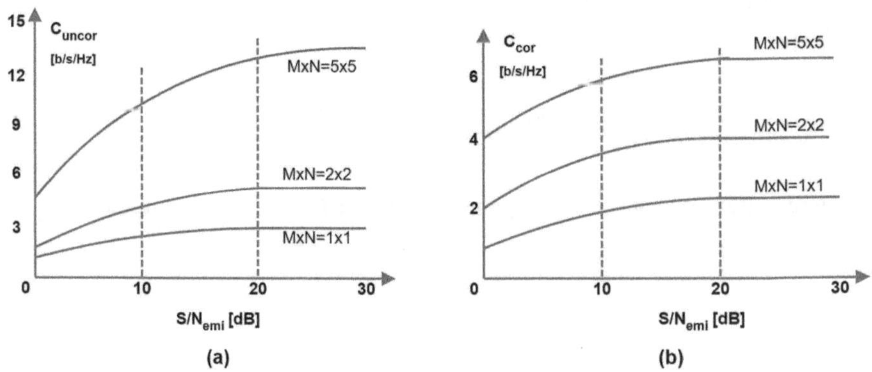

Fig. 10 Capacitance of (**a**) uncorrelated MIMO antenna lobs and (**b**) Correlated MIMO antenna lobes

These plots show the significant improvement in the MIMO system capacitance and spectral efficiency with increasing number of antenna array lobes, especially for an uncorrelated MIMO system. Additionally, a MIMO concept is effective against an EMI when used in conjunction with adaptive algorithms, which adopt the system parameters to real-time conditions on a radio communication channel with EMI and fading. This topic is considered later in this chapter. Additionally, the applications of MIMO systems to 5-G networks are discussed in chapter "Overview of 4G-LTE and 5G-NR Mobile Radio Systems from EMI Prospect".

5 Spread Spectrum Techniques and EMI Reduction

Some radio systems, stationary and mobile, use spread spectrum (SS) techniques to better exploit the radio spectrum and significantly increase the system's immunity to channel noise and EMI. An SS system works in such a way that a narrowband radio signal converts to a wideband signal at the SS transmitter before being emitted out in the transmitting channel. The SS signal looks like flat noise in the channel. A reason for the spread of operating signals is that narrowband radio signals are more vulnerable to selective fading than broadband (spread) signals are. Furthermore, the inverse process occurs at the RX side, where the received broadband signal is returned to the original narrowband signal.

Interestingly, the SS system can operate even when the operating signal level is below the level of the channel noise and EMI. This is because a narrowband EMI in the channel is spread by an SS demodulator at the receiver, resulting in the EMI level being below the level of the operating signal. Therefore, SS techniques significantly increase the system signal-to-noise/interference ratio (SNIR). In practice, three SS methods are used: *direct sequence* SS (DSSS), *frequency hopping* SS (FHSS), and the *hybrid* method, which combines DSSS and FHSS.

In the DSSS system, the digital signal is modulated (or correlated) by the TX encoder with a pseudorandom-coded digital sequence, which has a much higher transmission rate than the rate of the baseband digital signal. The result of the correlation process is broadening of the bandwidth of the operating signal before it is emitted in a transmission channel. At the RX side, the received broadband signal is decorrelated to the original narrowband signal. The decorrelation of the operating signal from the pseudocode signal in the RX is possible because the two received signals are orthogonal to each other.

5.1 DSSS Systems

The crucial element of the DSSS technique is the pseudorandom code sequence. As mentioned before, in DSSS, a narrowband input data signal correlates with a code sequence with a much higher bit rate, which results in broadening the data signal on the channel.

The common SS code sequences are the maximum-length sequence (M-sequence) and Gold codes. These coding methods are mainly used for the protection of SS systems against intentional and unintentional EMI and, ultimately, to enable multiple users to share the same frequency spectrum. The selection of the optimal code sequence for a given application is based on the requirements for good *autocorrelation* and *discrete-time periodic cross-correlation* between two discrete sequences. For example, when good autocorrelation is needed, the M-sequence code is employed, whereas for the *code division multiplex access* (CDMA), the cross-correlation method is employed. More details about the SS codes can be found in [12].

Figure 11a, b shows (a) the basic concept of the DS-SS system and (b) narrowband (baseband) and broadband (channel) SS spectra of the desired signal in the presence of EMI and additive medium (ground) noise.

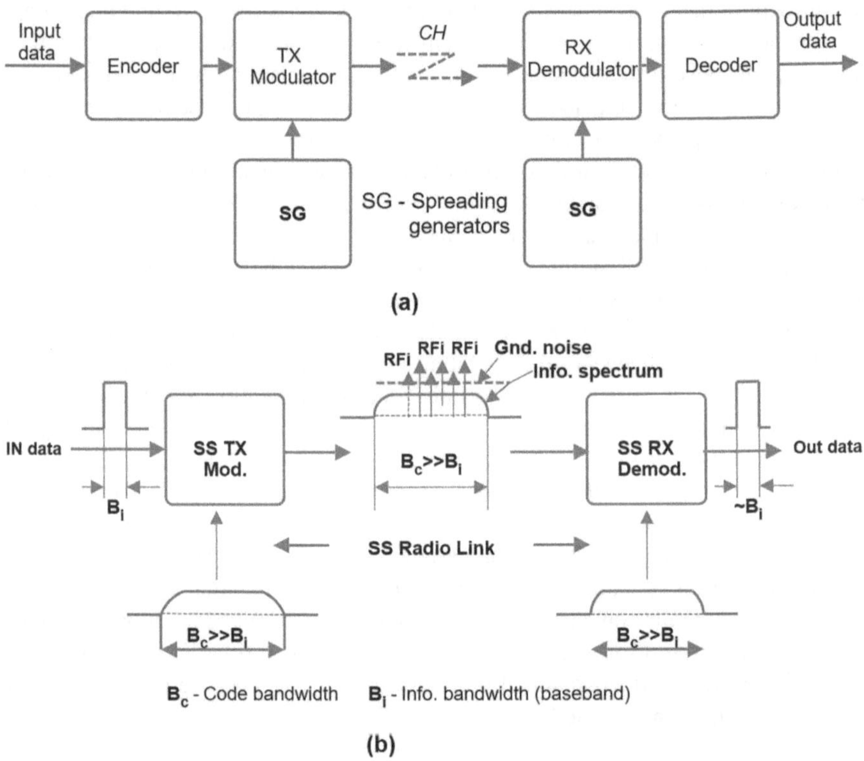

Fig. 11 (**a**) Basic block of the SS system, (**b**) Spreading and despreading of the original baseband B_i by an SS system

As shown in Fig. 11b, the DS-SS spreads the baseband B_i at the TX side via multiplication of the baseband signal with a pseudocode sequence generated by a pseudocode generator (SG). On the RX side, the broadband information signal is returned to the original baseband (narrowband) form with a bandwidth B_i. In the end, the higher level of the desired (operating) signal appears at the RX output, compared with the received low-level interfering signals and noise.

The important parameter of an SS system is the *processing gain* (G_P), which is defined as

$$G_P = \frac{B_c}{B_i} = \frac{R_c}{R_i} = \frac{P_i}{P_c} \quad (11)$$

where B_c, R_c, and P_c are the bandwidth, rate, and mid-power of the *coding digital signal*, respectively, and where B_i, R_i, and P_i are the bandwidth, rate, and mid-power of the *information (baseband) signal,* respectively. For example, the typical processing gain for a standard global DSSS GPS is approximately 47 dB, whereas for an FH terrestrial system, it is approximately 23 dB [7].

5.2 *FH-SS Systems*

Another SS technique is *frequency hopping* (*FH-SS*), which also provides significant improvements to a radio system's performance in terms of high immunity to channel EMI and noise. This technique is associated with frequency modulation (FM), where the rapid change in frequency is controlled by the same synchronized pseudocode employed at the TX and RX sides. For example, FH-SS synchronization is simpler than for TDMA systems. Additionally, FHSS systems are well resistant to fast fading effects because of a rapid frequency jump from one frequency to another before an alternative multipath signal of the first frequency reaches the RX decoder.

Figure 12 shows the functional block diagram of the FH-SS system and the channel with noise and EMI.

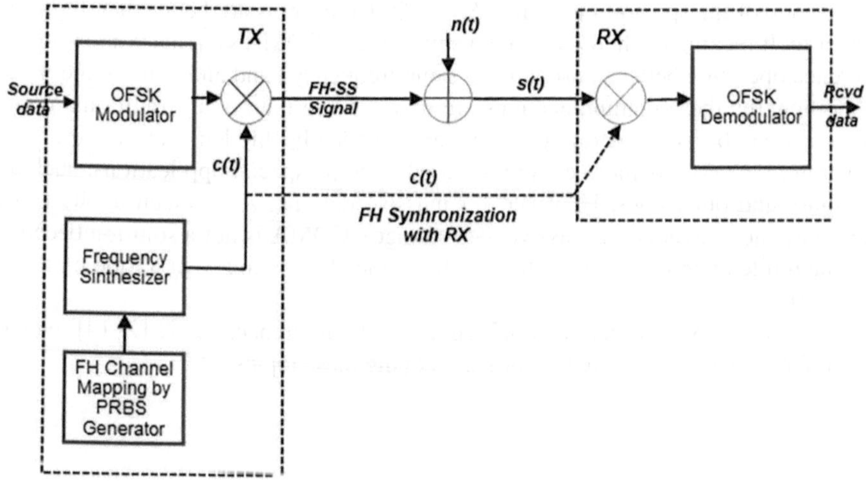

Fig. 12 Block diagram of the FH-SS system

Here, the FHSS operation is controlled by the generator of *pseudorandom bit sequences* (PRBS) in the TX section, which provides synchronization between the TX and the RX.

The FH-SS is classified into two categories: *fast* FHSS and *slow* FHSS. When the hopping rate is faster than the symbol rate (i.e., $T_h \leq T_{sym}$), the hopping frequency is related to *fast hopping*. Otherwise, the hopping frequency is *slow hopping*. In other words, in the *slow* FH, the transmitted signal does not change frequency during multiple symbols, whereas in the *fast* FH, the transmitted signal may change frequency more times during a symbol duration.

In addition to the discussed advantages of the SS systems, the disadvantages are that they are not as immune to broadband and flat Gaussian noise. Additionally, the capacity of the SS system is limited because of the increase in the amount of ambient noise produced by the imperfect coding of users from other radio networks at the same site, who operate with different pseudocodes. Accordingly, the maximum number of users in an SS network site is limited for a specified bandwidth BW and an acceptable BER. For example, the following equation gives the maximum number of users at an SS site [7]:

$$N_{\max} = \frac{P_G}{E_b / N_0} - \frac{1}{\mathrm{SNR}_i} + 1 \tag{12}$$

where P_G is the SS system processing gain, E_b/N_0 is the ratio between the signal power E_b and additive noise power N_0, both per bit, and SNR_i is the ratio between the powers of the spread desired signal and ambient noise with all spurious interference in the spread bandwidth B_c. For example, for a terrestrial FHSS link with $P_G = 23$ dB, $E_b/N_0 = +7$ dB, and $\mathrm{SNR}_i = -1$ dB, the maximum number of users at an SS site is $N_{max} \approx 40$.

The limit of the number of users can be increased by the optimal use of sectorally directive antenna arrays to minimize interference from users of other sectors and by adaptive control of the TX power level. However, the capacity and efficiency of the spectrum used by SS or CDMA systems are better than those of other multiplexing methods. The main advantage of CDMA systems is a low-interference operation between users in the same frequency band and at the same time. Additionally, these communications are more secure than other traditional communications because operating signals are practically ‘hidden’ below the noise. Therefore, CDMA is more commonly used in some special applications, such as military and other ones. However, for massive data networks (such as 4G-LTE, 5G-NR, and networks for massive data storage), CDMA is not a solution because of the limited number of users that can be serviced at a site and at a time by these networks.

The writing of this section is influenced by the references [5, 7, 12–14], which we refer to readers interested in further studying these topics.

6 Principles of Adaptive Software-Defined Radio (ASDR)

The software-defined radio (SDR) concept is becoming highly interesting in wireless cellular networks because of its great adaptability to real-time radio channel conditions and minimal hardware realization and integration complexity because most signal processing is preferred by software instead of hardware, as in traditional radio systems. Figure 13 shows the basic block diagram of an SDR transceiver.

Only analog parts of an SDR transceiver are at the user side (such as analog transducers, RX input and output analog amplifiers, and filters), and at the antenna side, TX power amplifiers and RX low-noise amplifiers are used in addition to analog filters.

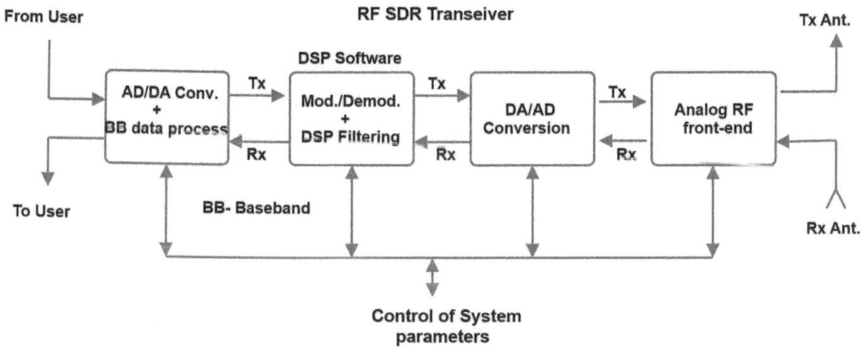

Fig. 13 SDR transceiver

All SDR parameters are constantly controlled and adaptively changed in real time to maintain the best system performance in the presence of fading, EMI-interfering signals, and additive noise in the communication channel. Additionally, distortions of operating signals caused by hardware and signal processing imperfections, which exist in conventional radio systems, are effectively minimized by a software.

The SDR concept is based on the idea of using a wideband analog-to-digital converter (ADC) and digital signal processing (DSP) softwares. Additionally, the use of the SDR significantly reduces the cost of base station hardware because only a single transceiver is required for multiple-channel transmission instead of a transceiver hardware for every single channel. Additionally, owing to the SDR, the operational flexibility of the base station is substantially increased because it can be just programmed for various existing and future communication requirements and standards.

Figure 14 shows an SDR receiver architecture for N user subchannels with centers at frequencies f_n and $n = 1, 2, \ldots, N$, where the bandwidths of different subchannels do not overlap.

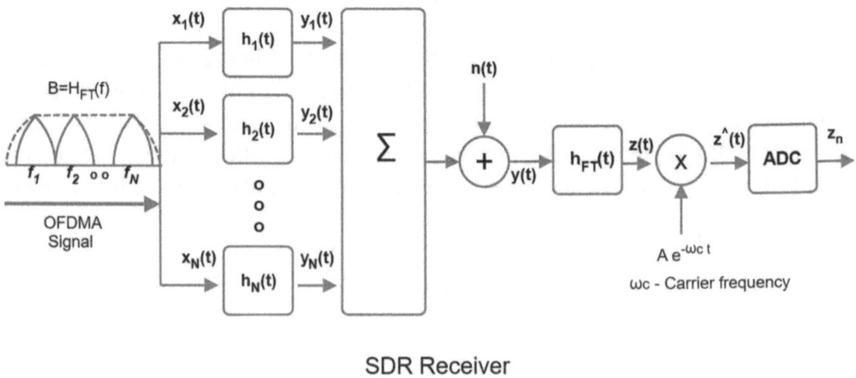

Fig. 14 SDR receiver block diagram

Per the present block diagram, the overall signal $y(t)$ with noise $n(t)$ can be expressed as

$$y(t)=\sum_{n=1}^{N}x_n(t)*h_n(t)+n(t)=\sum_{n=1}^{N}y_n(t)+n(t) \tag{13}$$

where $y_n(t)$ is the $x_n(t)$ signal filtered by an impulse-response subchannel filter $h_n(t)$. In Eq. (13), the operator $*$ denotes the convolution between the functions $x_n(t)$ and $h_n(t)$.

Additionally, the $y_n(t)$ signal of the *n-th* user can be expressed as

$$y_n(t)=\mathrm{Re}\left\{\tilde{y}_n(t)e^{i2\pi f_n t}\right\} \tag{14}$$

where $\tilde{y}_n(t)$ is the complex envelope associated with the center frequencies f_n of the n-th user's subchannel. Furthermore, the block h_{FE} in Fig. 14 is the final front-end filter with an impulse response that is transparent to the signal $y(t)$. After the h_{FE} filter, the signal $z(t)$ can be expressed as

$$z(t)=y(t)*h_{FE}+n(t) \tag{15}$$

Finally, the demodulated complex envelope $\tilde{z}(t)$ associated with the system channel carrier f_c can be expressed before entering the ADC as

$$\tilde{z}(t)=\sum_{n=1}^{N}\tilde{y}_n(t)e^{i2\pi(f_n-f_c)t}+n(t) \tag{16}$$

where $n(t)$ is the noise generated in the receiver.

Hence, an SDR station employs a broadband analog-to-digital converter (ADC) for sampling a wideband received signal that contains many user signals with different subbandwidths and powers. Given the need for a high sampling ratio and quantization, the ADC is a key part of the SDR receiver.

Furthermore, the implementation of a multicarrier modulation into an SDR system makes the SDR capable of adopting multicarrier parameters according to real-time channel conditions and significantly enhancing the overall system performance. The SDR adaptive signal processing includes the following steps:

1. The optimal radio frequencies of alternative bands for sending the desired signals are selected to obtain the best system performance. This method is very useful for radio systems that use random time-changing transmission channels, such as those used in *HF* communication and terrestrial radio mobile networks.
2. Radiochannel distortions are reduced by real-time adoption of RX digital filter characteristics until the signal-to-interference/noise ratio (SINR) is minimized to an acceptable level.
3. The constant *measures* the SINR or BER at the RX output and accordingly controls the TX output power via a feedback radio link to obtain reliable system performance in the presence of EMI from users of the same or other radio networks.
4. The *adaptive antenna array* technique is used to search for sectors with low interference.

Therefore, the SDR system can provide reliable and low-error signal reception because it can efficiently overcome channel fading, ambient noise, and EMI, as well as the natural electromagnetic disturbances commonly present in radio propagation media. Three main methods are implemented in SDR systems to achieve these goals. These methods are based on adaptive filtering and digital signal processing (DSP) [15–17], as well as adaptive ("smart") antenna arrays [15, 18–20]. The adaptive antenna arrays have the ability to adapt to unstable transmission channels with fading, ambient noise, and all kinds of EMI, which are discussed in chapter "Performance of Radio Digital Systems in Presence of EMI".

References

1. R.G. Gallager, *Principles of Digital Communication* (MIT, 2008), 6.1
2. S-M.M, Yang, *Modern Digital Radio Communications Signals and Systems* (Springer, 2020)
3. J. Speidel, *Introduction to Digital Communications*, 2nd edn. (Springer, 2021)
4. S.B. Weinstein, P.M. Ebert, Data transmission by FDM using the discrete Fourier transform. IEEE Trans. Commun. Technol. **19**(5), 628–634 (1971)
5. W. Stark, *Introduction to Digital Communications* (Cambridge University Press, 2023)
6. T. Araujo, R. Dinis, *Analytical Evaluation of Nonlinear Distortion Effects on Multicarrier Signals* (CRC Press, 2015)

7. J. Gavan, in *Radio Systems Parameters and Performance Criteria Concerning Computation and Mitigation of EMI Effects*, ed. by R. Perez. Chapter 20. Handbook of EMC (Academic Press, 1995)
8. J.G. Proakis, *Digital Communications* (McGraw-Hill, New York, 2008)
9. B. Clerckx, C. Oestges, *MIMO Wireless Networks* (Academy Press, Lagos, 2013)
10. B. Kumbhani, R.S. Kshetrimayum, *MIMO Wireless Communications over Generalized Fading Channels* (CRC Press, Boca Raton, 2017)
11. T.L. Marzetta, E.G. Larsson, H. Yang, *Fundamentals of Massive MIMO* (Cambridge University Press, 2016)
12. R.E. Ziemer, *Fundamentals of Spread Spectrum Modulation* (Springer, 2022)
13. D. Torrieri, *Principles of Spread-Spectrum Communication Systems*, 4th edn. (Springer, 2018)
14. R.L. Peterson, R.E. Ziemer, D.E. Borth, *Introduction to Spread Spectrum Communications* (Prentice-Hall, Hoboken, 1995)
15. N. Blaunstein, C.G. Christodoulou, *Radio Propagation and Adaptive Antennas for Wireless Communication Networks* (Wiley, Hoboken, 2014)
16. E. Grayver, *Implementing Software Defined Radio* (Springer, 2012)
17. S. Pollin, M. Timmers, L.V. der Perre, *Software Defined Radios; From Smarter to Cognitive* (Springer, 2011)
18. R.C. Hansen, *Phased Array Antennas* (Wiley, Hoboken, 2009)
19. R.J. Mailloux, *Phased Array Antenna Handbook* (Artech House, Norwood, 2005)
20. C.A. Balanis, *Antenna Theory*, 3rd edn. (Wiley, New York, 2005)

Overview of 4G-LTE and 5G-NR Mobile Radio Systems from EMI Prospect

1 Introduction

A new generation of cellular technology called the 5G New Radio (5G-NR) is being developed based on the existing 4G-LTE cell technology. It includes new and more complex network methods with a larger frequency spectrum, enhanced radio channel multiplexing, high-level of signal modulations, and highly efficient error correction techniques. The 5G-NR development responds to emerging and rapid demands for higher networks' capacity; massive user, computer, and machine connectivity; better area coverage; and high network reliability and low latency compared with the existing 4G-LTE networks. The 5G-NR technology is being developed and standardized in cooperation with different professional organizations, academies, and tech companies worldwide through the 3GPP (third Generation Partnership Project) organization. Therefore, the 5G-NR is a global and open network standard that enables anyone to contribute to its development. The main goal of this standard is that all 5G equipment and networks be compatible with each other and work together, regardless of where they are located.

A comparison between 4G and 5G technologies and their differences will be a subject in the next section. Hence, new wireless services, such as cellular networks, digital audio and video broadcasting, interactive video communications, video streaming, and e-commerce services, require larger and larger bandwidths due to constant demand for high-speed data access for end-users. To meet these requirements, the Multicarrier Modulation (MCM), Spatial Multiplexing (MIMO) techniques, and advanced error correction coding become very effective for these high-speed wireless communication systems and networks. These techniques are discussed in chapter "Overview of Broadband Radio Systems with High EMI Immunity".

M. Pajovic, *Fundamentals of EMI Phenomena in Radio Communication Systems*, https://doi.org/10.1007/978-3-031-88748-2_7

The writing of this chapter is motivated and influenced by the references [1–10], which we refer to interested readers for a further study of this new communication technology.

2 Concepts of 4G-LTE and 5G-NR Radio Cellular Networks

As noted in chapters "Performance of Radio Digital Systems in Presence of EMI" and "Overview of Broadband Radio Systems with High EMI Immunity", the channel capacity or throughput first depends on how the channel bandwidth is large, i.e., the greater channel bandwidth allows a greater amount of data to be transmitted through the channel. To achieve a channel bandwidth larger than the 4G bandwidth, new higher frequency spectra are needed for new high-data capacity 5G-NR networks. In accordance with the 3GPP recommendations, telecom regulators (from North America, Europe, and Asia) have allocated more international mobile telecommunication (ITM) bands to each global region. Additionally, network operators are doing all to share existing spectra with new radio services and acquire those used by abandoned old communication technologies. This includes adding the frequency-division duplexing (FDD) and the time-division duplexing (TDD) bands used in 3G mobile systems, as well as abandoning old analog broadcasting radio and TV spectra.

To further increase the 5G network capacity, several advanced techniques of 4G technologies have also been applied, such as high-order QAM modulations (64-QAM, 256-QAM, 512-QAM, etc.) and massive MIMO techniques in conjunction with antenna beamforming via adaptive sectorial antenna arrays. Additionally, the *carrier aggregation* (CA) technique from 4G is used in 5G technology, enabling larger aggregation bandwidths of up to 400 MHz in the mmWave spectrum. For example, 4G-LTE includes the CA method, which enables an increase in the usable bandwidth up to 140 MHz, combining multiple frequency carriers, each up to 20 MHz.

The 5G-NR frequency spectrum is divided into three main bands [1, 3]:

(a) The *low band* is between 410 MHz and 1 GHz with a limited data rate of up to 200 Mbps and with larger area coverage, including indoor penetration. This band is also used by 4G-LTE.
(b) The *midband* between 1 and 7 GHz is suitable for urban and rural deployment because of its propagation characteristics and increased capacity (data rate up to 2 Gbps). This band is also used up to 6 GHz by the 4G-LTE.
(c) The *high band* is between 24 and 100 GHz with very high capacity or data transmission (with a rate of up to 10 Gbps) and very low latency. However, these mm-Wave bands have limited area coverage because they have a significantly shorter transmission range and limited indoor propagation due to signal loss caused by objects and wet weather.

2.1 Basic Cellular Network Concept

Figure 1 shows a basic cellular network concept with hexagonal cells with their own base stations (BS), which are two-way connected to the mobile switching office (MSO). Furthermore, the MSO is connected to the Internet via the core network.

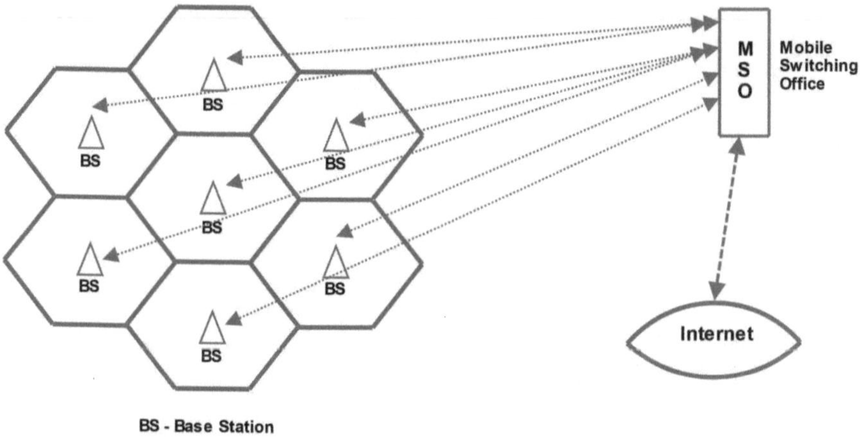

Fig. 1 Basic concept of a cellular network with hexagonal cells

2.2 5G Radio Access over the RF Layer and Cloud Internet Network

In general, the 5G-NR network includes two main access layers: the radio access network (RAN) layer and the core network (CN) layer, which connect users within mobile networks and with the global Internet. With core access, the coverage of the 5G-NR network is significantly improved, and the EMI effects in the network are reduced. To further improve the coverage of the 5G RAN, the network is divided into *macro cells* (where the low-frequency band *FR-1* is used for covering a larger aerial coverage) and *micro-* or *subcells* where a high-frequency band *FR-2* is used. Furthermore, massive MIMO systems are deployed in *microcells* with high-user density in conjunction with sophisticated antenna array beaming, called *space division multiplexing* (SDM). The MIMO concepts are discussed in chapter "Overview of Broadband Radio Systems with High EMI Immunity".

Figures 2 and 3 show the differences between the 4G-LTE and 5G-NR network structures, respectively: a 4G-LTE (Fig. 2) with RF links or a radio access network (RAN) between base stations, whereas a 5G-NR uses RAN and internet network layers (Fig. 3) for connecting users in the same and other networks and with the global Internet.

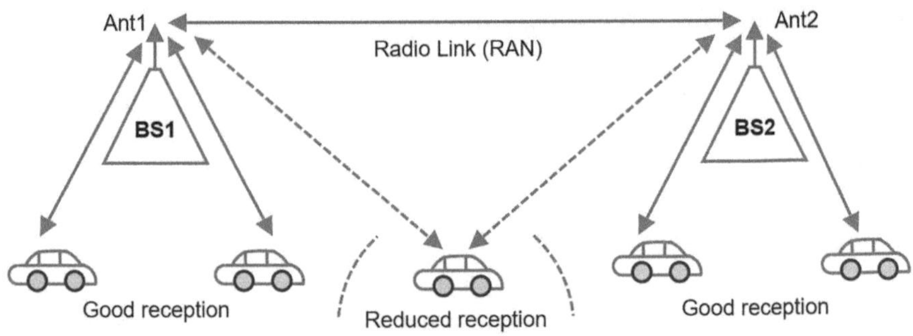

Fig. 2 4G cell RAN network concept

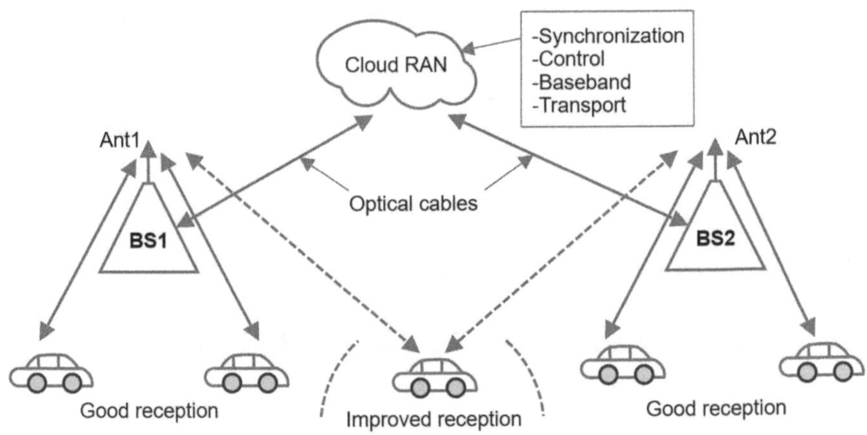

Fig. 3 New 5G-NR concept with core (cloud) network layers

Hence, cloud access is used in parallel with RF microwave links (RANs) for connecting 5G-NR base stations to mobile terminals.

2.3 4G-LTE vs. 5G-NR Network Performance

As noted earlier, one of the main reasons to use additional frequency spectra in 5G-NR networks is to increase the network's data capacity. As shown in Tables 1 and 2, the sub6 GHz frequency range is allocated for 4G-LTE, whereas two frequency bands, sub7 GHz (FR1) and mm-Wave (FR2), are allocated for 5G-NR [per the 3GPP proposal].

Table 1 4G-LTE/5G-NR (FR-1 and FR-2)

Parameters	5G-FR1 Sub7 GHz	5G-FR2 mmWave
Frequency range	410–7125 MHz (3GPP Rel.16)	24–49 GHz (3GPP Rel.16)
Standard channel BW for SC and MC	5–100 MHz	50–400 MHz
Carrier aggregation	Up to 16 carries per channel	Up to 16 carries per channel
Sub-carrier spacing	15, 30, 60 kHz	60, 120, 240 kHz
Channel multiplexing and modulation	DFT-S-OFD M, QPSK and 16, 64, 256 QAM	CP-OFDM, QPSK, π/2-BPSK, 16, 64, 256 QAM
MIMO for Downlink (DL) and Uplink (UL)	max. 8 Ch. in DL and max. 4 Ch. in UL	max. 8 Ch. in DL and max. 4 Ch. in UL
Deployment	Macro cells/long-range/more users	Microcells/short-range/fewer users
Spatial-antenna multiplex	Parallel delivery stream data of same data block	Beam steering for each user
Number of simultaneous users	Tens of users and larger covering area	Larger user density in small covering area
Chanel performance	High multipath propagation	A few multipath propagation
Spectral efficiency	High efficiency Due to spatial multiplex	Lower efficiency Due to high path loss

Table 2 5G (FR-2)

Freq. bands	Downlink/uplink (GHz)	Channel BW (MHz)
K-band	24.25–27.50	50–400
LMDS	27.25–27.50	50–400
Ka-band	27.50–28.35	50–400
Ka-band	37.00–40.00	50–400
V-band	37.50–43.50	50–400
V-band	47.20–48.20	50–400

Table 1 shows other 4G-LTE and 5G-NR network performances and parameters relevant for the network data capacity, power, and spectral efficiency, as well as enhanced resistance to noise and EMI effects. The single-carrier (SC) and multicarrier (MC) bandwidths of 4G-LTE reach 20 MHz and 140 MHz, respectively. The multicarrier 5G-FR1 bandwidth (7-GHz subband) is up to 100 MHz, whereas the 5G-FR2 bandwidth spreads up to 400 MHz in the mmWave bands. Table 2 also shows bands that belong to the 5G FR-2 spectrum.

Therefore, the maximum 5G-NR data capacity is significantly greater than the 4G-LTE capacity because a significantly wider bandwidth is allocated for the 5G network. In particular, the maximum data rate in the 4G (6-GHz subband) is up to 2 Gbps since, in the 5G FR2 band, the data rate is up to 10 Gbps. The 5G-NR capacity can be further increased by the use of new baseband waveforms, new channel modulation methods, and channel coding with digital compression techniques. Owing to these improvements and the greater bandwidth in the FR2 spectrum, the 5G-NR can also handle many more connected users at time than can the 4G-LTE network.

However, 5G-NR will not soon completely replace the 4G-LTE network, and one starts to work together with 4G-LTE because the new and independent network infrastructure for 5G-NR needs more time to be completely built. The two types of 5G projects are currently in development to transition smoothly from existing 4G-LTE to new 5G-NR cellular technology. They are the 5G *nonstandalone* network (5G-NSA), which works with the 4G-LTE partially using its infrastructure, and the 5G *standalone* network (5G-SA), which is being developed separately from a 4G-LTE. As a result, the 4G-LTE networks and *nonstandalone* 5G-NSA networks dynamically share the same spectra at the same time.

2.4 Summary of Performance and Dedicated Services of 5G Networks

Table 3 shows the 5G main parameters and allocated services.

Table 3 The 5G main parameters and services

	Main services	Core network	Air interface	Latency	Max. speed
4G	Full data, Phone, and Messaging services	IP	OFDM OFDMA MIMO	60–100 ms	1 Gbps
5G	e-MBB, m-MTC, URLLC	IP	AAS, FD-MIMO, OFDMA, SC-FDMA	<1 ms	>10 Gbps

The 5G primary services are *enhanced mobile broadband* (e-MBB), *massive machine type communications* (m-MTC), and *ultrareliable* and *low-latency communications* (URLLC). For example, the 5G latency is much shorter than the 4G latency, on the order of 60–100 times. As above, the maximum data speed of 5G is significantly above the 4G speed. Additionally, both 4G and 5G no longer support traditional circuit-switched telephony but only Internet protocol (IP) core networks.

The OFDM, OFDMA, and MIMO techniques are exploited as the 4G air interfaces, whereas the AAS, FD-MIMO, OFDMA, and SC-FDMA techniques are used for the 5G air interfaces. The AAS is an active antenna system, FD-MIMO is a full-dimensional MIMO, and SC-FDMA is a single-carrier FDMA.

In the next section, we describe the primary 5G services, e-MTC, e-MBB, and URLLC.

3 Applications of 5G-NR Networks

More information about the use and applications of 5G networks is given in Refs. [3–5].

3.1 Massive Machine Type Communications (m-MTC)

ITU-R M.2083 defines m-MTC as a very large number of connected devices that typically send a relatively low amount of nondelay-sensitive data. Devices are low-cost batteries with very long lifetimes. The m-MTC is deployed to support the Internet-of-Things, *smart cities*, *smart homes*, *energy monitoring*, *remote monitoring*, *smart agriculture,* etc.

The *Internet of Things* (IoT) is related to the expansion of interconnected smart devices from home appliances to tiny sensors. The IoT involves short-range mobile transceivers into a wide area, enabling a new form of communication between things themselves and people and things. For example, the Internet supports the interconnections of billions of industrial and personal devices through cloud systems.

Smart cities and *homes* are new city and home models that use the latest information and communication technologies to improve the quality of life and efficiency of urban services and operations, as well as efficient and optimal home functionalities, ensuring the needs of people in terms of social, economic, environmental, and cultural aspects.

Smart agriculture deploys the IoT to manage and optimize farm resources efficiently through sensors placed in ground and water, as well as in farm vehicles and machinery, to collect data such as crop health and soil moisture. For example, the IoT can support specific farm operations such as storage monitoring, weather condition monitoring, vehicle monitoring, and livestock monitoring.

In terms of the importance level, m-MTC is categorized by eight key characteristics:

- The high-importance connection density is expressed as the total number of devices available to satisfy a specific quality of services (QoS), measured in units of area (per km^2):
- Energy efficiency with medium importance, refers to the quality of data transfer with a sufficiently low BER sent and received from users.
- The following six m-MTC characteristics have low importance: area traffic capacity, peak data rate, user-experienced data rate, spectrum efficiency, mobility, and latency.

ITU-R Report M.2410 defines the performance requirements for enhanced mobile broadband (e-MBB), m-MTC, and ultrareliable and low-latency communications (URLCC). The performance requirements from m-MTC are defined only for connection density. The minimum requirement is set by M.2410 to 10^6 devices per km^2.

More about e-MTC and the IoT can be found in references [6–8] related to this chapter.

3.2 Enhanced Mobile Broadband (e-MBB) Services

ITU-R Report M.2410 gives three deployment scenarios for e-MBB: *indoor hotspots*, *dense urban areas*, and *dense rural areas*.

- *An indoor spot* is defined as *an indoor isolated area* at shopping stores and moles, offices, airports, train/bus stations, etc. This scenario includes relatively small spots with high user density in buildings or facilities. The main e-MBB characteristics exploited are the high data capacity needed for high user density and their consistent presence and ability to move into and out of the indoor area. As noted earlier, 5G has some challenges for indoor applications because the mmWave range is limited because of its high absorption in buildings with concrete walls. Therefore, interconnected 5G base stations are required for indoor applications, similar to outdoor 5G networks.
- *A dense urban area* is defined as *an urban area with high user density and heavy street traffic with car and pedestrian users*. Hence, the key characteristics of this deployment are covering heavy traffic, outdoor coverage, and outdoor-to-indoor coverage. The dense urban areas for 5G are specified by the use of densely collected small (micro) cells for supplementing macro cells. There are two reasons for this:
- The concentration of stationary, car, and pedestrian users that use a 5G network with a large data traffic load.
- 5G FR-2 bands (mmWave) are typically limited by high EMI, and consequently, only microcells can be covered sufficiently.

The rural area is defined as a rural zone where pedestrians, cars, and high-speed users are continuously serviced by 5G in larger areas. Therefore, the key ability of this 5G deployment is to cover wide areas continuously with high-speed motorized users. This deployment of microcells is limited by higher EMI and noise in the environment. The rural deployment actually provides the last access to Internet and telephone services for the users in the rural area, supplementing relatively expensive optical cabling in these remote zones. The way in which these remote zones are covered wirelessly is known as a wireless local loop (WLL). The WLL belongs to the category of fixed wireless access (FWA).

3.2.1 e-MBB Performance Requirements

Tables 4, 5, 6, 7, and 8 are based on ITU-R M.2083 indicating the key performance requirements for e-MBB. The following performance requirements with high importance are the *peak data rate requirements, user-experienced data rate, spectral efficiency requirements, mobility, network power efficiency,* and *area traffic capacity,* whereas *the latency and connection density requirements have medium importance for e-MBB communications.*

Table 4 e-MBB data rate requirements

Data rate	Downlink	Uplink	4G Downlink
Peak data rate	20 Gbps	10 Gbps	1 Gbps
User-experienced data rate	100 Gbps	50 Mbps	10 Gbps
Indoor hotspot area traffic capacity	10 Mbps/m^2	–	0.1 Mbps/m^2

Table 5 e-MBB spectral or bandwidth efficiency requirements

Parameters [bps/Hz]	Test environment	Downlink	Uplink
Peak spectral efficiency	All	30 bps/Hz	15 bps/Hz
5th percentile user spectral efficiency	Indoor hotspot	~0.3 bps/Hz	~0.2 bps/Hz
	Dense urban zone	~0.22 bps/Hz	~0.15 bps/Hz
	Rural zone	~0.12 bps/Hz	~0.05 bps/Hz
Average spectral efficiency	Indoor hotspot	~9 bps/Hz	~6.7 bps/Hz
	Dense urban zone	~7.8 bps/Hz	~5.5 bps/Hz
	Rural zone	~3 bps/Hz	~1.6 bps/Hz

Table 6 e-MBB requirements for latency

Parameters	Latency	4G latency
User latency	4 ms	10 ms
Control latency	20 ms	100 ms

Table 7 e-MBB mobility requirements

Test environment	Indoor hotspot	Dense urban area	Rural area	
Mobility classes	Pedestrian and stationary	Stationary, pedestrian, and cars	Pedestrian, Cars	High-speed cars
5G traffic data rate	1.5 bps/Hz	~1.1 bps/Hz	0.8 bps/Hz	0.45 bps/Hz
5G mobility	10 km/h	30 km/h	120 km/h	500 km/h
4G traffic data rate	1.0 bps/Hz	0.75 bps/Hz	0.5 bps/Hz	0.25 bps/Hz
4G mobility	10 km/h	30 km/h	120 km/h	350 km/h

Table 8 EMI-relevant parameters of BSs and associated cells

Cell type	Max. cell radius	Max. user number	Location	Max. TX power (W)/Band
Femtocell	100 m	30	Indoor	0.25/FR-2
Picocell	200 m	100	Indoor/Outdoor	1/FR-2
Microcell	2 km	2000	Indoor/Outdoor	10/FR-1
Macrocell	30 km	2000	Outdoor	>50/FR-1

Here, the substantial increase in the data rate enables 5G networks to obtain a broadband range of high-data traffic with high-user quality of experience (QoE).

Table 5 shows:

- *The peak spectral efficiency* is the maximum data rate normalized by a channel bandwidth in bps/Hz.
- The *5th percentile of the user spectral efficiency* is 5% of the CDF (the cumulative distribution function) of the normalized user throughput, which is defined as the number of correctly received bits over a certain time period.
- The average spectral efficiency is the average data throughput per unit of spectrum (bps/Hz) and per network cell.

3.2.2 e-MBB Latency Requirements

The user latency is the transmission time related to the wireless network to deliver a data packet from the sender user to the specific destination. Additionally, *the control latency* is the transmission time from the idle state to the start of continuous data transfer (the active state).

The last column in Table 6 shows the latency requirement for the advanced 4G network downlink. Like in the case of the data rate, the low latency requirements for 5G are more demanding.

3.2.3 e-MBB Mobility Requirements

The mobility requirements are related to a system's ability to effectively service users in motion, i.e., the maximum speed of user equipment (UE), to obtain a defined quality of service (QoS). Table 7 shows the classes of mobility. such as

- Stationary with a speed of 0 km/h
- Pedestrian movement at speeds ranging from 0 to 10 km/h
- Car movement at speeds of 10–120 km/h
- High-speed cars with speeds of 120–500 km/h

3.2.4 e-MBB System Requirements

The requirements listed in Tables 5, 6, and 7 concern the quality of the system parameters (QoS) is acceptable for 5G users. Generally, the system requirements can be two categorized into two categories: one is associated with the *air interface,* and the other is associated with the network functionality and architecture. The data rate and spectral efficiency (Tables 4 and 5) involve the design of the *air* or *radio interface*. These designs include the antenna design, signal coding methods, user protocol design, and spectrum allocation.

Mitigating the mobility and latency requirements (Tables 6 and 7) involves designing the core network and radio access network (RAN). One design strategy includes edge computing to achieve as much processing as possible to the network edge, i.e., closest to the user. Additionally, in this design process, virtualization methods that support the dynamic allocation of network resources to satisfy user needs are important.

In the end, 5G e-MBB is applied in organizing so-called "smart offices" (indoor hotspots) and in radio communication systems between trains and trackside (for rural use).

3.2.5 Ultrareliable and Low-Latency Communications (URLLC)

ITU-R M.2083 indicates the importance level of eight key performances of URLLC:

- *Mobility*, *latency*, and *reliability* with *high* importance.
- In addition, the *connection density, network energy efficiency, area traffic capacity, peak data rate, user-experienced data rate, and spectrum efficiency,* all have lower importance than the performance of the abovementioned three metrics.

Network mobility is related to the maximum user device (UD) speed at which the required quality of service (QoS) can be reached. The mobility requirement is defined and explained in the proceeding subsection related to e-MBB requirements.

The *latency requirement* is divided by ITU-R M.2410 into two parts: *user latency* and *control latency.* The *user latency* is defined as the time needed for successful delivery of a packet—message to the dedicated user, using the specific radio protocols for both network uplinks and downlinks. The minimum latency requirement is *1 ms* when considering a single user and small IP packets on both uplinks and downlinks. *The control latency* is related to the transition time for the most efficient battery state (idle state) to the start of continuous data transfer (active state). The minimum latency requirement is *20 ms.*

Cases of URLLC Applications

There are a wide variety of URLCC applications, such as emerging mission-critical applications, which have high reliability and latency requirements. They are applied in industrial automations, lend vehicles, robots, and drones.

Industrial automation is the area where URLLC supports smart factories and their automation. This application requires extremely reliable and low-latency 5G communication links between controllers, actuators, and sensors.

Lend vehicles, robots, and drones are remotely controlled over IoT and 5G links. These remotely controlled devices and equipment are commonly used in modern factories and smart agriculture, as explained in the proceeding subsection associated with e-MBB (enhanced massive broadband) communications.

Other areas of URLLC use include *emergencies*, natural *disasters*, and other *public safety* matters. The knowledge of an accurate location and very low latency of communication are critical requirements for rapid response to disasters and urgent needs. For example, urgent healthcare involves remote patient monitoring, remote intervention and surgery, and, in general, remote healthcare.

In addition to the abovementioned applications, URLLC is applied in *intelligent transportation*, which involves *assisted driving*, *autonomous driving*, and *tele-operated* driving for remote driver assistance in areas such as disaster areas and difficult terrains for manual driving. *Tele-operated* driving requires a reliable wireless network to support the communication of sound and video information as well as diagnostic data from the vehicle and environmental information to the remote driver.

4 Overview of 5G-NR Performance Concerning EMI

Techniques that improve the overall 5G-NR performance, including increasing the network immunity to EMI, are as follows [9]:

- Spectral carrier aggregation (CA)
- CP Orthogonal Frequency Division Multiplex (CP-OFDM)
- Multilevel baseband signal modulations (M-PSK and M-QAM)
- Massive MIMO and spatial division multiplexing (SDM)
- Adaptive Software Defined Radio Techniques with "Smart" Antenna Arrays

Carrier aggregation (CA) is a method that uses multiple frequency carriers for one data channel to increase the channel capacity. This technique was a game changer in 4G technology for enhancing network capacity and was found to play a significant role in the new 5G-NR technology. The spectral efficiency of the CA technique mainly depends on the types of deployed multiplexing methods and baseband signal modulation. As shown in Table 1, the multiplexing method used in the 4G network is *orthogonal frequency division multiple access* (OFDMA) for uplink data transported by a single frequency carrier (SC-FDMA), whereas for the downlink method, the 4G multicarrier orthogonal frequency division multiplex (OFDM) method is used. Both multiplexing methods are discussed in chapter "Overview of Broadband Radio Systems with High EMI Immunity".

The extended OFDM method is introduced to the 5G in the FR2 band with an added *cyclic prefix* (CP) to OFDM (resulting in CP-OFDM) for both uplink and downlink. This method is also implemented in Wi-Fi communication standards, as discussed in chapter "Overview of Broadband Radio Systems with High EMI Immunity". Notably, the CP-OFDM uses more parallel narrowband carriers instead of a single wideband carrier to transport more data. CP-OFDM is selected by 3GPP because it is as follows:

- Scalable with a lower RX complexity.
- It has the best compatibility with multiple antenna technology and MIMO *spatial* multiplexing, which results in higher *spectral* and *power* efficiency.
- Well controlled in the time domain and consequently suitable for time-division deployment (TDD) and low-latency applications.
- Less sensitive to the Doppler effect and phase noise (i.e., frequency and wavelength fluctuations) than other waveforms.
- This approach is suitable for uplink transmissions in macrocell deployments.

Table 1 also shows that the 5G-NR in the *FR1* band uses a variant of the OFDM format for uplinking (3GPP, Realize 16), such as *Discrete Fourier Transform Spread* OFDM (DFT-S-OFDM). The DFT-S-OFDM adopts a CP-OFDM waveform with a *low peak-to-average-power ratio* (PAPR). This multiplexing method increases the 5G uplink range and is used as a replacement for a higher-power link in situations when a user operates far from the base station tower. The DFT-S-OFDM is similar to the SC-FDMA used for 4G uplinks, whereas the CP-OFDM is similar to the OFDMA used for 4G downlinks.

Furthermore, the 5G-NR provides flexible subcarrier spacing beyond the fixed 15 kHz in the 4G-LTE. Owing to the wide FR2 band, the 5G carrier spacing ranges from 15 to 240 kHz, and this flexibility enables the optimal use of diverse bands needed for a 5G-NR network.

With respect to baseband modulation, the multiple-level QAM modulation is used for 4G and 5G networks, such as 64-QAM, 128-QAM, 256-QAM, and beyond. This baseband modulation is discussed in chapter "Introduction to Digital Communication Systems". Notably, multilevel QAM modulations are very robust to spurious interference and ambient noise. In addition to the QAM, the well-known Quadric Phase Shift Keying (QPSK) modulation is still used in 4G and 5G network links for low-speed and mid-speed data rates.

4.1 Massive MIMO and Spatial Division Multiplexing (SDM) for 5G-NR

As discussed in chapter "Overview of Broadband Radio Systems with High EMI Immunity", the beamforming technique used in MIMO systems exploits antenna arrays whose primary task is to focus radiated energy into narrow space regions. Standard MIMO technology effectively reuses the same bandwidth multiple times to transfer more data and provides more efficient use of the spectrum. As shown in Table 1, the massive MIMO in a 5G-NR base station (BS) contains up to 8 antennas for the downlink (DL) and up to 4 antennas for the uplink (UL). This enables the BS to transmit up to 8 signal streams simultaneously to 8 individual users in different sectors. Furthermore, the massive MIMO is an extension of the standard MIMO with arrays of 64, 128, 256, or even more antennas. Therefore, the primary task of an antenna array is to focus radiated energy into a specific narrow region via a beamforming technique. In general, the massive MIMO aims to

- Serve a larger group of users at a time.
- Minimizing radiation in undesired directions results in a reduction in EMI from users who share the same network site.
- A decreased in latency results in faster data transfer and high system reliability for real-time applications.
- This reduces channel fading and consequently reduces the decrease in the number of user connections.

Thus, a massive MIMO increases the cellular network *capacity* and *spectral efficiency* because of the ability to transmit and receive *RF* signal streams in parallel over narrow antenna beams. Consequently, EMI between the parallel RF streams can be significantly minimized because the RF streams travel by different antenna beams. In particular, antenna beams are created by arrays of multiple antennas whose direction and width are controlled by the appropriate magnitude and phase weightings of individual antenna signals. The beamforming technique is used in the mmWave spectrum above 24 GHz for very high-speed data transfer through a wide channel bandwidth (up to 400 MHz). Network operators also use this spectrum to deploy fixed wireless access (FWA) for high-speed connections to businesses and homes.

Figure 4 shows the concept of the *spatial division multiplex* (SDM) by the beamforming technique used in MIMO systems (per 3 GPP).

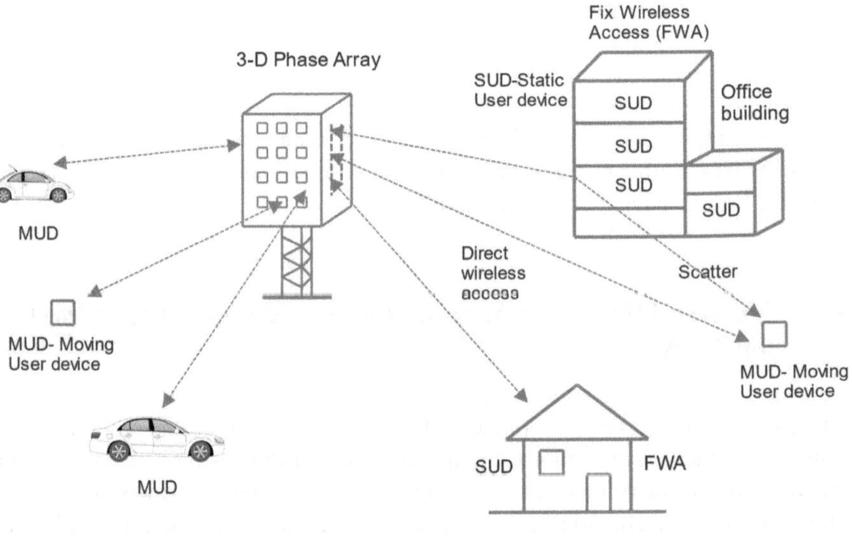

Fig. 4 Massive MIMO with antenna beamforming

Thus, data user streams do not interfere with each other at a MIMO base station because each stream is transferred by separate antenna beams, as shown in Fig. 4. Furthermore, the beamforming method is most efficient in the mmWave spectrum because more small antenna elements may comprise a larger phase array that provides higher gain and multiple narrow radiation beams. For example, the frequency of 30 GHz is related to the small wavelength $\lambda \approx 10$ cm, and because the size of array elements is in the range of the half-wavelength of a signal of interest, the number of array elements can be very large. Furthermore, by increasing the number of array elements in the 2D and 3D directions, the array radiation pattern may provide many narrow (sharp) beams. More information about adaptive antenna arrays can be found in Refs. [9, 10].

In addition to the mentioned advantage of mmWave bands, as noted earlier in Sect. 4, these bands have propagation difficulties due to signal attenuation and delays that may significantly limit the 5G-FR2 range. Millimeter waves are easily attenuated and distracted by traveling through constructions, trees, rain, fog, and foliage. Therefore, mmWaves can be used for short-range applications only, up to several hundred meters on average.

4.2 5G-NR Network Cell Sizes Concerning EMI

It is very important to consider base station (BS) parameters and their cell sizes with respect to potential EMI/RFI effects. These parameters include the maximum number of users the BS serves at a time, the cell size and location (indoor or outdoor), the deployed transmit power, and the spectrum used by the BS. Table 8 shows the EMI-relevant parameters of the BS and associated cells.

Here, 5G-NR cells are categorized from smallest to largest (such as femtocells, picocells, microcells, and macro cells) in terms of the cell radius and number of users, cell locations (indoor or outdoor), and employed TX power for specific cells and frequency bands.

Hence, as noted earlier, a larger difference between the 4G and 5G networks is that 5G uses a much wider frequency spectrum FR2. The 5G-FR2 standard is the first to use extremely high frequencies in the millimeter wave spectrum for public communications, up to ~50 GHz, and as expected in the future, up to 100 GHz. However, as mentioned earlier, the main problem with millimeter waves is their short propagation range, i.e., they cannot propagate as far as the low-frequency waves used in the 4G or 5G-FR1 networks. In addition to *intrinsic attenuation* in free space, 5G-FR2 waves are commonly blocked or highly attenuated through building constructions and other large obstacles. Therefore, to overcome this problem, 5G-FR2 networks must be divided into smaller cells since larger cells can be used for 4G/5G-FR1, as shown in Table 8.

5G-NR cell configuration concepts are not only interesting for user-to-user communications but also for the upcoming Internet of Things (IoT) and machine-to-machine communications, where many equipment and machines will connect each other with different wireless standards and techniques, such as short-range to mid-range wireless systems (Bluetooth, Wi-Fi, or ZigBee), low-power wireless area networks (WANs), and mid-range and long-range wireless networks (such as 4G-LTE, 5G-NR-FR1, and satellite communications).

Because most of the aforementioned communication techniques require a wideband radio spectrum and efficient coverage of areas with a constant increase in the number of wireless transmitters, a greater challenge arises for system network planners and network designers in terms of keeping potential EMI emissions at an acceptable level.

4.3 EMI Shielding for 4G and 5G Equipment and Devices

As mentioned earlier, excessive EMI can disrupt the functionality of base station electronic equipment and user handle devices, such as smartphones for 4G and 5G networks. The same can occur in industrial, automotive, medical, and aerospace environments. To prevent this EMI impact on electronic equipment and devices, strict EMC standards are established that electronic device and system designers and manufacturers must follow.

A first step to meet the EMC requirements for 4G and 5G devices and equipment is to properly shield these devices and equipment (discussed in chapter "EMI Shielding for Radio Equipment") against external interfering signals, whose frequencies fall into 4G and 5G operating bands that spread from 400 MHz to 7 GHz (for the FR-1 band) and from 24 to 50 GHz, and beyond in the future. As mentioned in chapter "EMI Shielding for Radio Equipment", openings or slots in device shielding can be efficient antennas if the largest dimension of the opening or slot is close to $\lambda/2$ (Babinet's principle), where λ is the wavelength of a potential interfering signal.

Table 9 shows the frequencies from 410 MHz to 50 GHz with corresponding λ and $\lambda/20$ values for the 4G/5G FR-1 and 5G FR-2 bands.

Table 9 λ values related to 4G/5G FR-1 and 5G FR-2 bands

4G/5G FR-1	λ	$\lambda/20$
410–7125 MHz	73–4.2 cm	3.65 cm to 0.21 cm
5G FR-2	λ	$\lambda/20$
24–50 GHz	1.3 cm–6 mm	0.65 mm–0.3 mm

Thus, an electrically large opening or slot on device conductive enclosures should be reduced such that the largest opening dimension is $\ll\lambda/2$. As discussed in chapter "EMI Shielding for Radio Equipment", EMI shielding materials used in the telecommunications and electronic information industry include conductive-coated plastics, conductive silicones, metal sheets and screens, conductive fabric gaskets, and ferrite wave-absorbing materials. Also, peripheral cables and wires of base station equipment often serve as entry and exit points for EMI. The strategy for reduction of this risk is also explained in chapters "Performance of RF Cables and IT Signal lines in Radio systems with EMI" and "EMI Shielding for Radio Equipment".

References

1. A.F. Molisch, *Wireless Communications: From Fundamentals to Beyond 5G*, 3rd edn. (Wiley-IEEE Press, 2022)
2. M. Sauter, *From GSM to LTE-Advanced Pro and 5G*, 4th edn. (Wiley, 2021)
3. W. Stallings, *5G Wireless: A Comprehensive Introduction*, 1st edn. (Addison-Wesley, 2021)
4. H. Tang, N. Yang, Z. Zhang, Z. Du, J. Shen (eds.), *5G NR and Enhancements—From R15 to R16* (Elsevier, 2021)
5. X. Lin, N. Lee (eds.) *5G and Beyond: Fundamentals and Standards* (Springer, 2021)
6. W. Chen, P. Gaal, J. Montojo, *Fundamentals of 5G* (McGraw-Hill, 2021)
7. C. Cox, *An Introduction to 5G: The New Radio, 5G Network and Beyond*, 1st edn. (Wiley, 2020)
8. S. Sorokin (ed.) *5G Radio Access Network Architecture: The Dark Side of 5G*, 1st edn. (IEEE Press, 2020)
9. E. Dahlman, S. Parkvall, J. Skold, *5G NR: The Next Generation of Wireless Technology*, 2nd edn. (Academy Press, 2020)
10. N. Blaunstein, C.G. Christodoulou, *Radio Propagation and Adaptive Antennas for Wireless Communication Networks* (Wiley, 2014)

Performance of RF Cables and IT Signals Lines in Radio Systems with EMI

1 Introduction

Cabling is a general term that involves not only cables themselves but also standards that regulate their quality and installation rules [1]. As discussed in Chapter "General Concept and Types of EMI Emissions", cables are common sources and victims of *EMI* conducted and radiated emissions, and their standardization at the domestic and international levels is very important. There are more types of cables, such as mains and DC power cables, low-voltage and high-voltage cables, and signal cables for communication, control, security, and safety systems.

The signal cables' quality installation rules are standardized mainly based on signal integrity and EMC standards. In this chapter, we focus on signal cabling for radio systems and related IT equipment. In *Appendix C*, we present the fundamental theory of signal transmission lines (*TLs*) and their electrical parameters and properties. In this chapter, we present different types of transmission lines that form the basis for developing practical signal cables.

The task of signal cables is to interconnect various radio and/or IT equipment. In terms of applications, there are two main groups of communication cables: RF signal cables and IT data cables. RF cables are typically coaxial cables, whereas IT data cables are typically twisted differential pairs (unshielded and shielded) in addition to fiber-optical cables. The EMI is not a concern for optical cables, as will be discussed at the end of this chapter.

In terms of the installation, there are external (outdoor) cables for interbuilding installations and indoor cables for in-building installations. The outdoor cables (especially long cables) are more prone to interact with the electromagnetic environment, and consequently, they are commonly placed in metallic conduits. Additionally, regarding EMI problems, the use of cable connectors as interfaces between cables and equipment is especially critical. Therefore, the important design and installation goal is that the connectors do not cause impedance mismatch or any

M. Pajovic, *Fundamentals of EMI Phenomena in Radio Communication Systems*, https://doi.org/10.1007/978-3-031-88748-2_8

discontinuity that may cause undesired cable radiation or cable coupling with the external EMI fields.

How shields of shielded cables are connected to the equipment enclosure is essential for the cable shielding efficiency and related equipment. This important topic will be discussed later in this chapter.

The writing of this chapter is motivated and influenced by references [1–14], which we recommend to readers interested in further reading these topics.

2 RF Coaxial Cables

As described in *Appendix C*, a coaxial cable is a cylindrical-concentric transmission line with two conductors, as shown in Fig. 1.

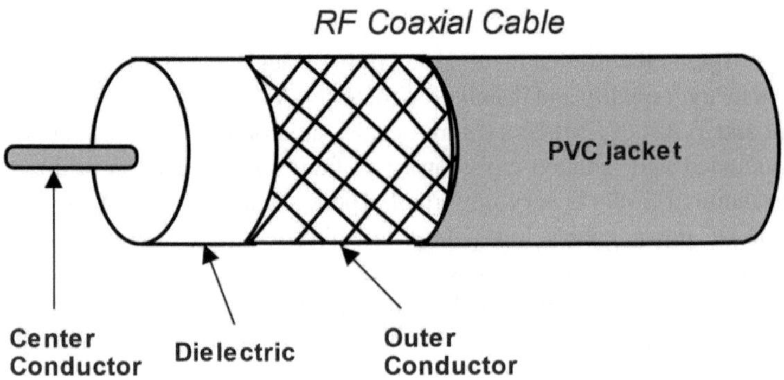

Fig. 1 Configuration of a coaxial TL

This *TL* structure is a wire (inner conductor) shielded all around by an outer conductor. A solid dielectric is placed between the inner and outer conductors to maintain fixed separation between the conductors. Notably, the dielectric has an essential influence on the cable's properties, such as its *characteristic* impedance and *propagation* constant, as discussed in *Appendix C*. Additionally, as noted in the same appendix, the coaxial line supports the TEM propagation mode up to a certain frequency (a cutoff frequency), and beyond this frequency, it becomes a waveguide that supports higher propagation modes. For instance, the lowest coaxial waveguide mode is TE_{11} with the cutoff frequency [2]:

$$f_c = \frac{c}{\sqrt{\varepsilon_r}} \cdot \frac{2}{(d+D)\pi} \tag{1}$$

where c is the speed of light and where ε_r is the relative permittivity of the dielectric. Additionally, d and D are the diameters of the inner and outer conductors, respectively. Thus, the coaxial cable only supports *TEM* propagation below the cutoff frequency f_c.

With respect to the current flow, the RF current flows from an RF source along the surface of the inner conductor to a TL load and returns to the source along the inner surface of the outer cable conductor (a shield). The signal attenuation along the coaxial cable directly depends on the resistances of the inner and outer conductors due to the skin effect, as well as the resistance of a TL lossy dielectric. As described in Chapter "Nonideal Passive Circuit Components", the skin effect can be the limiting factor in cable frequency performance, which may affect both the characteristic impedance and the propagation constant.

According to *Appendix C*, the *characteristic impedance* of any transmission line is defined as the ratio of voltage V and current I at any point on the *TL*. In terms of the per-unit-length parameters, series *impedance* $\Delta Z'$ and shunt *admittance* $\Delta Y'$, the *characteristic impedance* of a *lossy* coaxial *TL* can be expressed as

$$\tilde{Z}_c(\omega)=\sqrt{\frac{Z'(\omega)}{Y'(\omega)}}=\sqrt{\frac{R'_0+i\omega L'_0}{G'_d+i\omega C'_d}} \tag{2}$$

where R'_0 is the per-unit-length conductor resistance (including both inner and outer conductors) and where G'_d is the dielectric per-unit-length conductance. Additionally, L'_0 is the TL per unit-length loop inductance, and C'_d is the TL per-unit-lengtht capacitance between the conductors.

Figure 2 shows a lumped equivalent circuit for a lossy cable with lossy (parasitic) resistance and a lossy dielectric.

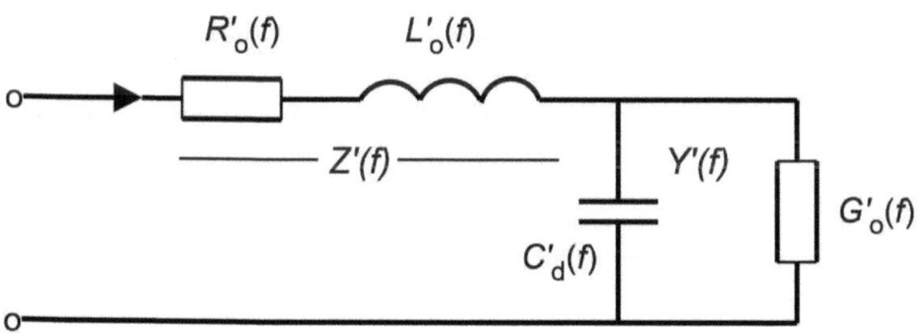

Fig. 2 Equivalent circuit for a lossy coax cable

The coaxial cable is *lossless* when $R'_0 = 0$ and $G'_d = 0$, and its *characteristic impedance Zc* can be simplified as follows:

$$Z_c=\sqrt{\frac{L'_0}{C'_d}} \tag{3}$$

The voltage and current are in phase along the cable for this case.

Additionally, the cable inductance L'_0 is related to the per-unit-length external (loop) inductance, which is created by the loop of the inner and outer conductors, whereas the internal inductance (described in Chapter "Nonideal Passive Circuit

Components") can be neglected because it is much smaller than the TL loop inductance.

Furthermore, Eq. (3) can be expressed in terms of the d and D diameters of a lossless cable when we substitute the following equations from *Appendix C* for per-unit-length inductance and capacitance, L'_0 and C'_d:

$$L_0' = \frac{\psi}{l} = \frac{\mu_0}{2\pi} \ln\left(\frac{D}{d}\right) \left(H/_m\right) \tag{4}$$

$$C_d' = \frac{\Delta q}{\Delta V} = \frac{2\pi\varepsilon_0\varepsilon_r}{\ln\left(\frac{D}{d}\right)} \left(\frac{F}{m}\right) \tag{5}$$

into Eq. (3) and obtain

$$Z_c = \frac{60}{\sqrt{\varepsilon_r'}} \ln\left(\frac{D}{d}\right) \quad (\Omega) \tag{6}$$

where ε'_r is the real relative dielectric constant.

Hence, any excessive mechanical or thermal deformation of the cable (conductors and/or dielectric) will change the cable axial symmetry and consequently change the cable dedicated *characteristic* impedance. Standard RF coaxial cables commonly have 50 and 75 Ω of impedance. The 50-Ω cables are mostly intended for RF applications and RF test equipment, whereas 75-Ω cables are commonly used in video and audio applications.

The *velocity of signal propagation* in the cable dielectric is defined in *Appendix C* as

$$v_d = \frac{1}{\sqrt{L_0' C_d'}} = \frac{c}{\sqrt{\varepsilon_r}} \tag{7}$$

where *c* is the speed of light.

Thus, any change in L'_0 and C'_d cable parameters due to cable mechanical or thermal deformation will also cause a change in the signal propagation velocity along the cable.

More information about RF coaxial cables can be found in [1–5].

3 EMI/RFI from Antenna Coaxial Cables

An RF cable is a standard line for connecting a radio system with an antenna or other RF equipment. Because coaxial cables are unbalanced transmission lines, they should be terminated by unbalanced loads or antennas with an impedance close to the cable's *characteristic* impedance. If this is not the case, differential (operating) mode currents inside the cable can be converted to common-mode currents on the cable shield such that the shield radiates EMI/RFI. Additionally, in this case, external fields may be coupled on the cable shield, inducing common-mode currents. The common-mode currents can be converted to differential-mode currents inside the cable in the form of EMI noise. Hence, the first case is associated with EMI/RFI emissions from the cable shield, whereas the second case is associated with low immunity to external EMI coupling.

Figure 3 shows a symmetric dipole antenna fed by a coaxial cable where the DM current flows inside the cable since the CM current flows outside the cable shield, causing RFI radiation.

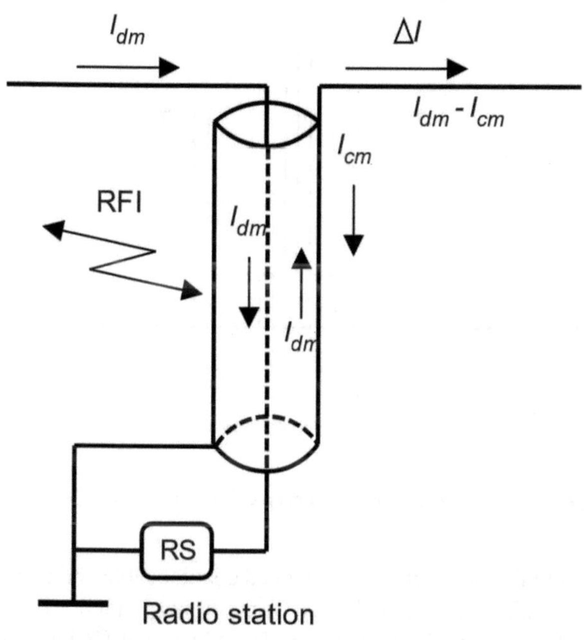

Fig. 3 Conversion of DM currents to CM currents at a coaxial cable

As shown in Fig. 3, the center-fed dipole has two identical legs whose impedances are equal with respect to the ground. If dipole terminals *A* and *B* are connected to a balanced (two-conductor) feed line with the same characteristic impedance to the input dipole impedance, then there is no common-mode current flow outside of the shield, and consequently, there is no RFI/EMI radiation from the shield. However, when the coaxial (unbalanced) cable is directly connected to the terminals of the center-fed dipole (as shown in Fig. 3), the situation is different: the common-mode currents flow outside the shield surface, causing shield radiation. The cable's immunity to the external RFI/EMI is also degraded in this situation.

Figure 3 shows that the differential-mode current I_{dm} flows from the antenna's terminal A to the radio station (RS) unit down over the inner conductor. When this current reaches the RS unit, I_{dm} flows back to the antenna's terminal B over the inner surface of the shield. At terminal B, the current I_{dm} mainly flows into the second dipole leg but also partially flows back as the CM current to the RS unit over the outer surface of the shield. This CM current causes undesired shield radiation.

For example, the bad scenario of RFI/EMI radiation also occurs when the antenna cable has a length that is a multiple of a half wavelength of the operating wavelengths. The cable shield then acts as a multiresonant wire antenna for RFI/EMI. Additionally, if the cable is vertically positioned to the ground and is long at multiple-quarters of the wavelength, it may act as a vertical resonant "antenna" for RFI/EMI. This case is shown in Fig. 4.

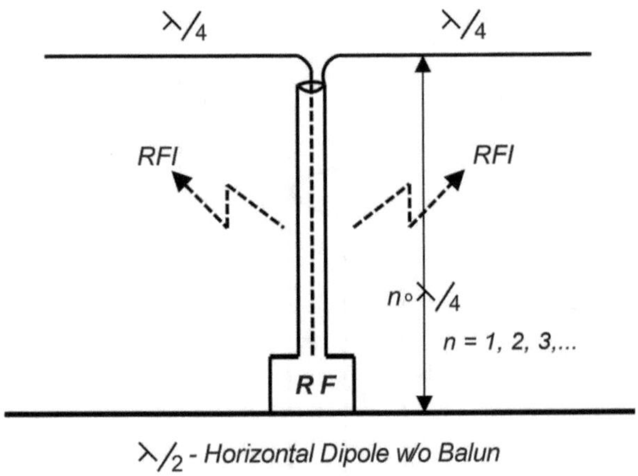

Fig. 4 Direct connection of coaxial cable to a center-fed antenna

3.1 *Balanced-to-Unbalanced Transformer (Balun)*

To avoid discontinuity between an unbalanced coaxial cable and a balanced input of the antenna, a balanced-to-unbalanced transformer called "balun" is usually employed. The term "balun" is an acronym for BALanced-UNbalanced. In addition to the balanced-to-unbalanced signal conversion, the baluns may be impedance transformers with an impedance ratio (*n: m*), where *n* and *m* are integer numbers. For the common cases, the impedance ratios are 1:1, 1:2, 1:4, …, 1:9, …, or 2:1, 4:1. 9:1, etc.

There are two electrical types of baluns: *voltage* and *current*. The *voltage* baluns are magnetic-flux-linked transformers. Figure 5 shows the concept of a *voltage* balun that produces equal voltage magnitudes with opposite polarity at antenna terminals *A* and *B* relative to the grounded shield and the transformer's center.

Therefore, if the impedance of the antenna legs is perfectly balanced with respect to the ground, then the current flow into the balun (from the coaxial output *C*) will be equal and with opposite polarity at terminals *A* and *B*. As a result, there is no undesired current flow outside the shield surface toward the RS unit, and consequently, there is no radiation from the shield.

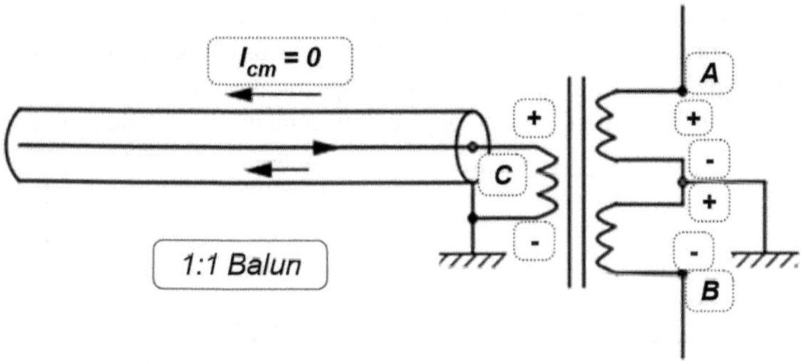

Fig. 5 Basic concept of a (1:1) voltage balun

Furthermore, Fig. 6a, b shows another topology of (1:1) balun, which functions in the same way as the balun shown in Fig. 5, but one is winded on a toroidal ferrite core.

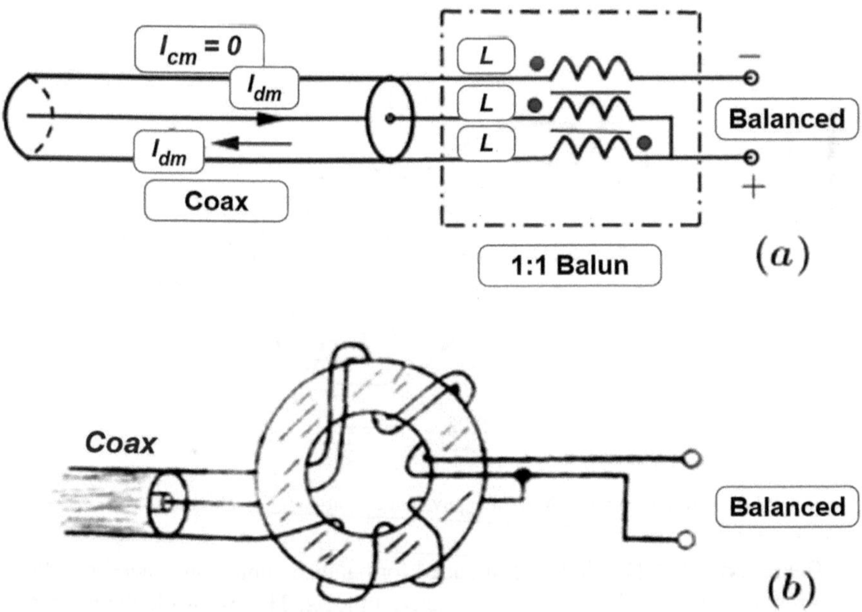

Fig. 6 (**a**) Circuit of the voltage balun and (**b**) the balun with a toroid core

Figure 6a shows the circuit of the *voltage* balun where three identical, well-coupled coils L are wounded on a ferrite core. The toroid ferrite core (shown in Fig. 6b) is a good solution because it is a closed magnetic circuit with high permeability and minimal magnetic *flux leakage*. As discussed in Chapters "Nonideal Passive Circuit Components" and "EMI Filters—Theory and Design", saturation of the toroid core by overrated currents should be avoided. Typically, manufacturers obtain power or current rates and frequency ranges for particular antenna baluns.

Another type of (1:1) balun is a *current-type* balun. Figure 7a, b shows (a) the balun equivalent circuit and (b) the physical placement on a ferrite toroid.

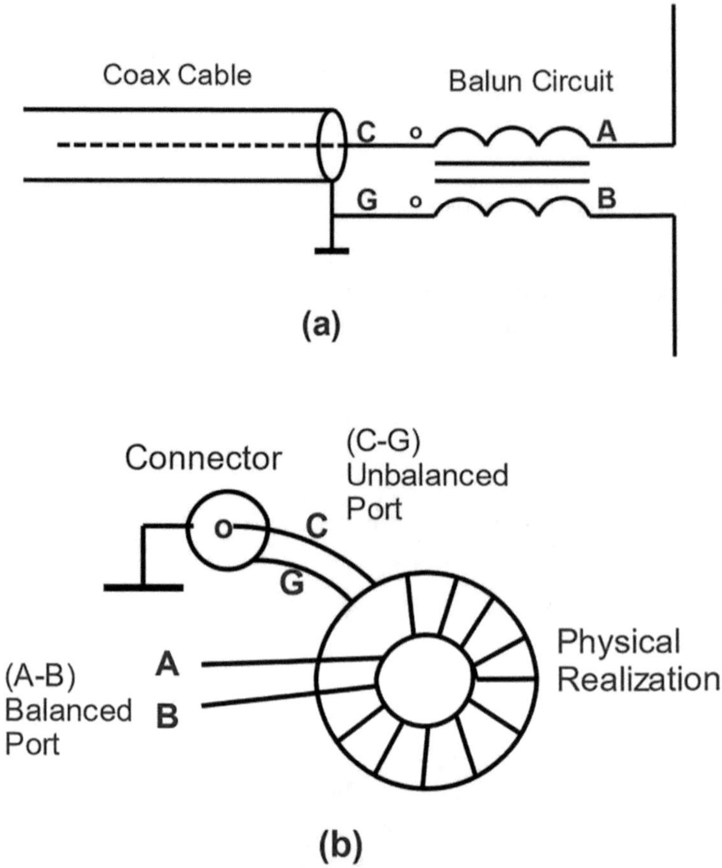

Fig. 7 (**a**) (1:1) Current balun with (**b**) an equivalent circuit

The described (1:1) baluns are intended for balanced antennas placed parallel to the ground and away from larger conductive objects. This is an idealized case. In real installations, one of the antenna legs is usually placed closer to the ground or larger conducting objects than the other leg. As a result, the antenna balance is distracted regardless of whether the proper balun is used. For this situation, we can use ferrite sleeves (chokes) on the coaxial cable instead of the impedance balun, as shown in Fig. 8.

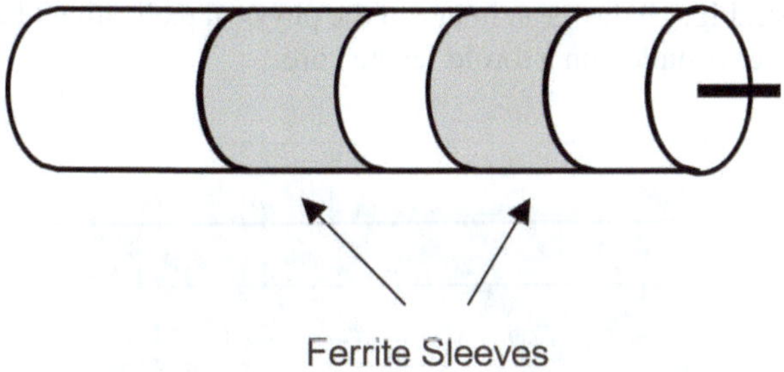

Fig. 8 CM ferrite sleeve choke instead of a balun

By using ferrite slaves or snap-on cable ferrites, the impedance of the outer shield surface significantly increases for CM currents, resulting in better suppression.

3.2 *Unbalanced-to-Balanced (1:4) Transformer*

When we need to feed an antenna with a different impedance than the characteristic impedance of a coax cable, for example, an off-center-fed dipole whose impedance is ~200 Ω and a (4:1) balun transformer can be used to obtain an impedance of 50 Ω. The circuit of this balun is shown in Fig. 9. Here, a ferrite sleeve (choke) is used for the suppression of common-mode currents. It is also shown.

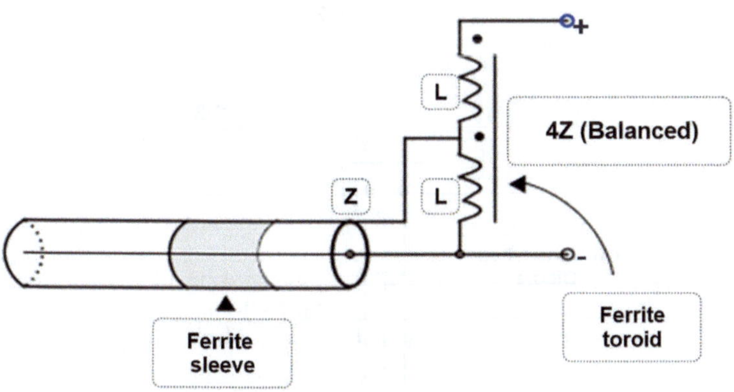

Fig. 9 (1:4) Impedance transformer

Furthermore, Fig. 10 shows an image of the physical realization of a (4:1) impedance transformer installed on a toroid ferrite core.

Fig. 10 4:1 Toroid impedance transformer (the author's construction)

For example, Fig. 11 shows the present impedance transformer applied for an off-center fed dipole with an input impedance of ~200 Ω.

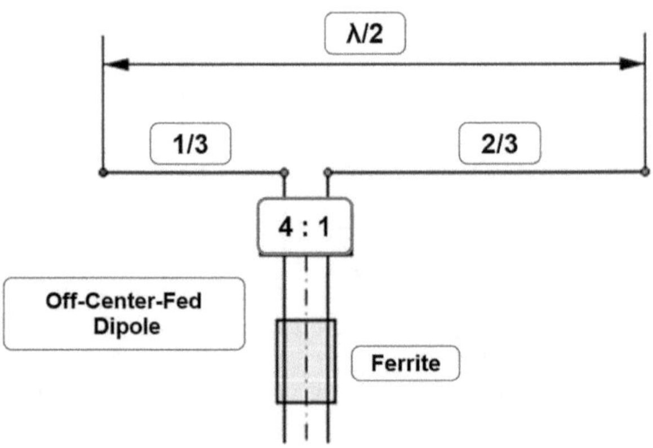

Fig. 11 An off-center-fed dipole

Additionally, Fig. 12 shows the case in which commercial snap-on ferrites are placed on the coaxial cable next to the (4:1) balun to increase the resistance of the shield to common-mode currents.

Fig. 12 Snap-on cable ferrites with a (4:1) balun (the author's construction)

Notably, the power or current rate are not critical parameters for snap-up common-mode ferrites because they manage significantly smaller CM currents on cable than the DM operating cable currents do.

3.3 *Unbalanced-to-Unbalanced (Un-Un) Transformer*

When we need to connect RF equipment to an unbalanced antenna, such as a monopole antenna, we do not need a balun but rather an unbalanced-to-unbalanced impedance transformer, shortly called the Un-Un transformer. Figure 13 shows a standard unbalanced monopole (Marconi) antenna with an input impedance of ~36 Ω fed by a 75-Ω coax cable.

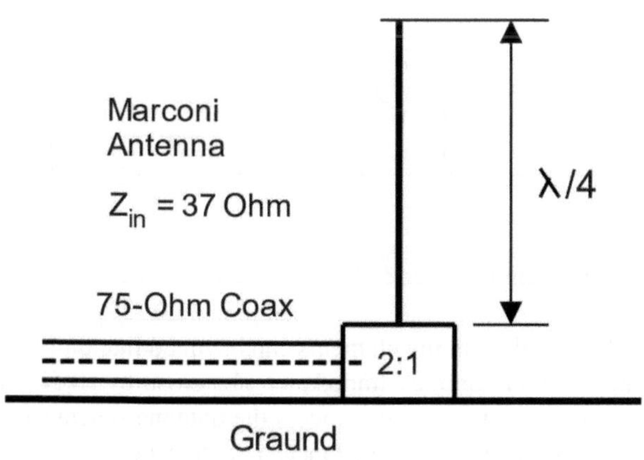

Fig. 13 (2:1) Un-Un transformer for the Marconi antenna

Here, a (1:2) Un-Un transformer is used. Un-Un transformers can also be built with other impedance ratios, such as 1:4, 1:6, and 1:9, or in oposite, 4:1, 6:1, and 9:1.

3.4 Antenna Tuners for Impedance Matching Broad-Band Antennas (Long Wires or Long Whips)

The methods of impedance matching of narrowband antennas are discussed in the the proceeding subsections, but broadband antennas, such as long wire antennas of arbitrary length, are commonly used in practice. We discuss these antennas in Appendix B.

The impedance matching for the broadband antennas is obtained by a manual or automatic antenna impedance tuner. The two common topologies of the antenna tuners are given in Fig. 14a, b.

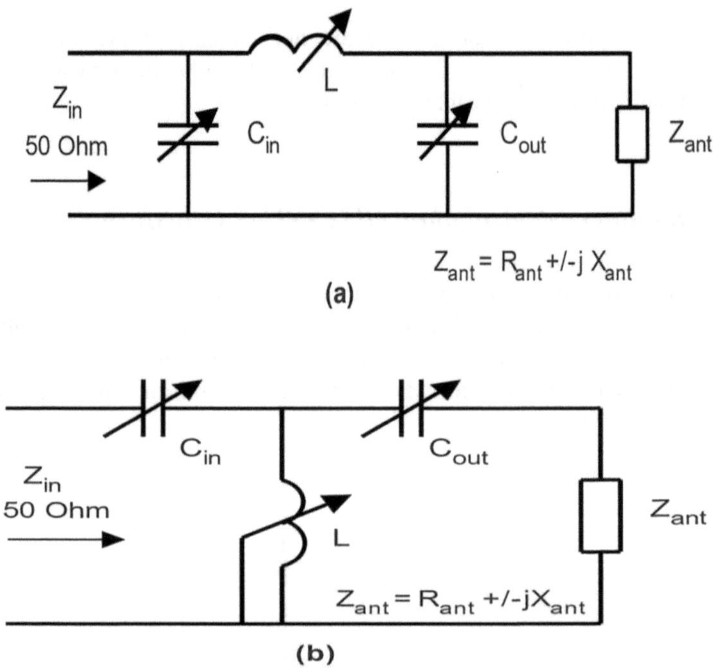

Fig. 14 (**a**) Pi- and (**b**) T-antenna tuner LC networks

The adjustable L- and C-tuning elements can be tuned manually or automatically to transform the antenna complex impedance Z_{ant} close to 50 Ω. The adjustment process is completed when the tuner cancels the antenna reactance X_{ant} and transforms the real part of the antenna impedance (R_{ant}) to 50 Ω.

Figure 15 shows the concept of an automatic antenna tuner where the process of the antenna matching to a radio station impedance is controlled and conducted by a microcomputer (CPU) to detect signals related to a phase, real part antenna impedance R, and VSWR. The phase detector and forward and reflected power detectors constantly monitor and send data to the CPU about the phase, forward, and reverse

RF signal powers. The process of antenna tuning is complete for frequencies when the antenna phase is close to zero, the antenna impedance Z_{ant} is close to 50 Ω, and the VSWR is close to 1.

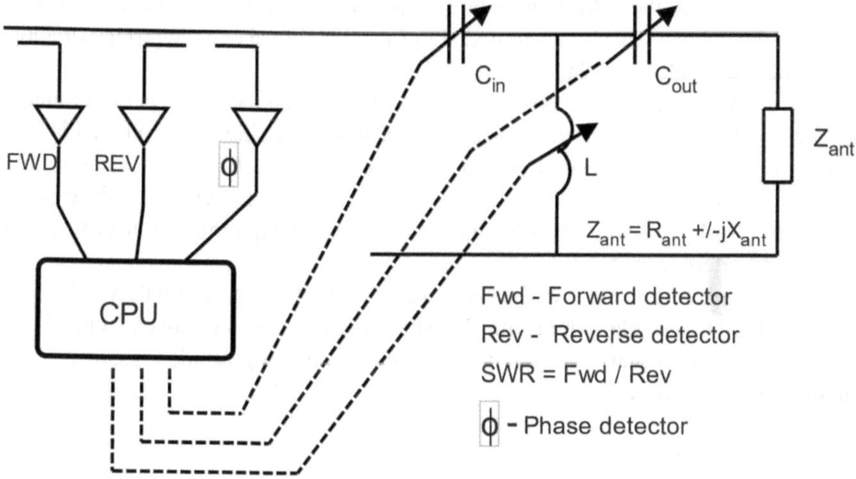

Fig. 15 The concept of an automatic antenna tuner

Interested readers are referred to references [3, 4] for more detail on the antenna impedance matching and coaxial cable radiation.

4 Coaxial Cables for Interconnecting RF Equipment

Coaxial cables are also used for connecting RF electronic equipment; for example, Fig. 16 shows a horizontal cable above the ground with induced common-mode currents at the shield due to an external field E_{inc}.

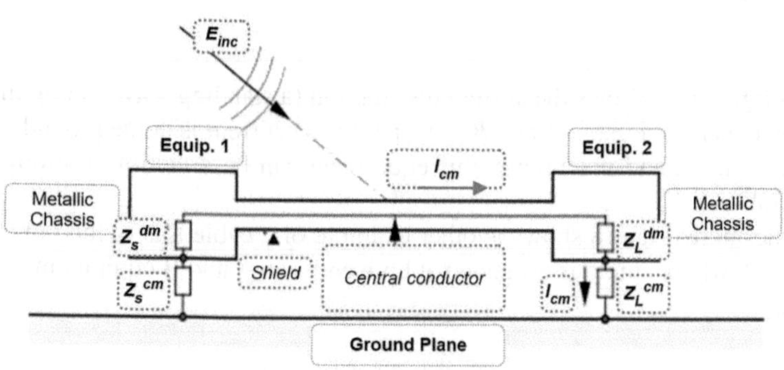

Fig. 16 Equipment 1 and 2 connected with a coaxial cable

We consider here electrically long cables whose shield forms a "transmission line" with the ground. If the shield is terminated at both ends by its common-mode *characteristic* impedance Z_c^{cm}, the external incident field will have less effect on the operating signal inside the cable. However, this termination is more theoretical than practical because the CM termination impedance is hardly predictable. This is because the cable is not always routed at the same height above the ground or above the same flat conductive plane.

The most critical situation for a coaxial cable is when the shield is not grounded at one end (the case of an open-ended TL) or when the shield is grounded at both ends (the case of a short-ended TL). However, both termination methods may cause shield-to-ground *TL* resonance, resulting in stronger EMI radiation, as discussed in *Appendix C*. More about cables for interconnecting RF equipment concerning EMI is given in references [5–8].

For example, Fig. 17 shows the coaxial cable length of approximately $\lambda/4$, where λ is the smallest wavelength of interest. The left end of the shield is grounded over the chassis of equipment A, whereas the other end is not grounded because equipment B is not grounded.

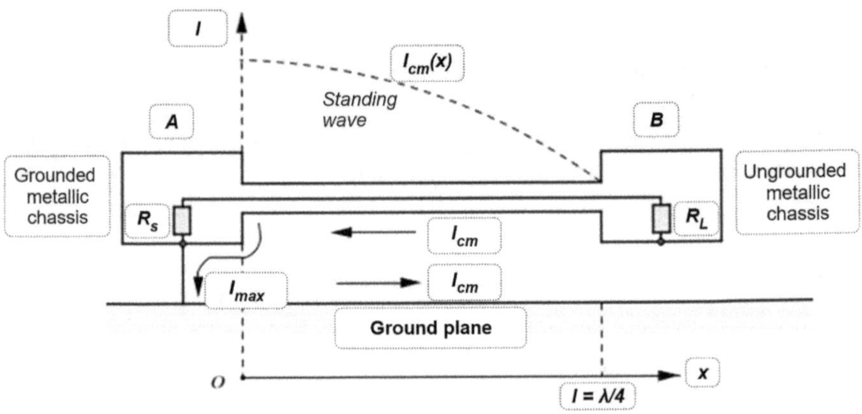

Fig. 17 Ungrounded shield at one end

This figure also shows the current distribution (a standing wave) along the cable shield between equipment *A* and *B*. The maximum current is at the ground terminal of equipment A, and potentially, this equipment can be a victim of strongly *conducted EMI*.

Furthermore, Fig. 18 shows another example of a cable that is long at approximately $\lambda/2$, whose shield is grounded at both ends via A and B equipment grounded chassis.

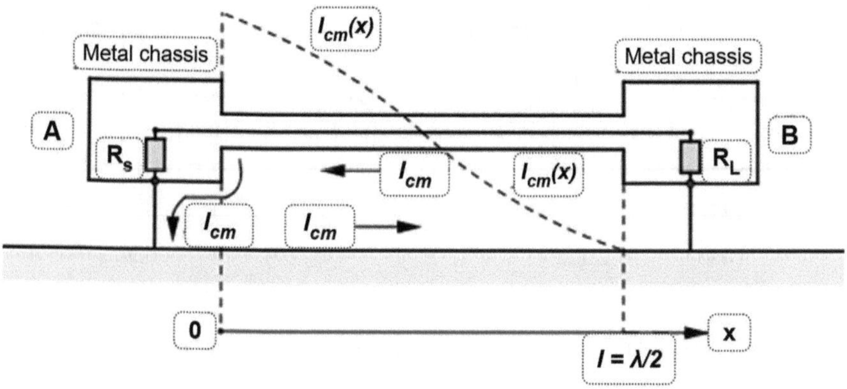

Fig. 18 Grounded shield at both ends

This figure also shows the current distribution (a standing wave) along the shield routed above ground. In this case, both pieces of equipment can be victims of EMI because the current maximum are at the equipment ground terminals.

In the end, it would be a better solution (not always practical) to ground the cable shield every $\lambda/10$ where λ is the smallest wavelength of interest. Additionally, to minimize EMI effects associated with shielded cables, placing the coaxial cable into a metallic conduit or on a duct or a conductive tray is good, as will be shown later in the next section. Additionally, this method is theoretically explained in Chapter "General Concept and Types of EMI Emissions".

4.1 *Coaxial Cable Terminations with Pigtails Versus Connectors*

Owing to its simplicity, the coaxial cable shield is sometimes connected to the equipment chassis at one point only. This is not always a good solution, especially for high-frequency signals. The perfect solution for the cable shield is to have a 360° connection to the equipment chassis. This concentric connection is commonly obtained by coaxial connectors standardized for a specific cable size, power, and frequency range. We will later show various types of standard coaxial connectors depending on their size and frequency range.

We first analyze the nonconnector cable connection when a cable shield is connected to an equipment chassis at one point only. This connection is called a "pigtail" and is shown in Fig. 19.

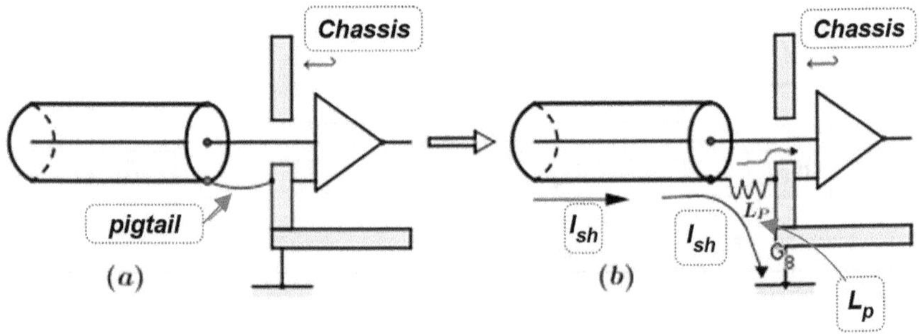

Fig. 19 (**a**) Cable shield grounding by "pigtail," (**b**) Equivalent circuit of "pigtail"

Figure 19a shows a "pigtail" connection of the cable shield to the equipment (conducted) chassis by a piece of wire long *l*. This simple connection can be satisfactory for low-frequency applications, but it can be critical for shielding radiation and EMI equipment immunity for high-frequency applications. Additionally, Fig. 19b shows the equivalent inductive impedance of the pigtail.

The voltage developed at the pigtail is proportional to the pigtail inductive reactance ωL_p and common-mode shield current I_{sh} that flows along the pigtail into the equipment. As a result, the developed voltage V_p on the pigtail usually drives the current noise into the equipment. The voltage can be expressed as a function of the pigtail inductive reactance ωL_p and external shield current I_{sh}:

$$V_p = (\omega L_p) \cdot I_{sh} \tag{8}$$

Additionally, the *transfer shield impedance* is here defined as the ratio between the voltage V_n of the noise penetrating the cable and the external (CM) shield current I_{sh}:

$$Z_T = \frac{V_n}{I_{sh}} = \omega L_p \tag{9}$$

Hence, the *transfer shield impedance* is associated with the pigtail inductive impedance. The EMI issue related to pigtail progressively increases with frequency. To minimize noise flow into equipment circuits, the inductance (i.e., length) of the pigtail should be minimized as much as possible. However, to eliminate the pigtail effect completely, the connection between the cable shield and the equipment chassis is shown in Fig. 20, where the external pigtail is physically eliminated. The shield is now connected to the chassis by a conductive clamp.

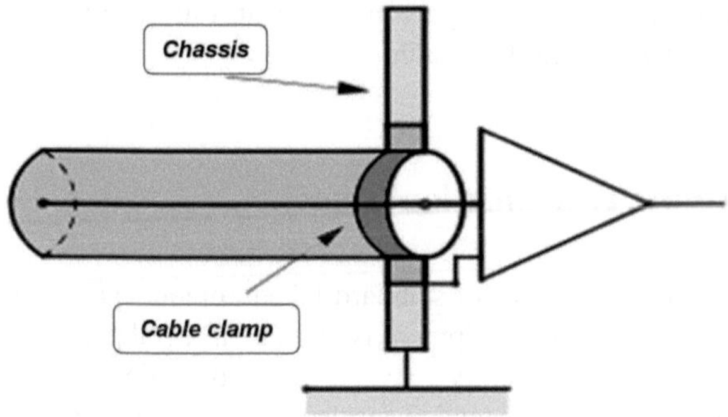

Fig. 20 Zero-pigtail connection of the shield by a clump

However, the best and most reliable connections of the coaxial cables to equipment are via standard coaxial connectors, which provide 360° shield connections to the equipment chassis.

Table 1 lists the sizes and maximal operating frequencies of common standard connectors.

Table 1 Standard coaxial connectors and their maximal operating frequencies

Connector type	Size	Maximum frequency
UHF/mini-UHF	Standard	1 GHz/4 GHz
C	Standard	4 GHz
BNC	Miniature	4 GHz
SMB	Subminiature	4 GHz
N	Standard	10 GHz
TNC	Miniature	10 GHz
SMA, SMC	Subminiature	10–20 GHz

Additionally, Fig. 21 shows images of some standard coaxial connectors available on the market.

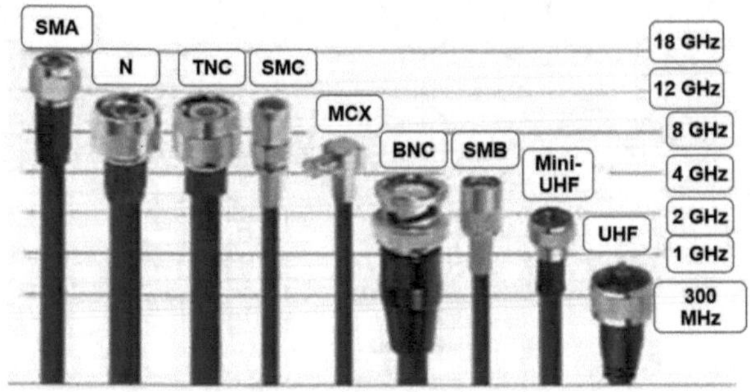

Fig. 21 Standard RF and microwave coaxial connectors

More information about the topics discussed in this section is given in [6, 8], which we recommend for further reading.

5 Twisted-Pair Transmission Lines

Here, we discuss the properties of standard IT equipment cables (such as twisted-differential pairs) because IT equipment is often an integral part of a radio communication system and consequently may be exposed to EMI and RFI emissions. Therefore, IT cables are also intended for data transfer between IT equipment and radio systems. Unlike unbalanced coaxial cables, as discussed in the preceding sections, twisted pairs are balanced transmission lines with two conductors tightly twisted. In practice, more balanced twisted pairs create a multitwisted differential-pair cable. The common case is a cable with four twisted differential pairs: unshielded twisted pairs (UTP) and shielded twisted pairs (STP).

5.1 Unshielded Twisted-Pair Cables (UTP)

We discuss the coupling (crosstalk) between two straight and adjacent differential lines in Chapter "General Concept and Types of EMI Emissions". To reduce the crosstalk between two or more adjacent differential lines, it is effective when differential line conductors are tightly twisted to each other. Furthermore, more twisted pairs can be twisted into each other, as in CAT-5, CAT-6, and CAT-7 cables [8].

Figure 22 shows two twisted differential pairs (next to each other) operating in differential mode and undesired common mode.

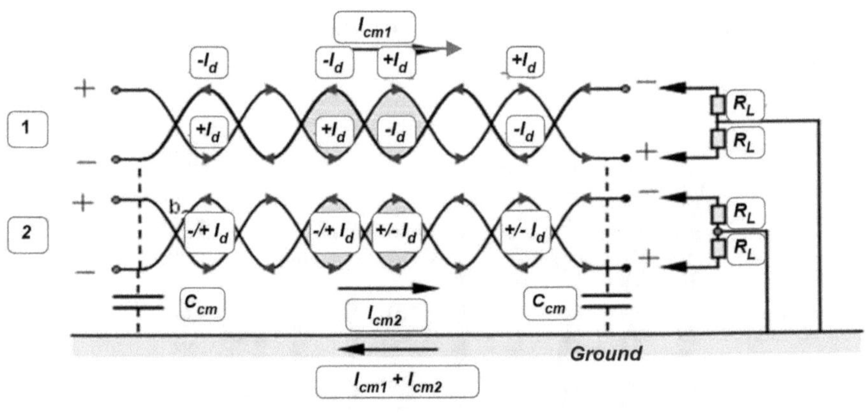

Fig. 22 Two adjacent balanced twisted pairs

Each twisted pair is divided into small loops (twists) where the differential currents I_d in adjacent loops have opposite directions, resulting in the significant minimization of inductive coupling between the two differential pairs. Thus, this is a very effective way to reduce inductive coupling between adjacent twisted pairs and their environment.

Additionally, for frequencies where twist loops are very small in terms of the signal operating wavelengths, the unshielded twisted wires have a uniform distribution of capacitance to support conductive structures, such as conductive ducts, conduits, or trays, leading to better control of capacitive coupling between adjacent twisted pairs. Additionally, if the untwisted termination areas at the cable ends are minimized enough and the external source and load circuits are well balanced, CM capacitive coupling between two twisted pairs becomes minimal.

Notably, the inductive coupling between two adjacent twisted pairs almost disappears if the differential pairs currents have opposite directions, as shown in Fig. 22.

5.2 *Mode Conversion at a Twisted-Pair Due to Different Lengths of Its Conductors*

We discuss possible sources of mode conversion at differential lines in Chapter "General Concept and Types of EMI Emissions". In this case, mode conversion is commonly produced because of the different lengths of conductors of a twisted pair created in the winding process, resulting in untwisted conductors with different lengths at the cable ends. In the ideal case, when the lengths of twisted conductors 1 and 2 are identical, the differential signals at the load end (receiver) are out of phase 180°. As a result of this ideal case, the desired differential mode exists only on the cable.

However, in reality, the differential line conductors have somewhat different lengths, as shown in Fig. 23, causing some differential-mode currents to be converted into unwanted common-mode currents.

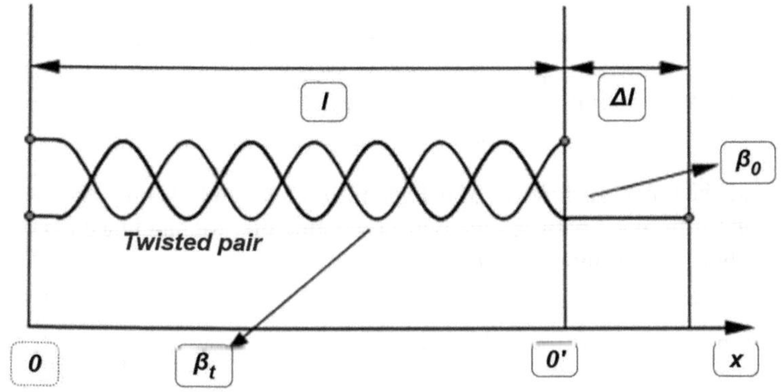

Fig. 23 Mode-conversion due to mismatch between the TL conductor lengths

Assuming that the present differential line is lossless, the voltages of the *direct* wave at the source end (for $x = 0$) are as follows:

$$\tilde{V}_1^+(\omega, x = 0) = V^+{}_1\, e^{i\omega t} \tag{10a}$$

$$\tilde{V}_2^+(\omega, x = 0) = V^+{}_2\, e^{i(\omega t + \pi)} \tag{10b}$$

and the voltages at the load end (for $x = l$) are

$$\tilde{V}_1^+(\omega, x = l) = V^+{}_1 e^{i(\omega t - \beta_t l)} \tag{11a}$$

$$\tilde{V}_2^+(\omega, x = l + \Delta l) = V^+{}_2\, e^{i(\omega t + \pi - \beta_t l - \beta_0 \Delta l)} \tag{11b}$$

where V^+_1 and V^+_2 refer to the voltage amplitudes of the *direct wave* at twisted conductors 1 and 2, respectively. Additionally, l is the length of the twisted pair and Δl is the length mismatch between conductors 1 and 2. Additionally, β_t is the propagation constant on the differential pair, and β_0 is the propagation constant at the Δl part of Conductor 2, which is split from Conductor 1.

Furthermore, starting from the definition of the DM-CM *conversion factor* as the ratio of voltages at the receiver and source ends [2]:

$$\tilde{K}_{dm-cm} = \frac{\tilde{V}_1^+(\omega, x = l) + \tilde{V}_2^+(\omega, x = l + \Delta l)}{\tilde{V}_1^+(\omega, x = 0) - \tilde{V}_2^+(\omega, x = 0)} \tag{12}$$

where the sign "+" denotes *direct* propagating waves along the *TL,* we derived the expression for the mode conversion factor at a differential TL with the mismatch between the TL conductor lengths.

Considering here a lossless differential pair, $V^+_1 = V^+_2 = V^+$. Furthermore, substituting Eqs. (10a), (10b) and (11a), (11b) into Eq. (12), the DM–CM conversion factor $\tilde{K}_{dm-cm}$ becomes

$$\begin{aligned} \tilde{K}_{dm_c m} &= \frac{V^+ e^{i(\omega t - \beta_t l)} + V^+ e^{i(\omega t + \pi - \beta_t l - \beta_0 \Delta l)}}{V^+ e^{i\omega t} - V^+ e^{i(\omega t + \pi)}} \\ &= \frac{e^{-i\beta_t l} - e^{-i(\beta_t l + \beta_0 \Delta l)}}{2} \rightarrow \\ \tilde{K}_{dm\_cm} &= \frac{1}{2}\left(1 - e^{-i\beta_0 \Delta l)}\right) e^{-i\beta_t l} \end{aligned} \tag{13}$$

Because the length difference Δl between two conductors is only relevant for the signal phase difference between the two TL conductors, we can place $l = 0$ into Eq. (13) and obtain the simplified form for $\tilde{K}_{dm\_cm}$:

$$\tilde{K}_{dm_{cm}} = \frac{1}{2}\left(1 - e^{-i\beta_0 \Delta l)}\right) \tag{14}$$

Furthermore, using Euler's identity, Eq. (14) becomes

$$\tilde{K}_{dm_{cm}} = \frac{1}{2}\left[1-\left(\cos\left(\beta_0\Delta l\right) - i\sin(\beta_0\Delta l)\right)\right] \tag{15a}$$

$$\left|\tilde{K}_{dm_{cm}}\left(\Delta l\right)\right| = \frac{1}{2}\sqrt{\left[1-\cos\left(\beta_0\Delta l\right)\right]^2 + \left[sin\left(\beta_0\Delta l\right)\right]^2} \tag{15b}$$

where $\left|\tilde{K}_{dm\_cm}\left(\Delta l\right)\right|$ is the magnitude of the DM-CM conversion factor.

Furthermore, we can reduce Eq. (15b) using the identity $cos(\alpha)^2 + sin(\alpha)^2 = 1$, and the magnitude of the mode conversion factor takes the following final form:

$$\left|K\left(\Delta l\right)\right| = \frac{1}{2}\sqrt{\left[2-2\cos\left(2\pi\frac{\Delta l}{\lambda_0}\right)\right]} \tag{15c}$$

where $\beta_0 = \frac{2\pi}{\lambda_0}$. Eq. (15c) shows that

$$\left|K\left(\Delta l\right)\right| \rightarrow 0 \;\; when \;\; \Delta l \rightarrow 0$$

Additionally, when the mismatch length Δl is much smaller than the wavelength of interest ($\Delta l \ll \lambda_{min}$), the DM-CM *conversion* factor approaches zero. In this case, the phase difference of the signals at the far end of the twisted pair approaches 180°.

Figure 24 shows the plot of the DM-CM conversion factor $|K(\Delta l)|$ as a function of the mismatch length Δl given in the wavelength λ.

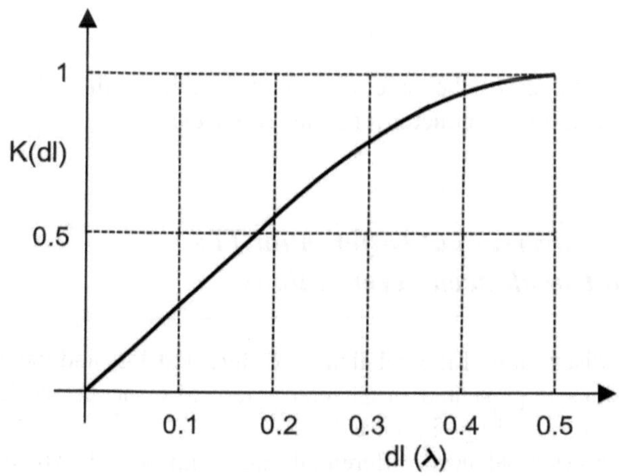

Fig. 24 DM-to-CM conversion factor K due to the mismatch of TL conductors' lengths

The mode-conversion factor ranges between *0* and *1* for $0 < \Delta l < 0.5\ \lambda$, where λ is the wavelength of interest. Thus, 100% of the DM signal energy will be converted into undesired CM energy when $\Delta l = 0.5\ \lambda$. Furthermore, by increasing the mismatch length Δl beyond 0.5 λ, the mode conversion decreases and reaches zero again when $\Delta l = \lambda$. Regardless of whether the mode conversion is now *zero* for $\Delta l = \lambda$, this case is not acceptable because the TL cannot perform correctly.

Table 2 shows the dependency of the DM-CM conversion factor on TL conductor mismatch length in terms of the signal wavelength, resulting in a signal phase difference $\Delta\varphi$ at the load end.

Table 2 DM-CM conversion factor dependent on a TL conductor mismatch length in terms of the signal wavelength and signal phase difference $\Delta\varphi$ at the load

Δl	$\Delta\varphi$	$\approx K_{dm-cm}$
$0 \cdot \lambda$	180°	0%
$0.01 \cdot \lambda$	183°	≈3%
$0.03 \cdot \lambda$	190°	≈9%
$0.05 \cdot \lambda$	198°	≈15%
$0.1 \cdot \lambda$	210°	≈31%

The mismatch length $\Delta l = 0.01\ \lambda_{min}$ contributes to ~3% of the DM-CM conversion. Furthermore, ~30% of the DM energy is converted into CM energy when the length mismatch Δl increases to 0.1 λ, which is not acceptable. It is recommended that the length mismatch be up to 0.01 λ. For example, the standard UTP Cat-6 cable is specified for frequencies up to 300 *MHz*, which corresponds to the wavelength of

$$\lambda_{\min} = \frac{c}{f_{\max}} \cong \frac{c}{300 \cdot 10^6} = 1m$$

Hence, for this case, the safe mismatch length of the TL conductors is $\Delta l < 0.01\ \lambda_{min}$, i.e., for a conductor long 1m, $\Delta l < 1\ cm$.

5.3 *Mode Conversion at Differential TLs Due to Unbalanced Terminations*

A preliminary discussion of the unbalanced differential TLs and the corresponding mode conversions is presented in Chapter "General Concept and Types of EMI Emissions". Here, we will go in more detail.

Figure 25 shows a balanced differential pair composed of two weakly coupled single-ended lines with the same *characteristic* impedance Z_c. Additionally, the pair lines share the same reference conductor and are terminated by different load impedances Z_{L1} and Z_{L2}.

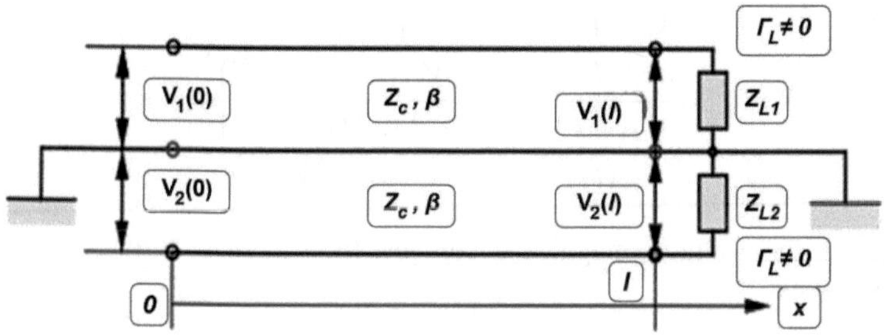

Fig. 25 Unbalanced-loaded differential pair

Here, the voltages of *direct* and *reflected* waves at arbitrary loads for lossless lines 1 and 2 are expressed as

– for *direct* (incident) waves:

$$\tilde{V}_1^+(l) = V_1^+(l) \cdot e^{-i\beta l} \tag{16a}$$

$$\tilde{V}_2^+(l) = V_2^+(l) \cdot e^{i(\pi-\beta l)} = -V_2^+(l) \cdot e^{-i\beta l} \tag{16b}$$

– for *reflected* waves:

$$\tilde{V}_1^-(l) = V_1^-(l) \cdot e^{i\beta l} \tag{16c}$$

$$\tilde{V}_2^-(l) = V_2^-(l) \cdot e^{i(\pi+\beta l)} = -V_2^-(l) \cdot e^{i\beta l} \tag{16d}$$

– where signs "+" and "–" represent the voltages of the *direct* and *reflected* waves, respectively, at loads Z_{L1} and Z_{L2}.

Furthermore, substituting Eqs. (16a), (16b) into the expression for the DM-to-CM conversion factor:

$$\begin{aligned} \tilde{K}_{dm-cm} &= \frac{V_1^-(l)e^{-i\beta l} - V_2^-(l) \cdot e^{-i\beta l}}{V_1^+(l)e^{i\beta l} + V_2^+(l)e^{i\beta l}} \\ &= \frac{V_1^-(l) - V_2^-(l)}{V_1^+(l) + V_2^+(l)} e^{-i2\beta l} \end{aligned} \tag{17a}$$

Equation (17a) can be expressed in terms of the *reflection coefficients* Γ_1 and Γ_2 at load Z_{L1} and Z_{L2}, respectively, as

$$\left|\tilde{K}_{dm-cm}\right| = \frac{\dfrac{V_1^-(l) - V_2^-(l)}{V_1^+(l)V_2^+(l)}}{\dfrac{V_1^+(l) + V_2^+(l)}{V_1^+(l)V_2^+(l)}} = \frac{\dfrac{1}{V_2^+(l)}\Gamma_1 - \dfrac{1}{V_1^+(l)}\Gamma_2}{\dfrac{V_{1+}(l) + V_{2+}(l)}{V_{1+}(l)V_{2+}(l)}} \tag{17b}$$

Because the voltage magnitude for low-loss TLs is $V_1^+(l) \approx V^+_2(l)$, Eq. (17b) becomes

$$\left|\tilde{K}_{dm-cm}\right| \cong \frac{1}{2}\left|\Gamma_1 - \Gamma_2\right| \tag{18a}$$

where the reflection coefficients at loads Z_{L1} and Z_{L2} are Γ_1 and Γ_2, respectively (*Appendix C*):

$$\Gamma_1 = \frac{V_1^-(l)}{V_1^+(l)} = \frac{Z_{L1} - Z_c}{Z_{L1} + Z_c};\ \Gamma_2 = \frac{V_2^-(l)}{V_2^+(l)} = \frac{Z_{L2} - Z_c}{Z_{L2} + Z_c} \tag{18b}$$

Furthermore, substituting Eq. (18b) into Eq. (18a), the DM–CM conversion factor becomes

$$\left|\tilde{K}_{dm-cm}\right| \cong \frac{1}{2}\left|\left(\frac{Z_{L1} - Z_c}{Z_{L1} + Z_c}\right) - \left(\frac{Z_{L2} - Z_c}{Z_{L2} + Z_c}\right)\right| \tag{18c}$$

Notably, if both lines of a differential pair are not perfectly terminated but have the same load impedance Z_L, the balance of the line pair is not disrupted, i.e., the mode conversion is 0. This does not mean that this is the correct impedance termination. The lines of the TL should also be terminated by the impedance equal to their characteristic impedance, as noted in *Appendix C*. In extreme situations, the impedance mismatch of a differential pair can enhance resonance effects, which can result in significant EMI radiation from the TL.

Example 1 Equation (18c) is used to find the DM-CM *conversion factor* for the differential pair with the characteristic impedance $Z_{dm} = 100\ \Omega$, whose single-ended lines have the characteristic impedance $Z_{c,\,\mathrm{sl}} = 50\ \Omega$ terminated as follows:

(a) By the impedance of 50 Ω equal to their characteristic impedance Z_c,.
(b) Both by the same impedance of 80 Ω.
(c) One line by the impedance $Z_{L1} = 50\ \Omega$ and another line by the impedance $Z_{L2} = 55\ \Omega$.
(d) One line by the impedance $Z_{L1} = 40\ \Omega$ and another line by the impedance $Z_{L2} = 60\ \Omega$.
(e) One line is perfectly terminated by 50 Ω, and the other line is unterminated ($Z_{L2} = \infty$).
(f) One line is perfectly terminated by 50 Ω since the other line is short-ended ($Z_{L2} = 0$).

The results of K_{dm-cm} factor are listed in Table 3 according to Eq. (18c).

Table 3 Results for K_{dm-cm} factor according to Eq. (18c)

$Z_{L1}(\Omega)$	$Z_{L2}(\Omega)$	$\lvert K_{dm-cm}\rvert$	$Z_{dm} = 100\ \Omega$ $Z_{c,\,sl} = 50\ \Omega$
50	50	0%	
80	80	0%	
52	55	1.4%	
40	60	10%	
50	∞	50%	
50	0	50%	

When both loads have identical impedances (even 0 or ∞), the mode conversion factor is zero, i.e., the differential *TL* remains balanced. However, as pointed out before, this is not enough for normal *TL* operation because the TL conductors should be loaded by an impedance equal to or close to their characteristic impedance $Z_{c,\,sl}$.

5.4 CM-DM Mode Conversion

Figure 26 shows the case of CM-to-DM conversion when the external field couples into a nonperfectly differential *TL*-producing CM currents on it, which can further be converted into TL DM currents in the form of noise. More information about this type of mode conversion can be found in Refs. [2, 9, 10].

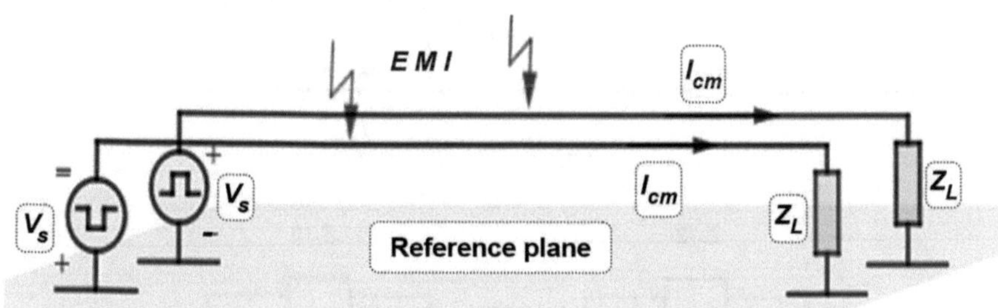

Fig. 26 CM current induced on a non-perfect differential TL by external field

Notably, induced CM currents on a perfectly balanced and impedance-matched *TL* are not converted into TL DM currents.

5.5 Longitudinal Conversion Loss (LCL) for UTP Cables

In terms of mode conversion, UTP (unshielded twisted pair) cables are also specified by *longitudinal conversion loss* (*LCL*). The LCL is usually measured at a network input port, and is required to be high. The LCL is the ratio between an injected *longitudinal common-mode* voltage V_{cm} at the network CM input (over the center CT of a symmetric transformer shown in Fig. 27a) and the developed *differential-mode* voltage V_{ab} at the network differential input, between terminals A and B. Therefore, the LCL is expressed as

$$LCL = 20\log\left(\left|\frac{V_{cm}}{V_{ab}}\right|\right) \quad (dB) \tag{19}$$

The goal of a balanced network is to have a small enough developed voltage V_{dm} at A-B input as a response of the voltage V_{cm}. The test setup for measuring the LCL at a network port, driven by a UTP cable, is shown in Fig. 27a, b [11].

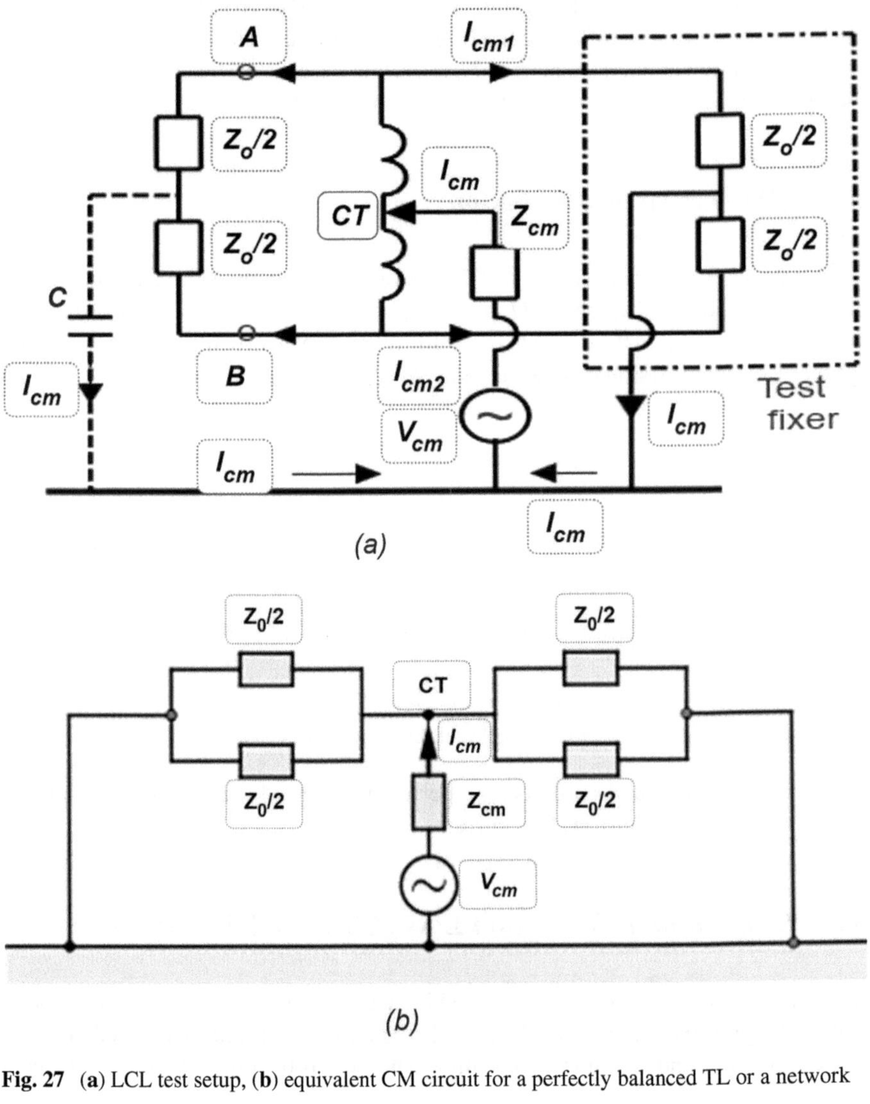

Fig. 27 (**a**) LCL test setup, (**b**) equivalent CM circuit for a perfectly balanced TL or a network

Figure 27a shows the differential-mode signal produced by an undesired common-mode signal, but this principle is reciprocal and can be used to determine the undesired common-mode signal at the transformer center (*CT*) as a response of a differential-mode voltage applied on the input differential network port A—B. In this circuit model, Z_{cm} is the common-mode impedance equal to 25 Ω, and Z_0 is the differential impedance of a network input equal to 100 Ω.

Based on the basis of the perfectly balanced CM circuit from Fig. 27b, the voltage V_{cm} is expressed as

$$V_{cm} = [Z_{cm} + Z_{ct}]I_{cm} = \left[Z_{cm} + \frac{Z_0}{4}\right]I_{cm} \tag{20}$$

where $Z_{ct} = Z_0/4$ is the input impedance from the center of the balanced transformer, and $Z_{cm} = Z_0/4$.

For example, we calculate the *differential-mode* voltage V_{ab} across the DM impedance $Z_0 = 100\ \Omega$ as a response to a *common-mode* excitation. Because the common-mode current limit for *Class B* is $I_{cm} = 31.6\ \mu A$ ($30\ dB\mu A$) per CISPR 22, the corresponding *LCL* is *70* dB for a Cat-6 cable.

Furthermore, per Eqs. (19) and (20), the differential voltage V_{dm} produced by $I_{cm} = 30\ dB\mu A$ is calculated as

$$\begin{gathered} V_{ab}(dB\mu V) = LCL(dB) + I_{cm}(dB\mu A) + 20\log\left(Z_{cm} + {}^{Z_0}\!/\!_4\right) \rightarrow \\ V_{ab}(dB\mu V) \approx 70\,\mathrm{dB} + 30\,\mathrm{dBuA} + 20\log(50) = 134\,dBuV \end{gathered} \tag{21}$$

where $LCL = 70\ dB$, $Z_{cm} = 25\ \Omega$ and $Z_0 = 100\ \Omega$.

This was the case when the tested network was perfectly balanced, i.e., when each leg of the network had the same impedance of $Z_0/2 = 50\ \Omega$ as shown in Fig. 27.

5.6 Shielded Twisted Pairs (STP)

To further reduce electromagnetic interactions between a UTP and its environment, shielded twisted pairs (STPs), such as Cat-7, can be used. Similarly to shielded coaxial cables, the shielding effectiveness of the STP depends on the shield quality and how the shield ends are grounded. The right solution is to connect both shield ends to a grounded equipment enclosure by a continuous 360°-connection with low impedance across the entire operating signal spectrum. However, this grounding scheme may not be efficient for installing excessively long STPs. This method is critical because of the potential for high common-mode voltage to develop between the STP shield and the ground. This is because the shield of an electrically long STP creates a "transmission line" with the ground, causing the cable to be an efficient EMI radiator or EMI receptor, as discussed earlier in the section about *RF* coaxial cables.

Thus, the developed voltage between an STP shield and the ground drives the common-mode current to flow along the ground and the cable shield, creating a large current loop. This case is shown in Fig. 28.

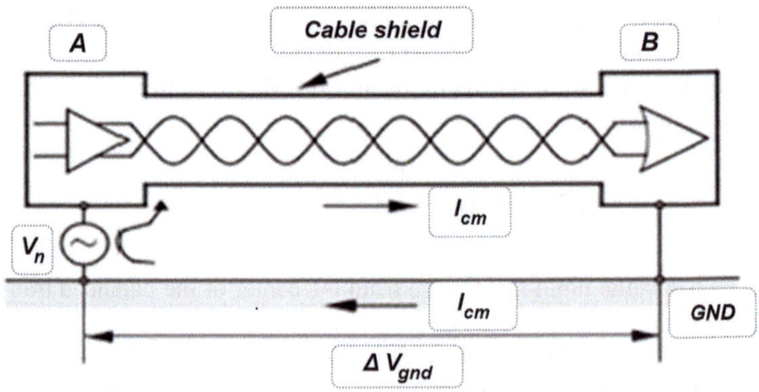

Fig. 28 STP shield grounded perfectly to the chassis of equipment A and B

Owing to this situation, it is important that two or more pieces of equipment connected by *STPs* share the same ground reference. Additionally, STP cables should be as close as possible to the ground or a larger conductive plane to minimize a common-mode current loop, whose *EMI* emission level is directly proportional to the loop area, as noted in *Appendix B* and Chapter "General Concept and Types of EMI Emissions". It is recommended that these cables (with other types of cables) be placed at metallic supports, such as conduits, ducks, or trays, as shown in Fig. 29a–e, according to the EN 61000-5-2 standard.

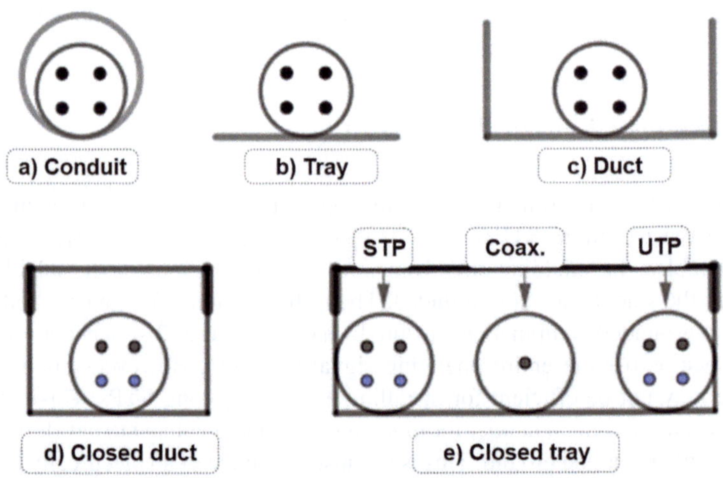

Fig. 29 Cross-sectional views of cable placement in conduits, ducts, and trays

Additionally, as pointed out earlier, the STP's shield should be grounded 360° around the twisted signal pairs to the grounded equipment chassis. If the

360°-connection is impossible, use at least a very short pigtail or, better, a cable metal clamp at the equipment chassis (see Fig. 20). For example, for fast data transfers (such as Gigabit Ethernet), the impedance between a cable shield and the ground is recommended to be less than 0.1 Ω, where a 360° shield grounding is the only solution.

5.7 *UTP Vs. STP Twisted-Pair Cables*

The most important and critical parameters of signal cables are their *characteristic* impedance, frequency bandwidth, immunity to *crosstalk*, and *attenuation*, which are all specified by the cabling standards. Unshielded twisted pairs (UTPs) usually meet these requirements for medium-speed data transfer, and as a result, these cables are widely used in local area networks (LANs) in commercial and residential buildings. The frequency content of operating signals at these cables is up to several hundred MHz.

Shielded twisted pairs are needed for higher-speed data links with broadband frequency content up to several GHz. Importantly, these cables have good shield quality to obtain minimal cable coupling with the electromagnetic environment. However, properly specified and installed UTPs are better and simpler solutions for widespread LANs, especially if equipment circuits possess a high *common-mode rejection factor* because of the circuits' good electrical balance and symmetric component layout.

An additional reduction in the common-mode current on a UTP cable can be obtained by loading both cable ends with commercial snap-on ferrite chokes next to the UTP connectors, as shown in Fig. 30.

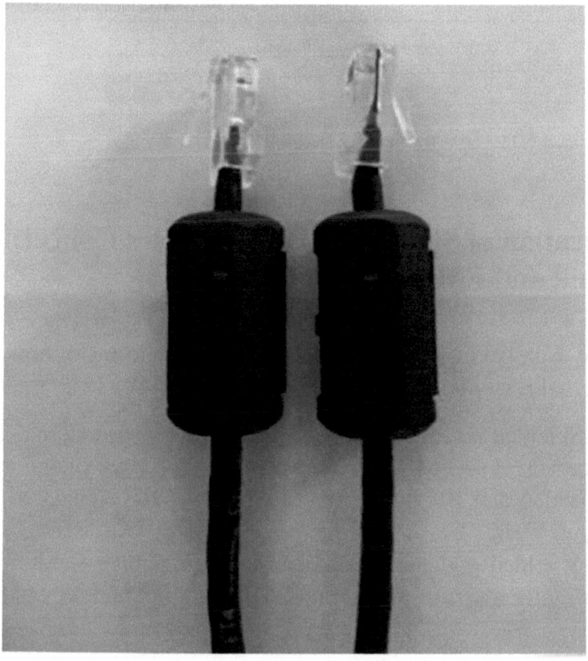

Fig. 30 Suppression of CM currents at UTP cable ends

In this way, the ferrite choke increases the high-frequency resistance to the common-mode currents that flow along the UTP cable. For example, particular chocks (by ferrite material 31) are efficient at suppressing CM currents on Cat-6 UTP cables up to 300 MHz. Suitable ferrite chokes for specific UTP cable sizes and frequency ranges can be found in the catalogs of ferrite manufacturers.

5.8 Unbalanced Twisted Pairs

Capacitive coupling is dominant for unbalanced twisted pairs when the connected circuits have *high impedance;* consequently, twisting of the cable does not significantly reduce capacitive coupling. However, when the connected circuits have *very low impedance*, the capacitive coupling between adjacent differential pairs decreases, but inductive coupling dominates. As pointed out earlier, unshielded twisting power lines should be close and parallel to the reference ground, which may significantly reduce *inductive* coupling. This approach provides good protection of power supply cables against the coupling of an external magnetic field.

Figure 31 shows a DC power line made from an unshielded twisted pair where the negatively polarized conductor is grounded at both ends.

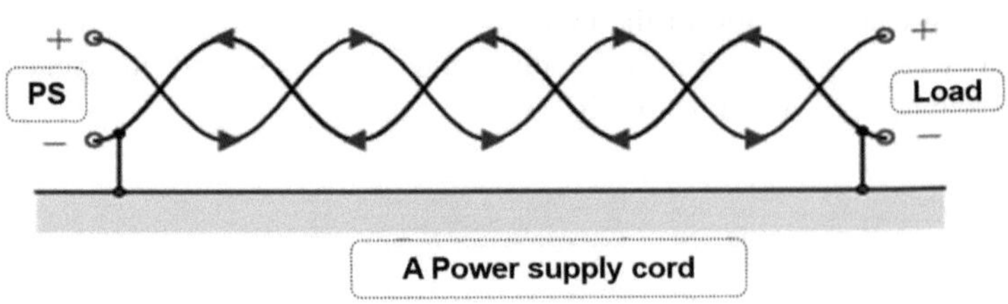

Fig. 31 Unbalanced UTP cable for the power supply cord

6 Classification of Signal and Power Cables Concerning Crosstalk and EMI

The cables are usually classified into four categories in terms of how sensitive they are to the external EMI and how much EMI producers are [11]:

Category 1—Shielded *coaxial cables* are *less sensitive* to external EMI coupling and produce very low EMI. They are intended for low-voltage analog signals, broadband video and *RF* signals, and high-power RF signals with frequencies greater than 1 MHz.

Category 2—Shielded and unshielded twisted-pair cables (with 2 and multiple pairs) are *slightly sensitive* to external EMI coupling, and they are mild producers of EMI. They are intended for low- and medium-speed data signals, differential analog instruments, or outputs for DC power supplies.

Category 3—These cables are usually single-pair unshielded cables that are *noisy* and commonly intended for external DC and AC filters that supply low-noise equipment.

Category 4—These cables are *very noisy*. They are AC 3-conductor or 5-conductor power cords, chassis grounding conductors, cords for welding, and other noisy equipment, such as noisy DC adjustable-speed motors.

The reason for classifying cables according to the types of signals they carry is to choose the correct terminations and safe distances between the cables from different categories. For example, the IEC (IEC 61000–5-2) suggests that signal and power cables and grounding conductors be placed on conductive trays at a distance depending on the type of signal they carry. The larger difference between the cable categories requires a larger distance between them.

Figure 32 shows the IEC proposal for separations between cables that belong to categories 1, 2, 3, and 4, according to the IEC 61000-5-2 standard.

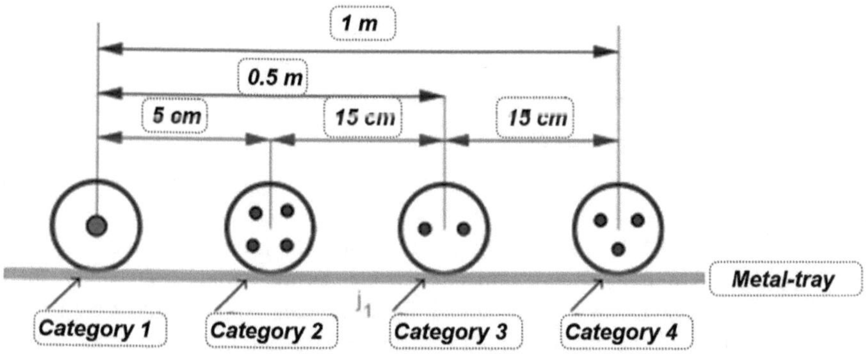

Fig. 32 IEC recommendation for cable placement and separations of different categories

More information about the cabling and cable installations can be found in [7, 12, 13].

7 Fiber-Optic Cables with Zero-EMI

Compared with traditional copper–wire cables, fiber optical cables have many advantages in communication systems, especially with respect to high-speed data transfers. Therefore, the main advantage of fiber cables is an increased data transfer capacity due to their wide bandwidth and absolute immunity to the external EMI. As a result, fiber cables are increasingly used in high-speed and low-power communications and control systems. With respect to EMI performance, fiber cables are not conductors and do not use electrons for signal transfer but rather photons—mostly infrared light.

As noted, the fiber cables do not couple with the electromagnetic environment, but some potential EMI problems may exist in optical-to-electronic converters, which are integral parts of fiber-optic communication systems. Notably, optical-to-electronic converters possess weak electronic signals, causing the optoelectronic receiver to be sensitive to external EMI fields and currents. Therefore, the standard strategy for improving the receiver's EMI immunity can be used as discussed in Chapter "General Concept and Types of EMI Emissions".

Figure 33 shows the concept of an optical-to-electronic receiver as an interface between the electronic and optical domains.

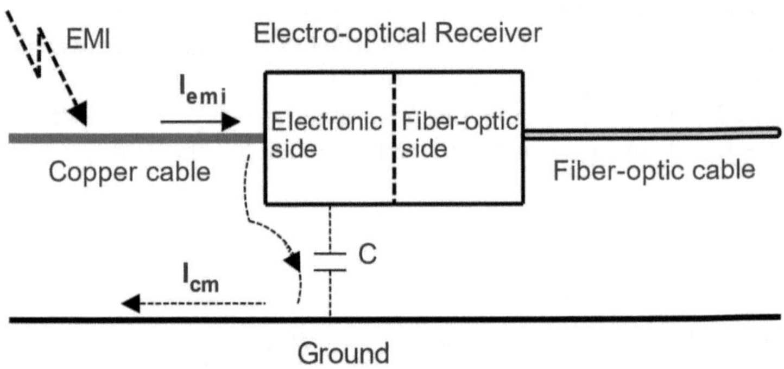

Fig. 33 Optical–electronic receiver

With respect to EMI, it is interesting to compare a high-quality coaxial cable for a low-power and high-speed link with a fiber cable. As noted for coaxial cables, there are two potential paths for a return signal that flows back to its source. One is over the inner surface of the coaxial shield (operating–DM case), and the other is over common ground (nonoperating–CM case). Both currents form current loops where the loop formed by the cable shield with the ground is significantly larger, resulting in greater mutual coupling between the cable and the external electromagnetic space.

Furthermore, if we use an optical cable for the above link, we do not need to care for any cable–ground loop because the return conductor does not exist in fiber cables, and consequently, a critical CM current loop does not exist. However, optical cables are designed solely for low-power and high-speed signals and cannot be used for high-power RF applications where coaxial cables are commonly used.

More about the topics discussed in this section can be found in Ref. [14].

References

1. H. Johnson, M. Graham, *High-Speed Signal Propagation* (Pearson, 2003) [1]
2. S.H. Hall, H.L. Heck, *Advanced SI for High-Speed Digital Design* (IEEE & Wiley, 2009) [2]

3. S.W. Ellingson, *Radio System Engineering* (Cambridge University Press, 2016) [3, 4]
4. H.W. Silver (ed.), *The ARRL Antenna Book for Radio Communications*, 22nd edn. (ARRL, 2013) [3, 4]
5. E.F. Kuester, *Theory of Waveguides and Transmission Lines* (CRS Press, 2021)
6. H.W. Ott, *Electromagnetic Compatibility Engineering* (John Wiley & Sons, Inc., NY, 2009)
7. T. Williams, *EMC for Product Designers* (Elsevier, Newnes, 2007)
8. C.R. Paul, R.C. Scully, *Introduction to Electromagnetic Compatibility* (Wiley, 2022)
9. F.M. Tesche, Plane wave coupling to cables, Chapter 4, in *Handbook of EMC*, ed. by R. Perez, (Academic Press, 1995)
10. E.F. Vance, Coupling to a long transmission line, Appendix C, in *Handbook of EMC*, ed. by R. Perez, (Academic Press, 1995)
11. ITU-T Recommendations G 117, ITU. (1996)
12. D. Groth, *Cabling: Guide to Network Wiring* (Sybex, 2000)
13. H. Hughes, *Telecommunications Cables: Design, Manufacture and Installation* (John Wiley, 1997)
14. A.H. Cherin, R. Perez, Chapter 25, in *Handbook of Electromagnetic Compatibility*, ed. by R. Perez, (Academic Press, 1995)

References

Conventional Methods for EMI Control in Radio Systems

EMI Filters: Theory and Design

1 Introduction

The main purpose of EMI filters is to prevent EMI-conducted currents from leaving electronic equipment or flowing into the equipment from other equipment. Excessive EMI currents may degrade equipment performance and even damage sensitive circuits, especially in the case of lightning or ESD occurrences, as discussed in Chapter "Lightning Physics and Impact on Radio Systems on the Earth". When *EMI* current is so excessive, the equipment may not also comply with mandatory requirements established by EMC domestic and international standards and, as such, cannot be placed on the market.

Filters are generally used in many electrical and electronics engineering areas, such as regular electric and electronic circuit designs, communication systems, automatic control, and signal processing. These filters are conventional, and their properties differ from those of EMI filters at some points. Conventional filters predominantly have *reactive* components from which noise energy reflects and dissipates somewhere in the circuit. These filters include various types of filters, such as Butterworth, Bessel, Chebyshev, and linear phase filters. Our primary focus in this chapter will be on passive filters with *reactive* and *lossy* components for EMI applications.

Hence, when a filter is built with lossy components (such as resistors and ferrites) in addition to reactive components, the noise energy dissipates mainly within these components. The filters with lossy components belong to the category of EMI filters. In the following section, we discuss the general properties of conventional and EMI filters.

The writing of this chapter is based on classical passive circuit and filter theories. We refer interested readers to references [1–10] for further reading of the filter topics, which are discussed in this chapter.

M. Pajovic, *Fundamentals of EMI Phenomena in Radio Communication Systems*, https://doi.org/10.1007/978-3-031-88748-2_9

2 Filter Parameters

The fundamental parameters of filters are the *frequency response,* or *transfer function,* and the *insertion loss* in the case of *passive* filters or the *gain* in the case of *active* filters. For communication systems, the essential filter parameters are the *amplitude, frequency,* and *phase* responses. The *linear phase* response is a significant parameter for the communication filters because all frequency components of the operating signal should have the same propagation delay through the filter's bandwidth. The *amplitude* response is more important for EMI filters than the *phase* response. Note that the main purpose of EMI filters is to suppress the amplitude of the noise frequency content associated with undesired EMI currents.

2.1 *Frequency Response (Transfer Function)*

Any filter can be represented by the 2-port network shown in Fig. 1, which introduces *attenuation* or *insertion loss* of a signal passing through the filter.

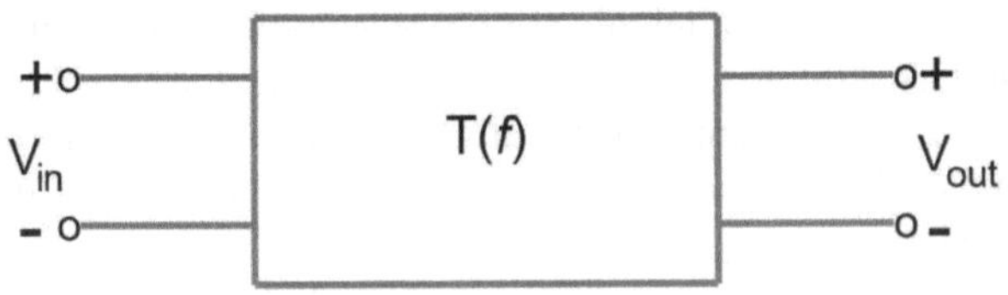

Fig. 1 Filter as a 2-port network

The filter *transfer function* is given as the ratio of two polynomials $T_n(i\omega)$ and $T_d(i\omega)$, or $V_{in}(i\omega)$ and $V_{out}(i\omega)$, where each function has the complex variable $i\omega$. ω denotes an angular frequency in radians per second:

$$\tilde{T}(i\omega) = \frac{\tilde{T}_n(i\omega)}{\tilde{T}_d(i\omega)} = \frac{\tilde{V}_{out}(i\omega)}{\tilde{V}_{in}(i\omega)} \tag{1}$$

where V_{in} and V_{out} are the filter *input* and *output* voltages, respectively. The roots of the polynomial numerator $T_n(i\omega)$ are referred to as the filter zeros since the roots of the polynomial denominator $T_d(i\omega)$ are referred to as the filter poles. For example, the *transfer function* for a standard low-pass filter shown in Fig. 2 is

$$\tilde{T}(i\omega) = \frac{\tilde{V}_{out}}{\tilde{V}_{in}} = \frac{R_L // \left(1/_{i\omega C} \right)}{R_s + i\omega L + R_L // \left(1/_{i\omega C} \right)} \rightarrow$$
$$\tilde{T}(i\omega) = \frac{1}{\tilde{T}_d(i\omega)} = \frac{1}{k_2 \cdot (i\omega)^2 + k_1 \cdot (i\omega) + k_0} \tag{1a}$$

where $k_2 = (LC)$, $k_1 = \left(\frac{L}{R_L} + CR_s\right)$, and $k_0 = \left(1 + \frac{R_s}{R_L}\right)$.

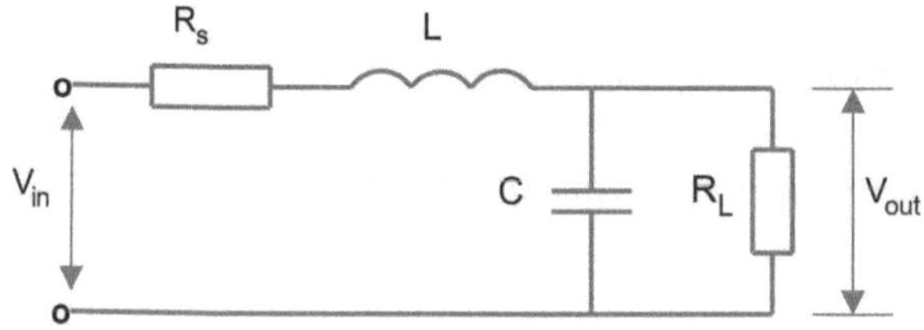

Fig. 2 Two-pole low-pass filter

The polynomial $\tilde{T}_d(i\omega)$ in Eq. (1a) has two roots, and accordingly, its transfer *function* $\tilde{T}(i\omega)$ has two poles.

We analyze the single-pole and multipole filter configurations in this chapter. Regarding the bandwidth BW of the filter, there are four main categories of filters: *low-pass*, *high-pass*, *bandpass*, and *bandstop* filters. Table 1 shows the *amplitude transfer functions* of the regular filters with the corresponding *corner* frequencies.

Table 1 Amplitude transfer functions for low-pass, band-pass, band-stop, and high-pass filters

	$\lvert T(i\omega)\rvert$	BW
Low-pass	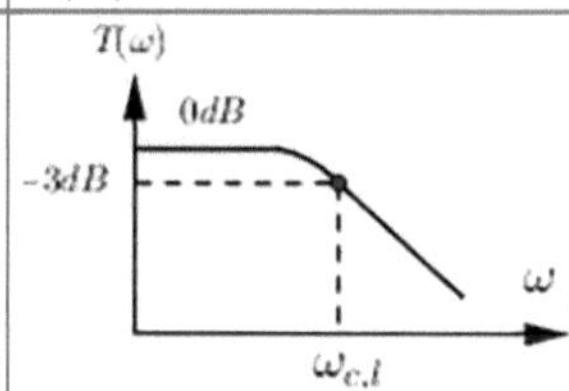	$\omega < \omega_{c,l}$
Band-pass	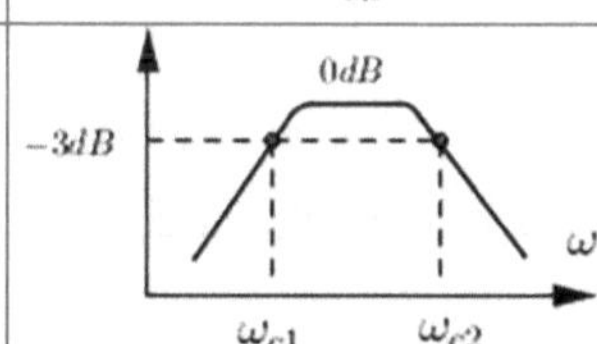	$\omega_{c1} < \omega < \omega_{c2}$
Band-stop	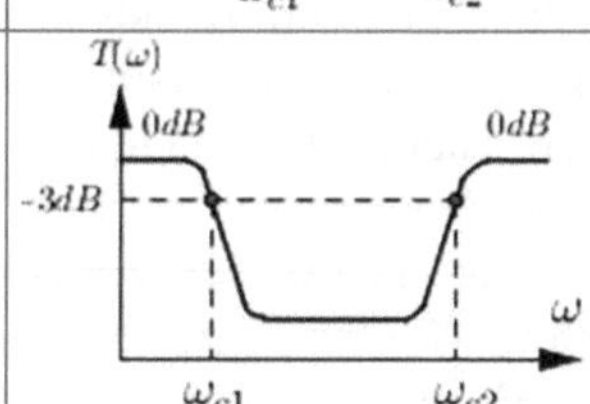	$\omega_{c1} < \omega < \omega_{c2}$
High-pass	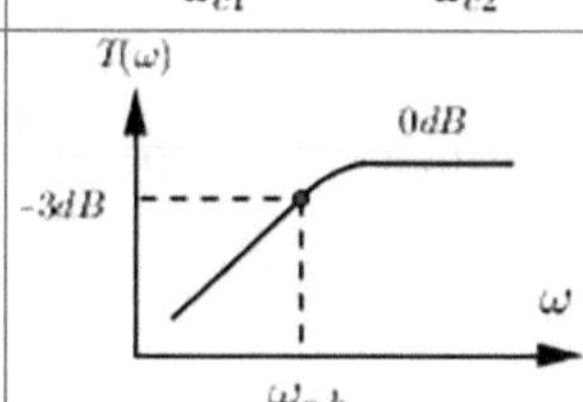	$\omega > \omega_{ch}$

The frequency at the end of the filter bandwidth (where the amplitude of the output signal is—3 dB) is called the *corner* frequency.

2.2 Definition of Insertion Loss

Figure 3a, b shows the circuits *without* and *with* the filter placed between the circuit source and load, respectively.

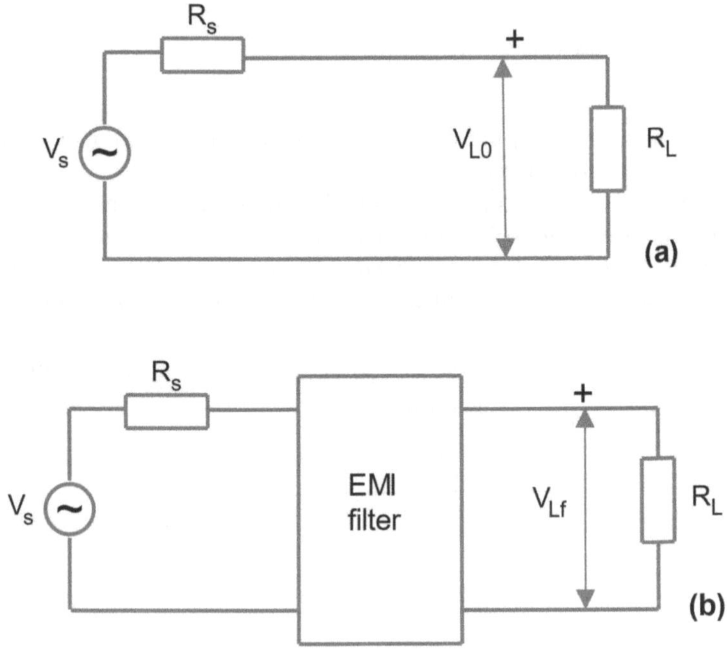

Fig. 3 (**a**) A circuit without the filter and (**b**)—a circuit with the filter

The load voltage $V_{L,0}$ corresponds to the circuit *without* a filter, whereas load voltage $V_{L,f}$ corresponds to the circuit *with* the filter. The source voltage V_S and the source and load resistances (R_S and R_L) are assumed to remain unchanged after filter insertion.

Per the definition, the *insertion loss* is the ratio of the signal *attenuations* of the circuit *without* and *with* the filter. This parameter is commonly expressed in logarithmic form (in *dB*) as

$$IL = 10\log\left(\frac{P_{L,0}}{P_{L,f}}\right) = 20\log\left(\frac{V_{L,0}}{V_{L,f}}\right) = 20\log\left(\frac{I_{L,0}}{I_{L,f}}\right) \tag{2}$$

where $P_{L,0}$, $V_{L,0}$, and $I_{L,0}$ denote the power, voltage, and current, respectively, at the load *without* the filter, while $P_{L,f}$, $V_{L,f}$, and $I_{L,f}$ denote the power, voltage, and current at the load when the filter is placed in the circuit.

For example, Fig. 4 shows the plot of the *insertion loss* for a low-pass filter from Fig. 2. As mentioned earlier, the filter bandwidth is defined from the *DC* (where the insertion loss is *0 dB*) to the *corner* frequency ω_c, where the *insertion loss* is *3 dB*.

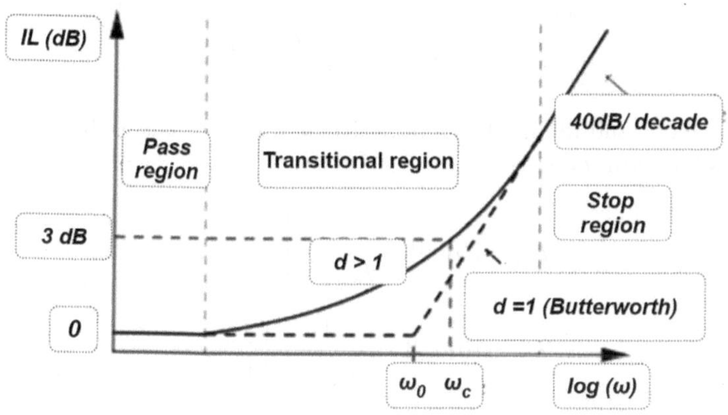

Fig. 4 Insertion loss and bandwidth of the LC low-pass filter

This plot shows three filter regions: the *pass*, *transitional*, and *stop* regions. Additionally, the dashed curve refers to Butterworth's filter, whose transfer characteristic is perfectly dumped down. The frequencies ω_0 and ω_c are the corner angular frequencies of the Butterworth filter and L-filter, respectively.

More information about filter theory can be found in the references [1–4], which we refer to interested readers for future reading.

3 Series Filters

Let us first consider the *insertion loss* for the simplest filter configurations with only one reactive element (a single-element filter) in series with the circuit source and load. The impedance iX_1 of the filter is shown in the circuit in Fig. 5.

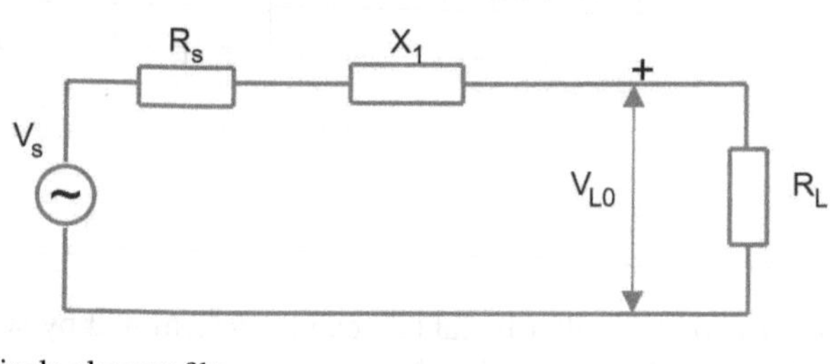

Fig. 5 Series single-element filter

For this case, the series reactance $\tilde{X}_1$ can be *inductive* or *capacitive.* The voltage attenuation A_0 *without* the filter and the attenuation A_f *with* the filter can be expressed as a function of the *source* and *load* resistances and reactance $\tilde{X}_1$, respectively, as

$$V_{L,0} = \frac{R_L}{R_S + R_L} V_S \rightarrow V_{L,0} = A_0 V_S \tag{3a}$$

$$\tilde{V}_{L,f} = \frac{R_L}{R_S + R_L + \tilde{X}_1} V_S = \tilde{A}_f V_S \tag{3b}$$

Furthermore, substituting Eqs. (3a), (3b) into Eq. (2) leads to the expression of the series filter *insertion* loss:

$$\begin{aligned} IL_{ser} &= 20\log\left|1 + \frac{\tilde{X}_1}{R_S + R_L}\right| \quad or \\ IL_{ser} &= 10\log\left|1 + \left(\frac{X_1}{R_S + R_L}\right)^2\right| \quad (dB) \end{aligned} \tag{4}$$

The subscript *ser* denotes a series filter. The typical *series* filters with the corresponding reactance are shown in Table 2.

Table 2 Series filters

Related to Fig. 4	Series-element filters	$\lvert\tilde{X}_1(\omega)\rvert$
(1) low pass	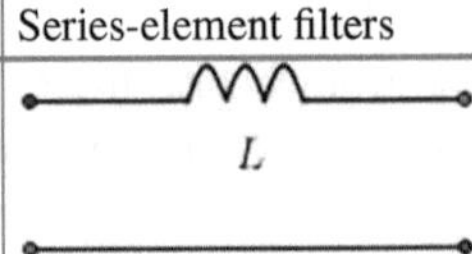	ωL
(2) high pass	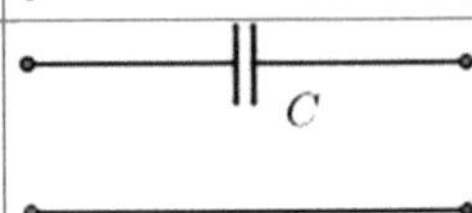	$\frac{1}{\omega C}$
(3) band pass	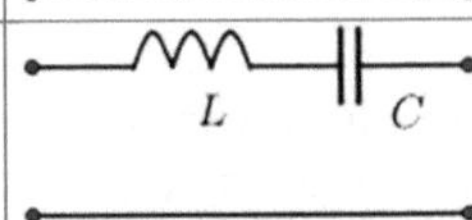	$\left(\omega L - \frac{1}{\omega C}\right)$
(4) band stop	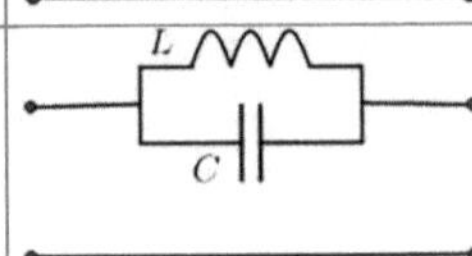	$\frac{1}{\left(\omega L - \frac{1}{\omega C}\right)} \cdot \frac{L}{C}$

The *insertion loss* of each filter in Table 2 can be determined by substituting the corresponding reactance $X_1(\omega)$ into Eq. (4).

3.1 Series Low-Pass Filter

The filter with the *series inductance L*, shown in Fig. 6, is a low-pass filter whose reactance is $\tilde{X}_1 = i\omega L$.

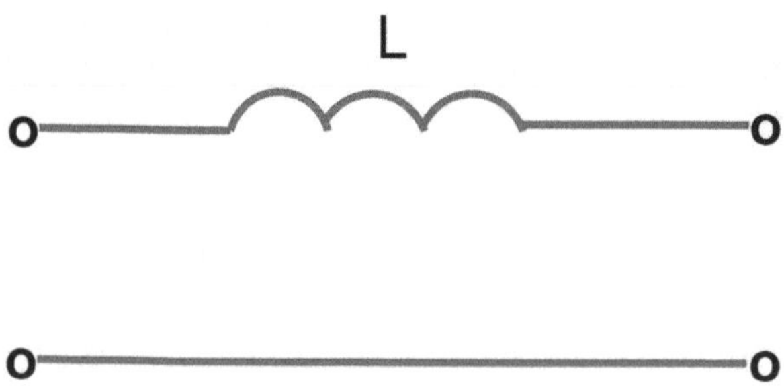

Fig. 6 Series single-element low-pass filter

On the basis of Eq. (4) and the data from Table 2, the *insertion loss* of the filter is given:

$$IL_{ser} = 10\log\left|1+\left(\frac{\omega L}{R_S+R_L}\right)^2\right| \quad \text{or} \qquad (5)$$

$$IL_{ser} = 10\log\left|1+(\omega\tau)^2\right|$$

where τ is the filter time constant equal to $\tau = L/(R_S + R_L)$.

Furthermore, based on Eq. (5), the *insertion loss* is equal to 3 dB when the time constant τ is equal to 1, resulting in $IL_{ser} = 10 \log 2 = 3\ dB$. Accordingly, the *corner angular* frequency ω_c of the low-pass filter is

$$\omega_c = \frac{(R_S+R_L)}{L} \qquad (6)$$

or the *corner* frequency f_c is

$$f_c = \frac{1}{2\pi}\frac{(R_S+R_L)}{L} \qquad (7)$$

where $\omega = 2\pi f \rightarrow f = \omega/2\pi$. Thus, the corner frequency of this filter exclusively depends on the circuit elements R_S, R_L, and L.

Furthermore, combining Eqs. (5) and (7), the *insertion loss* can be expressed in terms of the *corner* frequency f_c as

$$IL_{ser} = 10\log\left[1+\left(\frac{f}{f_c}\right)^2\right] \qquad (7a)$$

Example 1 numerically illustrates the *corner frequency* and the *insertion* loss of the low-pass filter with a series inductance *L*.

Example 1

(a) Find the *corner* frequency f_c ***and***
(b) the *insertion loss* at frequencies $f = 2f_c$, $3f_c$, $4f_c$ and $10f_c$ for the *low-pass* filter, as shown in Table 5, with an inductance of $L = 10\ \mu H$ and source and load resistances of $R_S = R_L = 50\Omega$.

Solution

(a) *The corner* frequency of the low-pass filter with the single-series *L*-element is expressed by Eq. (7):

$$f_c = \frac{1}{2\pi}\frac{(R_S + R_L)}{L} = \frac{1}{2\pi}\cdot\frac{(50+50)\Omega}{10\cdot 10^{-6}\,H} = 1.59\,MHz$$

(b) The *insertion* loss is expressed by Eq. (7a):

$$IL_{ser} = 10\log\left[1 + \left(\frac{f}{f_c}\right)^2\right]$$

Additionally, using this expression, the *insertion losses* at frequencies $2f_c$, $3f_c$, $4f_c$ and $10f_c$ are *7 dB*, *10 dB*, *12.3 dB*, and *20 dB*, respectively. Note that for $f \gg f_c$ (i.e., $f = 10 f_c$), the insertion loss increases to a maximum of 20 dB/decade.

This type of single-element series filter is commonly used to block AC, RF, and EMI currents from flowing into DC circuits.

3.2 *Series High-Pass Filter*

Figure 7 shows the high-pass filter with a series capacitance *C*.

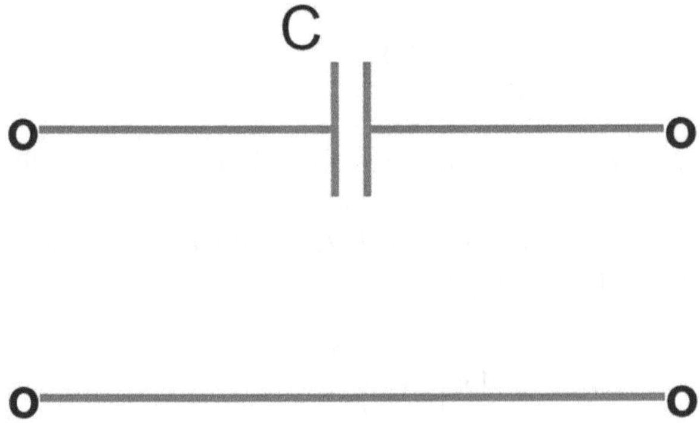

Fig. 7 Series single-element high-pass filter

Here, the series reactance is

$$\tilde{X}_1 = \frac{1}{i\omega C}$$

On the basis of Eq. (4), the filter *insertion* loss is given as

$$IL_{ser} = 10\log\left[1 + \left(\frac{1}{R_S + R_L} \cdot \frac{1}{\omega C}\right)^2\right] \quad (dB) \tag{8}$$

Additionally, per Eq. (8), the *insertion loss* is equal to 3 dB when

$$\frac{1}{(R_S + R_L)} \cdot \frac{1}{\omega C} = 1 \tag{9}$$

and on the basis of this identity, the *corner* frequency f_c of the high-pass filter is

$$f_c = \frac{1}{2\pi} \frac{1}{(R_S + R_L)C} \tag{10}$$

This type of single-element series filter is commonly used to decouple (isolate) DC circuits from AC and RF circuits.

4 Shunt Filters

In Table 3, two *shunt filters* for *low-pass* and *high-pass* filtering are shown.

Table 3 Topologies of single-shunt filters

Related to Fig. 5	Shunt filters	$\tilde{X}_2(\omega)$
Low pass	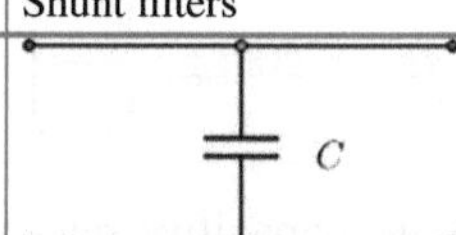	$\frac{1}{i\omega C}$
High pass	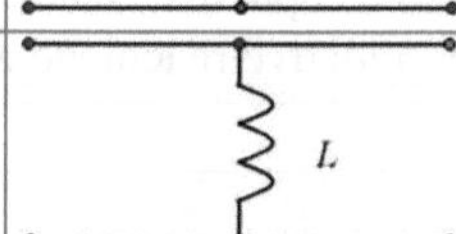	$i\omega L$

Additionally, Fig. 8 shows, in general, the shunt filter parallel to the circuit load R_L.

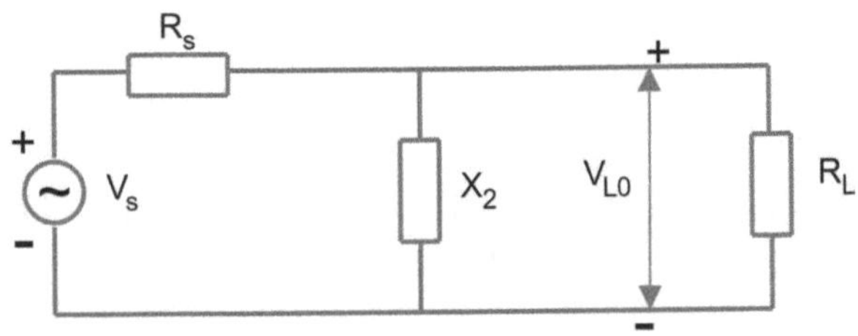

Fig. 8 Shunt single-element filter

The *insertion loss* for this filter will be derived similarly to before, i.e., as the ratio between the voltage attenuation *without* a filter $\left(\frac{V_{L,0}}{V_S}\right)$, and that *with* a filter $\left(\frac{V_{L,f}}{V_S}\right)$:

(a) *Without* the filter:

$$\frac{\tilde{V}_{L,0}}{\tilde{V}_S} = \frac{R_L}{R_S + R_L} \tag{11a}$$

(b) *With* the filter:

$$\frac{\tilde{V}_{L,f}}{\tilde{V}_S} = \frac{\tilde{X}_2 // R_L}{R_S + \tilde{X}_2 // R_L} \tag{11b}$$

Furthermore, by substituting Eqs. (11a), (11b) into Eq. (2), the *insertion loss* expression is obtained as

$$IL_{sh} = 20\log\left(\frac{V_{L,0}}{V_{L,f}}\right) = 20\log\left|\frac{R_L}{R_S + R_L}\frac{R_S + \tilde{X}_2 // R_L}{\tilde{X}_2 // R_L}\right| \ (dB) \tag{12}$$

Note that the capacitive reactance $\tilde{X}_2 = 1/_{i\omega C}$ is an element of a *low-pass* filter, whereas the inductive reactance $\tilde{X}_2 = i\omega L$ is an element of a *high-pass* filter.

4.1 Shunt Low-Pass Filter

A shunt low-pass filter is shown in Fig. 9.

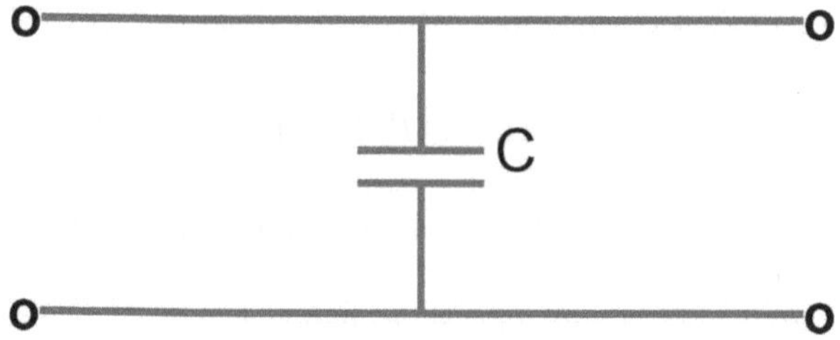

Fig. 9 A shunt low-pass filter

By substituting $\tilde{X}_2 = \frac{1}{i\omega C}$ into Eq. (12) and rearranging, the *insertion loss* for the *shunt low-pass* filter becomes

$$IL_{sh} = 10\log\left\{1 + \left[\left(R_S // R_L\right)\omega C\right]^2\right\} \quad (dB) \tag{13a}$$

$$IL_{sh} = 10\log\left|1 + \left[\omega\tau\right]^2\right| \quad (dB) \tag{13b}$$

where $\tau = (R_S//R_L)\ C$ is the filter *time constant*. Additionally, per Eq. (13b), the *insertion* loss is equal to 3 dB when the term $(R_S//R_L)\ \omega C = 1$; accordingly, the *corner* frequency f_c is expressed as

$$f_c = \frac{1}{2\pi\left(R_S // R_L\right)C} = \frac{1}{2\pi\tau} \tag{14}$$

Note that the *corner* frequency of this filter depends on the circuit elopements R_S, R_L, and C or the filter time constant τ. The next example numerically illustrates Eqs. (13a) and (14).

Example 2

(a) Find the shunt capacitance C for the low-pass filter with a corner frequency of f_c = ***1.59*** *MHz*, as shown in Table 2. The filter source and load resistance are equal at 50 Ω.

(b) Additionally, find the filter insertion loss at the frequencies $2f_c$, $3f_c$, $4f_c$, and $10f_c$.

Solution

(a) Per Eq. (14), the shunt capacitance C can be fined for f_c = 1.59 *MHz* as follows:

$$C = \frac{1}{2\pi\left(R_S // R_L\right)f_c} = \frac{1}{2\pi(25)1.59\cdot 10^6} = 4\cdot 10^{-9}\,F\left(4\,nH\right)$$

(b) The insertion loss is given by Eq. (13a) as.

$$IL_{sh} = 10\log\left[1+\left(2\pi f\left(R_S // R_L\right)C\right)^2\right] \quad (dB),$$

For the frequencies $2f_{c,\,lp}$, $3f_{c,\,lp}$, $4f_{c,\,lp}$, and $10f_{c,\,lp}$, this loss is 7 dB, 10 dB, 12.3 dB, and 20 dB, respectively.

Note that the *low-pass* filter with a the series inductance of 10 μH (Example 1) and the filter with a shunt capacitance of 4 nF (Example 2) have the same *corner* frequency and *insertion loss*. Additionally, we can see that *the insertion loss increases by 20 dB/*decade far beyond the corner frequency for the filter.

This simple filter is commonly used for decoupling EMI currents from a DC circuit to the filter's corner frequency, diverting EMI currents from the circuit and directing them to the circuit ground.

4.2 Shunt High-Pass Filter

Figure 10 shows a shunt high-pass filter.

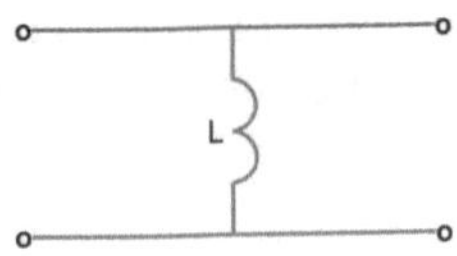

Fig. 10 Shunt high-pass filter

Substituting $\tilde{X}_2 = i\omega L$ into Eq. (12) and after its rearrangement, the *insertion loss* of this filter becomes

$$IL_{sh} = 10\log\left[1+\left(\frac{R_S // R_L}{\omega L}\right)^2\right] \quad (\text{dB}) \tag{15}$$

Additionally, the *corner* frequency can be derived when $\frac{R_S // R_L}{\omega L} = 1$, i.e., when the insertion loss is 3 dB:

$$f_c = \frac{1}{2\pi}\frac{\left(R_S // R_L\right)}{L} = \frac{1}{2\pi\tau} \tag{16}$$

where τ is the filter *time constant* equal to $L/\left(R_S // R_L\right)$.

Furthermore, combining Eqs. (15) and (16) gives us:

$$IL_{sh} = 10\log\left[1+\left(\frac{f_c}{f}\right)^2\right] \quad (\text{dB}) \tag{16a}$$

The next example numerically illustrates the use of Eqs. (15) and (16).

Example 3 The shunt capacitance C for the filter with a corner frequency of 1.59 MHz is found, as shown in Table 3. The filter source and load resistance are equal to 50 Ω.

(a) Additionally, find the insertion loss at frequencies $2f_{c\_l}$, $3f_{c\_l}$, $4f_{c\_l}$ and $10f_{c\_l}$.

Solution

(a) The *corner* frequency for the given filter is defined by Eq. (14), and the shunt capacitance C, for $f_{c,lp} = 1.59\ MHz$, can be defined as

$$C = \frac{1}{2\pi\left(R_S//R_L\right)f_{c\_l}} = \frac{1}{2\pi(25)1.59\cdot 10^6} = 4\cdot 10^{-9}F\left(4nH\right)$$

(b) The *insertion loss* is given by Eq. (13a), (13b) as

$$IL_{sh} = 10\log\left[1+\left(2\pi f\left(R_S//R_L\right)C\right)^2\right] \quad (dB),$$

and for the frequencies $2f_{c,lp}$, $3f_{c,lp}$, $4f_{c,lp}$ and $10f_{c,lp}$, the *insertion losses are* 7 dB, 10 dB, 12.3 dB, and 20 dB, respectively.
Note that the *low-pass* filter with a series inductance of 10 μH (from Example 1) and the filter with a shunt capacitance of 4 nF (Example 2) have the same *corner* frequency and *insertion loss*. Additionally, we can see that *the insertion loss increases by 20 dB/decade* far beyond the corner frequency for both filter configurations, *L-series* and *C-shunt*. In fact, this is a property of single-element filters.

5 Combination of Series and Shunt Filters

We considered series and shunt filters with a single element in the preceding sections. To increase the filter's insertion loss, we need to add at least one more reactive element to a single-element filter. Figure 11a, b shows the two-element filters made with a combination of series and shunt elements, whose maximal insertion loss is 40 dB/decade in the stopband.

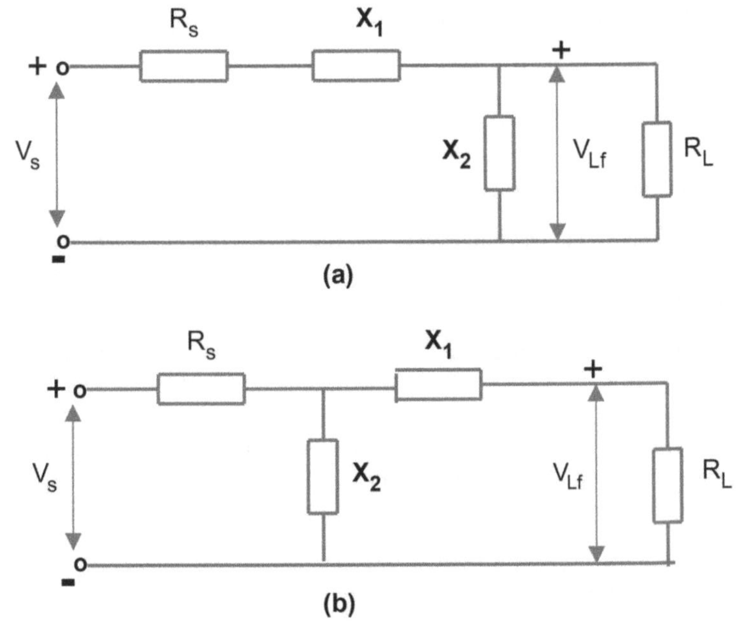

Fig. 11 L-filter topologies: (**a**) series-to-shunt and (**b**) shunt-to-series

The present filters may have different placements of series and shunt elements to the source and load. We find that the *shunt* reactance iX_2 in circuit (a) is placed closer to the source, whereas in circuit (b), it is placed closer to the load. As a result, the filters have different *input* and *output* impedances, which is important in terms of impedance matching the filter with associated circuits or TLs. Notably, both filter configurations have the same *insertion loss* and frequency bandwidth if they have the same components *R*, *L*, and *C* elements.

As mentioned earlier, the *insertion loss* for any filter is defined as the ratio between the voltage attenuations of a circuit *without* and *with* the filter. Applying the circuit theory for the *L*-filter shown in Fig. 11a, we obtain:

$$IL = 20\log\left|\left(\frac{R_L}{R_S + R_L}\right)\left[1 + \frac{(R_S + iX_1)}{iX_2 \, // R_L}\right]\right| \; (dB) \tag{17}$$

In summary, Table 4 shows *low-pass* and *high-pass* filters as a combination of series and shunt elements.

Table 4 LC low-pass and high-pass filters

LC Low pass	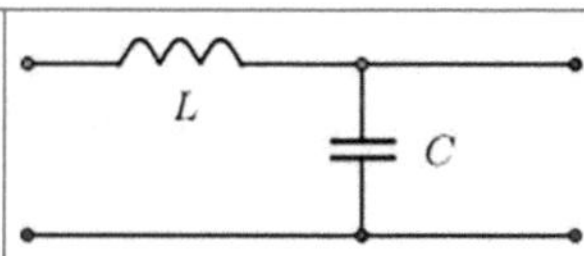
CL Low pass	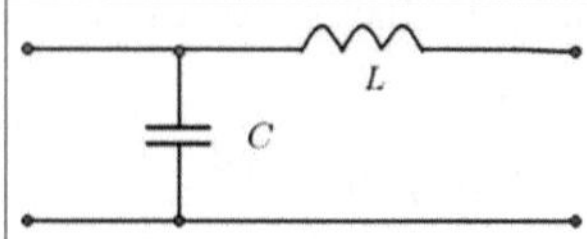
CL High pass	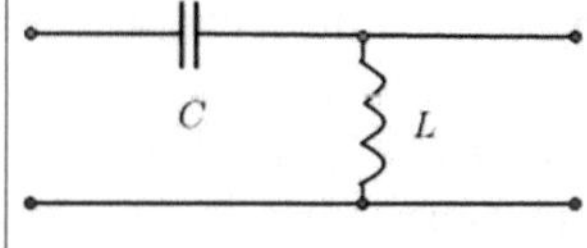
LC High pass	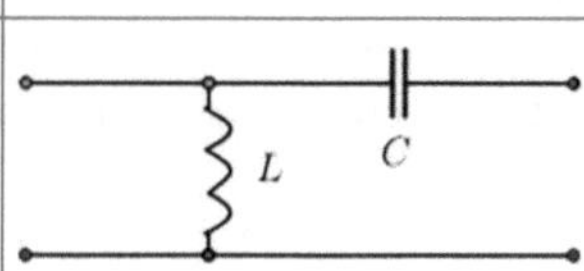

Hence, these filters have different *input* and *output impedances* that are important in the impedance filter's matching to its source and load. This important topic will be further considered in this chapter.

5.1 LC Low-Pass Filters

An LC low-pass filter is a common filter used in EMI applications, and we discuss the properties of this filter in this section.

The *insertion loss* for the LC low-pass filter from Table 4 can be derived by substituting the filter reactance:

$$X_1 = i\omega L \quad \text{and} \quad X_2 = \frac{1}{i\omega C}$$

into Eq. (17) and rearranging it, we obtain:

$$IL = 20\log\left|\left(\frac{R_L}{R_S + R_L}\right)\left(\left(1 + \frac{R_S}{R_L} - LC\omega^2\right) + i\omega\left(CR_S + \frac{L}{R_L}\right)\right)\right| \rightarrow$$
$$IL = 10\log\left\{\left[\left(\frac{1}{A_0} - \omega^2 LC\right)^2 + (\omega CR_S)^2(1+d)^2\right]A_0^{\,2}\right\} \tag{18}$$

where $A_0 = {R_L}/{(R_L + R_s)}$ is the circuit voltage attenuation without the filter and d is the damping factor expressed as $d = {L}/{(CR_S R_L)}$.

The *damping factor d* is an important filter parameter that significantly affects the filter response. The filter is in the optimal operating regime when perfectly damped, i.e., when its *damping factor is* close to 1. For example, this is the case for the response of the Butterworth filter shown in Fig. 4.

When $d < 1$, the filter is *underdamped,* causing *insertion gain* instead of *insertion loss.* When $d > 1$, the filter is *overdamped,* causing excessive low-frequency attenuation in its passband.

As noted, before, the *corner* frequency f_c for a filter is defined as the frequency where the *insertion loss* (expressed by Eq. 18) is equal to 3 dB*:*

$$f_c \cong \frac{0.113}{LC}\left\{\frac{2CL}{A_0} - \left[R_S C(1+d)\right]^2 + \left[\left(\frac{2CL}{A_0}\right)^2 + \left(\left[R_S C(1+d)\right]^2 - \frac{2CL}{A_0}\right)^2\right]^{\frac{1}{2}}\right\}^{\frac{1}{2}} \tag{19}$$

For an example, Fig. 12 graphically illustrates the *insertion loss* for the *L*-low-pass filter with the series inductance $L = 2.5\ \mu H$ and the shunt capacitance $C = 1\ nF$. Additionally, the filter's source and load resistances (R_S and R_L) are given below for these cases:

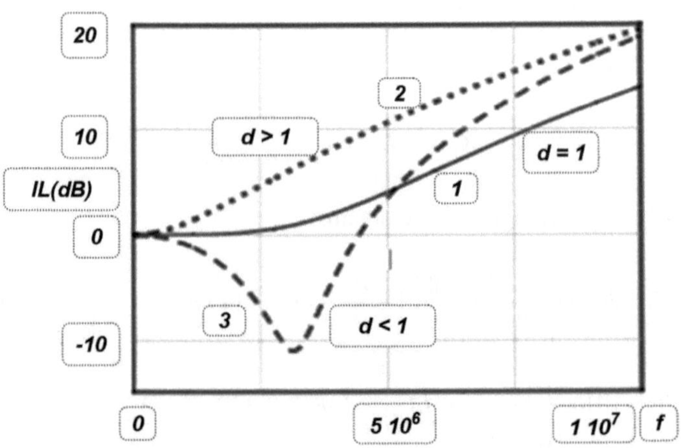

Fig. 12 R_s / R_L: 1—50/50 Ω, 2—5/20 Ω, 3—10/600 Ω

1. $R_S = 5\Omega,\ R_L = 20\Omega,$
2. $R_S = 50\ \Omega,\ R_L = 50\ \Omega,$
3. $R_S = 10\Omega,\ R_L = 600\Omega$

Curve 1 (when *d* = *1*) shows that the filter is perfectly damped with an *insertion* loss of 40*dB*/*decade* in the region far beyond the *corner* frequency.

Curve 2 (*d* > *1*) shows an overdamped filter with inefficient filtering inside the bandpass.

Curve 3 shows an underdumped filter with a *self-resonance* of ~3.18 MHz in its bandwidth. This resonance causes the filter to have an *insertion gain* instead of an *insertion loss*. The filter's self-resonance is expressed as

$$f_0 = \frac{1}{2\pi\sqrt{LC}} = \frac{1}{2\pi\sqrt{2.5\cdot10^{-6}\cdot1\cdot10^{-9}}} = 3.18\,MHz \tag{20}$$

The results of the damping factor calculations for cases 1, 2, and 3 are given as

$$d_1 = \frac{L}{CR_LR_S} = \frac{2.5\cdot10^{-6}(H)}{1\cdot10^{-9}(F)\cdot50\cdot50} = 1$$

$$d_2 - \frac{L}{CR_LR_S} = \frac{2.5\cdot10^{-6}(H)}{1\cdot10^{-9}(F)\cdot5\cdot20} = 25\ (d \gg 1)$$

$$d_3 = \frac{L}{CR_LR_S} = \frac{2.5\cdot10^{-6}(H)}{1\cdot10^{-9}(F)\cdot10\cdot600} \approx 0.4\ (d < 1)$$

Thus, EMI filter designers should be aware of the filter self-resonance because the *gain* may occur in the filter bandpass region instead of the *insertion* loss. Additionally, there are other circumstances when the EMI filter may be resonant, such as when the filter's source and load circuits are not purely resistive but reactive, which is often the case in EMI filtering. Additionally, per Eq. (19), the present filters have the *corner* frequencies:

1. $R_S = 50\ \Omega$; $R_L = 50\ \Omega$ $\quad f_{c1} \approx 4.5\ MHz.$
2. $R_S = 5\ \Omega$; $R_L = 20\ \Omega$ $\quad f_{c2} \approx 1.86\ MHz$
3. $R_S = 10\ \Omega$; $R_L = 600\ \Omega$ $\quad f_{c3} \approx 4.9\ MHz$

In conclusion, the filter's elements (*L, C,* R_S, R_L) influence the filter's *damping factor*, i.e., its *insertion loss* and *bandwidth*. The conventional filters have a *damping factor* of approximately *1* because they are well impedance-matched to TLs or circuits with poor resistive impedances. This is here case (1) when $R_S = R_L = 50\,\Omega$. Because the filter series inductance is $L = 2.5\ \mu H$ and the shunt capacitance is $C = 1\ nF$, the filter's characteristic impedance Z_c is

$$Z_c = \sqrt{\frac{L}{C}} = \sqrt{\frac{2.5\,\mu H}{1nF}} = 50\,\Omega$$

6 Multistage Filters: T- and Pi-Topologies

We considered *L*-filter topologies that belong to two-pole or one-stage filters in the preceding section. These topologies can be used to create multiple-stage filters. Very common multistage filters are *T*- and *Pi*-filters, which are created by two *L*-filters to obtain a higher *insertion* loss outside its bandwidth, greater than 40 dB/decade—up to 60 dB/decade. Figure 13a, b shows the composition of a *T*-filter by the two *L*-filters with the same L and C elements.

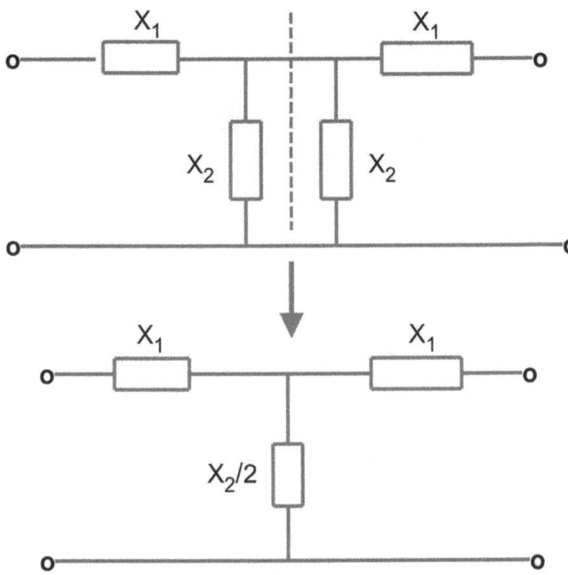

Fig. 13 Composition of the T-filter with two L-filters

Furthermore, Fig. 14 shows a π-filter created by two L-filters with the same L and C elements.

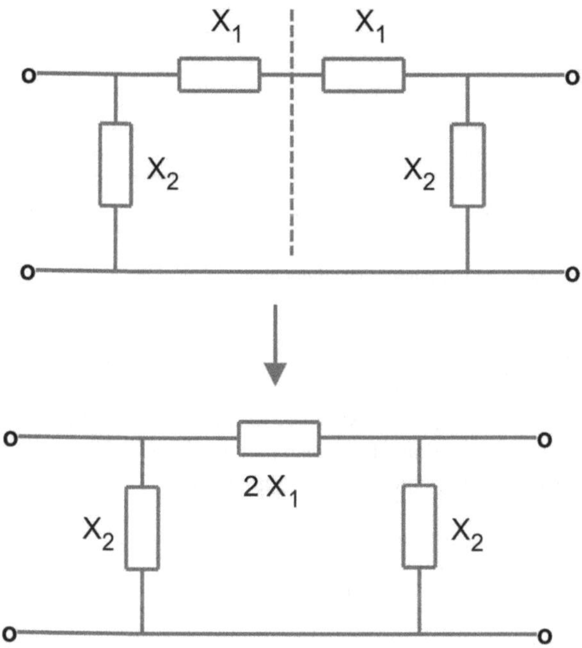

Fig. 14 Composition of a π filter with two L filters

Figure 15 specifically shows *high-pass T-* and *Pi*-filter configurations.

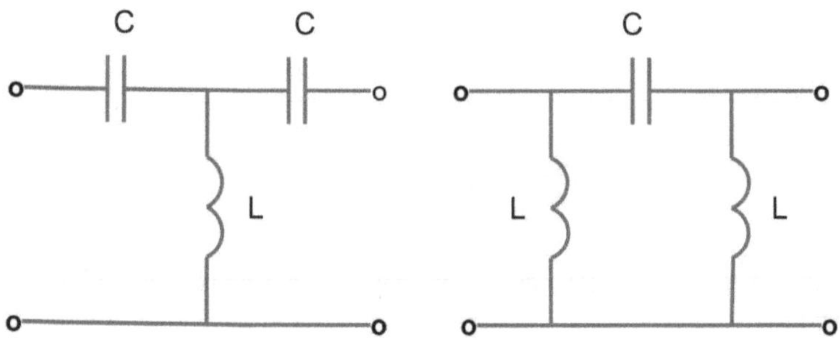

Fig. 15 T and π high-pass filters

Additionally, Fig. 16a, b shows *low-pass T* and *Pi* filters.

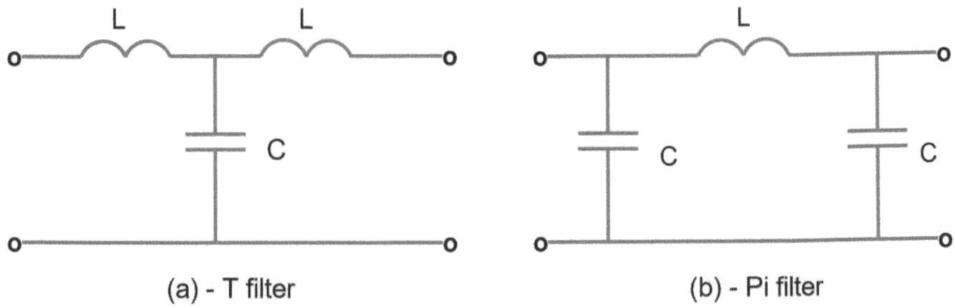

Fig. 16 (**a**) T low-pass and (**b**) π low-pass filters

The expressions of the *insertion loss* for these filter configurations are derived similarly to those of filter circuit theory, as is done for the single-element and two-element filter configurations. For example, when the *T* and π filters from Fig. 15 are terminated by equal source and load resistance ($R = R_S = R_L$), the *insertion* losses (IL_T and IL_π) for these filters are expressed as follows:

$$IL_T = 10\log\left(\left(1-LC\omega^2\right)^2+\omega^2\left(\frac{L}{R}+\frac{RC}{2}-\omega^2\frac{CL^2}{2R}\right)^2\right) \tag{21}$$

$$IL_\pi = 10\log\left(\left(1-LC\omega^2\right)^2+\omega^2\left(\frac{L}{2R}+RC-\omega^2\frac{RLC^2}{2}\right)^2\right) \tag{22}$$

Additionally, the *damping* factors d_T and d_π for these filters are as follows:

$$d_T = \frac{CR^2}{2L} \tag{23a}$$

and

$$d_\pi = \frac{L}{2CR^2} \tag{23b}$$

and their *resonance* frequencies when they are perfectly damped (d = 1) are expressed as

$$f_{0,T} = f_{0,\pi} = \frac{1}{2\pi}\sqrt{\frac{1}{LC}} \quad \text{for } d \approx 1 \tag{24}$$

For example, Fig. 17 graphically illustrates the *insertion losses* for the low-pass *T* filter per Eq. (21). This filter has the same two series inductance *L* and shunt capacitance *C*, respectively 2.5 μH and 2 nF. Additionally, this filter is terminated by equal source and load resistances R_S/R_L with values of 250/250Ω, 50/50Ω and 5/5 Ω.

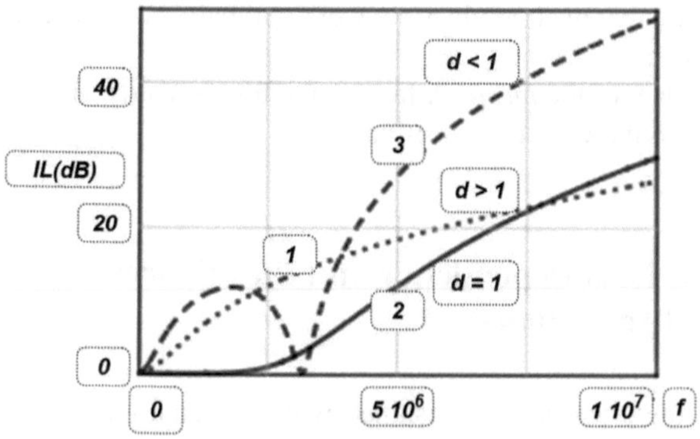

Fig. 17 T-Filter; R_S/R_L: (1) 250/250 Ω, d > 1; (2) 50/50 Ω, d = 1; (3) 5/5 Ω, d < 1

According to Eq. (23a), the *T*-filters with loads of 250/250 Ω, 50/50 Ω, and 5/5 Ω at both ends have the *damping factors of* $d > 1$, $d = 1$, and $d < 1$, respectively.

Curve 1 has a *higher insertion* loss (*d > 1*) in the filter's passband, resulting in filter inefficiency. *Curve* 2 shows that the filter works in the *optimal* regime (*d = 1*), whereas C*urve 3* shows that the filter does not work in the optimal regime (due to *d < 1*), having a large insertion loss bump in its passband.

As a further example, Fig. 18 graphically illustrates the *insertion losses* for the low-pass *Pi* filter expressed by Eq. (22). The filter has a series-centered inductance of 2.5 *μH* and two input and output shunt capacitances, each by 2 *nF*. The filter is loaded by equal source/load resistance pairs: R_S/R_L= 250/250 Ω, 50/50 Ω, and 5/5 Ω.

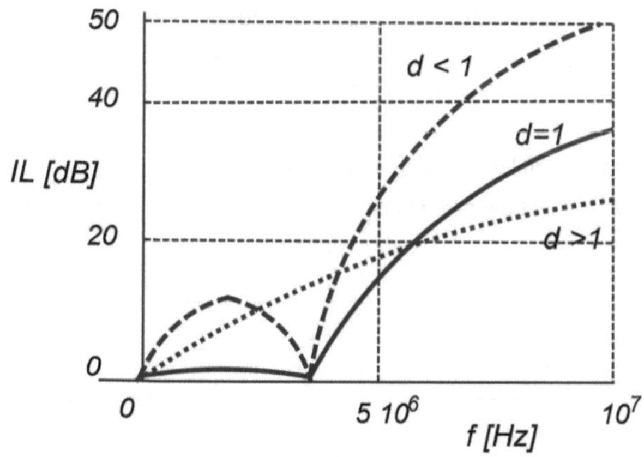

Fig. 18 Pi filter; R_S/R_L: 250/250 Ω, d < 1; 50/50 Ω, d = 0.25; 5/5 Ω, d > 1

Curve 1 here represents the *overdamped* filter with high *insertion loss* in the pass region, with a damping factor greater than 1 (*d* > 1). In this case, the filter does not work in an efficient mode.

Curve 2 represents the optimally damped filter with an *insertion* loss below 3 dB/ decade in the pass region and up to 60dB/decade in the filter stop region.

Curve 3 shows that the filter does not work optimally because it has a large bump of *insertion loss* in the pass region.

The dependency of the *insertion* loss on the filter's source and load impedances will be discussed in the next two sections.

7 Insertion Loss Dependency on Filter Source and Load Impedance

Conventional filters are usually designed with the same input and output impedances equal to some of the standard circuit impedances, such as 50, 75, 100, or 600 ohms, and as such, they do not have significant problems in terms of impedance matching. The situation of EMI filters is quite the opposite because the EMI source and load resistance may vary from low to high values, and in the worst case, they are commonly a reactive or complex impedance.

To ensure a fair enough impedance match between an EMI filter and its terminations, the filters should be categorized according to their input and output impedances, such as low-impedance and high-impedance filters. Table 5 shows the group of low-pass filters with low and high input/output impedances.

The conclusion from Table 5 is that if the filter element seen by a source and/or load is a series inductor, then the filter should be connected to a *low-impedance* source and/or load. However, if the input and/or output element is a shunt capacitance *C*, then the filter should be connected to a *high-impedance* source and/or load. This recommendation should be applied at frequencies where the filter is expected to operate. Additionally, "low" or "high" impedance does not indicate extremely low or extremely high impedance. In the case of impedance extremes, filters can be easily impedance-mismatched and, consequently, inefficient.

Table 5 Low-pass filters and source/load impedances

Source impedance	Low-pass filters	Load impedance
Low	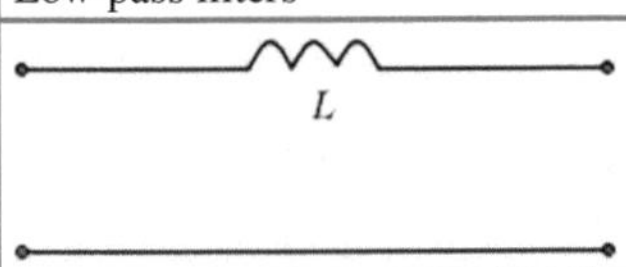	Low
High	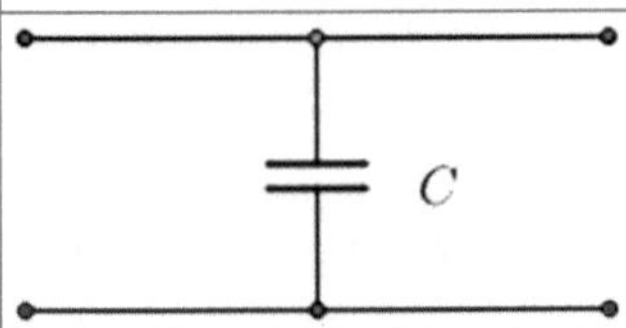	High
Low	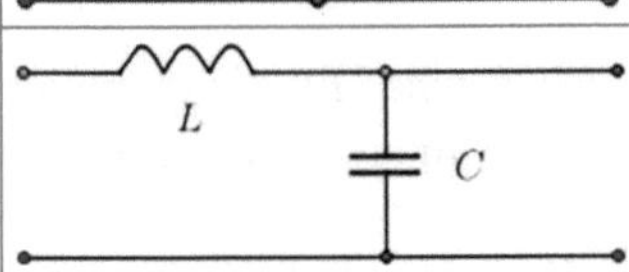	High

Table 5 (continued)

Source impedance	Low-pass filters	Load impedance
High	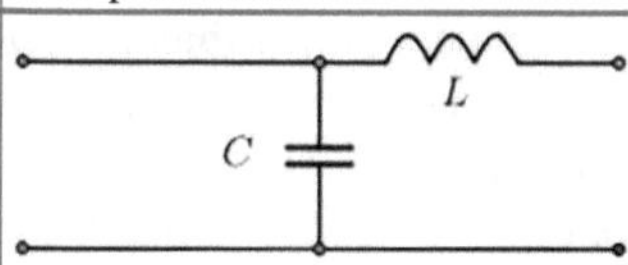	Low
Low	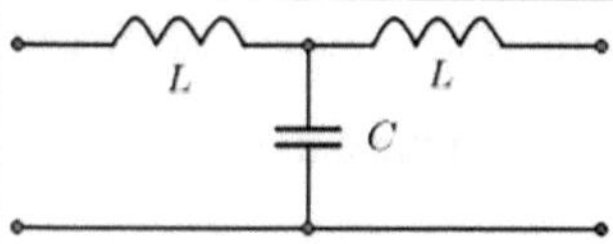	Low
High	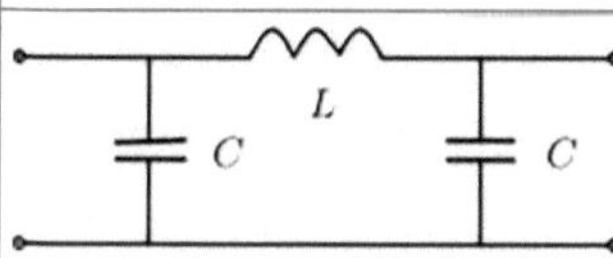	High
Low	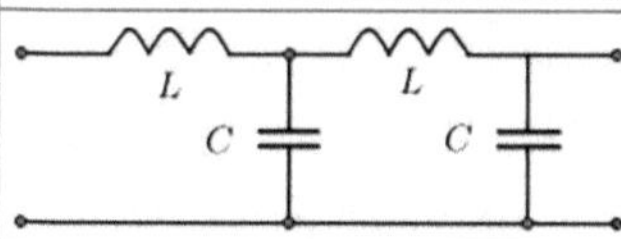	High
High	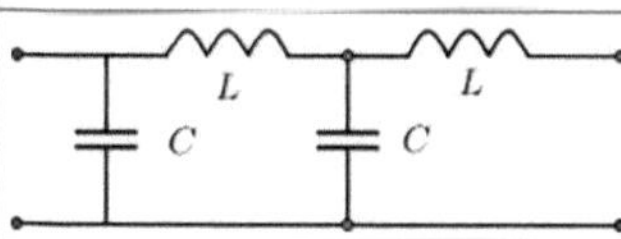	Low

As examples of the extreme filter impedance mismatch, Fig. 19 illustrates the problem when the CL low-pass filter is connected to a low-impedance source instead of a high-impedance source or to a high-impedance load instead of a low-impedance source.

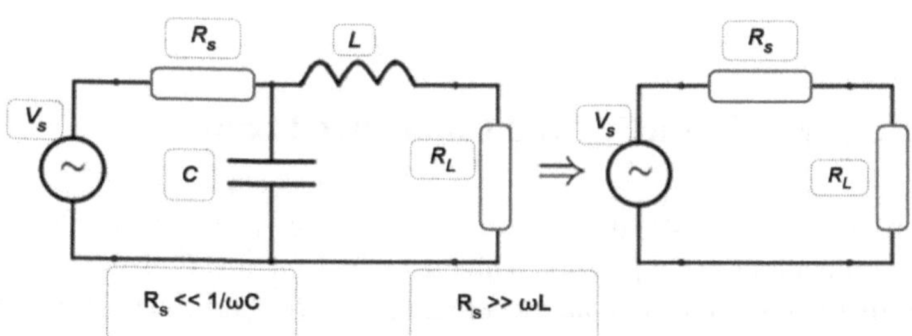

Fig. 19 Incorrect L-filter termination

The CL filter is transformed into a nonfilter, the so-called "all-pass" filter, when the source resistance R_s is much lower than the input capacitive reactance $\left(\frac{1}{\omega C}\right)$ and when the load resistance R_L is much greater than the output inductive reactance (ωL). In this case, the filter has an *insertion loss* in the stopband close to *0 dB/decade* instead of the expected *40 dB/decade*.

Furthermore, Fig. 20 illustrates the case in which a Pi-filter has a low-impedance *source* (R_s) much lower than the input capacitive reactance $\left(\frac{1}{\omega C}\right)$ and a low-impedance load (R_L) much lower than the output capacitive reactance $\left(\frac{1}{\omega C}\right)$.

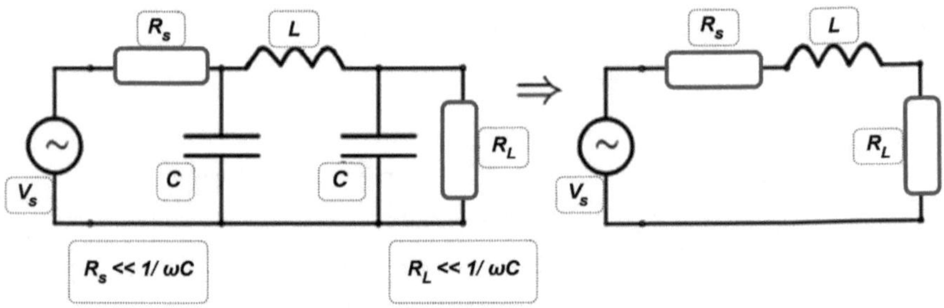

Fig. 20 Incorrect π-filter termination

The Pi filter is effectively reduced to a series single-element filter with an *insertion loss* in the stopband of *20 dB/decade* instead of the expected *60 dB/decade*. For good filter impedance math, the following equation should be used:

$$R_s \approx \left|\frac{1}{i\omega C}\right| \text{ and } R_L \approx |i\omega L| \tag{25}$$

More about the topic discussed in this section is given in [5, 6].

8 Filters with Inductive and Capacitive Loads

Other problems with EMI filters are their resonances and oscillations caused by reactive terminations in conjunction with their reactive L and C elements. For example, the filter's load may be *inductive* or *capacitive* (or both), which may enhance the filter's resonance with the *C* or *L* filter's element, causing significant degradation of the filter's performance.

For example, Fig. 21a, b shows the LC filter with the reactive elements $L = 2.5\ \mu H$ and $C = 1\ nF$ when $R_s \ll R_L$. This filter is alternatively terminated by the inductance $L_L = 1.6\ \mu H$ and the capacitance $C_L = 4\ nF$ parallel to the load resistance $R_L = 500\ \Omega$. Additionally, the source resistance is $R_s = 10\ \Omega$.

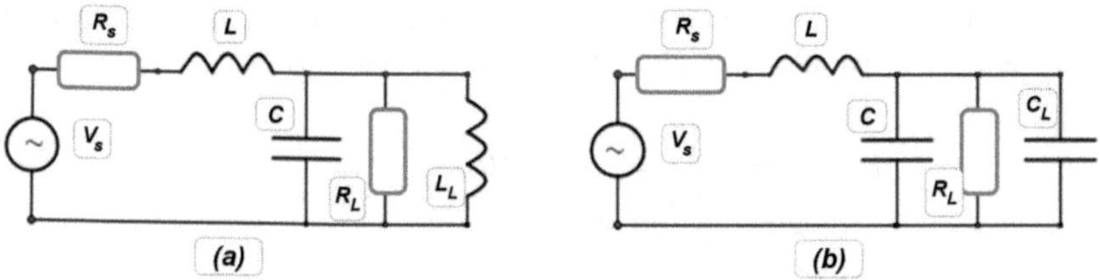

Fig. 21 Reactive terminations (**a**) with an inductive load and (**b**) with a capacitive load

Figure 22 shows the plot of the *insertion loss* with the filter's *self-resonance* when the filter is underdamped. Here, the three cases of filter self-resonance are shown: (a) with the purely resistive source $R_S = 10\ \Omega$ and the load $R_L = 500\ \Omega$, (b) with the parallel capacitance $C_L = 4\ nF$ to R_L, and (c) with the parallel inductance $L_L = 1.6\ \mu H$ to the R_L.

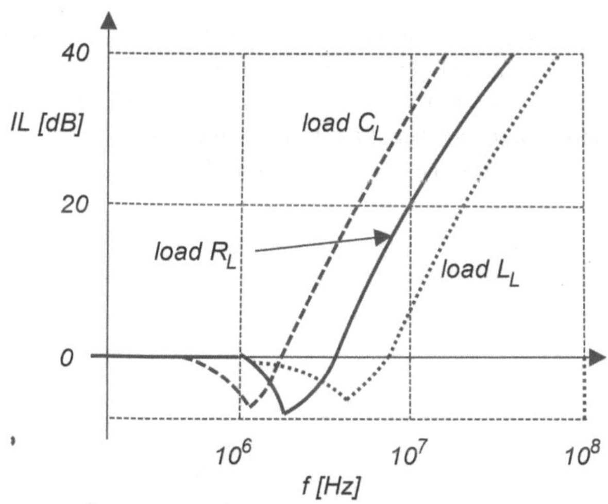

Fig. 22 The IL plots for the underdamped filter loaded with resistive, capacitive, and inductive loads

In the present plots, a solid curve (R_L) belongs to the filter with a purely resistive load, whereas curves (C_L) and (L_L) belong to the filter with *capacitive* and *inductive* loads, respectively. Therefore, the capacitive load causes the filter's bandwidth to narrow (C_L curve), whereas the inductive load causes the filter's bandwidth to widen (L_L curve).

Per Eq. (20), the self-resonance of the LC low-pass filter with $L = 2.5\ \mu H$ and $C = 1\ nF$ and with purely resistive terminations at both ends is

$$f_{0R} = \frac{1}{2\pi\sqrt{LC}} = 3.18\,MHz \tag{26}$$

as shown in Fig. 22 by curve (R_L).

Additionally, per circuit theory, the self-resonance for an *LC* filter terminated by the inductance L_L parallel to the load R_L can be calculated as

$$f_{0L} = \frac{1}{2\pi\sqrt{\left(\frac{L \cdot L_L}{L + L_L}\right)C}} \approx 6\,MHz \tag{27a}$$

where $L_L = 1\ \mu H$. Additionally, this resonance is shown by the curve (L_L) in Fig. 22. Thus, the inductive load L_L produces the resonance $f_{0L} \approx 6\ MHz$ in conjunction with the filter's *L* and *C* elements.

Additionally, per circuit theory, the self-resonance for an LC filter loaded by the resistance R_L in parallel with the capacitance C_L can be calculated as

$$f_{0C} = \frac{1}{2\pi\sqrt{L(C + C_L)}} \approx 1.4\,MHz \tag{27b}$$

where $C_L = 4\ nF$. Also, this resonance is shown in Fig. 22 by the curve (C_L).

Note that in both cases (Figs. 22 and 23), the *capacitive* load makes the filter bandwidth *narrow* (curve C_L), whereas the *inductive* load makes the filter bandwidth *wide* (curve L_L), regardless of the level of filter damping.

More about the topic discussed in this section is given in [6].

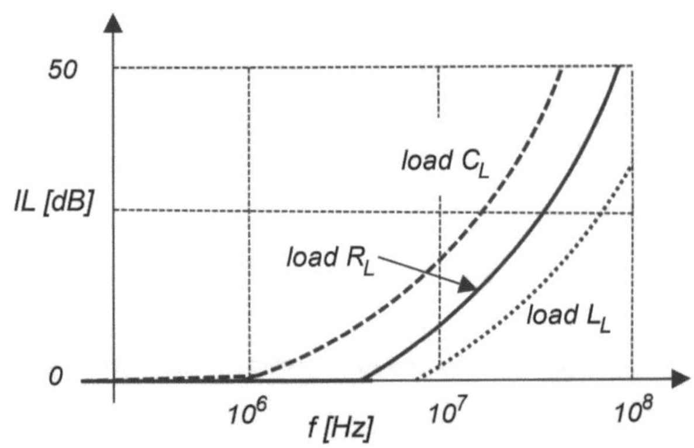

Fig. 23 A well-damped filter bandwidth with capacitive, resistive, and inductive load

9 Effects of Filter Parasites on Filter Performance

Real passive circuit components are not ideal, as noted in Chapter "Nonideal Passive Circuit Components". This is especially true at high frequencies. Consequently, a filter with nonideal *L* and *C* components will start to lose filtering efficiency at high frequencies, i.e., its *insertion loss* begins to fall at the frequencies where the filter's parasites dominate in the filter's impedance profiles. Thus, to minimize the loss of filter efficiency, the parasites of the filter elements should be controlled and minimized [5, 6].

9.1 High-Frequency Capacitors

Realistic ceramic capacitors perform well at relatively high frequencies because they have small *R* and *L* parasites due to their small packages and very short leads. In particular, capacitors in small packages with no leads, such as SMD capacitors, have very few parasites and high self-resonance. Hence, small capacitors with higher self-resonance are preferable for decoupling high-frequency EMI noise, as considered in Chapter "Nonideal Passive Circuit Components". To select proper decoupling capacitors for specific frequency spectra, manufacturing catalogs are commonly available on manufacturers' webs.

Furthermore, with respect to dielectric materials and EMI applications, it is better to have a capacitor with a somewhat lossy dielectric, such as in Y5V and X7R technology capacitors. Additionally, the capacitor size depends on the operating voltage and current because a higher voltage and current require a larger capacitor that has larger parasites. For example, high-voltage and high-current filters are needed to suppress lightning current effects. With respect to high frequencies, the lightning currents have a relatively low-frequency content, as discussed in Chapter "Lightning Physics and Impact on Radio Systems on the Earth", where the lightning physics is discussed.

9.2 LC-Filter with Nonperfect Components

Figure 24 shows the circuit of an LC low-pass filter with *imperfect* *L* and *C* elements.

Fig. 24 LC filter with nonperfect L and C elements

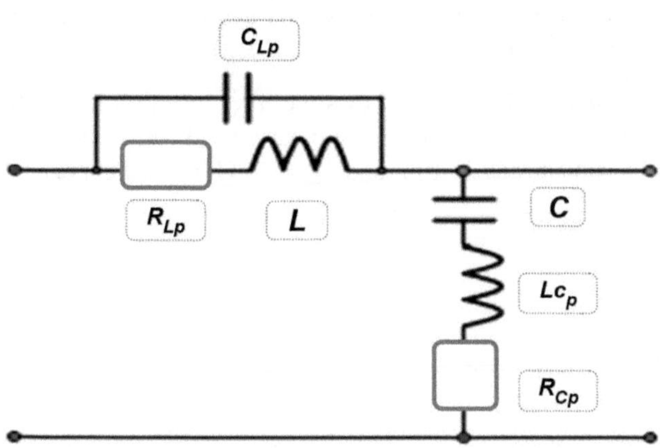

Because parasites C_{Lp} and L_{Cp} are more critical for EMI filter performance than parasitic resistance R_{Cp}, we consider filters with only *reactive parasites,* i.e., $R_{Lp} = R_{Cp} = 0$. Per Eq. (17), the *insertion loss* for an *LC* filter is expressed in terms of the filter reactance only as

$$IL = 20\log\left|\left(\frac{R_L}{R_S + R_L}\right)\left[1 + \frac{\left(R_S + \tilde{X}_1\right)}{\tilde{X}_2 // R_L}\right]\right| \tag{28}$$

where C_{Lp} and L_{Cp} are the parasites expressed as

$$\tilde{X}_1 = \left(i\omega L\right) // \frac{1}{\left(i\omega C_{Lp}\right)} \tag{29a}$$

$$\tilde{X}_2 = \left(\frac{1}{i\omega C} + i\omega L_{Cp}\right) \tag{29b}$$

Substituting Eqs. (29a), (29b) into Eq. (28), the *insertion loss* becomes

$$IL = 20\log\left|\left(\frac{R_L}{R_S + R_L}\right)\left[1 + \frac{R_S + \left(i\omega L\right) // \frac{1}{\left(i\omega C_{Lp}\right)}}{R_L // \left(\frac{1}{i\omega C} + i\omega L_{Cp}\right)}\right]\right| \tag{30}$$

or

$$IL = 20\log\left|\left(\frac{R_L}{R_S + R_L}\right)\left[1 + \frac{R_S + i\omega L / \left(1 - \omega^2 L C_{Lp}\right)}{R_L\left(1 - \omega^2 C L_{Cp}\right) / \left(R_L + 1 - \omega^2 C L_{Cp}\right)}\right]\right| \tag{30a}$$

The next example numerically illustrates the difference between the *insertion losses* of the LC filter with *perfect* and *imperfect* components on the basis of Eqs. (28) and (30).

Example 4 Figure 25 shows the filter insertion loss with the regular components L = 2.5 μH and C = *1 nF* (solid curve) and with *imperfect* L and *C* components with the parasites C_{Lp} = 1 *pF* and L_{Cp} = 1 *nH* (dashed curve).

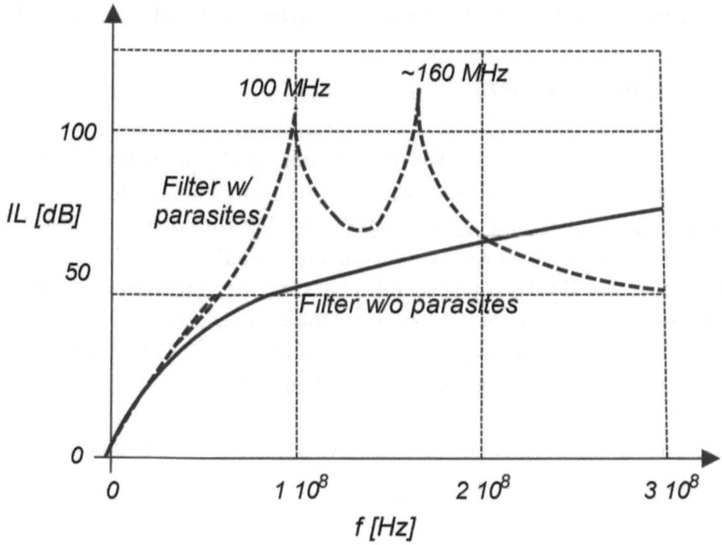

Fig. 25 Insertion losses for the filters without and with parasites

Note from the plot that the parasites of the filter's components introduce two resonance peaks above 100 MHz. In particular, the first resonance is associated with the parallel resonance of the inductor $L = 2.5\ \mu H$ and its parasitic capacitance $C_{Lp} = 1\ pF$:

$$f_{Lp} = \frac{1}{2\pi\sqrt{L\,C_{Lp}}} \approx 100\,MHz \tag{31}$$

The second resonance is associated with the series resonance of the main capacitor $C = 1\ nF$ with its parasitic inductance $L_{Cp} = 1\ nH$:

$$f_{Cp} = \frac{1}{2\pi\sqrt{C\,L_{Cp}}} \approx 160\,MHz \tag{32}$$

As shown in Fig. 25, the filter's insertion loss starts to fall after 200 MHz.

More information about the filters with nonperfect components can be found in reference [6].

10 Differential Mode (DM) and Common Mode (CM) Filters

10.1 6-Terminal Filters

We considered various types of 4-terminal filters in the preceding sections. Now, we analyze 6-terminal filters that include a reference conductor. As considered in Chapter "General Concept and Types of EMI Emissions", the differential mode (DM) and common-mode (CM) currents usually flow along the differential lines, and accordingly, 6-terminal filters are needed for effective suppression of both mode currents. The suppression of the CM currents is a priority for EMI applications because they contribute significantly more to EMI emissions than the DM currents do, as noted in Chapter "General Concept and Types of EMI Emissions".

Figure 26a shows a typical 6-terminal low-pass filter with *L*-topology as a combination of DM and CM filters.

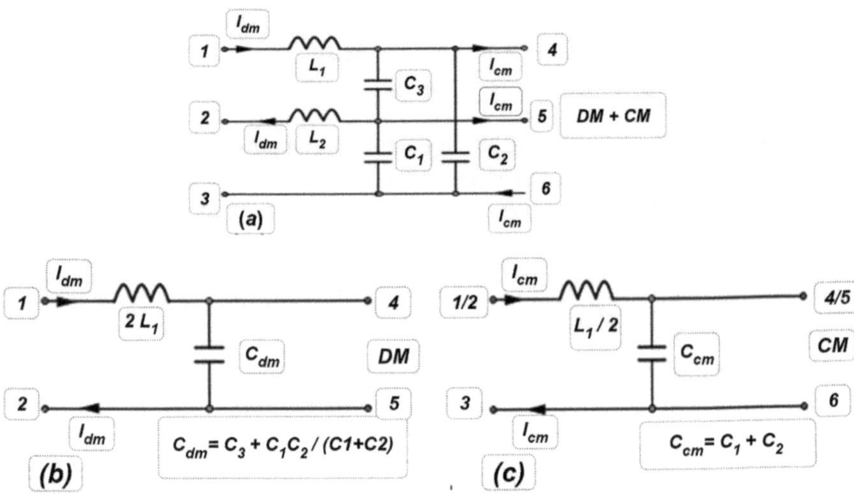

Fig. 26 (**a**) DM + CM composite filter, (**b**) equivalent DM circuit, (**c**) Equivalent CM circuit

The composite filter from Fig. 26a contains two identical and uncoupled coil inductors with an inductance *L* and three capacitors C_1, C_2, C_3. Figure 26b, c separately shows the equivalent circuits for the *differential-mode* and the *common-mode* filters as parts of the composite filter. The subscripts *dm* and *cm* denote the differential-mode and common-mode filter elements and currents, respectively. We discuss the main power filters in the next subsections.

10.2 A Common-Mode Choke Transformer

One of the most effective CM filters is the common-mode choke, which can effectively suppress undesired CM currents without significantly affecting the desired DM currents. For example, Fig. 27a, b, c shows (a) the perfect CM choke as a series filter, (b) its equivalent DM circuit, and (c) its equivalent CM circuit.

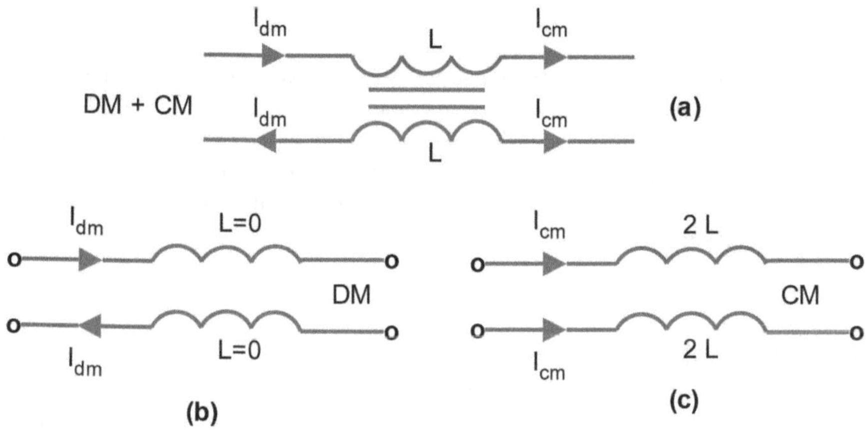

Fig. 27 (**a**) Perfect choke transformer, (**b**) equivalent DM circuit, (**c**) equivalent CM circuit

Figure 27a shows the two well-coupled inductors each per line, which are winded on a magnetic or ferrite core with high permeability. Hence, the choke circuit contains two identical self-inductances L and mutual inductance L_M. Owing to the *opposite directions* of the DM currents on signal lines 1 and 2, the effective DM inductance L_{dm} per line is given as

$$L_{dm} = L - L_M \tag{33}$$

Owing to CM currents having the *same directions* on lines 1 and 2, the resulting CM inductance L_{cm} per line is expressed as

$$L_{cm} = L + L_M \tag{34}$$

Thus, CM currents effectively flow through a larger inductance than DM currents. Because inductors 1 and 2 are well coupled, we can consider $L \approx L_M$, and Eqs. (33) and (34) result in $L_{dm} \approx 0$ and $L_{cm} \approx 2\ L$. Consequently, the impedances of the DM and CM circuits are as follows:

$$Z_{dm} = \frac{V_{dm}}{I_{dm}} = \omega(L - L_M) \approx 0 \tag{35a}$$

$$Z_{cm} = \frac{V_{cm}}{I_{cm}} = \omega\left(L + L_M\right) \approx 2\omega L \tag{35b}$$

Additionally, regarding filter theory, the coupling coefficient k between two well-coupled inductors with the same inductance L is close to 1:

$$k = \frac{L_M}{\sqrt{L_1 L_2}} = \frac{L_M}{L} \cong 1 \quad > \quad L \cong L_M \tag{36}$$

which confirms that the *mutual inductance* between two well-coupled inductors is approximately equal to the inductance of one inductor.

Note that here, we did not consider the inductance of the choke leads (called the *leakage parasitic* inductance) because this parasitic inductance is much smaller than the choke main inductance L. The choke parasites will be discussed later in the following section.

11 Power Mains Filters (PMF)

Now, we discuss the filters for suppressing EMI currents produced by power digital circuits and switching power supplies (PSs). These currents commonly flow into a power mains grid and contaminate other equipment connected to the grid if they do not have adequate mains filters.

There are two emission problems associated with switching PS: the rectification process of main currents and the switching processes in the *PS*. The switching emission is associated with fundamental PS switching frequencies and their harmonics. The approximate frequency spectra of the DM and CM emissions from the power supplies are shown in Fig. 28.

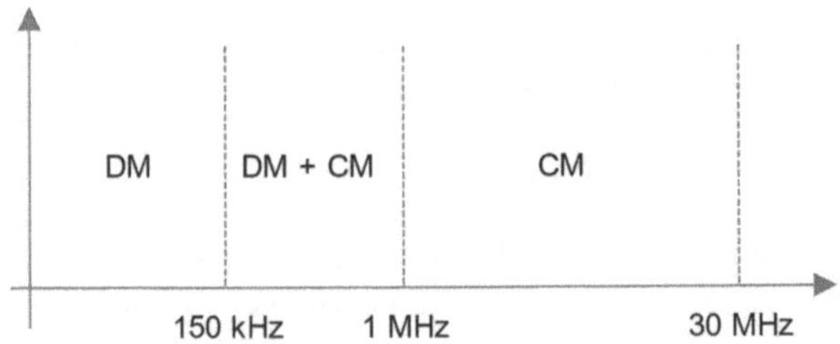

Fig. 28 Spectra of AC DM and CM emissions

A basic scheme of a main filter with a CM choke is shown in Fig. 29a–c.

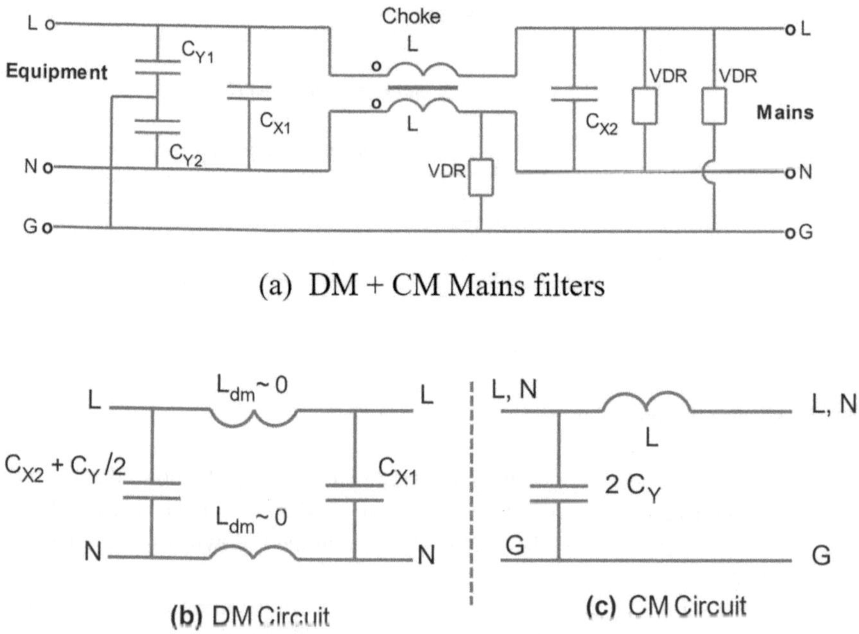

Fig. 29 (**a**) Basic mains filter, (**b**) its equivalent DM circuit, (**c**) its equivalent CM circuit

Here, the filter terminals *L, N*, and *G* denote a phase line, neutral line, and ground (or safety) line, respectively. Thus, the filter contains components to block both *CM* and *DM* currents flow from a PS into the power mains.

Additionally, Fig. 29b, c shows the equivalent DM and CM filter circuits in addition to the filter elements C_X, C_Y, and L. For higher accuracy, we also need to consider a *leakage* (*parasitic*) inductance in the DM circuit associated with the choke leads and magnetic flux *leakage* from the choke coil.

The following sections will describe the roles of the filter components C_X, C_Y, and L.

11.1 Surge Protection of Mains Filters

Regarding the surge protection of the main filters, Fig. 30 shows the response of the main filter to a surge transient without and with surge protection.

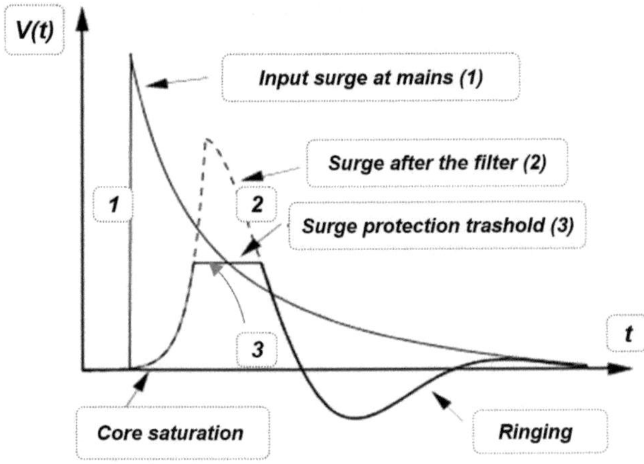

Fig. 30 Clump surge protection

In this figure, curve (1) shows the input surge pulse, curve (2) shows the response of the filter with a transient protector to the input transient with pulse delay and ringing, and curve (3) shows the surge pulse clumped down to the protector break-voltage.

Hence, the solution against incoming transients from the main lines is the use of nonlinear devices at AC equipment power terminals. The nonlinear devices are usually varistors (VDRs—voltage-dependent resistors), Zener diodes, gas discharge tubes, or spark gaps. Figure 29 shows the main filter with VRD transient protection for both the DM and CM transients.

Owing to potential surge problems, the L and C elements of the main filters are rated and standardized accordingly in terms of high voltage and currents. As a result, the equipment and its capacitive elements need to pass mandatory high-pot and leakage current tests to comply with safety standards. For example, test failure of capacitors C_X and C_Y may result in fire and shock hazards.

11.2 Nonperfect CM Chokes with Parasites

One more critical point of the CM choke should be noted. As discussed earlier in this chapter, LC filter elements are not perfect at higher frequencies because they contain parasites. As mentioned in Chapter "Nonideal Passive Circuit Components", a real inductor contains parasitic parallel capacitance, which causes a parallel resonance that degrades the filter performance, such as *insertion loss and bandwidth.* Therefore, in terms of the CM choke inductors, unintentional (parasitic) capacitance exists across the choke inductors because of the interwinding capacitance. This parasitic capacitance is shown in Fig. 31, where the choke is a two-line series filter.

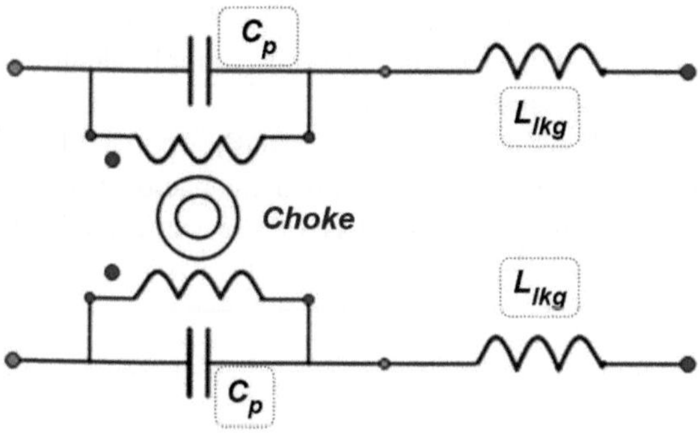

Fig. 31 Capacitive parasites of a choke

For example, a choke with an inductance of $L = 0.5\ mH$ and a *parasitic* capacitance of $C_p = 1\ pF$ has an antiresonance f_{ar} of ~7 MHz:

$$f_{ar} = \frac{1}{2\pi\sqrt{LC_p}} = \frac{1}{2\pi\sqrt{0.5mH \cdot 1pF}} \approx 7\ MHz$$

Because the parasitic capacitance forms a series bandstop filter (earlier considered in this chapter), we use Eq. (4) to plot the *insertion loss* of the choke:

$$IL_{ser} = 10\log\left[1+\left(\frac{X_1}{R_S+R_L}\right)^2\right]\ (dB) \tag{37}$$

where X_1 is given in Table 2 as

$$\tilde{X}_1 = \frac{i\omega L_{chk}\cdot\frac{1}{i\omega C_p}}{i\omega L_{chk}+\frac{1}{i\omega C_p}} = \frac{1}{i\left(\omega L_{chk}-\frac{1}{\omega C_p}\right)}\frac{L}{C_p} \rightarrow$$
$$\left|\tilde{X}_1\right| = \frac{1}{\left(\omega L_{chk}-\frac{1}{\omega C_p}\right)}\cdot\frac{L_{chk}}{C_p} = \frac{\omega L}{\left(\omega^2 C_p L_{chk}-1\right)} \tag{38}$$

and substituting Eq. (38) into Eq. (37) results in

$$IL_{ser} = 10\log\left[1+\left(\frac{\omega L}{\left(R_S+R_L\right)\left(\omega^2 C_p L_{chk}-1\right)}\right)^2\right]\ (dB) \tag{39}$$

Using Eq. (39), we plot the insertion losses for the choke *without* and *with* parasitic capacitance C_p, as shown in Fig. 31. This example includes a low-resistivity source and a load with resistance $R_S = R_L = 5\ \Omega$.

Therefore, Fig. 32 shows the insertion loss (dashed curve) of the imperfect choke *with* an inductance of $L = 0.5\ mH$ and a *parasitic* capacitance of $C_p = 1\ pF$, whereas the solid curve shows the insertion loss of the same choke without parasitic capacitance.

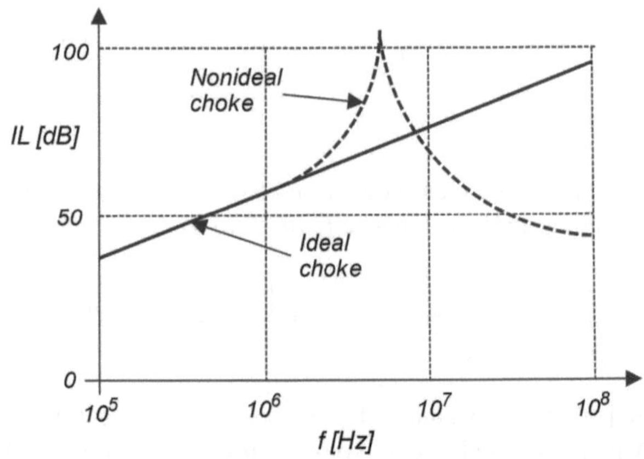

Fig. 32 Dash curve: a choke w/parasitic C, and solid curve: choke w/o parasitic C

The choke parasitic capacitance causes resonance at 7 MHz. Here, the insertion loss of the choke starts to drop at 9 MHz below the insertion loss of the perfect choke (without a parasitic capacitance).

Additionally, the choke has *parasitic* or *leakage* inductance that belongs to its leads. This inductance is denoted as L_{lkg} in the present circuit (Fig. 31). L_{lkg} is usually much smaller than the operating choke inductance L (i.e., $L_{lkg} \ll L$). Owing to this fact, the leakage inductance is usually neglected when chokes are intended for low-frequency applications.

Additionally, one more critical point of CM choke is the *saturation* of its ferromagnetic or ferrite core, which is related to the overrated intensity of the CM current flow through the choke. This topic is discussed in Chapter "Nonideal Passive Circuit Components".

More information about the main filters can be found in Ref. [5]. Additionally, interesting topics about EMI filters are in references [8–10], which we recommend to the readers interested in further reading.

References

1. H.G. Dimopoulos, *Analog Electronic Filters, Theory, Design and Synthesis* (Springer, 2012)
2. R. Smolenski, *Conducted EMI in Smart Grids* (Springer, 2012)
3. A. Williams, F.J. Taylor, *Electronic Filter Design Handbook*, 4th edn. (McGraw-Hill, 2006)
4. Editor-in-Chief W.K. Chen, Circuits and Filters Handbook, IEEE Press, 1995
5. T. Williams, *EMC for Product Designers* (Elsevier, Newnes, 2007)
6. L. Tihanyi, *Electromagnetic Compatibility in Power Electronics* (IEEE Press, 1995)
7. W.T. Rhoades, Conduction filters, in *Handbook of EMC*, ed. by R. Perez, (Academic Press, 1995)
8. A. Gerfer et al., *Trilogy of Magnetics: Design, Guide for EMI Filters Design, SMP RF Circuits* (Swiridoff Verlag, 2018)
9. R.L. Ozenbaugh, T.M. Pullen, *EMI Filter Design* (CRS Press, 2012)
10. A. Goldman, *Modern Ferrite Technology* (Springer, 2005)

References

EMI Shielding for Radio Equipment

1 General Concept of EMI Shielding

The purpose of EMI shielding is to achieve two goals: first, to reduce unwanted EMI emissions from electric and electronic devices and second, to protect devices against external EMI fields. Figure 1a, b shows two models for the purpose of shielding. Figure 1a shows a device as an EMI emitter (*T*) within a shielding enclosure, whose task is to reduce unwanted radiation from the device. In contrast, Fig. 1b shows a device as an EMI receptor (*R*) inside a shielding enclosure, whose task is to protect the device from external interfering fields.

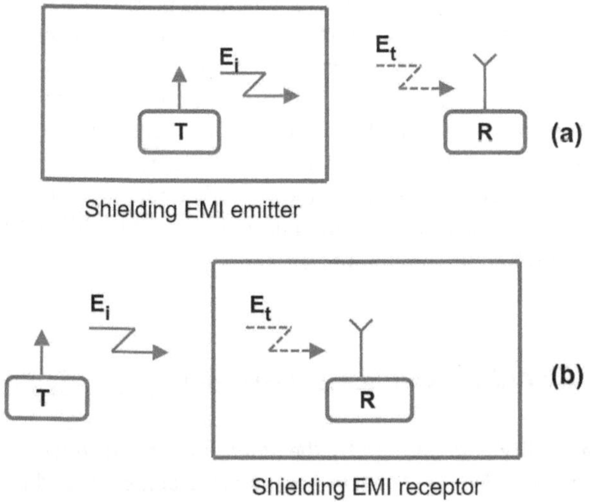

Fig. 1 (**a**) Reduction in unwanted radiation from equipment and (**b**) Shielding equipment against external fields

M. Pajovic, *Fundamentals of EMI Phenomena in Radio Communication Systems*, https://doi.org/10.1007/978-3-031-88748-2_10

Furthermore, Fig. 2 shows a large conductive barrier (wall) of an enclosure with thickness *d*, conductivity σ, and relative permeability μ_r. Additionally, the figure shows the *incident* wave and reflected waves R_1, R_2, … from the left surface of the barrier and transmitted waves T_1, T_2, …, through the right barrier surface.

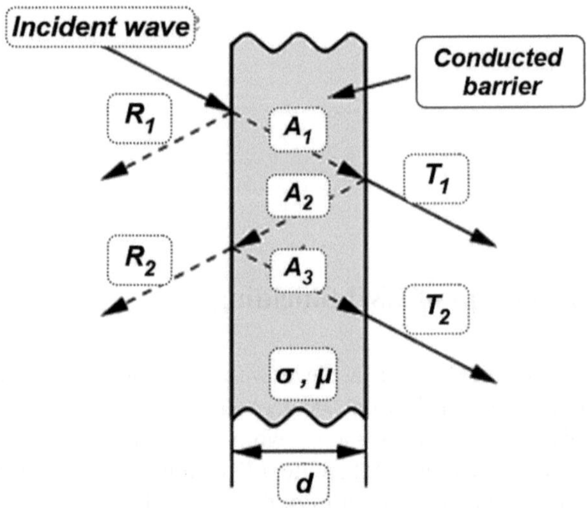

Fig. 2 Multiple reflections inside the shield wall

Owing to finite shield conductivity (the shielding barrier is not a perfect conductor), some wave energy is absorbed and converted into heat inside the shield wall. Additionally, multiple reflected waves occur when the waves travel back and forth in the wall, as shown in Fig. 2. In the following section, we consider the *shielding effectiveness* (SE) of an EMI shield with limited conductivity.

The writing of this chapter is motivated and influenced by the references [1–8], which we refer to readers interested in more detailed reading. Additionally, readers interested in further studying conventional and new EMI shielding concepts and materials are referred to the literature [10–17].

1.1 Definition of Shielding Effectiveness

The shielding quality is quantified by the *shielding effectiveness,* which is defined as the *ratio* of the *incident* field, electric or magnetic at the left surface of the barrier, and the field transmitted through the barrier:

$$SE(dB) = 20 \cdot \log\left|\frac{\tilde{E}_i}{\tilde{E}_t}\right| \equiv 20 \cdot \log\left|\frac{\tilde{\mathrm{H}}_i}{\tilde{\mathrm{H}}_t}\right| \tag{1}$$

where $\tilde{E}_i$ and $\tilde{H}_i$ are the incident electric and magnetic fields, respectively, since $\tilde{E}_t$ and $\tilde{H}_t$ are the transmitted electric and magnetic fields through the barrier, respectively. Additionally, we can simply define the *shielding effectiveness* of a device enclosure by ratio of field intensity at a point away from the device *without* and *with* the enclosure.

Furthermore, per the shielding model in Fig. 2, the *total shielding effectiveness* can be expressed as

$$SE_{tot}(dB) = R(dB) + A(dB) + M_R(dB) \tag{2}$$

where $R(dB)$ is the *reflection loss* due to wave reflection at the first wall surface, $A(dB)$ is the *absorption loss* inside the wall caused by the conversion of electromagnetic energy into heat, and $M_R(dB)$ is the *multiple-reflection loss* caused by the *multiple reflections* inside the wall. As shown in Fig. 2, *multiple reflection losses* also contribute to the transmission of waves through the wall barrier, degrading the shielding effectiveness at some level. This will be discussed later in this chapter.

2 Near-Field Shielding

In *Appendix B*, we discuss the difference between the *near-field* and *far-field* radiation regions associated with the distance from a radiation source. The structure of the field closer to the radiation source (up to the radial distance $\approx \lambda/6$) is more complex than the field structure in the *far-field* region ($>\lambda$), where uniform planar TEM waves exist. In this section, we discuss *near-field* shielding and, in the next section, *far-field* shielding methods. More details about these topics are given in Refs. [1–6].

In terms of *near-field shielding*, the two methods are, in practice, dependent on whether a source dominantly radiates *magnetic* or *electric* fields. As noted in *Appendix B*, the key difference between the *far-field* and *near-field* sources is associated with the wave impedance in these regions. In the *far-field* region, the wave impedance is the intrinsic impedance η_0 of free space, equal to $\approx 377\ \Omega$, whereas in the *near-field* region, the wave impedance depends on whether the field is dominantly *electric* or *magnetic*.

Additionally, as considered in *Appendix B*, the source of the *electric field* is the small electric dipole whose magnitude of wave impedance in the *very near-field* region ($r < 0.1\ \lambda_0$) is expressed as

$$\left|\tilde{Z}_w\right|_{ne} \cong \eta_0 \frac{\lambda_0}{2\pi r} = \frac{1}{\omega \varepsilon_0 r} \tag{3}$$

where λ_0 is the wavelength in free space, r is the radial distance of an observing point from the radiation source, and ω is the angular frequency ($\omega = 2\pi f$).

Furthermore, the magnitude of the wave impedance of the *magnetic field source* (a small current loop) in the *near-field* region is expressed as

$$\left|\tilde{Z}_w\right|_{nm} \cong \eta_0 \frac{2\pi r}{\lambda_0} = \omega \mu_0 r \tag{4}$$

Notably, the *electric field source* has a higher impedance magnitude in the near-field region than in the *far-field* region, whereas the impedance magnitude of the *magnetic field source* is lower in the *near-field* region than in the *far-field* region. Therefore, the *near-field electric source* is the *high-impedance* source, whereas the *near-field magnetic source* is a *low-impedance source*. In the next section, we discuss the effectiveness of *near-field* shielding.

2.1 Intrinsic Impedance of a Shield Barrier

Figure 3 shows the *reflected* and *transmitted* electric and magnetic fields at and inside a long conductive barrier with intrinsic impedance $\hat{\eta}_b$.

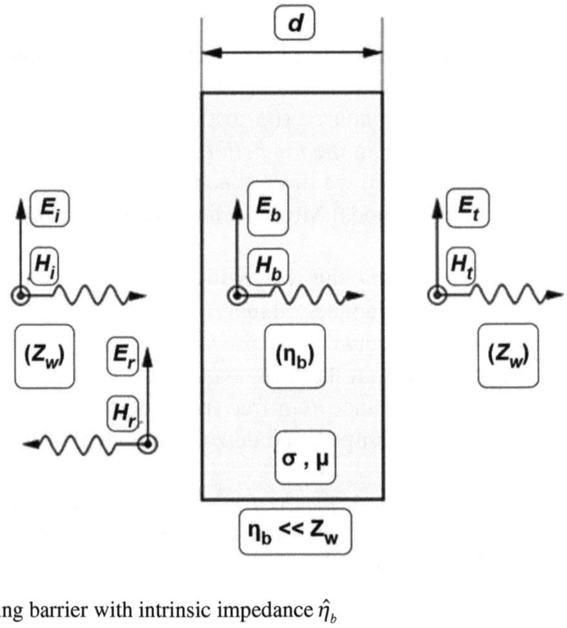

Fig. 3 A shielding barrier with intrinsic impedance $\hat{\eta}_b$

For conductors and dielectrics, per Eq. (6a) in Chapter "Nonideal Passive Circuit Components", the intrinsic impedance is

$$\tilde{\eta}_c = \tilde{\eta}_b = \sqrt{\frac{\tilde{Z}'}{\tilde{Y}'}} = \sqrt{\frac{i\omega\mu}{\sigma + i\omega\varepsilon}} \tag{5}$$

where $\tilde{Z}'$ and $\tilde{Y}'$ are a medium's per-unit-length impedance and admittance, respectively. Additionally, σ, ε, and μ are the conductivity, permittivity, and permeability of the medium, respectively. For a good conductor, where $\sigma \gg \omega\varepsilon$, Eq. (5) has a simplified form:

$$\tilde{\eta} \cong \sqrt{\frac{i\omega\mu}{\sigma}} \tag{6}$$

Owing to the *total shielding effectiveness* expressed by Eq. (2) as the sum of the *reflection* loss R, *absorption* loss A, and *multiple-reflection* loss M_R, we analyze these losses separately in the following subsections.

2.2 Reflection Loss (R)

By analogy with the transmission line theory from *Appendix C*, the *reflection* coefficient $\tilde{\Gamma}$ and *transmission* coefficient $\tilde{T}$ at the boundary between arbitrary mediums 1 and 2 are *generally* expressed as

$$\tilde{\Gamma} = \frac{\tilde{E}_r}{\tilde{E}_i} = \frac{\tilde{Z}_2 - \tilde{Z}_1}{\tilde{Z}_1 + \tilde{Z}_2} \tag{7}$$

$$\tilde{T} = \frac{\tilde{E}_t}{\tilde{E}_i} = \frac{2\tilde{Z}_2}{\tilde{Z}_1 + \tilde{Z}_2} \tag{8}$$

where $\tilde{Z}_1$ and $\tilde{Z}_2$ are the intrinsic impedances of media 1 and 2, respectively.

When Eq. (8) is applied for the case in Fig. 3, the transmission coefficients at the first and second boundary surfaces are expressed as

$$\tilde{T}_1 = \frac{\tilde{E}_b}{\tilde{E}_i} = \frac{2\tilde{\eta}_b}{\tilde{\eta}_b + \tilde{Z}_w} \tag{9}$$

$$\tilde{T}_2 = \frac{\tilde{E}_t}{\tilde{E}_b} = \frac{2\tilde{Z}_w}{\tilde{\eta}_b + \tilde{Z}_w} \tag{10}$$

where $\tilde{Z}_w$ is the wave impedance in the air outside the shield barrier.

Furthermore, the part of the shielding effectiveness SE_R associated with the *reflection loss R* can be expressed as

$$R = SE_R = \left|\frac{\tilde{E}_i}{\tilde{E}_t}\right| = \left|\frac{\tilde{E}_i}{\tilde{E}_b}\cdot\frac{\tilde{E}_b}{\tilde{E}_t}\right| = \left|\frac{1}{\tilde{T}_1}\cdot\frac{1}{\tilde{T}_2}\right| \tag{11}$$

and substituting Eqs. (9) and (10) into Eq. (11), the shielding effectiveness (i.e., the *reflection loss R*) for *near-field* sources is expressed as

$$R = SE_R = \left| \frac{\left(\tilde{\eta}_b + \tilde{Z}_w\right)^2}{4\tilde{\eta}_b \tilde{Z}_w} \right| \tag{12}$$

Because $\tilde{\eta}_b \ll \tilde{Z}_w$ for a good conductor, Eq. (12) has the simplified form

$$R \cong \left| \frac{\left(\tilde{\eta}_b + \tilde{Z}_w\right)^2}{4\tilde{\eta}_b \tilde{Z}_w} \right| = \frac{Z^2{}_w}{4\tilde{\eta}_b \tilde{Z}_w} = \frac{\tilde{Z}_w}{4\tilde{\eta}_b} \tag{13}$$

Additionally, the *reflection loss* expressed by Eq. (13) can be expressed in *dB* as

$$R(dB) \cong 20\log \left| \frac{\tilde{Z}_w}{4\tilde{\eta}_b} \right| \tag{14}$$

Furthermore, with Eq. (13), the *shielding effectiveness* against the *near-electric* field can be expressed as

$$R_{ne}(dB) \cong 20\log \left| \frac{\tilde{Z}_{w,ne}}{4\tilde{\eta}_b} \right| \tag{15}$$

Since the shielding effectiveness against the *near-magnetic* field is given as

$$R_{nm}(dB) \cong 20\log \left| \frac{\tilde{Z}_{w,nm}}{4\tilde{\eta}_b} \right| \tag{16}$$

Per Eqs. (3) and (4), $\tilde{Z}_w$ has an imaginary form in the *very-near-electric* and *magnetic* fields for $r < 0.1\ \lambda_0$. Their magnitudes are as follows:

$$\left|\tilde{Z}_w\right|_{ne} \cong \eta_0 \frac{\lambda_0}{2\pi r} = \frac{1}{\omega \varepsilon_0 r} \tag{17}$$

$$\left|\tilde{Z}_w\right|_{nm} \cong \eta_0 \frac{2\pi r}{\lambda_0} = \omega \mu_0 r \tag{18}$$

Per Eq. (6), the wave impedance $\tilde{\eta}_b$ for a good conductive barrier is expressed as

$$\tilde{\eta}_b \cong \sqrt{\frac{i\omega\mu}{\sigma}} \tag{19}$$

For $\omega = 2\pi f$ and the barrier parameters $\sigma = \sigma_r \sigma_0$ and $\mu = \mu_r \mu_0$, where σ_r and μ_r are the relative barrier conductivity and permeability, respectively, the following equation is obtained:

$$|\tilde{\eta}_b| \cong 3.710^{-7}\sqrt{f\frac{\mu_r}{\sigma_r}} \tag{20}$$

Furthermore, substituting Eqs. (17), (18), and (20) into Eqs. (15) and (16), we can express the *reflection losses $R_{ne}(dB)$* and $R_{nm}(dB)$ for *very near electric* and *magnetic* fields, respectively, as

$$R_{ne}(dB) \cong 320 + 10\log\left(\frac{\sigma_r}{\mu_r r^2 f^3}\right) \tag{21a}$$

$$R_{nm}(\text{dB}) \cong 15 + 10\log\left(\frac{\sigma_r \text{r}^2 f}{\mu_\text{r}}\right) \tag{21b}$$

According to Eqs. (21a) and (21b), both *electric* and *magnetic reflection losses* (R_{ne} and R_{nm}) *increase* with the barrier conductivity σ_r, and they decrease when the barrier permeability μ_r increases. Additionally, regarding the frequency, the situation is the opposite for these fields: the *electric reflection loss R_{ne}* rapidly decreases with increasing frequency, Eq. (21a), whereas the *magnetic reflection loss R_{nm}* progressively increases with increasing frequency, Eq. (21b).

2.3 *Absorption Loss (A)*

In Chapter "Nonideal Passive Circuit Components", we observed that the field decreases exponentially through the conductive barrier by a factor of $e^{-\alpha d}$, where α is the attenuation constant and *d* is the barrier thickness. This attenuation is the *absorption loss* (*A*) expressed as

$$A = \frac{|\tilde{E}_i|}{|\tilde{E}_i|e^{-\alpha d}} = \frac{1}{e^{-\alpha d}} = e^{d\alpha} = e^{\frac{d}{\delta}} \tag{22}$$

where α is expressed as $\alpha = 1/\delta$. Additionally, δ is the *skin depth* expressed as

$$\delta = \sqrt{\frac{2}{\omega\mu\sigma}} = \frac{1}{\sqrt{\pi f(\mu_r\mu_0)(\sigma_r\sigma_0)}} \tag{23}$$

where $\omega = 2\pi f$, $\sigma = \sigma_r\sigma_0$, and $\mu = \mu_r\mu_0$. Additionally, $\sigma_0 = \sigma_{cu} = 5.8 \cdot 10^7$ *S/m* and $\mu_0 = 4\pi \cdot 10^{-7}$ *H/m*.

Furthermore, substituting Eq. (23) into Eq. (22), the *absorption loss* can be expressed in *dB* as

$$A(dB) = 20\log\left(e^{d\sqrt{\pi f \mu_r \sigma_r}}\right) \rightarrow$$
$$A(dB) \cong 8.69\frac{d}{\delta} \cong 130d\sqrt{f\mu_r\sigma_r} \quad (24)$$

where μ_r and σ_r are the relative permeability and conductivity of the shield material, respectively. As shown in Eq. (24), the *absorption loss* increases with the square root of the frequency.

2.4 Multiple-Reflection Loss (M_R)

We considered only the wave reflection from the first surface of the shield barrier in the previous section. However, multiple reflections can occur inside the barrier when the wave propagates back and forth inside the barrier wall. This can occur when the skin depth δ is equal to or greater than the barrier thickness *d*. As seen later, the *multiple-reflection loss* is more significant for low-frequency magnetic shielding, and it can even be a gain instead of a loss.

Figure 4 shows the multiple reflections inside the shielding barrier and transmitted waves through the barrier, degrading the barrier *shielding effectiveness*.

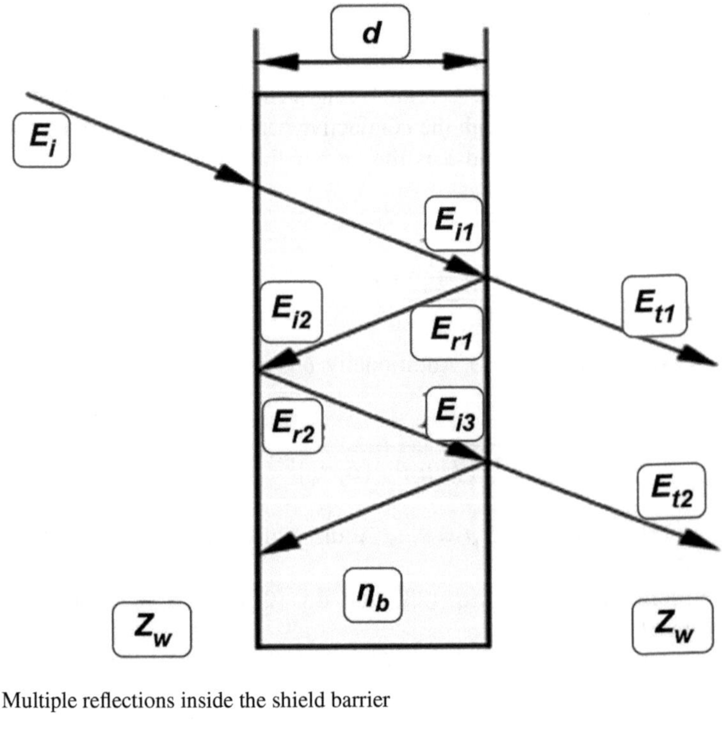

Fig. 4 Multiple reflections inside the shield barrier

Therefore, the first wave $\tilde{E}_{t1}$ transmitted through the right wall boundary can be expressed via Eq. (10):

$$\tilde{E}_{t1} = \left(\frac{2\tilde{Z}_w}{\tilde{Z}_w + \tilde{\eta}_b} \right) \tilde{E}_{i1} \tag{25}$$

where $\tilde{E}_{i1}$ is the incoming (incident) wave that travels inside the wall toward the right boundary. The reflected wave $\tilde{E}_{r1}$ from the right boundary can be expressed via Eq. (7):

$$\tilde{E}_{r1} = \left(\frac{\tilde{Z}_w - \tilde{\eta}_b}{\tilde{Z}_w + \tilde{\eta}_b} \right) \tilde{E}_{i1} \tag{26}$$

Furthermore, the reflected wave from the right boundary propagates back toward the left boundary, is attenuated and phase-shifted by the factor $e^{-\tilde{\gamma}d}$, and can be expressed at the left boundary as

$$\tilde{E}_{i2} = \tilde{E}_{r1}\, e^{-\tilde{\gamma}d} \tag{27}$$

Furthermore, considering Eqs. (26) and (27), the second reflected wave $\tilde{E}_{r2}$ from the left boundary and propagating toward the right boundary can be expressed as

$$\tilde{E}_{r2} = \left(\frac{\tilde{Z}_w - \tilde{\eta}_b}{\tilde{Z}_w + \tilde{\eta}_b} \right) \tilde{E}_{i2} = \left(\frac{\tilde{Z}_w - \tilde{\eta}_b}{\tilde{Z}_w + \tilde{\eta}_b} \right)^2 \tilde{E}_{i1}\, e^{-\tilde{\gamma}d} \tag{28}$$

Additionally, via Eq. (26), $\tilde{E}_{i3}$ can be expressed as

$$\tilde{E}_{i3} = \tilde{E}_{r2}\, e^{-\tilde{\gamma}d} = \left(\frac{\tilde{Z}_w - \tilde{\eta}_b}{\tilde{Z}_w + \tilde{\eta}_b} \right)^2 \tilde{E}_{i1}\, e^{-\widetilde{2\gamma}d} \tag{29}$$

Finally, the second transmitted wave $\tilde{E}_{t2}$ can be expressed via Eq. (10), and Eq. (29) becomes

$$\tilde{E}_{t2} = \left(\frac{2\tilde{Z}_w}{\tilde{\eta}_b + \tilde{Z}_w} \right) \tilde{E}_{i3} = \left(\frac{2\tilde{Z}_w}{\tilde{\eta}_b + \tilde{Z}_w} \right) \left(\frac{\tilde{Z}_w - \tilde{\eta}_b}{\tilde{Z}_w + \tilde{\eta}_b} \right)^2 \tilde{E}_{i1}\, e^{-\widetilde{2\gamma}d} \tag{30}$$

Owing to Eq. (25), Eq. (30) has the form

$$\tilde{E}_{t2} = \left(\frac{\tilde{Z}_w - \tilde{\eta}_b}{\tilde{Z}_w + \tilde{\eta}_b} \right)^2 \tilde{E}_{t1} e^{-2\tilde{\gamma}d} \tag{31}$$

For a good conductor, when $\tilde{\eta}_b \ll \tilde{Z}_w$, Eq. (31) can be approximated as

$$\tilde{E}_{t2} \cong \tilde{E}_{t1} e^{-2\tilde{\gamma} d} \tag{31a}$$

Furthermore, substituting $\tilde{F} = e^{-\tilde{\gamma} d}$ into Eq. (31a), one becomes

$$\tilde{E}_{t2} = \tilde{E}_{t1} F^2 \tag{32}$$

Next, the considering n transmitted waves through the right shield boundary, the total transmitted field E_t through the shield wall takes the form:

$$\tilde{E}_t = \tilde{E}_{t1} \left(1 + F^2 + F^4 + \ldots + F^{2n}\right) \tag{33}$$

Based on the approximate relation:

$$\left(1 + F^2 + F^4 + \ldots + F^{2n}\right) \cong \frac{1}{\left(1 - F^2\right)}$$

for $|F| < 1$, the last Eq. (33) has the form:

$$\tilde{E}_t = \frac{\tilde{E}_{t1}}{\left(1 - F^2\right)} \tag{34}$$

and the *shielding effectiveness* M_R, due to multiple reflections, is expressed as

$$M_R = \frac{\tilde{E}_{t1}}{\tilde{E}_t} = \left|1 - F^2\right| \tag{35}$$

Because $\tilde{F} = e^{-\tilde{\gamma} d}$, the multiple-reflection loss has the form

$$M_R = \left|1 - e^{-2\tilde{\gamma} d}\right| \tag{36}$$

Owing to the complex propagation constant $\tilde{\gamma}$ for good conductors ($\sigma \gg \omega\varepsilon$),

$$\begin{aligned} \tilde{\gamma} &= \alpha + i\beta = \sqrt{i\omega\mu\left(\sigma + i\omega\varepsilon\right)} \approx \sqrt{i\omega\mu\,\sigma} \\ &= \sqrt{\pi f \mu \sigma}\left(1 + i\right) = \frac{1}{\delta} + i\frac{1}{\delta} \end{aligned} \tag{37}$$

where $\delta = 1/\sqrt{\pi f \mu \sigma}$, and substituting Eq. (37) into Eq. (36), the final expression for M_R in dB becomes

$$M_R = 20\log\left|1 - e^{-\frac{2d}{\delta}} e^{-i\frac{2d}{\delta}}\right| = 20\log\left|1 - e^{-2d\sqrt{\pi f \mu \sigma}} e^{-i2d\sqrt{\pi f \mu \sigma}}\right| \tag{38}$$

Additionally, after rearrangement of Eq. (38) via Euler's identity and $\sigma = \sigma_r\, \sigma_0$ and $\mu = \mu_r\mu_0$, the last expression is given in terms of σ_r and $\boldsymbol{\mu_r}$ as

$$M_R\left(dB\right) \approx 10\log\left|1-2e^{-30d\sqrt{f\mu_r\sigma_r}}\cos\left(30d\sqrt{f\mu_r\sigma_r}\right)+e^{-60d\sqrt{f\mu_r\sigma_r}}\right| \qquad (39)$$

Notably, the *multiple-reflection* loss is 0 dB when the skin depth δ is much smaller than the barrier thickness d. This is the case for high frequencies when $\delta \ll d$. However, for low frequencies when $\delta \geq d$, the multiple-reflection loss is negative, i.e., the M_R is effectively a gain. Therefore, M_R may degrade the shielding effectiveness at very low frequencies.

3 Far-Field Shielding

We can derive the *reflection loss* when the shield is in the f*ar-field region* of an EMI source by substituting the wave impedance $\tilde{Z}_w$ with the wave impedance η_0 in Eq. (14). Thus, the *reflection loss* in the *far-field* region is

$$R_{ff}\left(dB\right) \cong 20\log\left|\frac{\eta_0}{4\tilde{\eta}_b}\right| \qquad (40)$$

and per Eq. (19): $\tilde{\eta}_b \cong \sqrt{\dfrac{i\omega\mu}{\sigma}}$ and $\eta_0 = 120\pi\ \Omega$, Eq. (40) becomes

$$R_{ff}\left(dB\right) \cong 40+10\log\left(\frac{\sigma}{2\pi f\mu}\right) \qquad (41a)$$

or in terms of the shielding barrier relative parameters σ_r and μ_r:

$$R_{ff}\left(dB\right) \cong 168+10\log\left(\frac{\sigma_r}{f\mu_r}\right) \qquad (41b)$$

Furthermore, the derived expressions for the *absorption loss* and *multiple-reflection loss* in the *near-field region,* Eqs. (24) and (39), are also valid for the far-field region. They are derived before as

$$A\left(dB\right) \cong 130d\sqrt{f\mu_r\sigma_r} \qquad (41c)$$

$$M_R\left(dB\right) \cong 10\log\left|1-2e^{-30d\sqrt{f\mu_r\sigma_r}}\cos\left(30d\sqrt{f\mu_r\sigma_r}\right)+e^{-60d\sqrt{f\mu_r\sigma_r}})\right| \qquad (41d)$$

In the following examples, using Eqs. (41b), (41c) and (41d), we numerically illustrate the shielding effectiveness associated with a *near-field* and *far-field* source for two shields made of copper and stainless steel (430), each with a thickness of $d = 1$ mm.

Example 1 The plots of the *reflection loss* with frequency for the near fields (electric and magnetic) are made when a radiated source is located at $r = 0.1\ m$ from:

(a) The copper shield thickness $d = 1\ mm$, whose relative parameters are $\sigma_r = 1$ and $\mu_r = 1$.
(b) Stainless steel (430) is $d = 1\ mm$ thick, with the relative parameters $\sigma_r = 0.02$ and $\mu_r = 500$.

Solution

(a) *Solution for the cooper*:

Based on Eqs. (21a) and (21b) yield:

$$R_{ne}(dB) \cong 320 + 10\log\left(\frac{\sigma_r}{\mu_r r^2 f^3}\right)$$

$$R_{nm}(dB) \cong 15 + 10\log\left(\frac{\sigma_r r^2 f}{\mu_r}\right)$$

Figure 5 shows the plots of the *reflection losses* R_{ne} **and** R_{nm} **for the copper shield.**

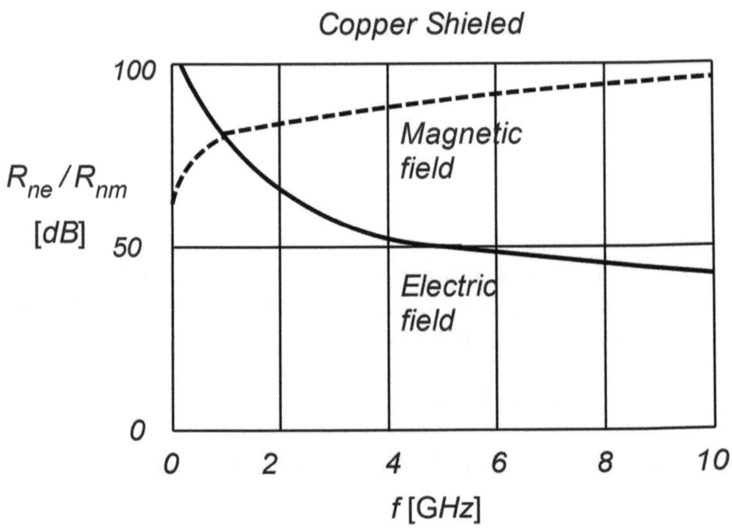

Fig. 5 Near-field reflection losses at the cooper shield: (electric field) R_{ne} and (magnetic field) R_{nm}

Here, the *reflection loss* for the *electric* field is greater than that for the *magnetic* field for frequencies up to 478 MHz. The frequency of 478 MHz is associated with $\lambda/2\pi$ where $\lambda = r = 0.1\ m$ and where r is the distance of the radiation source from the shield. Therefore, the frequency of 478 MHz corresponds to the boundary between the *reactive* and *radiating near-field* regions where the two losses (R_{ne} and R_{nm}) are equal. We also obtain the boundary frequency of 478 MHz via the following calculation:

$$\lambda_0 = r = 0.1m \rightarrow f_0 = 3GHz$$

$$\frac{3GHz}{2\pi} \rightarrow f_b = 478\,MHz$$

(b) *Solution for the stainless steel (430).*
Stainless steel (430) has the relative parameters $\sigma_r = 0.02$ and $\mu_r = 500$, and via Eqs. (8) and (39), the corresponding plots of the reflection losses R_{ne} and R_{nm} are shown in Fig. 6.

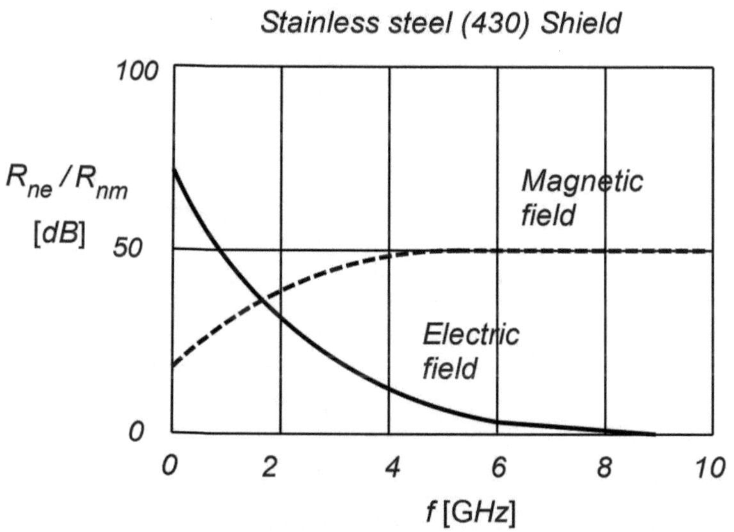

Fig. 6 Nier-field reflection losses at a stainless steel (430) shield: (electric field) R_{ne} and (magnetic field) R_{nm}

Here, the boundary frequency between the near-field and far-field regions of 478 MHz is also shown.
Furthermore, a comparison of the plots of the reflection losses of the copper shield (Fig. 5) and the stainless steel (430) shield (Fig. 6) reveals that copper has greater reflection loss for both electric and magnetic fields, i.e., it has better shielding effectiveness by ~45 *dB*.

Example 2 This example shows the calculated plots for the *absorption losses* with frequency for the shielding materials in the previous example (copper and stainless steel 430), which are 1 mm thick.

Solution Here, we use Eqs. (24) or (41c), $A(dB) \cong 130 d \sqrt{f \mu_r \sigma_r}$, to create the plots shown in Fig. 7, where the two absorption losses are shown for the two shields: (1) the stainless-steel 430 shield and (2) the copper shield.

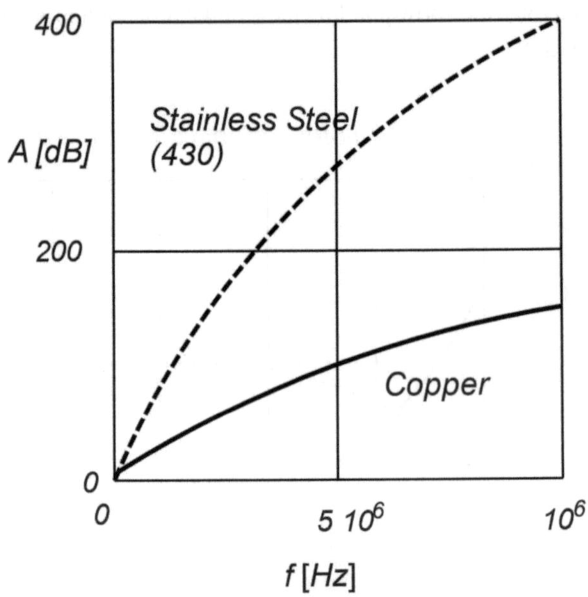

Fig. 7 Absorption losses for stainless-steel 430 and copper shields

Note that *absorption losses* progressively increase with frequency for both shields, but the stainless-steel (430) shield introduces greater absorption loss at higher frequencies than does the copper shield with the same thickness. Hence, against low-frequency magnetic fields, stainless steel has better shielding effectiveness than copper.

Example 3 For the shields in the previous examples (the copper and stainless-steel examples), we find multi-reflection losses.

Solution Using Eq. (41d), the plot shown in Fig. 8 is made for the same shields from the previous examples.

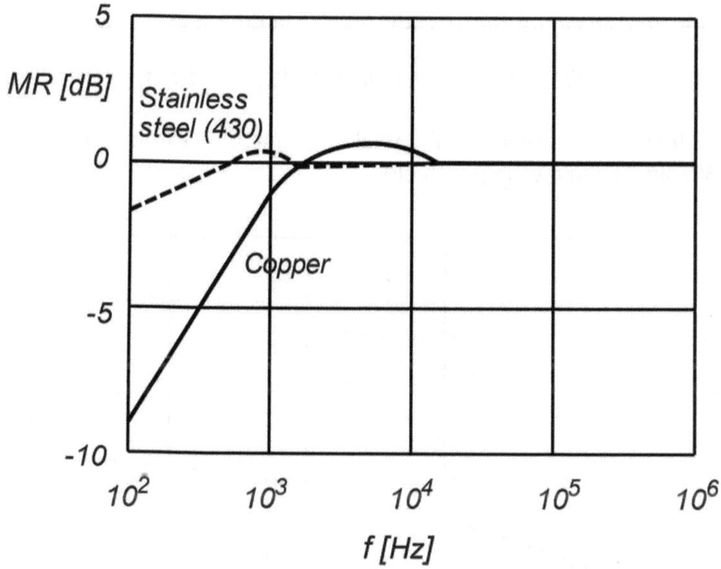

Fig. 8 Multiple-reflection loss for the shields: (1) stainless-steel (430) and (2) copper

The present plots show that the reflection losses at low frequencies have negative values, meaning that the shields do not introduce losses but rather some gains. Note that stainless steel has a "gain" at lower frequencies (between approximately 0.8 kHz and 1.2 kHz), whereas copper has a "gain" at higher frequencies (between approximately 3 kHz and 12 kHz).

Example 4 Next, we find the plots for the *total shielding effectiveness* (the sum of losses: *reflection*, *absorption*, and *multiple reflection*) for a *far-field* source distance of $r = 1\ m$ from the *copper* and the stainless steel (430) shields, both of which are thick by $d = 1\ mm$.

Solution Using Eqs. (41b), (41c) and (41d),

$$R_{ff}(dB) \cong 168 - 10\log\left(\frac{\sigma_r}{f\mu_r}\right);\; A(dB) \cong 131d\sqrt{f\mu_r\sigma_r}$$

$$M_R(dB) \cong 10\log\left|1 - 2e^{-30d\sqrt{f\mu_r\sigma_r}}\cos\left(30d\sqrt{f\mu_r\sigma_r}\right) + e^{-60d\sqrt{f\mu_r\sigma_r}}\right|$$

We find the *total shielding effectiveness* $SE_{tot,ff}$ via Eq. (2) as

$$SE_{tot,ff}(dB) = R_{ff}(dB) + A(dB) + M_R(dB).$$

Figure 9 shows two plots for the total shielding effectiveness for the stainless steel (430) shield and the copper shield.

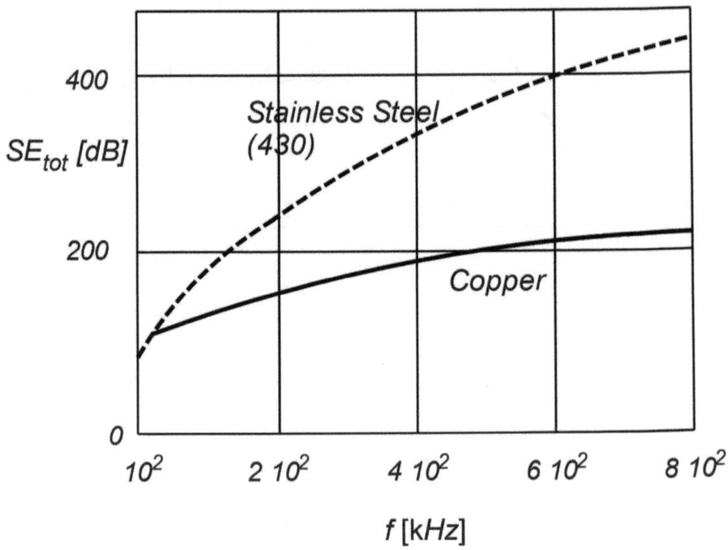

Fig. 9 Total shielding effectiveness for copper and stainless-steel shields for the far field

Thus, for the *far field,* the *total shielding effectiveness* of the stainless-steel shield is better than that of the copper shield.

The writing of this section is influenced by the references [1–3].

4 Low-Frequency Magnetic Shielding

As seen in the previous section, shielding with copper or stainless-steel results in a low shielding effectiveness against low-frequency magnetic fields. To solve this problem, other shielding techniques for *low-frequency magnetic* fields are applied in practice, such as (a) the generation of an opposite magnetic flux per Faraday's law to cancel unwanted magnetic fields at a specific space, (b) diverting magnetic flux from the shielded space bounded with high-permeability materials, and (c) the use of multilayer shielding combining copper and steel layers. We discuss all these techniques in this section, which are described well in [1, 2].

4.1 Magnetic Shielding by Closed Conductive Structures

Figure 10 shows a conductive (nonmagnetic) ring that bounds a surface through which an *incident* magnetic flux penetrates, inducing current flow in the ring.

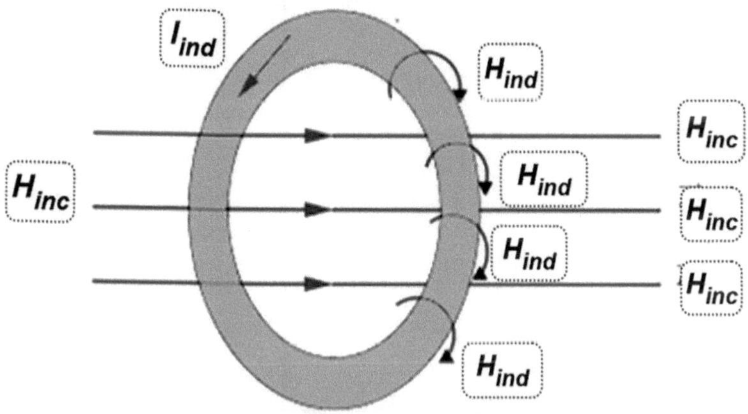

Fig. 10 Low-frequency magnetic shield by opposite magnetic field

Furthermore, the induced current produces a magnetic field in the opposite direction to the incident field $\vec{H}_{inc}$, tending to cancel this field inside the ring. As a result, electronic circuits (sensitive to low-frequency magnetic fields) can be protected inside a conductive and nonmagnetic ring. Instead of a conductive ring, the shielding structure can be any shaped conductive loop or a tube because the resulting magnetic field tends to be zero inside these closed structures.

As another example, Fig. 11 shows a finite-length conductive and nonmagnetic circular tube. The figure also shows the axial external and internal magnetic fields, $\vec{H}_{ext}$ and $\vec{H}_{\mathrm{int}}$, respectively, and the surface current $\vec{J}_S$ induced by $\vec{H}_{ext}$.

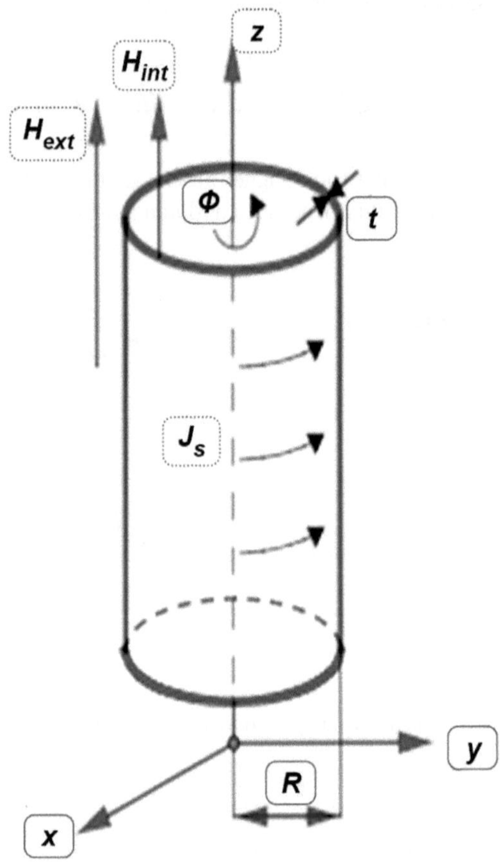

Fig. 11 Copper tube as a low-frequency magnetic shield

In this case, the induced surface current J_S produces an internal (inside the tube) magnetic field $\vec{H}_{\mathrm{int}}$ in the opposite direction to the field $\vec{H}_{ext}$ (per *Ampere's law*), resulting in cancelation of the magnetic field inside the tube. This shielding method is efficient against low-frequency magnetic fields when the skin depth is $\delta > t$, where t is the thickness of the tube wall, the tube radii $R \gg t$ and $R \ll \lambda_0$, and the tube length $l \gg R$.

4.2 *Magnetic Shielding by Diverting Magnetic Flux from a Shielded Space*

The next example, shown in Fig. 12, is another magnetic shielding method, but for very low-frequency magnetic fields, where the external (incident) magnetic flux is diverted from the shielded area by a low-reluctance shield (with $\mu_r \gg 1$), assuming that the external environment is a nonmagnetic space. Therefore, owing to the magnetic field's tendency to concentrate in high-permeability materials, the external magnetic field can be diverted by closed high-permeable magnetic material placed around a shielded interior, where circuits sensitive to a magnetic field (such as electronic sensors) can be safely placed.

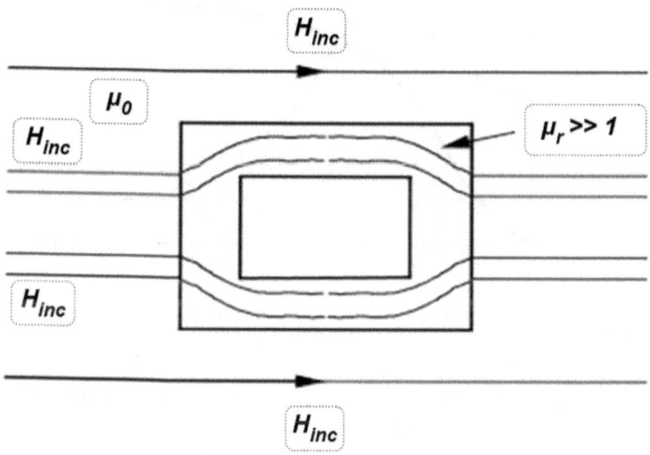

Fig. 12 Diverting a magnetic field with a magnetic shield

The ratio of the magnetic flux ϕ_{sh} inside the shield wall and the magnetic flux ϕ_0 in the air (inside the shielded space) defines the shielding effectiveness:

$$SE = \frac{\phi_{sh}}{\phi_0} = \frac{\mu_r \mu_0 \phi_0}{\mu_0 \phi_0} = \mu_r \tag{42}$$

where μ_0 is the air permeability and where μ_r is the relative permeability of the magnetic shield, usually $\mu_r \gg 1$. Thus, with increasing shield permeability μ_r, the magnetic shielding effectiveness is better.

4.3 Degradation of Low-Frequency Magnetic Shielding

Degradation of the effectiveness of magnetic shielding is usually caused by decreasing the shield permeability due to saturation of a magnetic material with the excessive intensity of the magnetic field *H*, i.e., the excessive current *I* that produces the magnetic field. This is discussed in Chapters "Nonideal Passive Circuit Components" and "EMI Filters—Theory and Design".

Figure 13 shows curves of the magnetic induction *B* and the relative permeability μ_r of the ferromagnetic material with respect to the intensity of the magnetic field *H*.

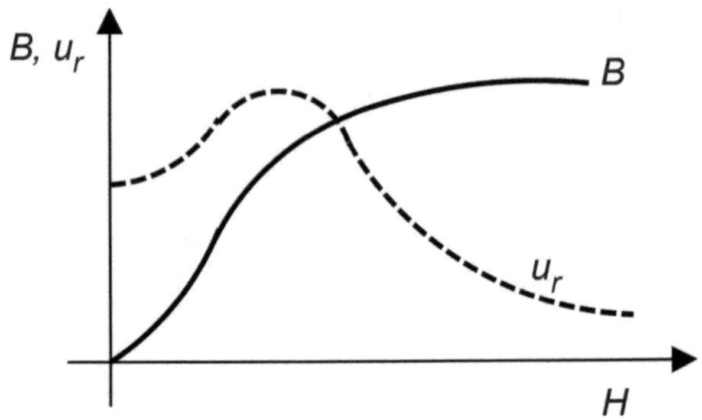

Fig. 13 Magnetic induction B and relative permeability μ_r with the intensity of the magnetic field H

Regarding this plot, an operating region for a magnetic material is the region where curve *B* is linear, roughly up to μ_{max}.

Additionally, permeability is constant up to some frequency, after which it starts to decrease, as shown in Fig. 14.

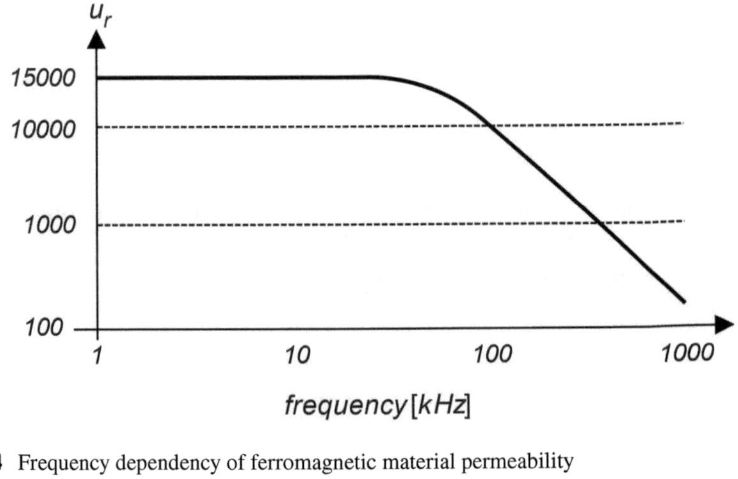

Fig. 14 Frequency dependency of ferromagnetic material permeability

Typically, manufacturers specify the relative permeability of ferromagnetic materials at low frequencies, such as *1 kHz*. For example, Permalloy and Mumetal have very high relative permeabilities (over 10,000) up to *100 kHz*. Above *100 kHz*, the permeability of these materials rapidly decays with frequency, and above several *1 MHz*, the permeability is like that of cold-rolled steel.

5 Improvement in EMI Shielding by Multilayer Shielding

Hence, high currents produce a high magnetic field that lowers the permeability of ferromagnetic materials due to saturation. For example, two-layer shielding can be applied to reduce the effects of saturation of a ferromagnetic material. Figure 15 shows the double shield where a low-permeability material (copper) is placed in parallel with the higher-permeability material (e.g., stainless steel 430).

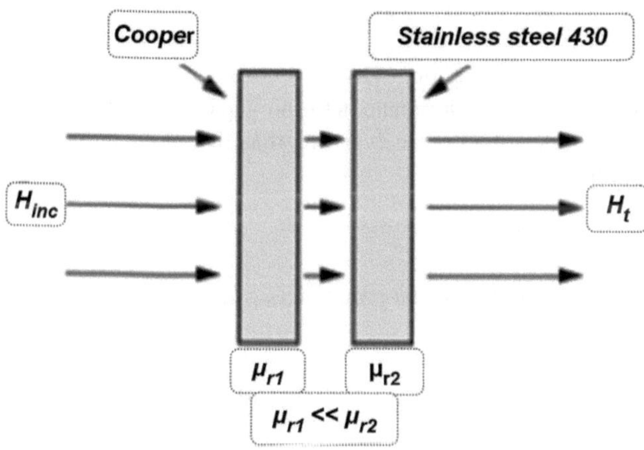

Fig. 15 Double-layer shielding against EMI fields

The first shield layer, which has a low μ_r (= *1*) and a low tendency to saturate, is employed to reduce the incident magnetic field penetrating the second magnetic layer, which has a high tendency to saturate. Therefore, the copper layer reduces the incident electric field mainly because of the higher reflection loss at low frequencies, but at higher frequencies, absorption loss is dominant in the Cooper layer, as discussed earlier in this chapter.

According to Eq. (45), the *far-field reflection loss* and *absorption loss* related to the copper layer (shown in Fig. 15), are expressed via Eqs. (41b) and (41c) as

$$R_{ff1}(dB) \approx 168 + 10\log\left(\frac{\sigma_{r1}}{f\mu_{r1}}\right) \tag{43a}$$

$$A_1(dB) \approx 130\, d\sqrt{f\mu_{r1}\,\sigma_{r1}} \tag{43b}$$

$$SE_1(dB) = R_{ff1}(dB) + A_1(dB) \tag{43c}$$

Additionally, we can express *reflection* and *absorption* loss, related to the second (magnetic) shield, as

$$R_{ff2}(dB) \approx 168 + 10\log\left(\frac{\sigma_{r2}}{f\mu_{r2}}\right) \tag{44a}$$

$$A_2(dB) \approx 130\, d\sqrt{f\mu_{r2}\sigma_{r2}} \tag{44b}$$

$$SE_2(dB) = R_{ff2}(dB) + A_2(dB) \tag{44c}$$

The *multiple-reflection loss* inside of the double shield is expressed as the sum of multiple-reflection losses in the three shielding layers: copper ($M_{R,\,1}$), air ($M_{R,\,air}$), and stainless-steel-430 ($M_{R,\,2}$):

$$M_{R,tot} = M_{R,1} + M_{R,\,air} + M_{R,2} \tag{45}$$

where $M_R(dB)$ is expressed by Eq. (41d).

Multiple reflection losses dominate in the air gap between the two metals because the intrinsic impedance of air is much greater than that of the metals. As a result, Eq. (45) has the simplified form:

$$M_{R,tot} \approx M_{R,\,air} \tag{46}$$

Additionally, substituting the air parameters, such as $\mu_r = 1$, $\sigma_r = 1$, and $s = d$, into Eq. (39) results in the following:

$$M_{R,air}(dB) \approx 10\log\left|1 - 2e^{-30s\sqrt{f}}\cos\left(30s\sqrt{f}\right) + e^{-60s\sqrt{f}}\right| \tag{47}$$

In the end, using Eqs. (43c), (44c), and (47), the *total shielding effectiveness* $SE_{tot}(dB)$ of the double-layer shield is

$$SE_{tot}(dB) \approx SE_1(dB) + SE_2(dB) + M_{R,air}(dB) \tag{48}$$

where Eqs. (43c) and (44c) yield

$$SE_1(dB) = R_{ff1}(dB) + A_1(dB) \tag{49a}$$

$$SE_2(dB) = R_{ff2}(dB) + A_2(dB) \tag{49b}$$

The following example numerically illustrates the difference between the shielding effectiveness of the magnetic single-layer shield made by the stainless-steel 430 and the double-layer shield made by the stainless-steel 430 and the copper layer.

Example 5 Here, we plot the shielding effectiveness for a single-layer shield (stainless-steel 430) with parameters μ_{r2} = 500 and σ_{r2} = 0.02) and a double-layer shield with a stainless-steel 430 and copper with parameters $\mu_{r1} = 1$ and $\sigma_{r1} = 1$. Both layers have the same thickness: $d_1 = d_1 = 1\ mm$. Additionally, separation between the metallic layers is $s = 2\ mm$.

Solution On the basis of Eqs. (46a), (41b), (41c) and (47a), (47b), (47c), Fig. 16a, b shows the two curves of the shielding effectiveness (SE_{cu} and $SE_{cu/ss}$) at lower frequencies (< $1MHz$) and at higher frequencies (>$1MHz$), respectively, which are related to the magnetic single-layer and double-layer shields.

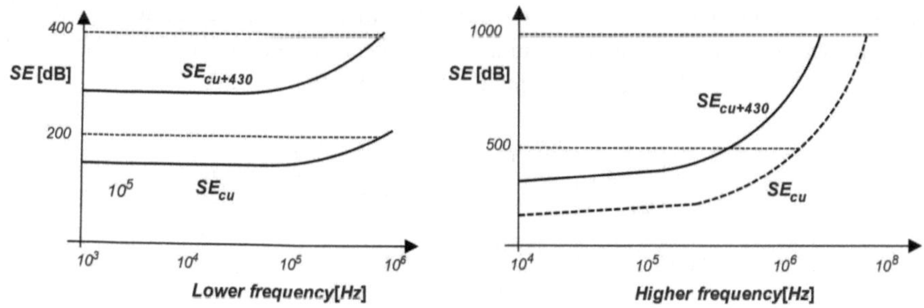

Fig. 16 Shielding effectiveness improvement by the double-layer shield: (**a**) below 1 MHz and (**b**) above 1 MHz

Therefore, the copper layer effectively shields against the electric field, reducing the intensity of the incident field before one hits the second (magnetic) layer; consequently, the saturation of the second layer is effectively lowered. Additionally, Fig. 16a, b show the significant improvement (theoretically by ~240 dB) in the shielding effectiveness at low frequencies (below 1 MHz) and at higher frequencies in the higher MHz range. In fact, SE values >120 dB are not usually achievable in practice, as we are not able to use infinitely large metallic sheets.

6 EMI Impact of Apertures in Equipment Enclosures

6.1 Openings and Slots in EMI Shields

Openings or apertures in equipment enclosures are among the common causes of the reduction in shielding effectiveness due to EMI radiation through apertures. This event is theoretically explained by Huygens's principle, which we discuss in *Appendix B.*

Therefore, per Huygens's principle, any aperture in a conductive plane on the way of incident electromagnetic waves acts as a secondary radiation source. Additionally, this phenomenon is known as field scattering, and the waves leaving the aperture are called scattering waves, as discussed in *Appendix B.*

When a circular hole in a large conductive plane is electrically small, the scattered electric field decays with distance r from the shield via the approximate equation [6]:

$$\left|\vec{E}_s\right| \cong \left(\frac{d}{r}\right)^3 \frac{E_{inc}}{6\pi} \quad \left(\frac{V}{m}\right) \tag{50}$$

where E_{inc} is the magnitude of the incident electric field before one hits the hole. Additionally, d is the hole diameter, and r is the distance of an observing point at the right side of the plane. Owing to the term $1/r^3$ in Eq. (50), the scattered field E_s rapidly decays with increasing distance from the hole. For example, at distance $r = 3\,d$, the incident electric field is attenuated by 54 dB:

$$A = \frac{\left|\vec{E}_s\right|}{E_{inc}} \cong \left(\frac{d}{r}\right)^3 \frac{1}{6\pi} = \left(\frac{d}{3d}\right)^3 \frac{1}{6\pi} = \frac{1}{48\pi} \quad \rightarrow$$

$$A(dB) \cong 20\log\left(\frac{1}{48\pi}\right) = -54\,dB$$

This knowledge is useful when we place sensitive circuits close to *electrically small* holes in the shield. Therefore, placing a sensitive circuit according to the distance from the hole to several hole diameters is not critical. Hence, this is valid when $r \ll \lambda$, where λ is the minimal wavelength of interest.

When the aperture is a thin slot, as shown in Fig. 17a. This slot can be an efficient radiator depending on its length in terms of the wavelength of interest. Per Babinet's principle, a thin slot antenna is a complementary electromagnetic structure to a wire antenna with the same length and shape, as shown in Fig. 17b.

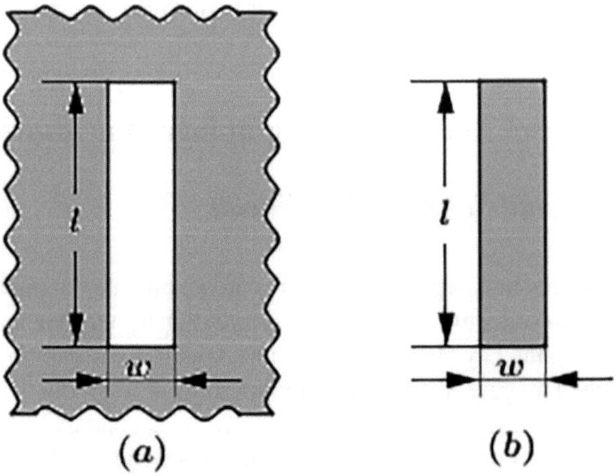

Fig. 17 Babinet principles: (**a**) slot antenna and (**b**) linear antenna

In general, an arbitrary aperture in a conductive plate can be an effective radiator, such as a conductor antenna with the same shapes and dimensions. For example, a slot long $l = \lambda_0/2$ is a resonant radiator as a half-wave dipole. Additionally, the slot width w does not significantly affect slot radiation when $w \ll \lambda_0$. Even if a slot is less visible or invisible, it can efficiently radiate, causing a significant reduction in the enclosure's shielding effectiveness.

Many intentional openings cannot be avoided, such as ventilation openings or openings for device peripheral cables or displays. Potential radiating circuits or cables inside the enclosure should be placed far from the openings. Additionally, with respect to regarding ventilation openings, it is better to use many small holes rather than large holes. This will be confirmed later in this chapter.

6.2 Perturbation of Surface Currents at a Shield Wall by a Slot

Figure 18a–c shows a solid shield panel (case *a*) and two panels with different slot orientations (cases *b* and *c*) with a wire conductor with current *I* routed next to the panel.

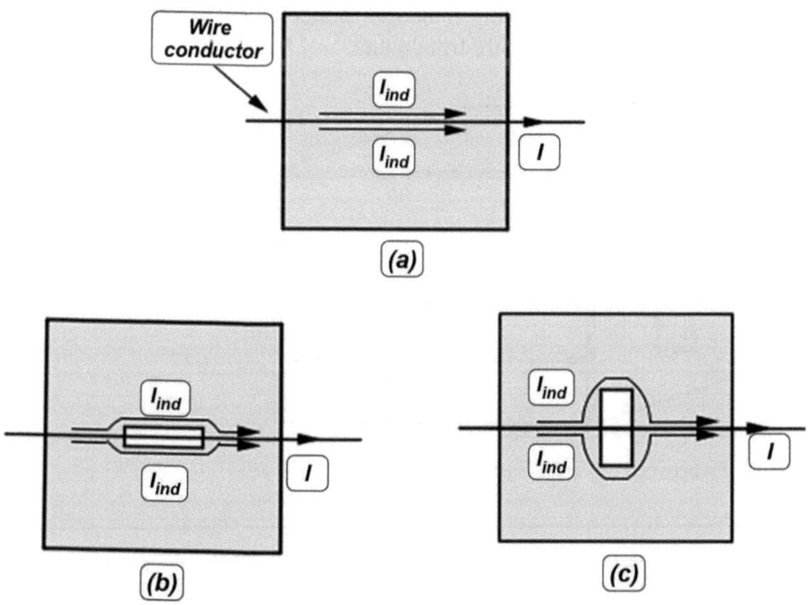

Fig. 18 (**a**) Perfect shielding, (**b**) somewhat good shielding, (**c**) bad shielding

For the panel without a slot (case *a*), the *incident* field from the wire induces a current I_{ind} on the panel surface that produces a reflected field that tends to cancel the *incident* field. This is in accordance with the boundary condition, where the resulting tangential electric field must be zero on the good conductive panel surface, as discussed in *Appendix A*.

Furthermore, Fig. 18b, c shows the situations in which the surface-induced currents at the panel are perturbed (diverted) around slots, creating current loops. When an induced current on the panel is less perturbed by the slot, causing a smaller current loop, the slot radiates less. In contrast, when an induced current is more perturbed by a slot, which results in a larger current loop, the slot radiates more.

Thus, to avoid larger current perturbations, the wire conductor with the current I should be routed along the larger size of the slot (*case b*), not perpendicular to the larger size (case c). As mentioned above, a larger current loop, formed around a larger slot, is a more efficient EMI radiator, as discussed in *Appendix B*.

6.3 Reduction in Shielding Effectiveness by Slots

As mentioned above, a slot may radiate like a wire conductor with the same length and shape as the slot. To reduce the effects of the slot radiation, the slot length should be reduced to move the slot resonance frequencies above the frequency bandwidth of interest.

Figure 19 shows an electrically long slot (nonconductive gap air or dielectric) between two parts of the enclosure front panel.

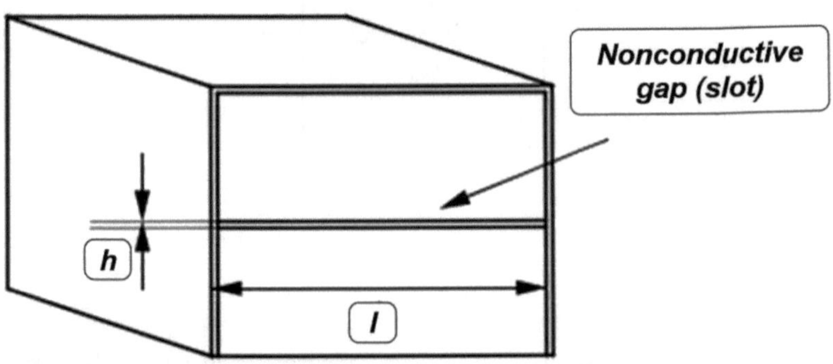

Fig. 19 A slot between two chassis panels

If the slot is long $\lambda/2$, then it acts as an efficient resonant dipole. In general, when a slot has a length less than the half-wavelength, $l < \lambda/2$, the shielding effectiveness can be expressed as

$$SE(dB) = 20\log\left(\frac{\lambda_0/2}{l}\right) \quad (51)$$

Therefore, for $l = \lambda/2$, the shielding effectiveness (SE) is zero and by decreasing the slot length, the SE increases.

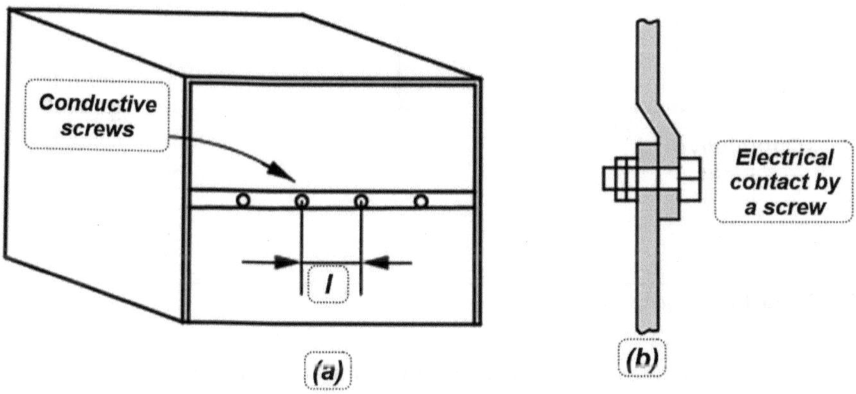

Fig. 20 (**a, b**) Slot reduction length by conductive screws

To reduce the effect of the slots, Fig. 20 shows the enclosure from Fig. 19 when the long slot is divided into smaller slots by conductive screws. These smaller slots should have the first resonance frequency much beyond the resonance frequency of the long slot.

Therefore, if a slot is long, then the critical frequency for the chassis *shielding effectiveness* is when $l > \lambda_0/10$. For some applications that are more sensitive to EMI, the slot length must be smaller than $\lambda_0/20$. For example, if the potential EMI has a frequency spectrum up to 1 GHz ($\lambda_0 = \frac{c}{f} = 30\,cm$), it is recommended that the minimal spacing between conductive screws along a slot be less than 1.5 cm (30 cm/20 = 1.5 cm).

However, the better solution to minimize EMI leakage from an electrically long slot is to fill the slot with a conductive gasket, as shown in Fig. 21a, b.

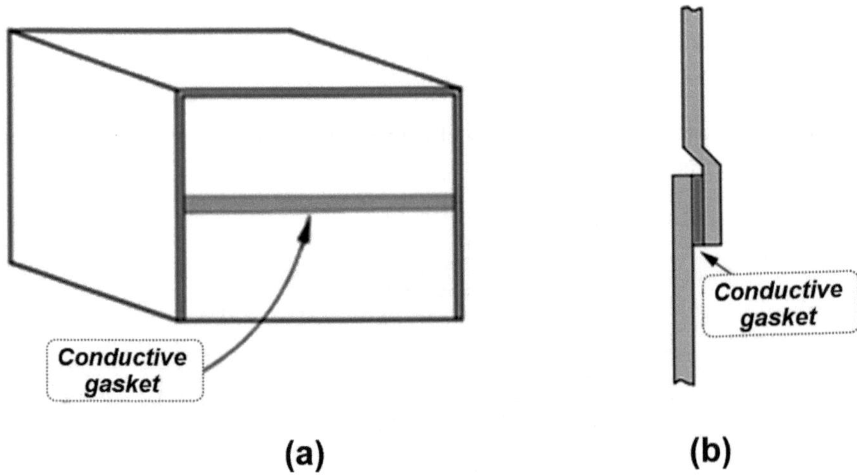

Fig. 21 (**a, b**) Conductive gasket

Figure 22 shows several commercial EMI conductive gaskets and foams.

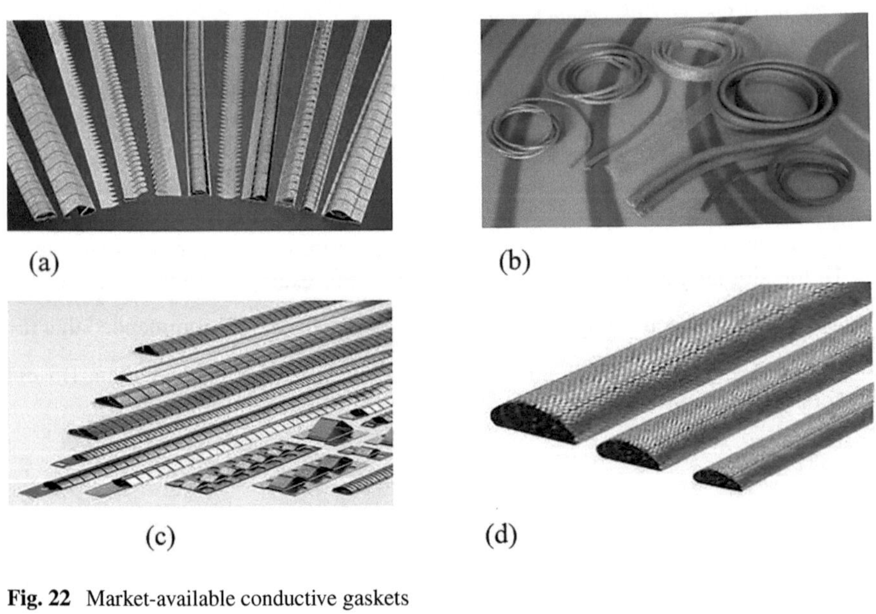

Fig. 22 Market-available conductive gaskets

6.4 Ventilation Panels and Openings

A common problem of EMI radiation is ventilation openings in the equipment chassis and device enclosures. As a solution, more electrically small square or circular holes close to each other should be used for the ventilation panel instead of one or several larger openings.

Figure 23 shows a standard perforated ventilation panel with small holes whose dimensions are much smaller than the wavelengths of potential EMI inside the device.

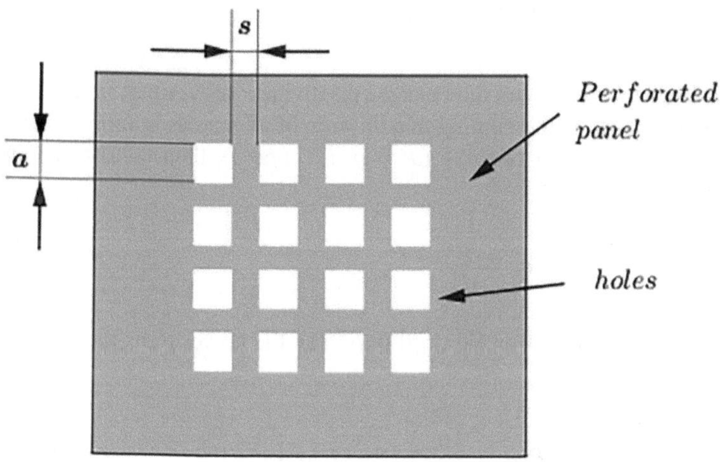

Fig. 23 Perforated ventilation panel

The square hole size in the ventilation pattern is *a*, whereas the separation between adjacent holes is *s*. In general, the panel shielding effectiveness decreases with increasing hole number. Therefore, in addition to the hole size, the number of holes determines the shielding effectiveness of the vent panel.

The ratio of the incident power P_i and transmitted power P_t through a single opening represents the degree of *radiation attenuation* or *gain*:

$$G = \frac{P_t}{P_i} \tag{52a}$$

since the power density of the incident wave (P_{di}) is

$$P_{di} = \frac{P_i}{S} = \frac{P_i}{a^2} \tag{52b}$$

where S is the area of the square opening equal to a^2. Furthermore, according to aperture antenna theory, the gain of an opening illuminated by a uniform plane wave is [9]

$$G = \frac{4\pi}{\lambda^2} S = \frac{4\pi}{\lambda^2} a^2 \tag{53}$$

and per Eqs. (51), (52a), and (53), the transmitted power P_t through the hole is expressed as

$$P_t = P_i G = P_{di}\, a^2 \left(\frac{4a}{\lambda}\right)^2 \tag{54}$$

If the large opening (with dimension a) is divided into n small square holes, each with dimensions a_n, and assuming that the sum of all n areas is approximately equal to the area of the larger opening ($a^2 \approx n\, a_n^2$; $a \gg s$), then the dimension of one small hole is

$$a_n = \frac{a}{\sqrt{n}} \tag{55}$$

Furthermore, substituting Eq. (55) into (54), the power transmitted through one of the small holes is

$$P_{tn} = P_d\, a_n^2 \left(\frac{4a_n}{\lambda}\right)^2 = \frac{1}{n^2} P_d\, a^2 \left(\frac{4a}{\lambda}\right)^2 \tag{56}$$

Because the dimension a is small in terms of the wavelength of the incident wave, from Eq. (56), the total power transmitted through all small holes is as follows:

$$nP_{tn} = \frac{P_t}{n} \quad \rightarrow \quad P_{tn} = \frac{P_t}{n^2} \tag{57a}$$

Equation (57a) shows that by dividing a large hole into n of small holes, the shielding effectiveness is improved by n:

$$\frac{P_t}{n\,P_{tn}} = n \tag{57b}$$

or in dB

$$\Delta_{SE}(dB) \cong 10\log\left(\frac{P_t}{nP_{tn}}\right) = 20\log\left(\sqrt{n}\right) \tag{57c}$$

Furthermore, per Eqs. (51) and (57c), the total attenuation of the ventilation panel, when the large hole is divided into n small holes, is

$$A_{tot}(dB) = 20\log\left(\frac{c}{2af}\right) + 20\log\left(\sqrt{n}\right) \tag{58}$$

where $c = 3\ 10^8\ m/s$.

Thus, the attenuation of the perforated panel with more small holes of size a_n, instead of a large hole of size $a = n\ a_n$, is improved by $20\log\left(\sqrt{n}\right)(dB)$.

Additionally, from the opposite perspective, per Eqs. (51) and (55), the $A_{tot}(dB)$ panel with one small hole with dimension a_n decreases when the number of holes increases:

$$A_{tot}(dB) = 20\log\left(\frac{\lambda_0}{2a_n}\right) - 20\log\left(\sqrt{n}\right) \tag{59}$$

We numerically illustrate the last expression (59) in Example 6.

Example 6 The shielding effectiveness for the ventilation panels (Fig. 24) is determined for frequencies up to *1* GHz with hole patterns 4 × 4 = 16, 6 × 6 = 36, and 10 × 10 = 100. All the squared holes have equal sizes, $a = 5$ mm, and the separation between the holes is $s \ll a$.

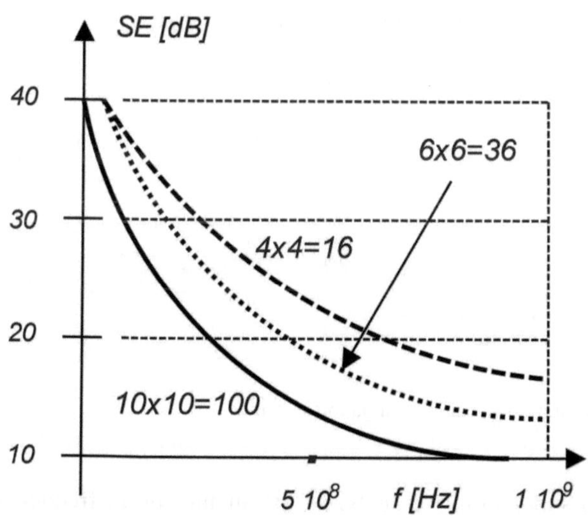

Fig. 24 SE vent panel dependency of the number of vent holes

Solution Here, we use Eq. (59) to plot the shielding effectiveness for vent panels with hole patterns of $4 \times 4 = 16$, $6 \times 6 = 36$, and $10 \times 10 = 100$ holes. This solution is shown in Fig. 24.

As expected, by increasing the number of vent holes and incident wave frequency, the shielding effectiveness of the ventilation panel decreases. Additionally, with 16, 36, and 100 small holes, instead of 1 small hole, the shielding effectiveness is reduced by 12 dB, 15.5 dB, and 20 dB, respectively.

However, during equipment design, the compromise between the acceptable level of EMI leakage through vent holes and the equipment thermal requirement (airflow) should be reached. Additionally, the EMI level from vent holes can be reduced by careful component layout inside the enclosure, placing noisy parts far enough from the vent panel.

The writing of this and the following section is influenced by Refs. [1–9].

7 Waveguides for Vent Panels

Another more efficient method for ventilation shielding, introduced in *Appendix C*, uses a larger number of adjacent waveguides, instead of small vent holes, with cutoff frequencies below which *EMI* radiation is attenuated.

Figure 25a shows a screen of rectangular waveguides in a conductive panel (front view), whereas Fig. 25b shows the waveguide side view.

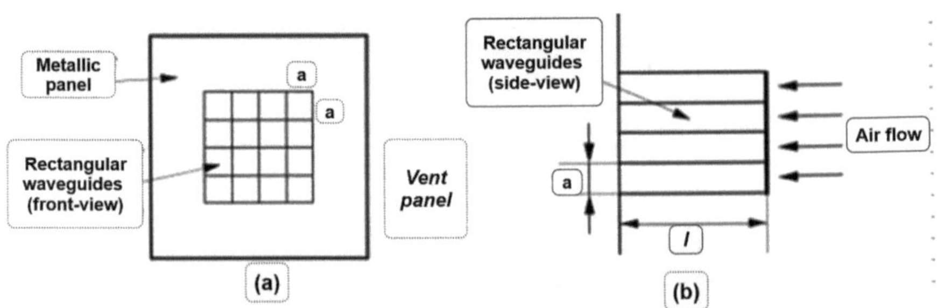

Fig. 25 (**a, b**) Multiple-waveguide ventilation system

In the next subsection, we analyze EMI attenuation at frequencies below the cutoff frequency for single- and multiple-rectangular waveguides.

7.1 *Attenuation of a Single-Rectangular Waveguide*

A single rectangular waveguide, shown in Fig. 26 and discussed in *Appendix C*, has length l and cross-sectional size a by a.

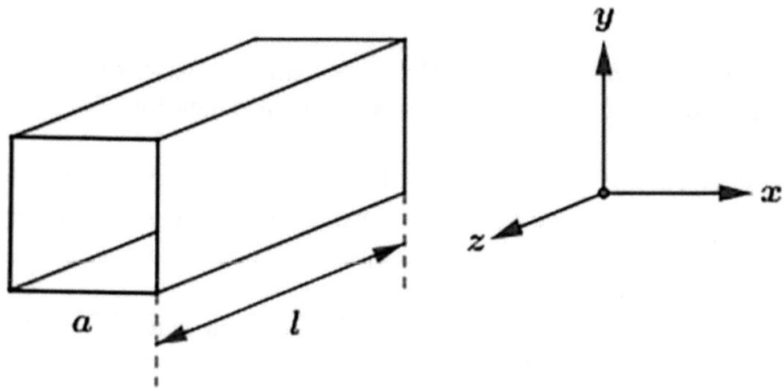

Fig. 26 Rectangular waveguide

Per Eqs. (C.38b) and (C.38c) in *Appendix C*, the waveguide attenuation of the dominant propagation mode TE_{10}, with the lowest cutoff frequency $f_c = {}^{c}\!/_{2a}$, is derived as

$$A(dB) \cong 8.7l\left(\frac{\omega}{c}\right)\sqrt{\left(\frac{\omega_c}{\omega}\right)^2 - 1} \tag{60a}$$

or for $\omega \ll \omega_c$:

$$A(dB) \cong 27l\left(\frac{\omega_c}{c}\right) \cong 27\left(\frac{l}{a}\right) \tag{60b}$$

Additionally, in the case of a rectangular waveguide with cross-sectional dimensions a and b, where $b > a$, Eq. (60b) becomes

$$A(dB) \cong 27l\left(\frac{\omega_c}{c}\right) \cong 27\left(\frac{l}{b}\right) \tag{60c}$$

By analogy with Eq. (59), the attenuation of the system of n rectangular waveguides, which are closely placed to each other ($s \ll a$), can be expressed as

$$A(dB) = 27\left(\frac{l}{a}\right) - 20\log\left(\sqrt{n}\right) \tag{61}$$

In the next example, using Eqs. (59) and (61), we numerically illustrate the attenuation of the panel with different numbers of rectangular waveguides, as shown in Fig. 26.

Example 7 We find:

(a) *Shielding attenuation* (*A*) at 1 GHz of the ventilation panel with a (4 × 4) pattern of squared waveguides, whose cross-sectional dimensions are (0.5 × 0.5) *cm* and length *l* = *2 cm.*
(b) Additionally, we compare the SE of this vent waveguide panel with the 4 × 4 pattern of a square-hole panel from Example 6 (*n* = 16, *a* = 0.5 cm).

Solution

(a) First, we find the cutoff frequency for a single squared waveguide with *a* = 0.5 *cm* and *l* = 2 *cm* via Eq. (C.34a):

$$f_c = \frac{c}{2a} = \frac{3{\cdot}10^8}{2{\cdot}5{\cdot}10^{-3}} = 30\,GHz,$$

Next, we calculate the attenuation of the given (4 × 4) waveguide panel (*n* = *16*) via Eq. (61):

$$A(dB) = 27\left(\frac{l}{a}\right) - 20\log\left(\sqrt{n}\right) = 96\,\text{dB}$$

(b) The SE of the (4 × 4) square-hole panel (*n* = *16, a* = *0.5 cm*) is calculated via Eq. (59):

$$SE(dB) = 20\log\left(\frac{\lambda_0}{2a_n}\right) - 20\log\left(\sqrt{n}\right) = 17.5\,\text{dB}$$

The above results reveal a large difference between the SE of the waveguide vent panel and the SE of the square-hole panel (by 78.5 dB) in favor of the waveguide panel.

7.2 *Attenuation of a Vent Panel with Honeycomb Waveguides*

In addition to rectangular waveguides, a circular waveguide structure, and more commonly, a honeycomb waveguide structure, is used for EMI suppression at ventilation panels. The attenuation of a honeycomb waveguide can be approximated with the attenuation of a circular waveguide with considerable accuracy. The attenuation of a circular waveguide (with a diameter *d*) far below its cutoff frequency is considered in *Appendix C*, where the cutoff frequency f_{c11} is expressed by Eq. (C.43):

$$f_{c11} \cong \frac{c}{2\pi}\frac{1.84}{r} \cong \frac{c}{2\pi}\frac{3.7}{d} \rightarrow \omega_c \cong \frac{3.7}{d}c \tag{62}$$

where $\omega_c = 2\pi f_c$ is the angular cutoff frequency and d is the waveguide diameter equal to $d = 2\ r$. Substituting Eq. (62) into Eq. (60b), the attenuation becomes

$$A(f) = 8.7l\left(\frac{2\pi f}{c}\right)\sqrt{\left(\frac{3.7c}{d2\pi f}\right)^2 - 1} \qquad (dB) \tag{63a}$$

or

$$A(dB) \cong 8.7l\frac{\omega_c}{c} = 8.7l\frac{\left(3.7\frac{c}{d}\right)}{c} \cong 32\left(\frac{l}{d}\right) \tag{63b}$$

when the frequency ω is much below the waveguide cutoff frequency ω_c ($\omega \ll \omega_c$).

Therefore, Eq. (63a), (63b) with Eq. (57c) can be used with a solid approximation of the attenuation of honeycomb waveguides. The system of multiple-honeycomb waveguides is shown in Fig. 27.

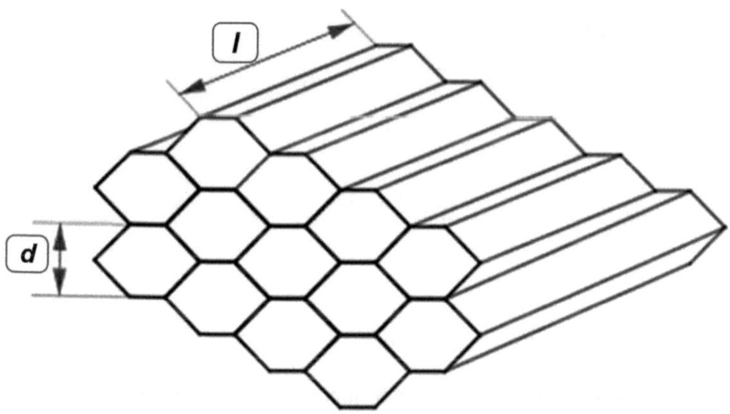

Fig. 27 Hanycom waveguides

Thus, using Eq. (63b) and by analogy with Eq. (59), the total attenuation (shielding effectiveness) of the system with n honeycomb waveguides is

$$A(dB) \cong 32\left(\frac{l}{d}\right) - 20\log\left(\sqrt{n}\right) \qquad \omega \ll \omega_c \tag{64}$$

where d is the waveguide cross-sectional dimension (the equivalent diameter) and l is the waveguide length.

Example 8 In Example 10, we graphically illustrate the attenuation of three honeycomb panels by 16 (4 × 4), 36 (6 × 6), and 100 (10 × 10) waveguides. The equivalent diameter d of a single waveguide is *5 mm* since the length is 20 mm. The attenuation plot is made up of the waveguide cutoff frequency f_c, expressed via Eq. (62) as $f_c \cong \frac{3.7}{d}\frac{c}{2\pi}$.

Solution Therefore, the cutoff frequency for this waveguide structure is

$$f_c \cong \frac{3.7}{d}\frac{c}{2\pi} = \frac{3.7}{5 \cdot 10^{-3}\, m}\frac{3 \cdot 10^{8}\,\frac{m}{s}}{2\pi} \approx 35\,GHz.$$

Furthermore, the attenuation plots shown in Fig. 28 were obtained via Eq. (63a):

$$A(f) = 8.7l\left(\frac{2\pi f}{c}\right)\sqrt{\left(\frac{3.7c}{d 2\pi f}\right)^2 - 1} - 20\log\left(\sqrt{\mathrm{n}}\right)(\mathrm{dB}).$$

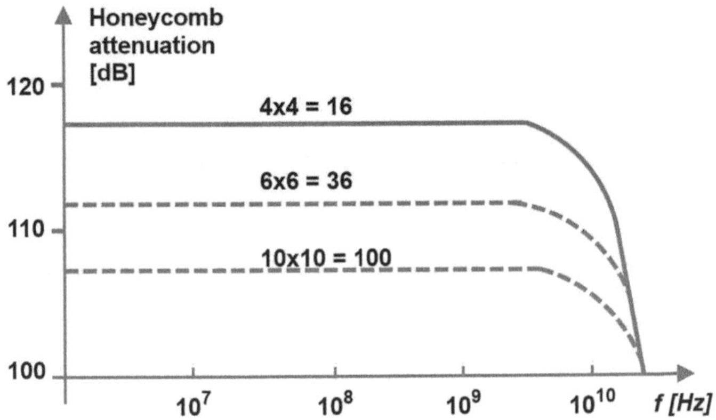

Fig. 28 Attenuation of honeycomb panels with different numbers of waveguides

A single waveguide has a constant attenuation with frequency and approaching its cutoff frequency, starting to decrease as expected. Additionally, by increasing the number of waveguides, the total attenuation of this panel decreases, as shown in Fig. 28. At a frequency of $f = 30\ GHz$, the attenuations of all waveguides fall to 0 *dB*. This is because the waveguide dimension a = 5 mm resonates at 30 GHz ($\frac{\lambda}{2} = 5\,\mathrm{mm} \rightarrow f = 30\,\mathrm{GHz}$).

7.3 *Comparison of Square-Holes, Rectangular Waveguides, and Honeycomb Waveguides Vent Panels in Terms of EMI Attenuation*

Previous equations (59), (61), and (64) yield:

$$A(dB) = 20\log\left(\frac{\lambda_0}{2a_n}\right) - 20\log\left(\sqrt{n}\right) \quad \omega \ll \omega_c \tag{65a}$$

$$A(dB) = 27\left(\frac{l}{a}\right) - 20\log\left(\sqrt{n}\right) \quad \omega \ll \omega_c \tag{65b}$$

$$A(dB) \cong 32\left(\frac{l}{d}\right) - 20\log\left(\sqrt{n}\right) \quad \omega \ll \omega_c \tag{65c}$$

Per these equations the EMI attenuations of the (4 × 4), (6 × 6), and (8 × 8) square-hole, rectangular-waveguide, and honeycomb-waveguide vent panels at *1 GHz* are listed in Table 1.

Table 1 The attenuations of vent panels with square-holes, square waveguides, and honeycomb waveguides

	EMI antennation (dB)			
Vent pattern	4 × 4 (n = 16)	6 × 6 (n = 36)	8 × 8 (n = 64)	Dimensions
Square holes	17.5 dB	14 dB	11.5 dB	a = 5 mm
Square waveguides	96 dB	92.5 dB	~90 dB	a = 5 mm, l = 20 mm
Honeycomb waveguides	116 dB	~112 dB	110 dB	d = 5 mm, l = 20 mm

Table 1 shows that the highest EMI attenuation is obtained by the honeycomb-waveguide vent panel at f = *1 GHz*. The cutoff frequency for the present panels is approximately f_c = 30 *GHz*,, which means that $f \ll f_c$.

8 Degradation of Shielding Effectiveness by a Cable's Penetration into a Device Enclosure

The shielding effectiveness of a device enclosure is strongly affected when an unshielded wire enters the enclosure, as shown in Fig. 29.

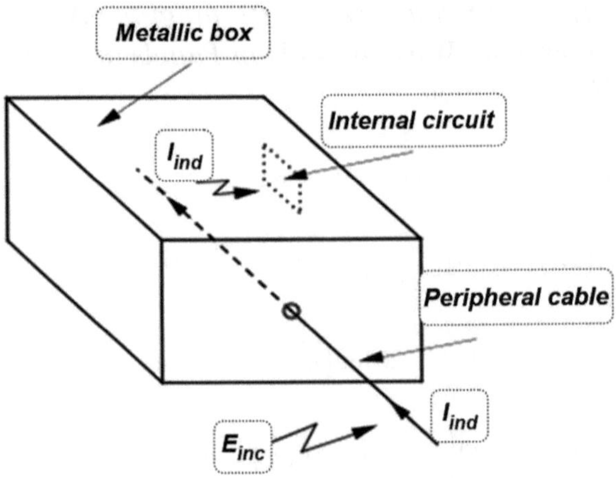

Fig. 29 Nonregular cable access in a device enclosure (the worst-case scenario)

Here, a penetrated EMI over the wire may easily affect electronic circuits inside the enclosure.

Additionally, connecting a cable with a pigtail, as discussed in Chap. 8: "Performance of RF Cables and IT Signal lines in Radio systems with EMI" is not a perfect solution because the shielding effectiveness is also degraded to some level, especially at higher frequencies. For example, Fig. 30 shows the mechanism of how the pigtail may introduce an external EMI into the enclosure.

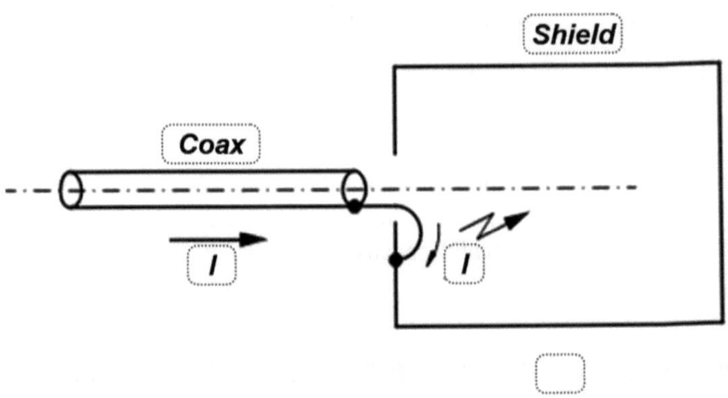

Fig. 30 Pigtail shield connection and degradation of a device SE

Figure 31 shows an example of the best solution for connecting the shielded wire (coaxial cable) via a coaxial connector to a device conductive chassis.

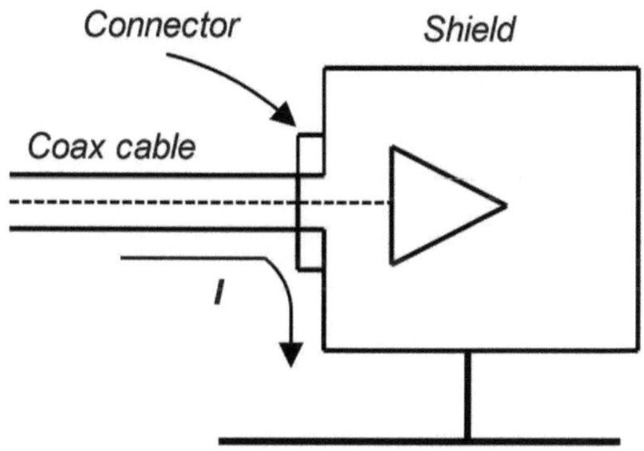

Fig. 31 Regular connection of a coaxial cable to a grounded chassis

References

1. G.M. Kunkel, *Shielding of Electromagnetic Waves* (Springer, 2020)
2. K.L. Kaiser, *Electromagnetic Shielding* (Taylor & Francis, CRC, 2006)
3. R.B. Schulz, Chapter 9, in *Handbook of Electromagnetic Compatibility*, ed. by R. Perez, (Academic Press, 1995)
4. H.W. Ott, *Electromagnetic Compatibility Engineering* (John Wiley & Sons, Inc., NY, 2009)
5. T. Williams, *EMC for Product Designers* (Elsevier, Newnes, 2007)
6. F.M. Tesche, M.V. Ianoz, T. Karlsson, *EMC Analysis Methods and Computational Models* (John Wiley & Sons, Inc., NY, 1997)
7. D. White, *Shielding Design–Methodology and Procedures* (Interference Control Technologies, Virginia, 1986)
8. C.R. Paul, R.C. Scully, *Introduction to Electromagnetic Compatibility* (Wiley, 2022)
9. C.A. Balanis, *Antenna Theory*, 3rd edn. (John Wiley & Sons, Inc., New York, 2005)
10. S. Thomas, S.M. Zachariah, *Nanostructured Materials for EMI Shielding* (CRC Press, 2022)
11. R.S. Yadav, I. Kurtika, J. Vilcakova, *Advanced Spinel Ferrite Nanocomposites for EMI Shielding Applications* (Elsevier, 2020)
12. K. Joseph, R. Wilson, G. Gejo (eds.), *Materials for Potential EMI Shielding Applications*, 1st edn. (Elsevier, 2019)
13. R. Morrison, *Grounding and Shielding: Circuits and Interference*, 6th edn. (Wiley-IEEE Press, 2016)
14. S. Celozzi et al., *Electromagnetic Shielding* (Wiley-IEEE Press, 2008)
15. G. Vijayaraghavan, M. Brown, M. Barnes, *Practical Grounding, Bounding, Shielding and Surge Protection* (Elsevier, 2004)
16. L.H. Hemming, *Architectural EM Shielding Handbook: Design and Specification Guide* (Wiley-IEEE Press, 2000)
17. D. White, *A Handbook on EM Shielding Materials and Performance* (Don White Consultants, Inc., 1980)

Grounding for Radio Systems and Facilities

1 Introduction to the General Concept of Grounding

The term "ground" for circuits and electrical equipment refers to a zero-ohm or zero-voltage reference. These are conventional definitions for the ground that are more appropriate for DC and very low-frequency applications (below 1 MHz). For higher frequencies, neither zero-ohm nor zero-voltage ground exists. This is true because the overall grounding impedance also includes the impedance of a grounding conductor (cable) for connecting equipment to the grounding system. This impedance is highly dependent on frequency, as we discuss later in this chapter.

Hence, a grounding system is a network of ground electrode(s) buried into the earth and connected to equipment over a grounding conductor. Figure 1a shows a simple and typical grounding installation with the equivalent circuit model.

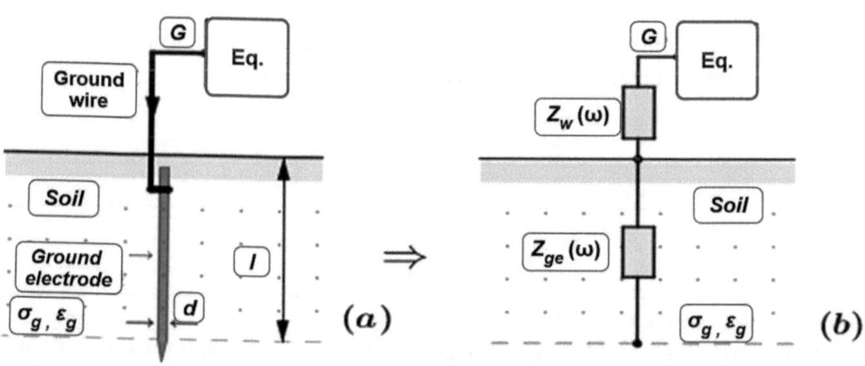

Fig. 1 Typical grounding: (**a**) physical realization, (**b**) an equivalent circuit

M. Pajovic, *Fundamentals of EMI Phenomena in Radio Communication Systems*, https://doi.org/10.1007/978-3-031-88748-2_11

This grounding involves a buried rod with length l and a grounding cable (a thick wire or strap) that connects equipment to the grounding system. Figure 1b shows the equivalent circuit scheme of the ground installation, where Z_w and Z_{ge} are the impedances of the grounding conductor and ground electrodes, respectively. Additionally, this figure shows the ground (soil) with conductance σ_g and a dielectric constant ε_g; as such, the ground can be treated as a conductive dielectric. Notably, the conductance of a grounding electrode is much greater than the conductance of the ground.

The writing of this chapter is motivated and influenced by the references [1–7] that we recommended to interested readers for future reading.

2 Resistance and Capacitance of a Vertical Conductor in a Lossy Ground

Here, we can find the self-capacitance of one conductor (electrode) by applying *Gauss's* law in the integral form (discussed in *Appendix A*):

$$\varepsilon \oint \vec{E}_n d\vec{S} = Q' \tag{1}$$

where ε is the permittivity of the ground around the electrode and where Q' is the total per-unit-length charge at the electrode enclosed by surface S. Furthermore, we can find the electrostatic field normal at the electrode surface from Eq. (1) as

$$E_n = \frac{Q'}{\varepsilon \oint dS} = \frac{Q'}{\varepsilon (2\pi r)} \quad \left(\frac{V}{m}\right) \tag{2}$$

The voltage potential V between the electrode surface and an observing point in space at a radial distance R from the electrode is expressed as

$$\begin{aligned} V &= -\oint \vec{E}_n d\vec{l} = \int_r^R \frac{Q'}{2\pi\varepsilon r} \\ &= \frac{Q'}{2\pi\varepsilon} \ln\left(\frac{R}{r}\right) \quad (V) \quad R \geq r \end{aligned} \tag{3}$$

where r is the conductor's radius.

Furthermore, based on the capacitance definition $C' = Q'/V$ and for $R = l_{el}$ (where l_{el} is the length of the buried electrode), the self-per-unit-length capacitance of the conductor becomes

$$C' = Q'/V = \frac{2\pi\varepsilon}{\ln\left(\frac{l_{el}}{r}\right)} \quad \left(F/m\right) \tag{4}$$

or the capacitance of the buried electrode long l_{el}

$$C_{gv} = l_{el} \frac{2\pi\varepsilon_0\, \varepsilon_g}{\ln\left(\frac{l_{el}}{r}\right)} \quad \left(\text{F}/\text{m}\right) \tag{4a}$$

The resistance R_{ge} of the present vertical electrode is given by Eq. (5) as

$$R_{ge} = \frac{\varepsilon_g}{\sigma_g\, C_{gv}} \tag{5}$$

where σ_g is the ground conductance. Eq. (5) is derived by combining Gauss's and Ohm's laws (described in *Appendix A*) given here by Eqs. (6a) and (6b), respectively, as

$$Q = \varepsilon \oint \vec{E}\, \mathrm{d}\vec{S} \;\rightarrow\; \frac{Q}{\varepsilon} =.\ \vec{E}\, \mathrm{d}\vec{S} \tag{6a}$$

$$\vec{J} = \sigma\, \vec{E} \tag{6b}$$

In accordance with Ohm's law, the current I at the electrode is equal to the flux of the surface current $\vec{J}$ as follows:

$$I = \int \vec{J} \mathrm{d}\vec{S} \tag{6c}$$

Substituting Eq. (6a) into Eq. (6c) results in

$$\mathbf{I} = \int \vec{J} \mathrm{d}\vec{S} = \sigma \oint \vec{E}\, \mathrm{d}\vec{S} \;\rightarrow\; \frac{I}{\sigma} = \oint \vec{E}\, \mathrm{d}\vec{S} \tag{7}$$

and, on the basis of Eqs. (6a) and (7), there is

$$\frac{Q}{\mu} = \frac{I}{\breve{A}} \tag{8}$$

Furthermore, substituting the following expressions from Appendix A, $Q = CV$ and $I = V/R$, into Eq. (8), the resistance R_{ge} of an electrode in a lossy dielectric (ground) is given as

$$R_{ge} = \frac{\varepsilon_g}{\sigma_g\, C_{gv}} \tag{9}$$

Finally, substituting Eq. (4a) into Eq. (9), the resistance of a vertical electrode in the ground is given as

$$R_{gv} \cong \frac{1}{2\pi l_{el}\sigma_g}\ln\left(\frac{2l_{el}}{d_{el}}\right) \quad \text{for } l_{el} \gg d_{el} \tag{10}$$

where l_{el} and $d_{el} = 2\, r_{el}$ are the length and diameter of the electrode, respectively.

Example 1 The resistance and capacitance of a vertical electrode of length $l_{el} = 2.4$ m (≈8 feet) with a diameter of $d_{el} = 2.5$ cm (≈1 inch) are determined when the electrode is buried in a well-conductive ground with $\sigma_g = 0.03$ S/m and a relative dielectric constant of $\varepsilon_g = 20$.

Solution According to Eqs. (4a) and (8), there is

$$C_{gv} = \frac{2\pi l_{el}\varepsilon_g\varepsilon_0}{\ln\left(\frac{2l_{el}}{d_{el}}\right)} \approx 50\,\text{nF}; \quad R_{gv1} = \frac{1}{2\pi l_{el}\sigma_g}\ln\left(\frac{2l_{el}}{d_{el}}\right) \approx 11\,\Omega \tag{10a}$$

We can see here that the electrode resistance R_{ge} is much smaller than the magnitude of the capacitive impedance $1/2\pi f C_{gv}$ for frequencies below 1 MHz, meaning that the grounding impedance is predominantly resistive, not capacitive:

$$\frac{1}{2\pi f C_{gv}} = \frac{1}{2\pi f \cdot 50 \cdot 10^{-9}} \approx 318\,\Omega \gg R_{ge} = 11\,\Omega \;\; at \; 1\,\text{MHz} \tag{10b}$$

3 Examples of Other Grounding Configurations

Figure 2 shows two vertical rods, rods 1 and 2, whose tops are bonded by horizontal conductor 3. For this situation, the grounding resistance depends on separation between the two rods: when the distance L between the rods is smaller than the electrode length ($L < l$), the ground resistance is greater than the ground resistance of the two rods when the distance between them is greater than the electrode length ($L > l$).

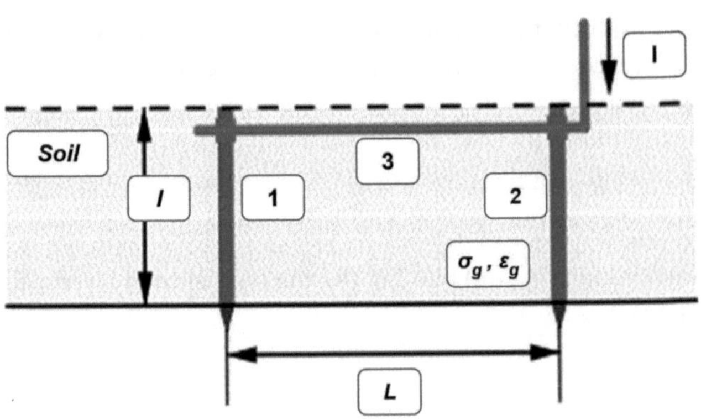

Fig. 2 Grounding with two vertical electrodes

The total resistance R_{gt} of this grounding, with a larger space between the electrodes ($L > l$), can be approximately represented as a simple parallel connection between the resistances of vertical electrodes 1 and 2 (R_{gv1} and R_{gv2}) and a horizontal strap (R_{gh3}). Therefore, this grounding has a significantly lower resistance than the single-electrode grounding considered in the previous section.

Hence, this expression of the resistive parallel connection is valid when the current fields of the electrodes in the ground do not overlap, i.e., when the vertical rods are sufficiently distant from each other. For practical opinions, $L > 5\ l$ is good. However, the most reliable method for obtaining accurate grounding resistance is measurement at the location of interest.

The grounding can also be made by buried horizontal electrodes, as shown in Fig. 3a–c.

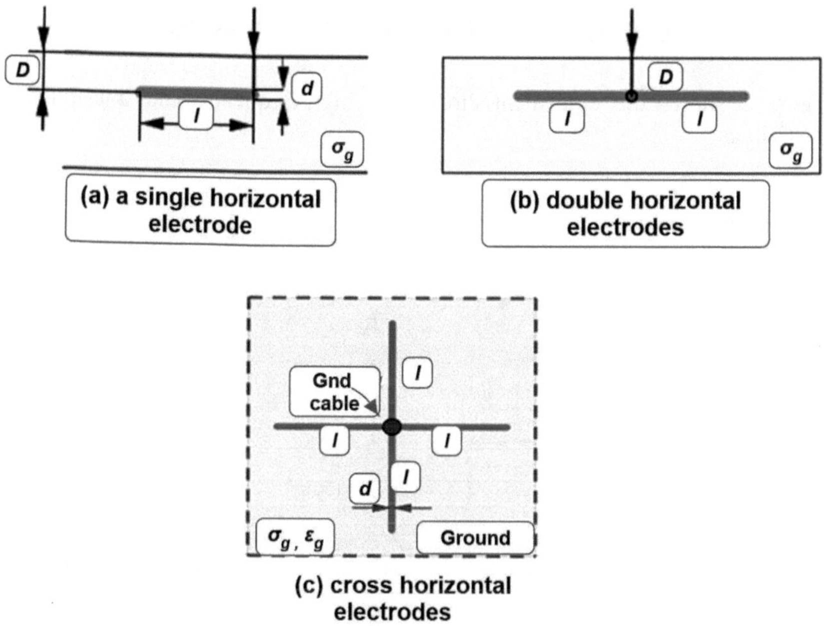

Fig. 3 Configurations of horizontal groundings. (**a**) Single—horizontal electrode. (**b**) Double—horizontal electrode. (**c**) Cross configuration of horizontal electrodes

The resistances of these ground grounding configurations depend on the length of the electrodes, as well as on the depth h where the electrodes are buried. Larger electrode lengths and deeper burial lead to lower resistance.

Grounding can also be achieved by various shapes of 2D or 3D electrodes. Figure 4a, b shows the two shapes of 2D grounding electrodes.

Here, a metallic solid disk and a conductive ring buried in the ground are shown. The ring is usually made of a thick wire conductor or straps. As in the previous cases, the grounding resistance of these groundings depends on the size of the electrodes and the depth h where the electrodes are buried.

More grounding configurations can be found in the literature [1–4].

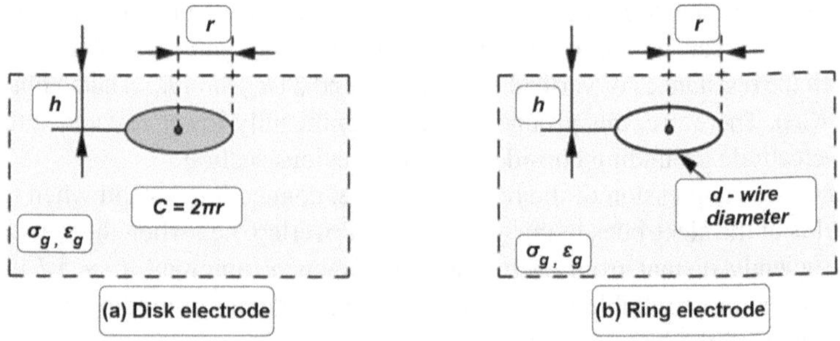

Fig. 4 2D electrodes: (**a**) disk and (**b**) ring

4 Low-Frequency Vs High-Frequency Grounding

Figure 5 shows the equivalent circuit for high-frequency and low-frequency groundings.

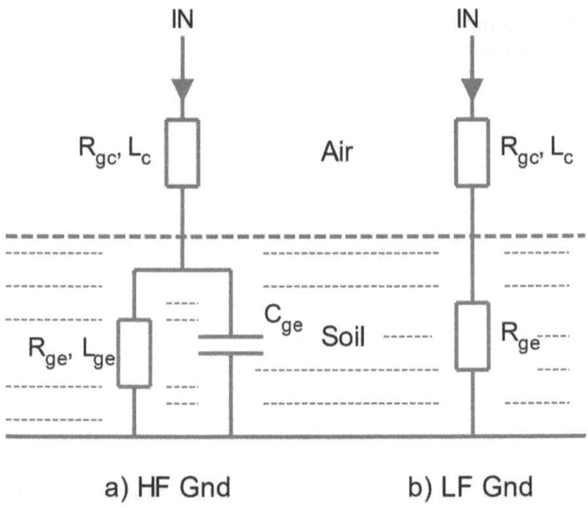

Fig. 5 Equivalent scheme for high-frequency grounding

Where R_{gc} and L_c are the resistance and inductance of the grounding cable (conductor), respectively, and where R_{ge}, L_{ge}, and C_{ge} are the resistance, inductance, and capacitance of the electrode, respectively. Hence, the total impedance of the grounding system includes the impedance of the grounding cable Z_w with *resistive* and *inductive* components (R_w and $i\omega L_w$) and the impedance of the underground electrode system with resistive and reactive components (R_{ge}, $i\omega L_{ge}$, and $1/i\omega C_{ge}$). For low frequencies, such as AC frequencies of 50/60 Hz or their harmonics in the kHz range, Z_{ge} is dominantly resistive and mainly depends on the soil conductance σ_g and the size of the grounding electrode system (the case related to Fig. 5b).

The ground *conductance* σ_g and *relative dielectric* constant ε_{gr} vary from poor soil conductance ($\sigma_g = 0.001$ S/m with $\varepsilon_{gr} = 3$) to very good conductance (for pastoral and rich soil: $\sigma_g \approx 0.03$ S/m with $\varepsilon_{gr} = 20$). The average ground has a conductance and dielectric constant of approximately $\sigma_g \approx 0.005$ S/m and $\varepsilon_{gr} = 10$, respectively. The ground conductance and dielectric constants from poor to rich soil are listed in Table 1 [3–6].

Table 1 The ground conductance and dielectric constants for grounds from poor to rich

Type of ground	Conductance (S/m)	Resistance (Ω/m)	Dielectric constant
Extremely poor: Cities with high buildings	0.001	1 k Ω/m	3
Very poor: Cities and industrial zone	0.001	1 k Ω/m	5
Sandy and dry terrain	0.002	500 Ω/m	10
Poor, rocky, mountains	0.002	500 Ω/m	12
Average: Pastoral, heavy soil	0.005	200 Ω/m	13
Pastoral, med hills, forest	0.006	~170 Ω/m	13
Flat, densely wooded	0.008	125 Ω/m	12
Pastoral, rich soil	0.01	100 Ω/m	14
Very good: Pastoral, rich soil	0.03	~33 Ω/m	20
Fresh water	0.001	1 k Ω/m	81
Salt water	5	0.2 Ω/m	80
Copper as the reference	$5.8 \cdot 10^7$	$\sim 1.7\ 10^{-6}$ Ω/m	1

We can see here that seawater (saltwater) has the best "grounding" properties, i.e., high conductance. For example, seawater with a low conductance with a conductive plate (such as bronze) provides a very low-impedance "grounding."

5 Grounding Conductor

As noted in the previous sections, grounding systems consist of a system of buried electrodes and a grounding conductor, such as a wire or strap conductor for connecting electronic equipment to the underground electrode system. The impedance of the grounding conductor can dramatically increase the overall impedance of the grounding system, which depends on the conductor's length and frequency of interest.

Figure 6 shows the impedance of a grounding system buried in the ground including the impedance of a grounding conductor above the ground.

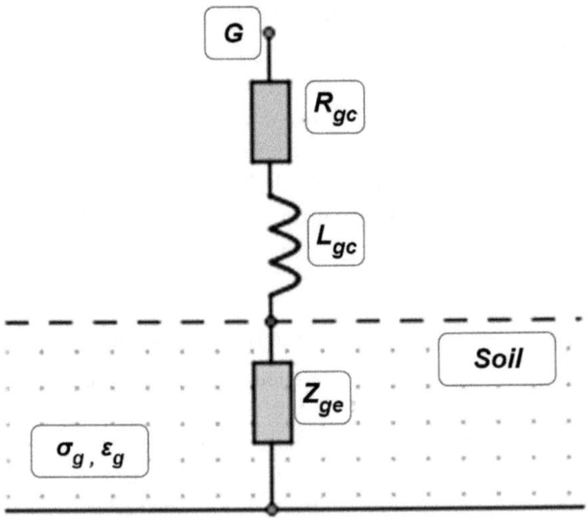

Fig. 6 Circuit model of a complete grounding

The impedance elements of the ground conductor are the resistance R_{gc} and the inductance L_{gc}, while Z_{ge} is the impedance of the electrode system in the ground. Additionally, per Chapter "Nonideal Passive Circuit Components", R_{gc} contains DC and AC resistance, i.e., $R_{gc} = R_{dc} + R_{ac}(\omega)$. Therefore, the grounding conductor impedance $\tilde{Z}_{gc}$ is

$$\tilde{Z}_{gc} = R_{dc} + R_{ac}(\omega) + i\,\omega\,L_{gc} \tag{11}$$

The main goal is for the grounding conductor to have as low an impedance as possible over the frequency spectrum of interest. For example, the frequency range is mostly below MHz for safe AC and lightning grounding. The final goal of the *low-impedance* conductor is to equalize the equipment chassis potential with the ground potential.

Hence, the grounding conductor should be short and thick to achieve this goal. Additionally, a grounding conductor should be made of a nonmagnetic material.

Figure 7a, b shows the round wire and flat strap conductors, which have dimensions important to the grounding impedance and the currents they should carry.

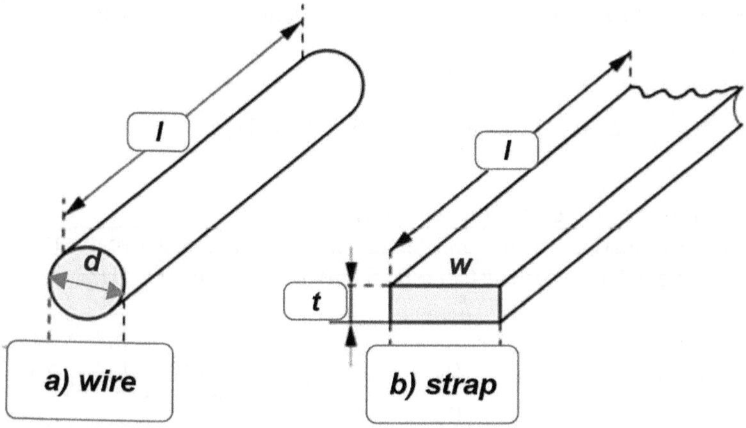

Fig. 7 Grounding conductor: (**a**) a wire and (**b**) a strip

The per-unit resistances R'_{dc} and R'_{ac} are given in Chapter "Nonideal Passive Circuit Components" by the expressions:

$$R'_{dc} = \frac{1}{\sigma}\frac{1}{\pi r^2} = \frac{1}{\sigma}\frac{1}{S} \qquad \left(\frac{\Omega}{m}\right) \tag{12a}$$

$$R'_{ac} = \frac{1}{(2\pi r)\sigma\delta} = \frac{1}{O\sigma\delta} \qquad \left(\frac{\Omega}{m}\right) \tag{12b}$$

where $S = \pi r^2$ and $O = 2\pi r$ are the area and circumference of the wire cross-section, respectively. Additionally, σ and δ are the wire conductance and the skin depth, respectively, expressed as (Chapter "Nonideal Passive Circuit Components"):

$$\delta = \frac{1}{\sqrt{\pi f \mu_0 \sigma}} \tag{12c}$$

The resistances R_{dc} and R_{ac} for a flat strap are

$$R'_{dc} = \frac{1}{\sigma}\frac{1}{wt} \qquad \left(\frac{\Omega}{\mathrm{m}}\right) \tag{13a}$$

$$R'_{ac} = \frac{1}{\sigma\delta C} = \frac{1}{2\sigma\delta(w+t)} \qquad \left(\frac{\Omega}{\mathrm{m}}\right) \tag{14}$$

The approximate expressions for the *self-inductance* of a round wire L_w are given as

$$L_w^{'} \approx 2l\left(2.3 \cdot \log\left(\frac{4l}{d}\right) - 0.75\right) \quad (\mathrm{nH}) \tag{15}$$

where d is the diameter of the wire in *cm.*

Table 2 summarizes the per-unit-length resistance and inductive reactance of the copper wire conductor AWG #1 ($d = 0.75$ cm) at DC up to 100 MHz.

Table 2 Round copper wire with $l = 1\ m$ and $d = 0.75\ cm$

$L'_{wg} \approx 1.1$ µH/m	$\left(R_{dc}^{'} + R_{ac}^{'}\right)(m\Omega / \mathrm{m})$	$X_{wg} = 2\pi f L'_{gc}$
DC	0.39	0
50 Hz	0.39	≈0.35 mΩ/m
100 kHz	0.39	≈0.7 Ω/m
1 MHz	0.39	≈7 Ω/m
10 MHz	0.39	≈70 Ω/m
100 MHz	0.39	≈690 Ω/m

The wire has a much larger *inductive* reactance X_{wg} than the resistance $(R_{dc}^{'} + R_{ac}^{'})$ above *1 MHz*. Consequently, the *inductive* reactance X_{wg} of the ground conductor introduces the most impedance to the grounding system above 1 MHz, which should be considered during the planning of a grounding system.

Thus, grounding with a long grounding conductor is not a low-impedance grounding solution at frequencies above 1 MHz. Low-frequency grounding is more realistic for DC, AC power, and lightning groundings, where the main frequency content of the electric energy lies below 1 MHz. However, the thickness of a grounding conductor is also important because of the current intensity that the grounding conductor needs to carry. This is especially important for lightning ground where the conductor should support very high current flows, as discussed in Chapter "Lightning Physics and Impact on Radio Systems on the Earth", which is related to lightning phenomena.

5.1 *Electrically Short Grounding Conductor Parallel to the Ground*

A grounding conductor is sometimes routed parallel and close to the ground or a larger conductive plane. These conductors can be electrically short and long. An electrically short conductor is shown in Fig. 8.

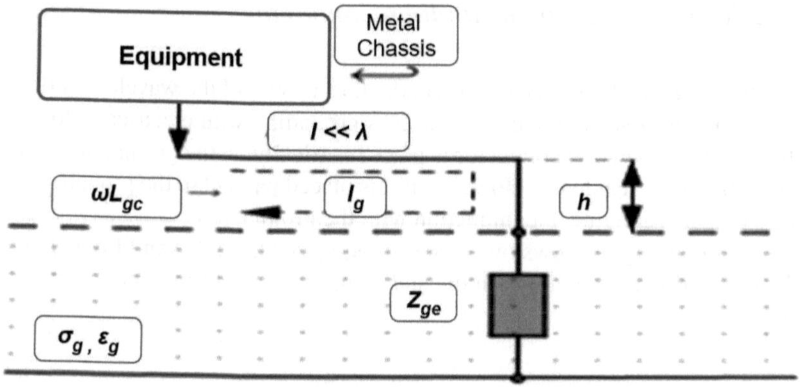

Fig. 8 Electrically short grounding conductor parallel to the ground

Here, the grounding conductor forms an inductive loop with the ground whose inductance is derived in *Appendix C* as

$$L_{wg} \approx l \frac{\mu_0}{2\pi} \ln\left(\frac{2h}{r_w}\right) \quad \left(\frac{H}{m}\right) \qquad for\, h \gg d \tag{16}$$

where l, d, and h are the length, diameter, and height of the conductor above the ground, respectively. For example, Table 3 gives a list of the *inductive* reactance values calculated via Eq. (16). The wire conductor with the length $l = 1\ m$ and radius $r_w = 0.5\ cm$ is routed parallel to the ground at $h = 0.1\ m$.

Table 3 Inductive reactance of the wire conductor

$L'_{wg} \approx 0.5$ µH/m	$X_{wg} = 2\pi f L'_{wXg}$
DC	0
50 Hz	≈0.23 mΩ/m
100 kHz	≈0.46 Ω/m
1 MHz	≈4.6 Ω/m
10 MHz	≈46 Ω/m
100 MHz	≈460 Ω/m

Per Eq. (16), the loop inductance increases with increasing height and length of the conductor. For example, according to Table 3, the per-unit-length impedance is approximately 46 Ω/m at 10 MHz. At higher frequencies, the impedance of the grounding conductor exceeds 460 Ω/m at 100 MHz. At this frequency, the grounding is completely nonfunctional.

Thus, the impedance of the ground conductor, which connects equipment with the grounding electrode system, is one of the most critical points in every grounding.

5.2 Electrically Long Grounding Conductor

We discussed grounding with a short conductor in terms of the wavelength of interest in the previous section. Now, we discuss grounding with electrically long conductors. If this grounding conductor is placed vertically on the ground, it can act as an EMI/RFI vertical antenna. However, if it is placed parallel to the ground, one can act as a transmission line in conjunction with the ground.

For example, Fig. 9 shows two pieces of equipment, Eq. (1) and Eq. (2), placed on the first and second building floors and connected to the same buried grounding electrodes.

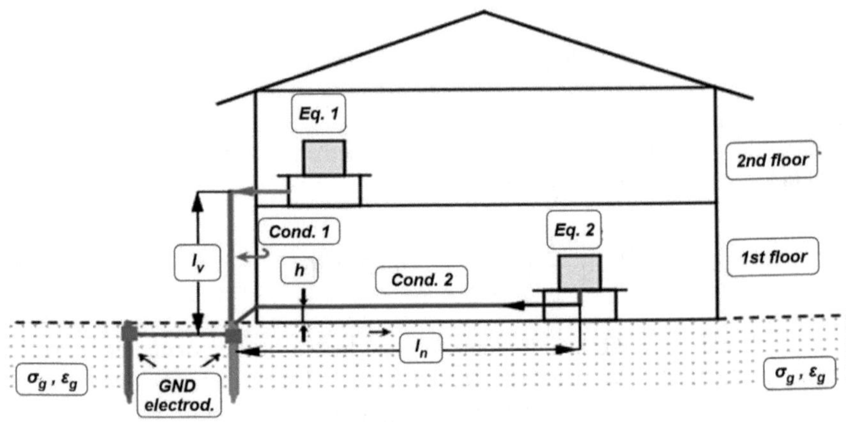

Fig. 9 Building ground installation

If the lengths of the grounding conductors are comparable to a quarter of the wavelength of the potential EMI, the vertical conductor may act as an EMI vertical antenna, whereas the horizontal conductor may act as a resonant transmission line with the ground surface. Figures 10 and 11 show the cases where vertical conductor 1 acts as an EMI radiator since horizontal conductor 2 acts as a TL.

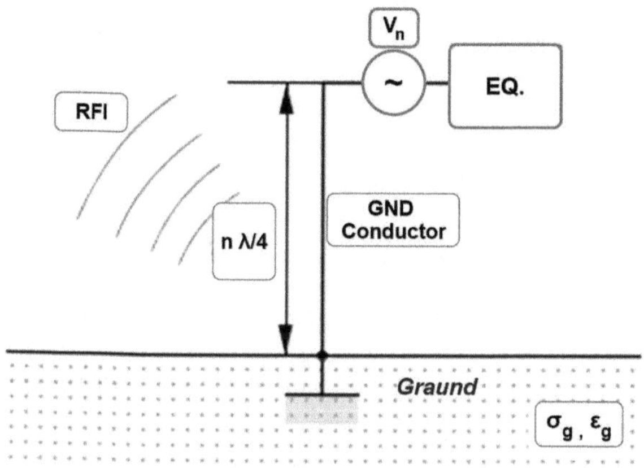

Fig. 10 Ground conductor as a vertical antenna

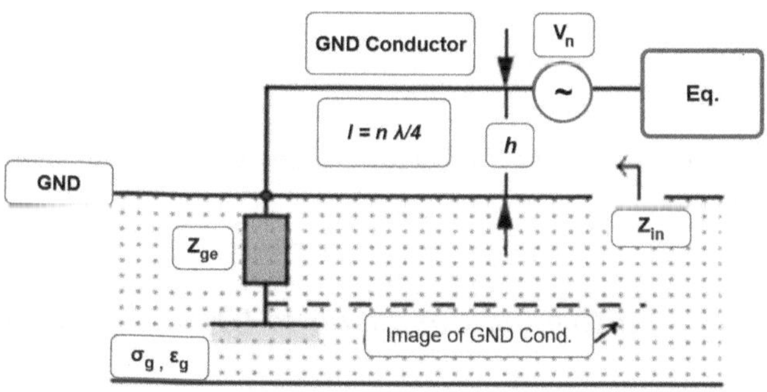

Fig. 11 Ground conductor as a transmission line

As discussed in Appendix B and Chapter "Performance of RF Cables and Signal Lines in Radio Systems with EMI", the maximum radiation of a resonant vertical conductor occurs when one is long by multiple quarters of the wavelength. Also, a slope grounding conductor is critical because may act as a slope wire antenna. Additionally, in addition to unwanted EMI radiation, an electrically long conductor may significantly increase the input grounding impedance at that resonance (discussed in *Appendix C*), where the grounding becomes completely inefficient. Thus, the grounding can be efficient at frequencies much below the conductor's first resonance when its length is much smaller than $\lambda/4$.

Another example of grounding involves a long conductor parallel to the ground, as shown in Fig. 11.

Here, the conductor forms a transmission line with the ground whose impedance "seen" by the equipment may be much greater than expected. Figure 12 shows the frequency profile of conductor impedance when the conductor forms a transmission line with the ground.

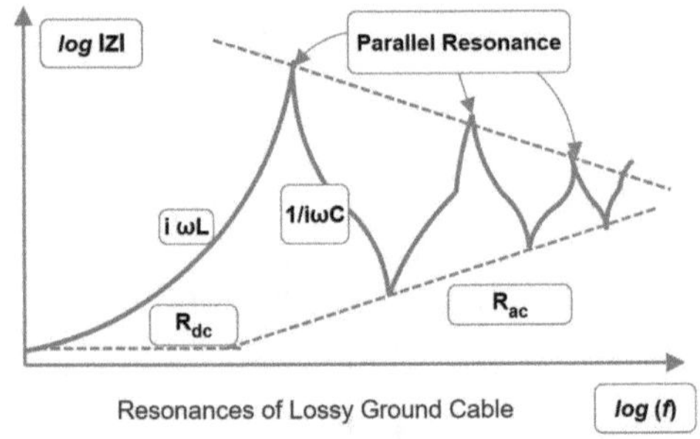

Fig. 12 Input impedance of an electrically long grounding cable parallel or vertical to the ground

This plot shows a series of unwanted high-impedance peaks related to a ground conductor parallel to the lossy ground, whose length l is expressed as a function of wavelength λ:

$$l = \left(\frac{2n+1}{4}\right)\lambda \qquad n = 0,1,2\ldots \tag{17}$$

Additionally, high-frequency peaks progressively decay with conductor length because of ground loss.

5.3 Bonding of a Ground Conductor to Ground Electrodes

Bonding the grounding conductor to grounding electrodes is a critical point for grounding quality. If the bonding is not good, there is an unexpectedly high impedance in the grounding system. Figure 13 shows an example of a ground conductor's proper and reliable bonding, which is placed in a building foundation.

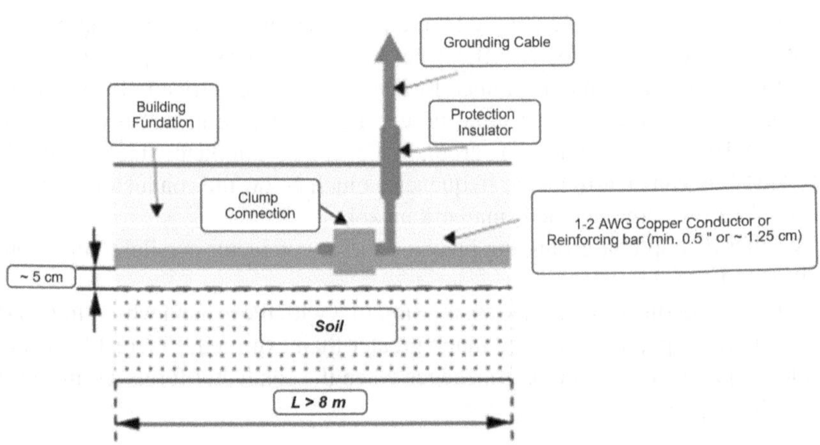

Fig. 13 Bonding ground cable to the building foundation

Hence, in addition to the low impedance of the grounding conductor and the ground electrode system, the low impedance *bonding* between these two parts of the grounding system is essential. The common cause of bonding degradation is corrosion between the materials of ground conductors and electrode systems. When good electrical contact is lost, the entire grounding system becomes unfunctional.

The writing of this section is influenced by the references [1–5], which we recommend to interested readers for more detailed reading.

6 Single-Point Vs. Multi-Point Grounding

Generally, two grounding concepts or topologies exist: *single-point* and *multipoint*. The first concept includes more pieces of equipment being grounded on a single ground point by their own conductors. The second concept includes each piece of equipment being grounded at different ground points.

6.1 Single-Point Grounding

By a *single-point grounding* or *a star topology,* each piece of equipment is grounded to the same ground point, as shown in Fig. 14a, b.

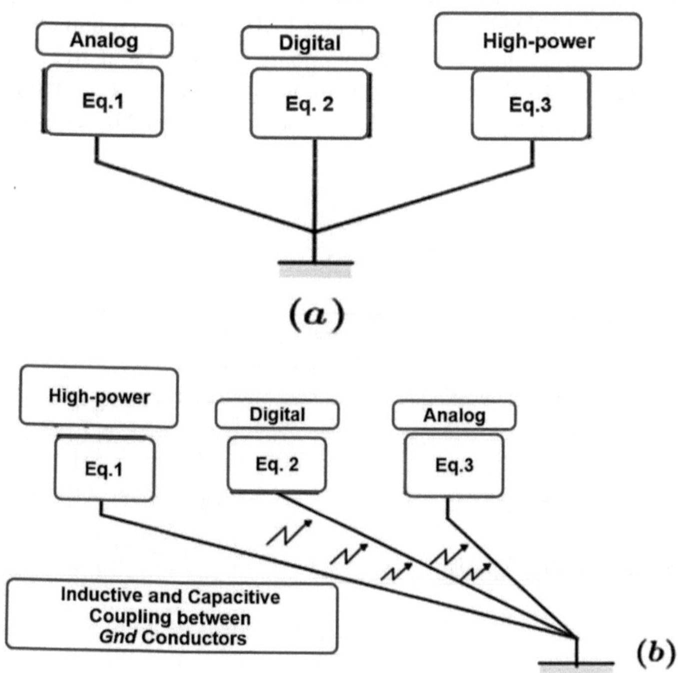

Fig. 14 (**a, b**) Ground cable routing: (**a**) good and (**b**) poor

However, owing to the potential *inductive* and *capacitive* coupling between grounding conductors of high-noise and low-noise equipment, their grounding conductors should be effectively separated. For example, regarding minimized conductor coupling, the topology in Fig. 14a is a *correct* solution since the grounding concept shown in Fig. 14b is *not correct.*

Another variant of single-point grounding, shown in Fig. 15, is called *chain* grounding.

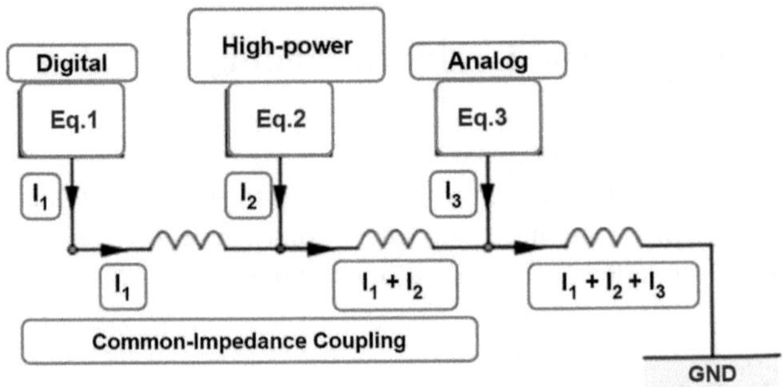

Fig. 15 A chain grounding

Here, different pieces of equipment are grounded in series at the same ground point sharing the grounding conductor, which may result in strong EMI coupling between the equipment with different sensitivities to EMI. Thus, this grounding concept should be avoided when grounding EMI-sensitive equipment and noisy *digital* and *high-power* equipment that commonly act as EMI "aggressors."

Additionally, Fig. 16 shows another example of an *incorrect* grounding scheme where a noisy digital unit is grounded over the chassis of EMI-sensitive (analog) equipment.

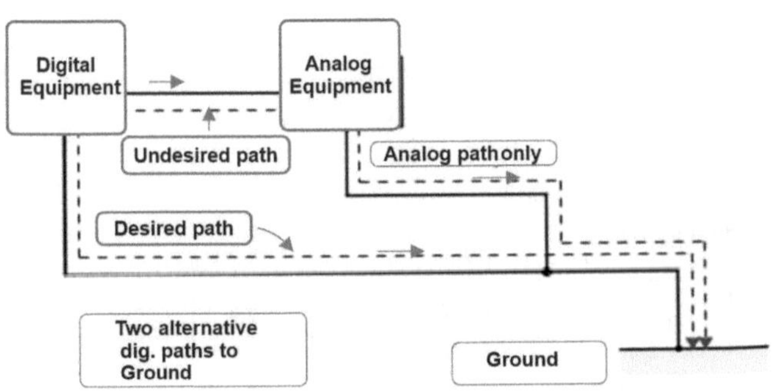

Fig. 16 Incorrect grounding topology

Here, we can see two current paths to the ground. The path via the enclosure of sensitive Equipment 2 is not good since the noise from Equipment 1 may penetrate Equipment 2. Thus, the bonding of two or more equipment chassis grounded to the same ground point should be avoided.

The alternative to grounding equipment (noisy and sensitive to noise) to the same grounding point is a hybrid single-point grounding scheme, as shown in Fig. 17.

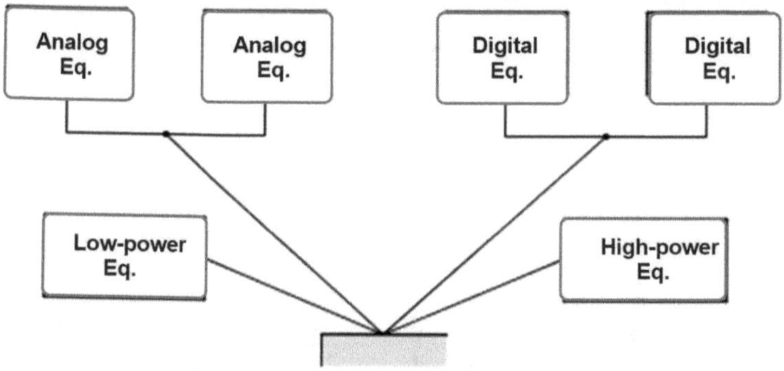

Fig. 17 Correct hybrid grounding scheme

Here, noisy and sensitive equipment are grouped into different groups, and each group is grounded by its own conductor to the same grounding point.

6.2 *Multipoint Grounding*

To reduce undesired coupling between the grounding conductors, multipoint grounding is the solution. Figure 18a shows a multipoint grounding concept where some grounded equipment coupling occurs through the ground having impedance Z_g.

The coupling in the multipoint grounding is proportional to the magnitude of the impedance Z_g, i.e., proportional to the developed voltages ΔV_{12} and ΔV_{23} between electrodes 1 and 2 and electrodes 2 and 3 (see Fig. 18a). By bonding the tops of the ground electrodes with thicker conductors (shown in Fig. 18b), the voltages developed between the electrodes will also be significantly minimized, and consequently, the coupling between equipment will also be minimized.

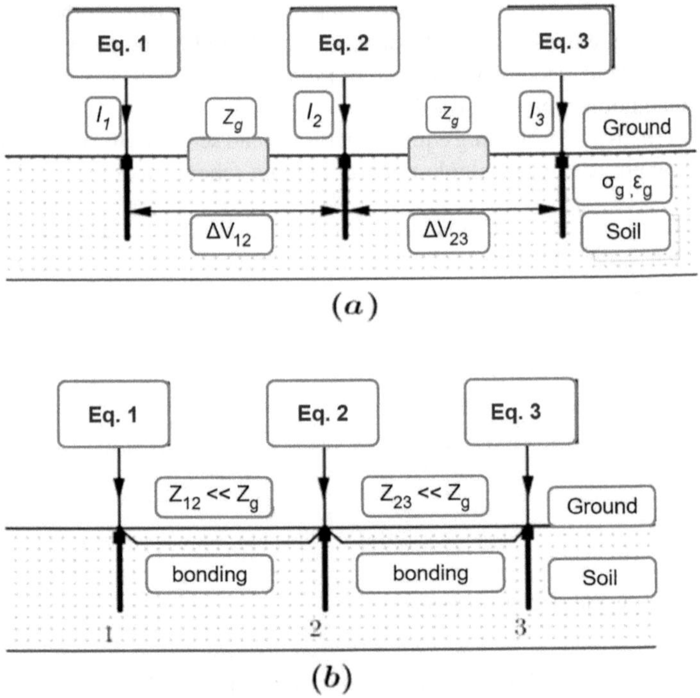

Fig. 18 (**a, b**) Multipoint grounding

6.3 High-Frequency Grounding

As mentioned above, if single-point grounding is improperly shared between noisy high-power and/or digital devices and noisy-sensitive low-power devices, the noise from noisy devices may disrupt the normal functions of low-power devices. Digital devices are relatively immune to external EMI noise but not to EMI produced at their reference conductor. However, the multipoint grounding concept with a solid and larger conductive plane, as a reference plane, is commonly used for these circuits.

Figure 19 shows a *decoupling* concept for a single-point grounding, which is shared by low-power (analog) and noisy digital devices, to use a unique single-point grounding for low-power analog devices and digital or other noisy devices.

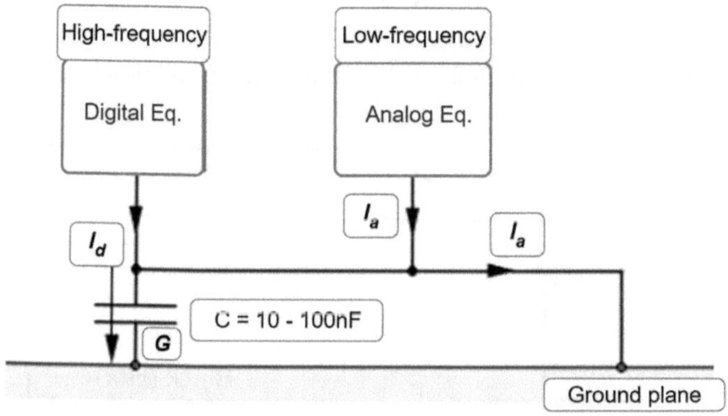

Fig. 19 Decoupling of EMI digital currents from low-power analog equipment

We can see here that noisy digital current I_d flows to the ground plane via the decoupling capacitor *C*, bypassing the grounding conductor used for the grounding of the analog-sensitive device.

The writing of this section is influenced by the references [1, 2, 5, 6], which we recommend to readers interested in more detailed reading.

7 Safety and RF Signal Grounding

With respect to applications, there are two major categories of grounding systems: (a) safety grounding (AC and lightning) and (b) RF grounding (low-frequency and high-frequency). Because of their low-frequency content, *AC* and lightning groundings belong to the category of low-frequency power groundings.

AC grounding is important because it protects people and equipment against shock hazards caused by AC-powered equipment faults. Hence, AC grounding provides a safe path for unwanted currents when a fault occurs in AC-powered equipment.

Lightning grounding is important because it keeps the equipment close to the ground potential during lightning strikes, directing strikes to the ground. In this way, lightning grounding also keeps equipment and personnel safe. More data about the grounding for lightning are given in Chapter "Lightning Physics and Impact on Radio Systems on the Earth", which the considers physics of *lightning* phenomena.

Another type of grounding is high-frequency grounding (or RF management grounding), whose task is to direct unwanted high-frequency EMI to the ground, avoiding impact on equipment functions. More information about RF grounding will be given later in this chapter.

7.1 AC Safety and Lightning Grounding

Figure 20a, b shows a typical 3-phase AC grounding installation in a building.

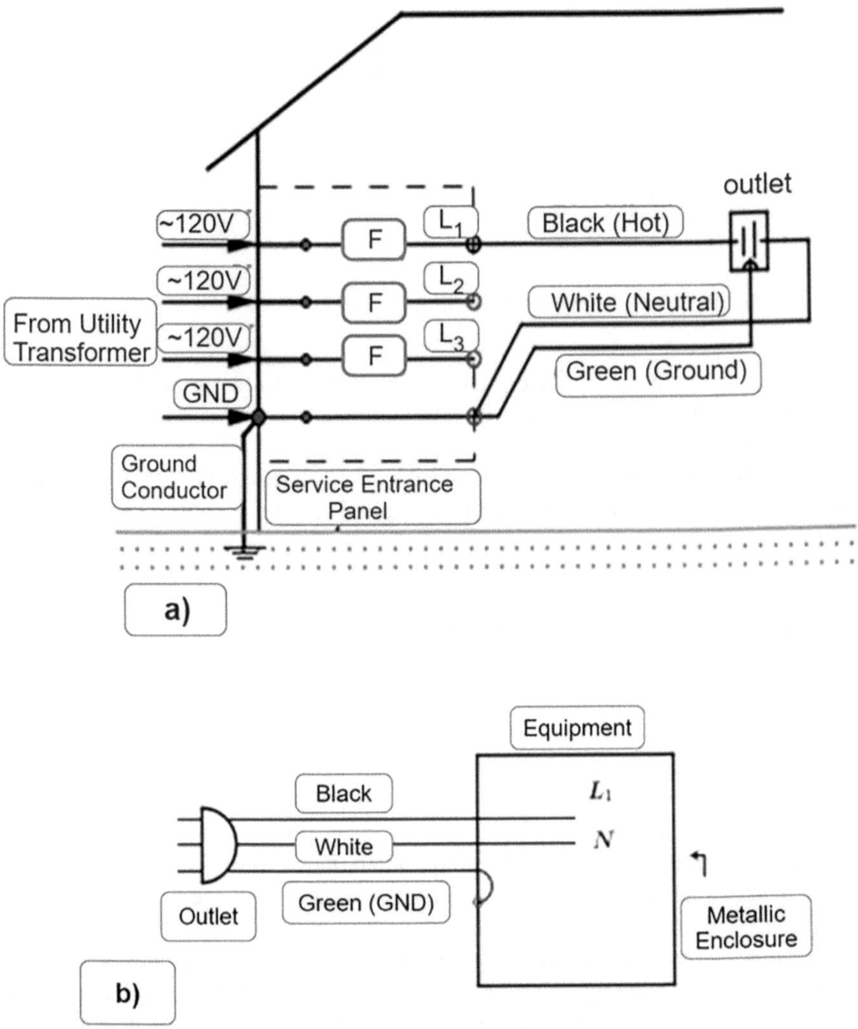

Fig. 20 (**a**) Building the main rounding scheme in the service entrance panel, (**b**) Equipment enclosure grounding

A 3-phase AC main system (3 × 120 V or 3 × 220 V) is used in commercial and residential buildings. For residential buildings in the United States, the AC main system is mainly a 2-phase system of 120V. However, the concept of AC safety grounding is similar to that of 2- and 3-phase main systems. Importantly, a service ground wire is grounded to the dedicated ground point in the entrance (service) panel. Additionally, note that the neutral line (white) of an outlet is connected to the ground (green) wire in the entrance panel [1, 2, 7].

Figure 20b shows that the green wire grounds the equipment enclosure via a room outlet and the service entrance panel to the outside-building grounding. If an equipment hotline (~120 V or ~220 V) touches the equipment enclosure, a corresponding fuse in the service panel instantly reacts by disconnecting AC power from the equipment, keeping the user safe.

8 Grounding for Multistory Buildings

Most of today's buildings are multistore structures ranging from several to tens and even hundreds of floors. As a result, a vertical distance from *real* ground becomes a serious problem because a very long grounding conductor leads to ineffective and useless grounding. Additionally, large buildings have large horizontal dimensions, and consequently, an additional grounding problem regarding horizontal bonding conductors, which should connect the floor equipment to a dedicated service panel at the current floor.

To overcome this problem, building a steel structure is exploited as a large 3-D grounding system. The steel structures (vertical members) are driven deep into the building foundation, and the horizontal greed of the steel is placed in the building foundation and bonded by welding to all vertical steel members, as shown in Fig. 21 [13].

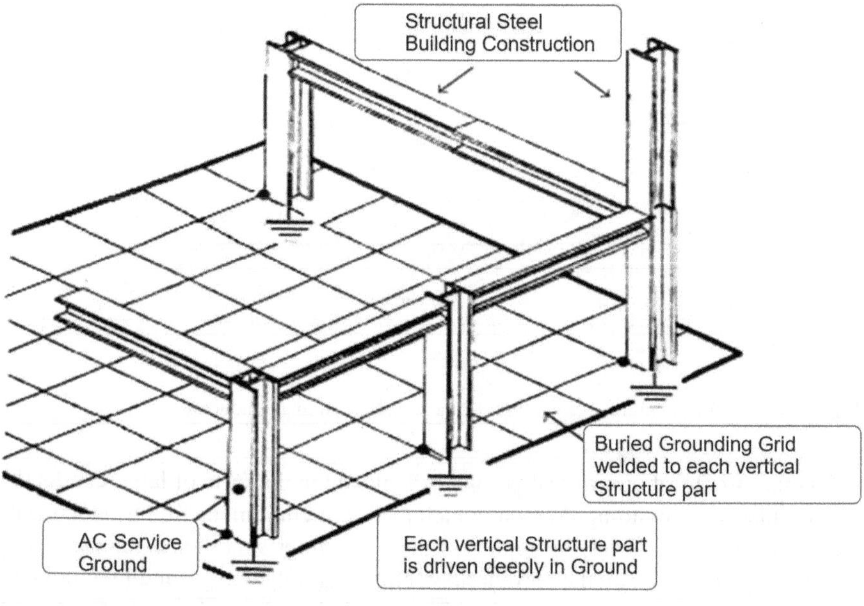

Fig. 21 Steel-building structure used for grounding

For a building structure, *I*-profiled steel bars are commonly used with a cross size of ~0.3 m (12 inches). There are concerns about the high impedance of steel bars, but interestingly, the *per-unit-length impedance* of the present steel bar is four times smaller than, for instance, a 4/0-copper cable (with $d \approx 11.5$ mm), which is commonly used in grounding systems. The lower per-unit-length impedance of the steel bar is due to its *larger cross-sectional* size than that of a 4/0 copper cable and consequently results in a smaller per-unit-length resistance and inductance. Hence, this 3-D grounding structure satisfies the criteria of *safe* grounding for AC mains and *lightning* protection.

8.1 Local Safety and Local RF Grounding for Multistory Buildings

Figure 22 shows the local grounding concept for grounding equipment on high floors, which are far from real ground [12].

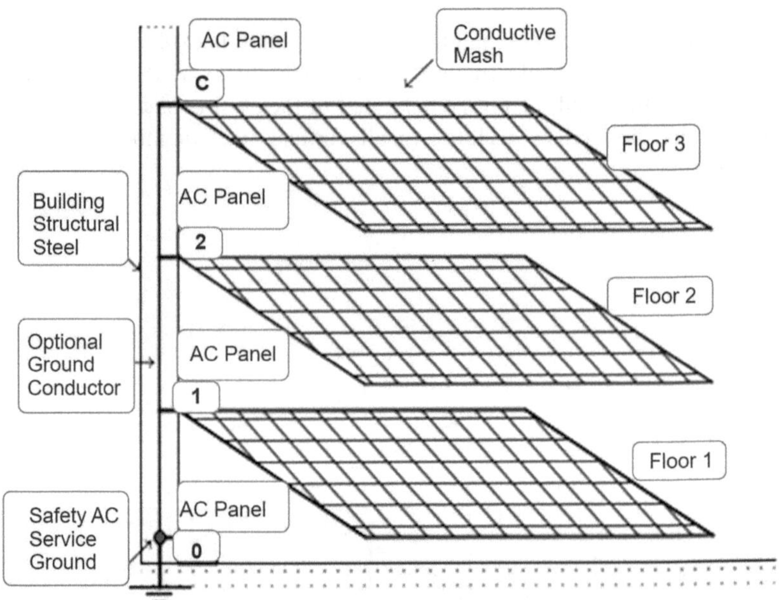

Fig. 22 Local (floor) building grounding with a single grounding conductor

In this case, local or artificial ground is deployed in the form of large conductive meshes placed on building floors to which pieces of equipment are grounded. This mesh is also used for safety grounding, where each floor's mesh is connected to the AC safety panel block on the current floor. Note that only one grounding conductor connects a floor mash down to the building foundation via the floor's service panel to avoid the creation of large inductive loops in the building grounding scheme. Figure 23 shows several acceptable grounding schemes for multistory buildings.

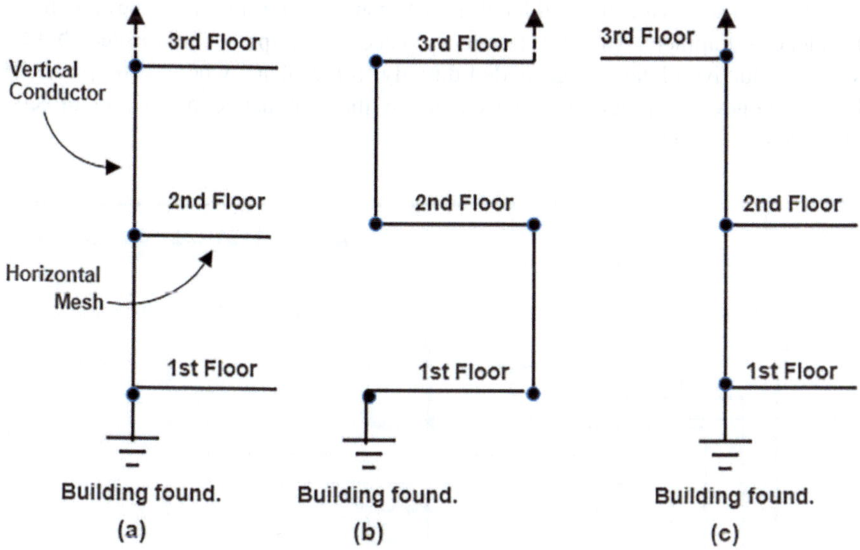

Fig. 23 Noninductive loops in a multistory building grounding

Some examples of undesired grounding, where grounding conductors form large current loops, are shown in Fig. 24.

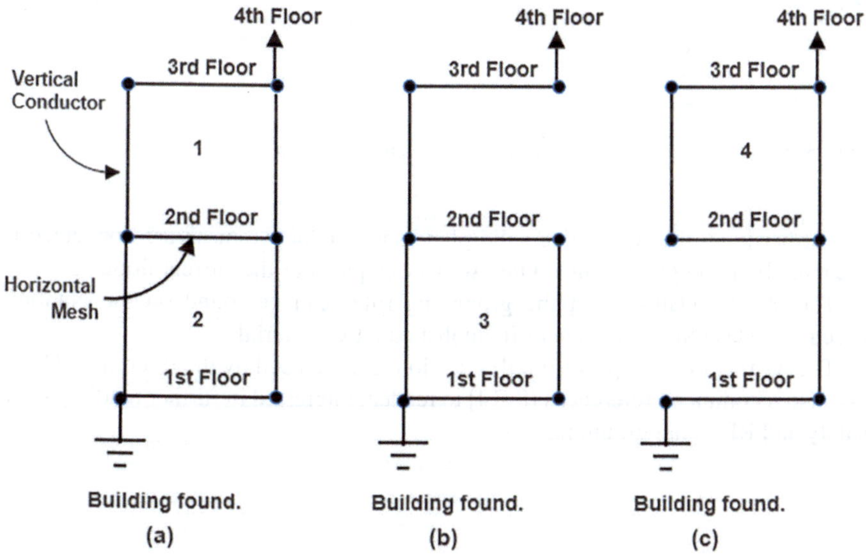

Fig. 24 Undesired current loops in a building grounding

Existing current loops 1, 2, 3, and 4 are especially critical during lightning strikes when high currents induced in the loops can harm the building's equipment and personnel.

Figure 25 shows a combined local (floor) grounding for low-frequency and high-frequency equipment. The high-frequency pieces of equipment are located above a solid conductive plane and grounded directly to the plane, whereas the pieces of low-frequency equipment are placed above the conductive mesh and directly grounded to the mesh.

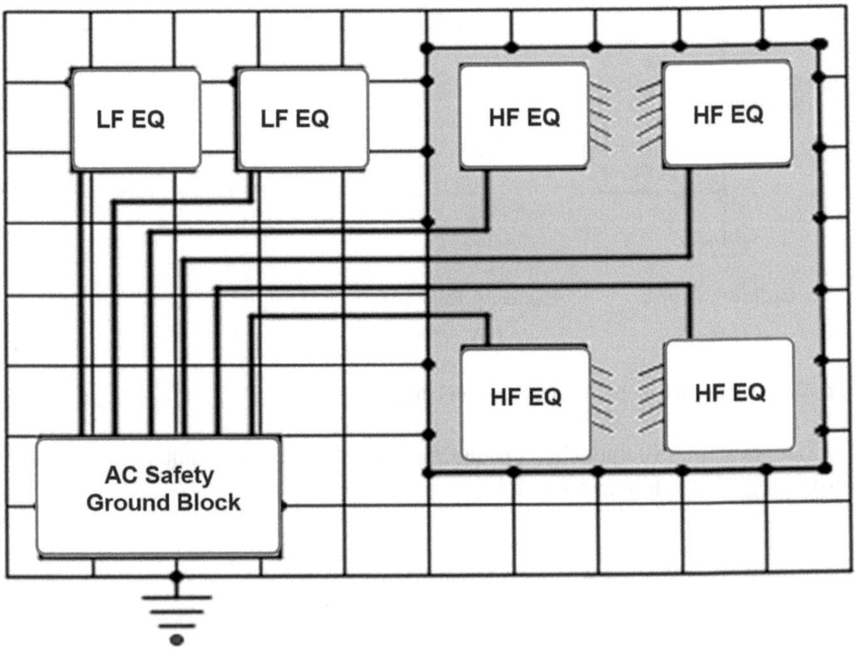

Fig. 25 Local high-frequency a groundings in an equipment room

Furthermore, the conductive solid plane and conductive mesh are connected to the main building grounding via the AC service panel of the current floor.

The latest updates about the grounding rules can be found on the National Electric Codes (NEC) website or in published NEC materials.

The writing of the topics from this section is influenced by the references [1–7]. We also recommend references [8–14] to readers interested in further reading about safety and RF signal grounding.

References

1. W.H. Lewis, Grounding and Bonding, in *Handbook of EMC*, ed. by R. Perez, (1995)
2. T. Williams, *EMC for Product Designers* (Elsevier, Newnes, 2007)
3. F.M. Tesche, M.V. Ianoz, T. Karlsson, *EMC Analysis Methods and Computational Models* (John Wiley & Sons, 1997)
4. H.W. Silver, *N0AX, Grounding and Bonding*, 2nd edn. (ARRL, 2019)
5. H.W. Ott, *Electromagnetic Compatibility Engineering* (John Wiley & Sons, Inc., NY, 2009)
6. C.R. Paul, R.C. Scully, *Introduction to Electromagnetic Compatibility* (Wiley, 2022)
7. MIL-HDBK-419A, *Grounding, Bonding, and Shielding for Electronic Equipment and Facilities* (Department of Defense, 1982)
8. M.A. Salam, Q.M. Rahman, *Power Systems Grounding* (Springer, 2016)
9. E.B. Joffe, K.-S. Lock, *Grounds for Grounding: A Handbook from Circuits to Systems*, 2nd edn. (Wiley IEEE Press, 2023)
10. M. Mitolo, *Analysis of Grounding and Bonding Systems* (CRC Press, 2023)
11. T. Henry, Grounding Workbook, Based on National Electric Code, 2020
12. R. Morrison, *Grounding and Shielding: Circuits and Interference*, 6th edn. (Wiley-IEEE Press, 2016)
13. K. Bryant, *Charmaz* (SAGE Publication Ltd, The SAGE Handbook of Grounding Theory, 2010)
14. G. Vijayaraghavan, M. Brown, M. Barnes, *Practical Grounding Bounding, Shielding, and Surge Protection* (Elsevier, 2004)

Natural Sources of EMI and EM Disturbances

Physics of Lightning and Its Impact on Radio Systems on Earth

1 Introduction

Lightning is a natural phenomenon during thunderstorms associated with high-energy electrostatic discharge (strikes or flashes) initiated from highly electrified thunderclouds. Electrostatic cloud energy is developed by a warm air stream that moves from the ground surface upward and enters the thundercloud during its formation from a regular cumulus cloud. In addition to the formation of cloud raindrops or ice particles, positive and negative charges also develop in the thundercloud. The positive charges are concentrated in the cloud-top region since the negative charges are concentrated in its lower region. Figure 1 shows typical thundercloud development and the charge distribution in the cloud volume [1–5].

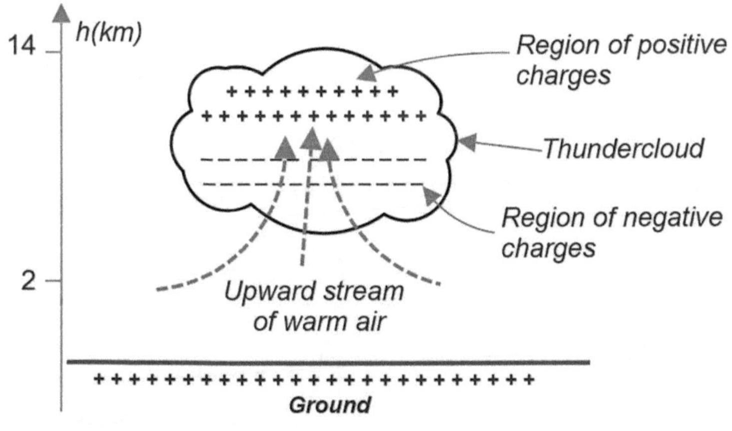

Fig. 1 Cloud charging process

M. Pajovic, *Fundamentals of EMI Phenomena in Radio Communication Systems*, https://doi.org/10.1007/978-3-031-88748-2_12

When the cloud is charged to a sufficient level, the discharging process begins toward the ground by a slow-moving ionized air stream followed by more intense and fast-moving negative charges from the cloud's lower part. Furthermore, as the cloud's negative charges approach the ground, the positive charges start to accumulate on the ground. When the strength of the *electric* field between *ground positive* and *cloud negative charges* reaches a sufficient level, the *ground-charged* stream (or leader) starts to move toward the downmoving cloud stream (or leader). This process is called the *step leader*. When the two streams (leaders) collide, a conductive low-impedance filament (a lightning channel) is created in the streams' centers, enabling the cloud charges to move faster toward the ground.

As the negative cloud charges are neutralized by the positive *ground* charges, a strong flesh strikes back from the ground toward the cloud at a speed of up to 100,000 km/s. This strike is called the *first return* strike, producing an intense flash typical for lighting. The *first return strike* is followed by several subsequent *return strikes* with a total flesh duration of approximately 2 s. Additionally, in addition to the high temperature in the lightning channel, which can reach 20,000 K, the lightning current peak can sometimes reach 200 kA. The more typical values of the current peaks are 20–50 kA [1, 2].

Figure 2 shows the *first* and a few subsequent *return* strikes with the corresponding current peaks on the ground, as well as the related current peaks on the ground.

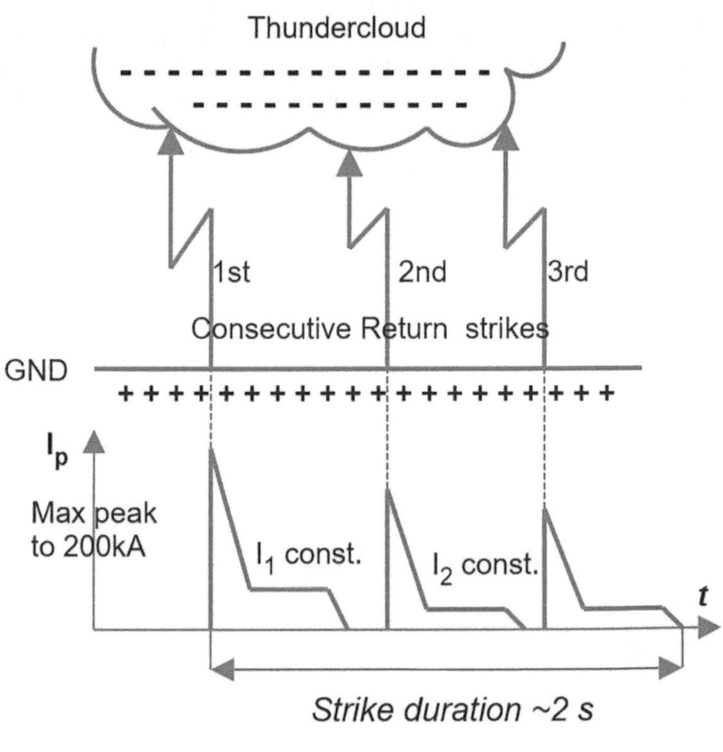

Fig. 2 First return strike followed by the next return strike with the corresponding current peaks on the ground

Hence, the current peak of the *first return* strike can reach 200 kA since the subsequent current peaks reach 100 kA. In general, the lightning process involves the exchange of electrostatic energy between charged thunderclouds and charged ground.

Additionally, Fig. 3 shows the return strike from the positively electrified ground over an electrified aircraft toward a negatively electrified thunderstorm cloud.

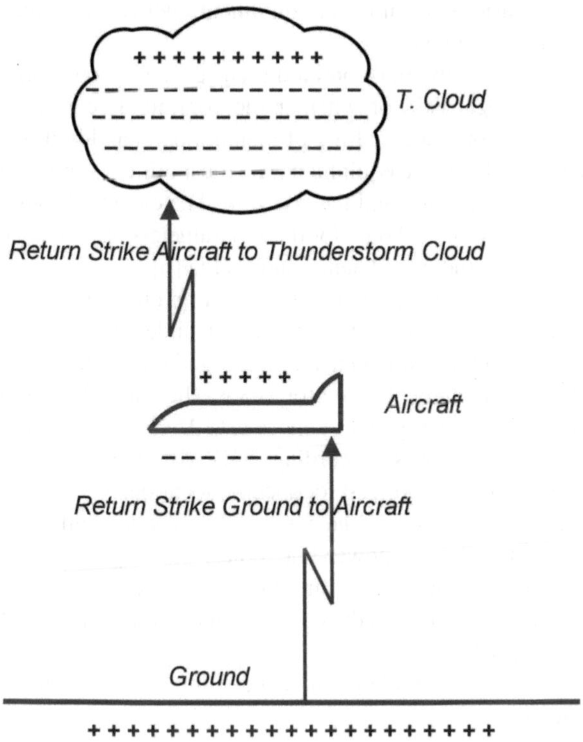

Fig. 3 The return strike from the positively electrified ground over an aircraft toward a negatively electrified Cloud

Therefore, positive charges are on the ground surface and upper side of the aircraft, whereas negative charges are at the lower side of the aircraft and lower side of the cloud, as shown in Fig. 3. Because the aircraft body is metallic (by aluminum), it acts as an efficient EMI shield for the aircraft interior.

2 Electrical Nature of the Lightning Strike

Lightning strikes associated with high currents, as shown in Fig. 2, can be highly destructive. These strikes usually generate temperatures and overpressures that can break nonmetallic objects, such as wood, glass, and concrete. In addition to mechanical destruction, strike currents induce a high voltage between conducting objects, which usually results in harmful EMI effects.

On the other hand, strike currents cause a rapid change in magnetic field strength, which can have a negative impact on electric and electronic equipment near a lightning strike. The equipment can be damaged by the surge voltage induced by the magnetic field that can penetrate the system equipment via peripheral wires and cables or through various openings at equipment enclosures in the form of high-voltage surges and transients.

Additionally, Fig. 2 shows that continuous currents with lower amplitudes follow the current peaks. Regardless of whether the continuous currents possess lower energy, they may still produce melting of metallic parts and high-temperature arcing. Associated with electronic equipment, the most critical parameters of the lightning current are the *amplitude* and the *rise rate*. The electrical lighting parameters are probabilistic, and statistical data for these parameters are obtained from empirical studies such as the studies of Cianos and Pierce [6].

Lightning or thunderstorm parameters depend on climate and differ from location to location worldwide. Most observations concluded that three lightning strikes or flashes occur per minute from an average charged cloud that may even cover the ground area of approximately 500 km^2 for 1–3 h. The data related to lightning have been accumulated over many years by the *World Meteorological Organization* for specific locations and countries. According to these data, lighting protection methods are established and standardized for specific locations.

Typically, lightning strikes exposed conductive objects or tall constructions, such as rooftops, radio towers, or power-line poles. An exposed conducting object absorbs most of the lightning energy and, in that way, reduces the lightning effects on neighboring objects within a radius equal to its double height (2H). This case is known as a *cone of protection.*

For example, Fig. 4 shows a large 3D conducting construction and the 2-H protection area around it.

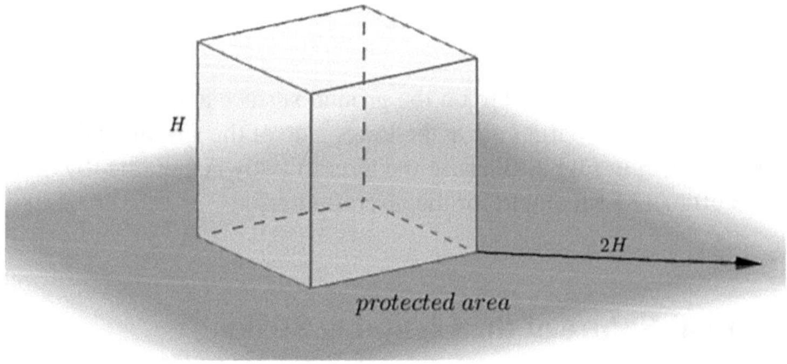

Fig. 4 3D metallic construction as a strike absorber

Outside of the 2H area, the protection effectiveness decreases. Hence, moving away from the exposed object by 2H steps, the lightning effectiveness obtained by the exposed object gradually vanishes. When the exposed metallic object is a thin tower or pole, Fig. 5 shows circular protection areas with 2H and 4H radii around the pole toll *H* [6].

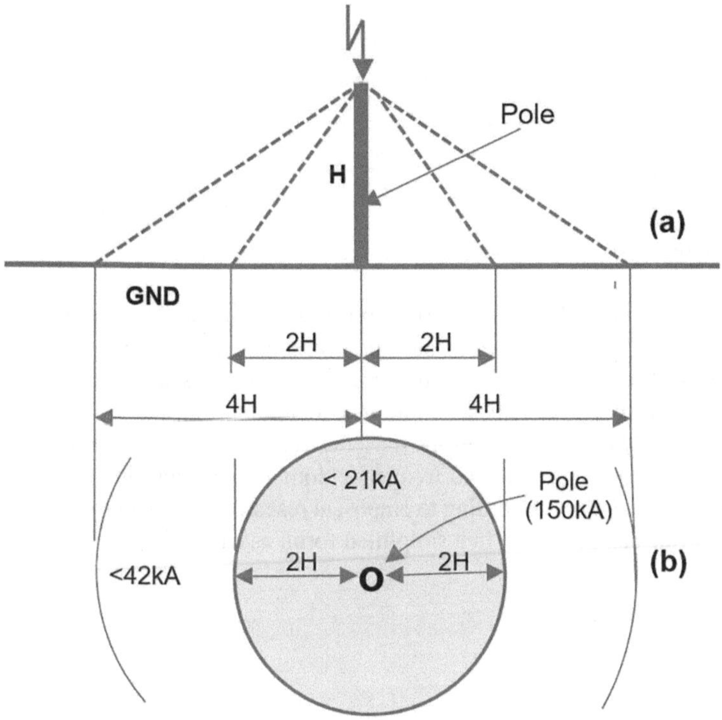

Fig. 5 Protection of the lightning area by a conductive pole

Figure 5b shows that the lightning strike hits a pole tall by *H* with a current peak of 150 kA. The original current peak away from the pole is reduced from 150 kA to ≈21 *kA* in the *2H*-radius area and to ≈42 *kA* in the *4H*-radius area.

2.1 Estimation of the Lightning Frequency Spectrum

As explained in the previous section, lightning strikes are a series of *return* strikes from the ground toward charged clouds. The approximative lightning current waveforms were proposed by [7] and are shown in Fig. 6.

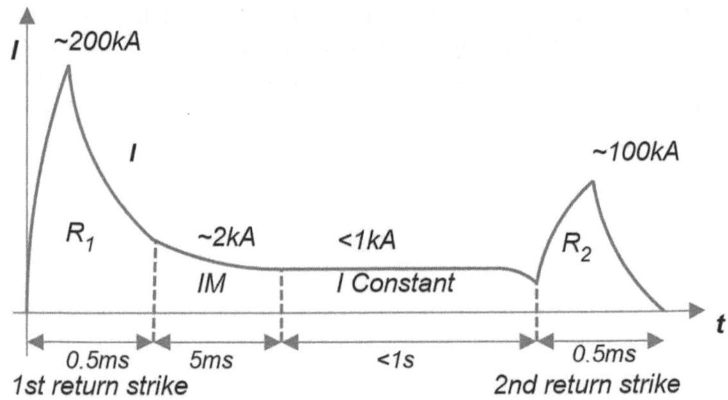

Fig. 6 Waveforms of lightning currents

Where R_1 and R_2 denote the current pulse of the *first* and *second return* strikes. IM and CONT denote intermediate and continuing current pulses, respectively, between the two *return* strikes. The present plot is not scaled but approximately shows the current pulse shapes, amplitudes, and durations.

Because the *first* strike carries the most destructive energy, we consider this pulse with more detail in the time and frequency domains to determine the approximate spectrum of this pulse. According to empirical research [2], the current waveform of the *first* strike can be shown in a simplified form, as shown in Fig. 7.

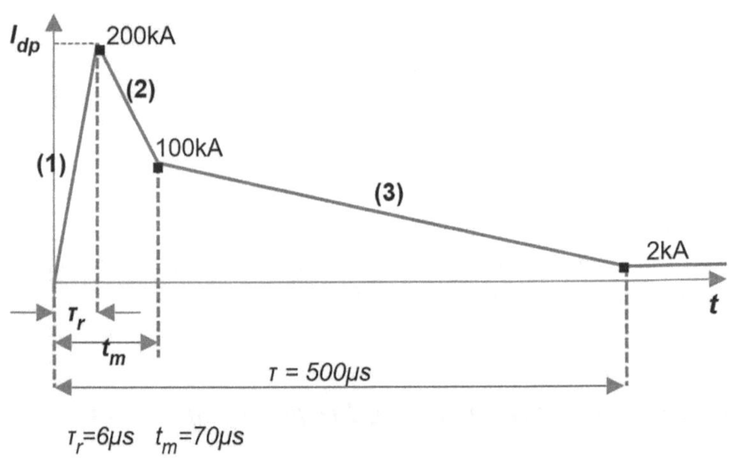

Fig. 7 Time-shape of the current produced by the first strike-up clouds

Here, we have the pulse that belongs to the category of nonperiodic signals that we considered in Chapter "Digital and Pulse Signals, Random Variables, and Stochastic Processes". The presented plot shows the current pulse $I(t)$ of the first return strike with a peak of 200 kA, the duration $\tau = 500\ \mu s$, the *rise*-time $\tau_r \approx 6\ \mu s$, and $t_m \approx 70\ \mu s$.

To simplify the analytical analysis, we approximated the current pulse with a composition of three linear functions: (1), (2), and (3), as shown in Fig. 7. Based on the present plot, we derived the corresponding linear function $I_1(t)$, $I_2(t)$, and $I_3(t)$ as

$$I_1(t) = I_p \frac{t}{\tau_r} \tag{1a}$$

$$I_2(t) \approx -\frac{I_p}{2}\frac{t}{(t_m - \tau_r)} + \frac{I_p}{2}\left(1 + \frac{t_m}{(t_m - \tau_r)}\right) \tag{1b}$$

$$I_3(t) \approx -\frac{I_p}{2}\frac{t}{(\tau - t_m)} + \frac{I_p}{2}\frac{\tau}{(\tau - t_m)} \tag{1c}$$

Furthermore, to find the spectral content of the current pulse, we use the Fourier transforms in normalized form as

$$F(f) = \left|\int_0^{\tau_r} I_1(t)\cdot e^{-i\omega t}dt + \int_{\tau_r}^{t_m} I_2(t)\cdot e^{-i\omega t}dt + \int_{t_m}^{\tau} I_3(t)\cdot e^{-i\omega t}dt\right| \tag{2}$$

By substituting Eqs. (1a), (1b) and (1c) into Eq. (2) and solving the given integrals for the given periods, the spectral bound of the lightning pulse from Fig. 7 is plotted in Fig. 8.

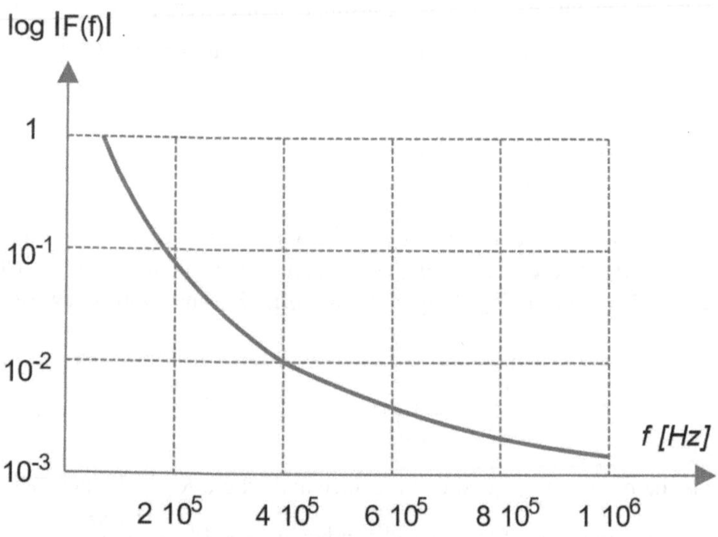

Fig. 8 Spectral bound of the current pulse from Fig. 6

For $\tau = 500\mu s$ and $\tau_r = 6\mu s$, the spectrum envelope frequencies are

$$f_1 = \frac{1}{\tau} = \frac{1}{500\,\mu s} \approx 2\,kHz,\; f_2 = \frac{1}{\tau_r} = \frac{1}{6\mu s} \approx 160\,kHz$$

Thus, the resulting plot shows the current spectrum when the current increases to the peak of 200 kA and decays to approximately 2 kA. Notice that most of the energy from the *first* strike exists below 1 MHz. However, lighting effects beyond several MHz and beyond should not be ignored because they can also significantly affect normal equipment operations depending on the equipment proximity to lightning strike.

2.2 *Magnetic Field and Voltage Developed on the Ground by Strike Currents*

Furthermore, the strike currents contain enough energy to produce a strong magnetic field and to develop high voltage on the ground. Owing the *magnetic field* and *ground high voltage* being destructive for equipment, it would be helpful to estimate (to some level) these parameters.

The following example illustrates the situation in which lightning strikes the ground with conductance $\sigma = 1 {mS}/{m}$, causing a high-current peak I_p of 150 kA [8]. The intensity of the produced magnetic field H can be expressed based on Ampere's law as

$$H = \frac{I_p}{2\pi d} \left(\frac{A}{m}\right) \tag{3}$$

where H is the magnetic field intensity at the ground at the radial distance d from the strike point. Furthermore, the developed ground voltage V' at the same distance d can be derived by combining Ampere's law, Eq. (3), and Ohm's law ($J = \sigma E$) to obtain

$$V'^{(d)} = \frac{I_p}{2\pi d\sigma} = \frac{I_p}{2\pi d} R'_g \tag{4}$$

where σ is the per-meter ground conductance and where R'_g is the per-meter resistance, expressed in terms of the ground conductance as $R'_g = \frac{1}{\sigma}\left(\frac{\Omega}{m}\right)$.

Figure 9 shows the shape of the circular magnetic field produced by the current with peak $I_p = 150\,kA$ and voltage V developed in the ground with $\sigma = 1 {mS}/{m}$.

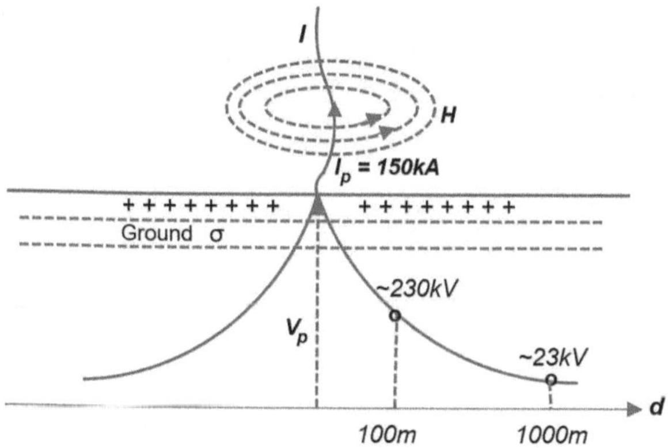

Fig. 9 Ground voltage at distances of 100 m and 1 km

Additionally, per Eq. (4), the ground voltages are approximately 230 kV and 23 kV at distances of 100 m and 1000 m, respectively, from the strike point, as shown in Fig. 9.

For example, according to Eq. (4), Table 1 shows the calculated voltages developed on the ground from the strike distance d (m) and different ground resistances R_g' (in Ω/m). More detailed data of various types of ground (soil) with the resistance are given in Chapter "Grounding for Radio Systems and Facilities", Table 1, which are related to the RF and safety grounding systems.

Table 1 The voltage developed on the ground from the different strike distance d (m) and for different ground resistances R_g'

$V(d)=\frac{I_p R_g'}{2\pi d}$ (4)	Ground voltage V with the distance d	
R_g' (Ω/m)	$d = 100\ m$	$d = 500\ m$
0.2 (salt water)	$\approx 48\ V$	$\approx 9.5\ V$
33 (very good pastoral, rich soil)	$\approx 7.8\ kV$	$\approx 1.5\ kV$
100 (pastoral, rich soil)	$\approx 24\ kV$	$\approx 4.8\ kV$
200 (average pastoral)	$\approx 48\ kV$	$\approx 9.5\ kV$
500 (sandy, dry, or rocky terrain)	$\approx 120\ kV$	$\approx 24\ kV$

Thus, the ground voltage can be in the order of tens of kV or even over 100 kV for some R_g' and strike distances.

Additionally, estimating the difference between two voltage potentials at two ground points would be helpful. The distances of those points from the strike point are d and $d_1 = d + \Delta s$. The voltage difference is called the *step-ground voltage*. According to Eq. (4),

$$V(d) = \frac{I_p}{2\pi\sigma d} \quad \text{and} \quad V(d_1) = \frac{I_p}{2\pi\sigma d_1}$$

and the step voltage ΔV is the difference between potentials $V(d)$ and $V(d_1)$:

$$\Delta V = V(d) - V(d_1) = \frac{I_p}{2\pi\sigma d} - \frac{I_p}{2\pi\sigma d_1}$$
$$\Delta V = \frac{I_p}{2\pi\sigma}\left(\frac{1}{d} - \frac{1}{d_1}\right) = \frac{I_p}{2\pi\sigma}\left(\frac{1}{d} - \frac{1}{d + \Delta s}\right) \tag{5}$$

where I_p is the current peak at the strike point and where Δs is the separation between the two points distant from the strike point by d and d_1, respectively.

For example, per Eq. (5), the voltage difference potential at two points separated by $\Delta s = 1\ m$ on the ground with $\sigma = 1 \, {}^{mS}\!/_{m}$ and where the first and second point are farther from the strike point by 200 m and 201 m is

$$\Delta V = \frac{I_p}{2\pi\sigma}\left(\frac{1}{d} - \frac{1}{d + \Delta s}\right) = \frac{150 \cdot 10^3 (A)}{2\pi} 200\ Ohm\left(\frac{1}{200m} - \frac{1}{201m}\right) \approx 590\ V$$

From this example, the developed step voltage between the two ground points separated by 1 m (1 step) is approximately 590 V when the strike occurred 200 m from the first ground point.

2.3 *Classification of Lightning Effects and Protection Against Lightning*

With respect to the effects of lightning on radio equipment and facilities, there are *direct* and *indirect* effects. The *direct* effects are more destructive and usually damage equipment and building structures, especially flammable structural materials, such as fuel and gas storage tanks that are exposed during lightning. Another common case of *direct* effects is related to *strike* currents that can cause significant damage to lightning-exposed objects or facilities on the ground.

The *indirect* effects are less destructive but can still damage electronic and electric equipment if the lightning currents penetrate facilities and their equipment over wires and cables. The following sections present several methods to protect equipment and facilities against *indirect* lightning impacts.

Hence, the lightning effects are categorized into two main groups, namely, *direct* and *indirect effects,* and equipment protection against these effects can be performed via the three strategic approaches that we consider in Chapter "General Concept and Types of EMI Emissions":

1. The *first* approach is to minimize lightning effects at a point where the original lightning strike can occur by using an efficient grounding system capable of absorbing most of the strike energy. Thus, the lightning effects at the lightning *source* are reduced.
2. The *second* approach is to protect external facility cables (power and signal cables) against penetration of the voltage and current surges into a facility and its equipment. This approach includes shielded cables whose shields are properly grounded, as discussed in Chapters "Performance of RF Cables and Signal Lines in Radio Systems with EMI" and "Grounding for Radio Systems and Facilities". This approach is a method for increasing the *resistance* of *lightning current paths.*
3. The *third* approach is to increase the *equipment's immunity* to voltage transients and current surges via effective surge-protection elements and high-power EMI filters at equipment terminals (also discussed in Chapters "EMI Filters—Theory and Design" and "EMI Shielding for Radio Equipment").

In the *first* approach, we consider several methods of lightning grounding in this chapter. Regarding the *second* approach, it should be noted that external cables are the main paths for incoming voltage transients and current surges into facilities and their equipment. Thus, the increase in the *path resistance* to a current surge should be reached as much as possible. Additionally, proper cable routing and shielding can help reduce the magnetic field coupling between cables to achieve this goal. This matter is discussed in more detail in Chapters "Performance of RF Cables and Signal Lines in Radio Systems with EMI" and "EMI Filters—Theory and Design", where system cabling and grounding are considered. Rigid metal conduits are commonly used for power outdoor and indoor cables, whereas flexible foils or braids may shield indoor signal cables.

With respect to the *third strategy approach*, enhancing the *immunity* of equipment to the voltage surge can be obtained by well-balanced input/output circuitry with high common-mode rejection, various surge protectors, series inductance and ferrite beads, decoupling (shunt) capacitance, isolation transformers, etc. This process is considered in Chapter "EMI Filters—Theory and Design", which is related to EMI filtering.

However, during a preliminary design, the lightning problem should be recognized before equipment and systems are built and installed. Improving and modifying lightning protection when systems are built and installed is impractical because employing surge suppressors, shielding, grounding, and bonding schemes is more costly and difficult to implement.

2.4 Protection Against First and Second Strikes by Lightning Grounding

The goal of lightning grounding is to divert lightning currents from objects and equipment to the ground via a low-impedance path. The low-impedance path is a dedicated lightning ground conductor or other conductive structure, such as suitable metallic walls or large metallic parts of building constructions connected directly to the ground. The path impedance depends on the *ground resistance* of the buried rods and more on the inductance of the grounding conductor, which are primary quality parameters of any low-impedance grounding, as noted in Chapter "Grounding for Radio Systems and Facilities". Excessive conductor *inductance* is the most critical part of the overall grounding because it is directly related to voltage development on the conductor when the strike current flows into the ground.

Therefore, as mentioned in Chapter "Grounding for Radio Systems and Facilities", the *ground resistance* and *ground conductor inductance* should be minimized as much as possible. However, the quality of a lightning grounding system also depends on the bonding between the grounded equipment and the grounding conductor, as well as between the grounding conductor and systems of the underground electrodes. As noted in Chapter "Grounding for Radio Systems and Facilities", if the bounding resistance is unexpectedly high, then the impedance of the overall grounding impedance is high. This situation usually leads to an inefficient or even useless grounding system.

2.5 Magnetic Fields Associated with Lightning Strikes

In *Appendix A*, we describe how the electric current in a linear conductor generates a circular magnetic field around the conductor per Ampere's law. Figure 10 shows the simple case of a single wire that carries the current *I*.

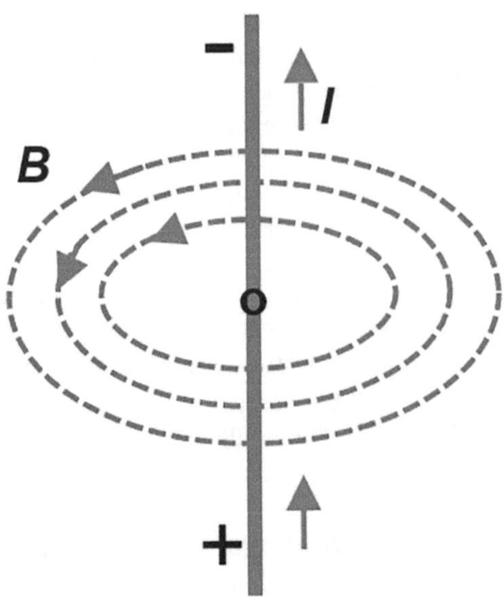

Fig. 10 Magnetic flux B produced by the current I

The magnetic flux B is expressed in terms of the magnetic field intensity H in free space as

$$B = \mu_0 H \tag{6}$$

where μ_0 is the magnetic permeability for free space equal to $4\pi 10^{-7}\left(\frac{H}{m}\right)$. Furthermore, substituting Eq. (3) ($H = \frac{I_p}{2\pi r}$) into Eq. (6) leads to the Biot–Savart law for the magnetic flux B:

$$B = \mu_0 \frac{I_p}{2\pi r} \quad \left(\frac{Wb}{m^2}\right) \tag{7}$$

where I_p is the peak lightning current in a grounding wire and where R is the radial distance (in meters) from the wire. Additionally, the total *magnetic flux* ϕ around the wire is represented by the integral from the wire's surface ($r = r_0$) to some distance R as

$$\phi = \int \vec{B} d\vec{S} = \mu_0 \int_{r_0}^{R} \frac{I_p}{2\pi r} dr = \mu_0 \frac{I_p}{2\pi} \ln\left(\frac{R}{r_0}\right) \quad \left(\frac{Wb}{m}\right) \tag{8}$$

The next example numerically illustrates Eq. (8).

Example 1 Find the magnetic flux around the stranded grounding AWG 4/0 wire (with radius $r_0 \approx 6.5\ mm$), which carries the lightning current with a peak of $I_p = 100\ kA$, up to a distance of 1500 m from the wire surface. Additionally, find a per-unit-length conductor inductance L'.

Solution Using Eq. (8) with the above parameters, the magnetic flux can be calculated as

$$\phi = \mu_0 \frac{I_p}{2\pi} \ln\left(\frac{R}{r_0}\right) = 4\pi 10^{-7}\left(\frac{H}{m}\right)\frac{100 \cdot 10^3 (A)}{2\pi} \ln\left(\frac{1500\, m}{6.5 \cdot 10^{-3}\, m}\right) \approx 0.25\left(\frac{Wb}{m}\right) \tag{8a}$$

Furthermore, by using the definition of the inductance as the ratio of the magnetic flux ϕ and current I_p, we can find the per-unit-length inductance L' related to I_p as

$$L' = \frac{\phi}{I_p} = \frac{0.25\left(\frac{Wb}{m}\right)}{100 \cdot 10^3 (A)} \approx 2.5\, \mu H / m \tag{9}$$

2.6 Lightning Current and Developed Voltage on a Grounding Conductor

Hence, the current flow in a conductor produces a circular magnetic field around the conductor, but it cannot occur instantaneously. Building the magnetic field requires a finite time or time rate. The rate of the current build-up to some level is proportional to the voltage (the current driving force) and inversely proportional to the conductor inductance L (as storage of the magnetic energy). This relation is given as

$$\frac{dI}{dt} = -\frac{V(t)}{L} \quad \left(A/s\right) \tag{10}$$

where $V(t)$ is the time-varying voltage and t is the time in seconds. During the lightning strike, the voltage along the grounding conductor increases at the same rate as the current rate, and per Eq. (10), the voltage $V(t)$ is given in *Appendix A*:

$$V(t) = -L\frac{dI}{dt} \quad (V) \tag{11}$$

Equation (11) shows that *the voltage* can even arise along a conductor with zero resistance because it is proportional to the conductor inductance L and the current rate $\frac{\mathrm{dI}}{\mathrm{dt}}$. $V(t)$ is called the *inductive* voltage, whereas the voltage developed on the resistance is the *resistive* voltage. Notably, the *inductive* voltage on the conductor is more critical than the *resistive* voltage.

Thus, the *conductor inductance* and the developed voltage are important parameters of any lightning ground installation. For instance, the per-unit-length voltage developed along the conductor with an inductance of $L' = 2.5\ \mu H/m$ and with a current change by rate of $\frac{dI}{dt} = 100\, kA / \mu s\left(10^{11}\ A/s\right)$ is calculated via Eq. (11) as

$$V(t) = (2.5\,\mu H/m) \cdot \left(10^{11}\frac{A}{s}\right) \approx 250\ (kV/m)$$

Other grounded conductors, which are close to the lightning ground conductor, can be metallic conduits, water pipes, or other nonlightning grounded conductors (wires). Therefore, harmful lightning arcs between the lightning conductor and these grounded conductors are possible and directly depend on how large the separation between the conductors is.

For example, common project practice considers that arcing between a lightning conductor and adjacent grounded conductors can occur when the voltage between them reaches a level of up to 100 kV per meter. On the basis of the functional form for a typical lightning current waveform [1 – cos (ωt)] that reaches its peak at 2 μs

[8], the *strike current* and per-meter *voltage* along the grounding conductor are given as

$$I(t) = I_p\left[1 - \cos(\omega t)\right] \quad (\mathrm{A}) \tag{12}$$

$$V(t) = L'\frac{dI}{dt} = L' \cdot I_p\, \omega \sin(\omega t) \quad \left(\frac{\mathrm{V}}{\mathrm{m}}\right) \tag{13}$$

where $I_p = 200kA$ and $L' \approx 2.5\ \mu H/m$ (from Example 1).

When the strike waveform reaches the maximal voltage in $t \approx 2\ \mu s$, i.e., when ωt is

$$\sin(\omega t) = 1 \rightarrow \omega t \approx \frac{\pi}{2} \rightarrow \omega \approx \frac{\pi}{2t} \tag{13a}$$

Accordingly, ω is $7.85 \cdot 10^5 \left(\frac{rad}{s}\right)$ at t ≈ 2 μs. Therefore, per Eq. (13), the maximum per-unit-length voltage developed along the conductor is

$$V_{\max}(t) = L' I_p\, \omega$$
$$\approx 2.5\left(\frac{\mu H}{m}\right) \cdot 200 \cdot 10^3 (A) \cdot 7.85 \cdot 10^5 \left(\frac{rad}{s}\right) \approx 390 \frac{kV}{m} \tag{13b}$$

Figure 11 shows examples of a protected building. The developed *inductive* voltage V_L on a grounding conductor is shown in this drawing.

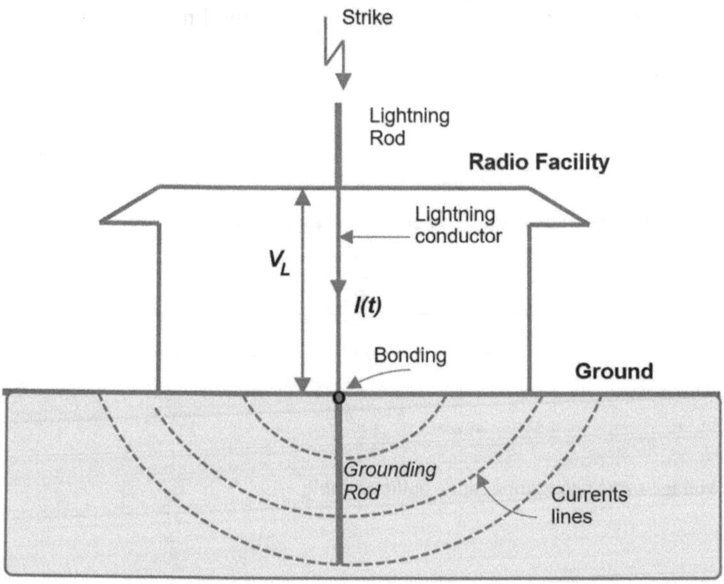

Fig. 11 Lightning grounding for a facility

Notably, the *inductive* voltages along the lightning conductor can produce second-strike effects (as mentioned above) when the lightning energy jumps (arcs) from the lightning conductor to next grounded conductors that are not parts of the lightning grounding system. For example, an inductive voltage with a peak of 100 kV on a grounding conductor can spark across 1 m or longer in air, depending on the conductor's length.

Thus, when the main strike occurs, some lightning current escapes the grounding conductor and restrikes to neighboring metallic objects and grounded wires. If some of these structural parts are not parts of the grounding system, damage is likely to occur. Additionally, as mentioned before, the *inductive* voltage produces current surges in cables of electrical and electronic equipment, and these events are designated *indirect* lightning effects, which may be harmful to equipment and personnel. These situations will be considered in the following subsection.

3 Indirect Lightning Effects

Hence, regardless of whether building facilities or equipment are shielded well, a voltage surge from neighboring lightning strikes can induce high currents into external power and signal cables and further enter a facility and its equipment. Therefore, filtering and surge protection methods should also be applied in addition to proper cable shielding. This is considered in more detail in Chapter "EMI Filters—Theory and Design".

Figure 12 shows the mechanism by which voltage transients penetrate facility buildings via an external cable interconnecting two buildings.

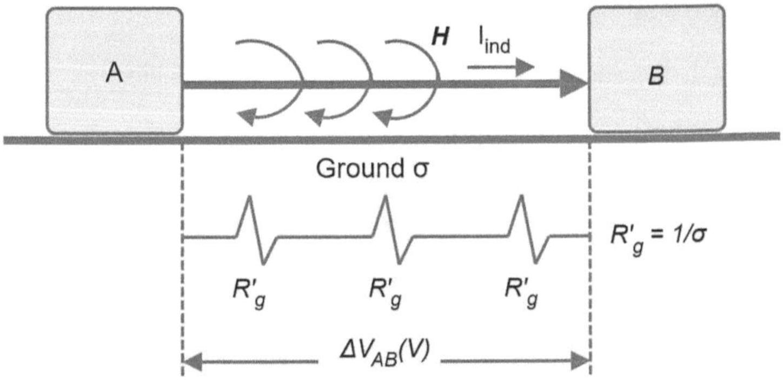

Fig. 12 Voltage surge penetration into a building cable

Here, how the magnetic flux produced by the lightning current from a neighboring strike induces a current in the loop created by a cable that connects facilities A and B in conjunction with the ground is shown. When the ground is represented by distributed (per-meter) resistances R'_g, the voltage $\Delta V'_{AB}$ on the ground can be expressed per Ohm's law as

$$\Delta V'_{AB} = I_{ind} R'_g = \frac{I_{ind}}{\sigma} \left(\frac{V}{m}\right) \tag{14}$$

where σ is the per-unit length ground conductance. The voltage produced by changing the magnetic field is considered in the next subsection.

3.1 Developed Voltage in the Loop Formed by Ground Cable and Ground by a Time-Changing Magnetic Field

On the basis of the model in Fig. 13, the distributed voltage $V(t)$ developed on the cable by a magnetic field $H(t)$ is expressed as

$$V(t) = \frac{d\phi(t)}{dt} = \mu_0 A \frac{dH(t)}{dt} \quad (\mathrm{V/m}) \tag{15}$$

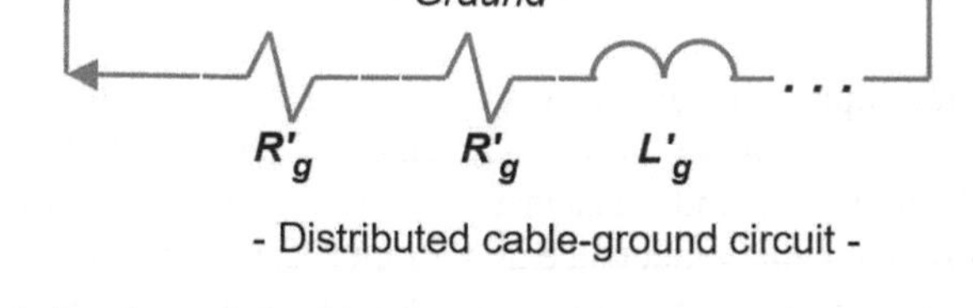

Fig. 13 The equivalent lumped circuit of the cable-ground loop

where $\phi(t)$ is the time-changing magnetic flux (in Weber) that crosses through the cable-ground loop, μ_0 is the magnetic permeability in free space equal to $4\pi \bullet 10^{-7}$ *H/m*, and A is the area of the cable-ground loop in m^2.

Because the magnetic field has the same waveform as the lightning current per Eq. (12), we can represent the magnetic field with the function $[1 - \cos(\omega t)]$, which usually reaches a peak at $t \approx 2\ \mu s$ [8]. Therefore, when the magnetic field has the waveform

$$H(t) = H_0\left[1 - \cos(\omega t)\right] \quad (16)$$

then the time derivative of $H(t)$ is

$$\frac{dH(t)}{dt} = \omega H_0 \sin(\omega t) \quad (17)$$

Furthermore, when Eq. (17) is substituted into Eq. (15), the $V(t)$ distribution along the cable and the ground is expressed as

$$V(t) = \mu_0 \omega H_0 A \sin(\omega t) \qquad \left({}^{\mathrm{V}}\!/_{\mathrm{m}}\right) \quad (18)$$

where A is the cable-ground area of 1 m^2.

The equivalent lumped RL circuit of the cable-ground loop is shown in Fig. 13, where L' and R' are per-unit-length elements of the cable inductance and resistance, respectively, whereas R'_g and L'_g are the ground resistance and inductance, respectively. Herein the ground acts as a return conductor for the current induced in the cable or a reference conductor for the distributed voltage $V(t)$ along the cable

3.2 *The Current Induced by Neighboring Lightning Strikes in a Conductor Above the Ground*

According to circuit theory, the current $I(t)$ in the RL circuit can be represented in the exponential form as

$$I(t) = \frac{V(t)}{R_g}\left(1 - e^{-\frac{R_g t}{L_g}}\right) \quad (A) \quad (19)$$

where $V(t)$ is the input voltage for an RL circuit. Rg is the sum of the cable and ground resistance, whereas Lg is the inductance of the grounding conductor. For lightning applications, ground resistance is more significant than the cable resistance. Therefore, the cable inductance and ground resistance are the dominant elements in the cable-ground loop. Substituting Eq. (18) into Eq. (19) leads to

$$I(t) = \frac{\mu_0 \omega H_0 A \sin(\omega t)}{R_g}\left(1 - e^{-\frac{R_g t}{L_g}}\right) \quad (A) \quad (20)$$

Equation (20) is valid when the current $I(t)$ of a moderately strong strike reaches the peak in 2 μs.

Furthermore, by calculating the peak magnetic field intensity at the lightning strike point (I_p = 150 kA), currents at different *distances* from the strike point and for different *ground resistances* can be found via Eq. (3):

$$H = \frac{I_p}{2\pi d} \quad \left(\frac{A}{m}\right)$$

Substituting Eq. (3) into Eq. (20) yields the final form:

$$I(t) = \frac{\mu_0 \omega I_p}{2\pi d\, R_g}\left(1 - e^{-\frac{R_g t}{L_g}}\right) \quad (A) \tag{21}$$

where $\omega = 7.85 \bullet 10^5$ *rad/s* per Eq. (13).

Per Eq. (21), Table 2 shows the induced peak current in the cable from different strike distances d and different ground resistances $R'_g\left(\frac{\Omega}{m}\right)$, when the current peak at a strike point is $I_p = 150\ kA$.

Table 2 The produced peak current in a cable from different strike distances d and different ground resistances R'_g when the current peak at a strike point is $I_p = 150\ kA$

$I(t) = \frac{\mu_0 \omega I_p}{2\pi dR}\left(1 - e^{-\frac{Rt}{L}}\right)$		Peak-induced current at ground with different ground resistance R'_g		
$d(m)$	$H_p(A/m)$	$1\left(\Omega/m\right)$	$10\left(\Omega/m\right)$	$20\left(\Omega/m\right)$
100	~240	~200 *A*	~23 *A*	~12 *A*
1000	~24	~20 *A*	~2.3 *A*	~1.2 *A*
5000	~5	~4 *A*	~0.5 *A*	~0.2 *A*

Under some conditions, the induced currents in the cable caused by neighboring lightning strikes can be tens or even hundreds of amperes for low ground resistance (~1 Ω/m) and a distance d from the strike point of 100 m. If such high voltage is developed on the ground, a high current is induced in unshielded equipment cables; most likely, electronic equipment will burn out. Properly shielded cables and grounded cable shields are common solutions for lightning problems (Chapter "Performance of RF Cables and Signal Lines in Radio Systems with EMI").

The writing of this section is influenced by [2], which we refer to interested readers for further study.

4 Lightning Grounding and Protection for Radio Systems

As mentioned earlier, *lightning* grounding is needed to protect personnel and equipment against *lightning* strikes. High-profile objects, such as towers and tall buildings, are most exposed to *lightning* strikes; as such, lighting grounding, along with other lightning protection methods, should be implemented around and within radio system objects. For example, Fig. 14 shows one concept of lightning grounding for an RF building facility with an antenna tower.

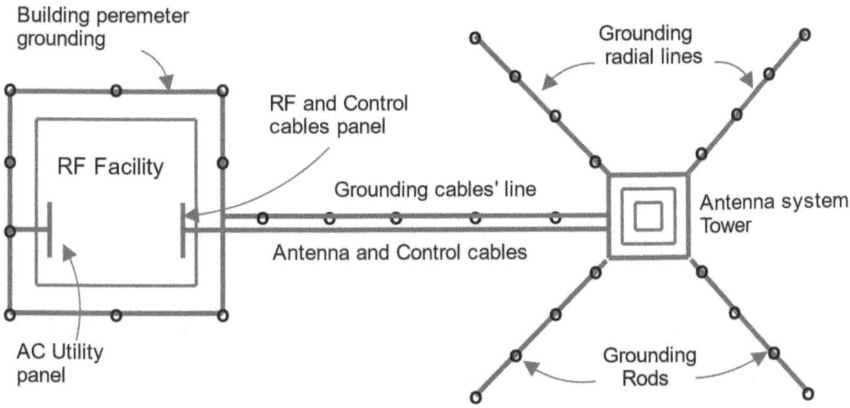

Fig. 14 "Star" grounding for an antenna tower and a perimeter grounding for an RF facility (top view)

This is a *star* configuration grounding for an antenna tower and the *perimeter* grounding around an RF facility building. The grounding is created by buried vertical rods bonded by buried horizontal thick conductors, usually straps. In addition to protection against lightning, this grounding is also used as AC grounding.

Additionally, Fig. 15 shows another facility with two antenna towers next to a two-story building, grounded and protected by lightning protectors (LPs) against lightning.

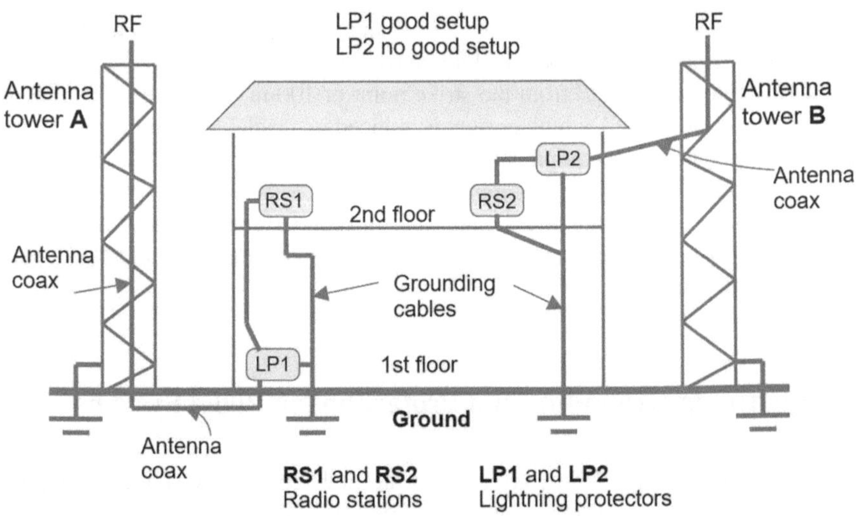

Fig. 15 (**a**) Good LT placement, (**b**) no good LT placement

The routing of the antenna coaxial cable and LP placement for tower A is correct, whereas for tower B, the coax cable routing and LP placement are incorrect. Hence, an LP should be placed closer to the ground level than too high from the ground. This is because the antenna cable and the LP grounding cable in case B are more exposed to direct and indirect lightning effects.

Furthermore, Fig. 16 shows a lightning protector commonly installed in addition to lightning grounding to bypass lightning currents from equipment to the ground.

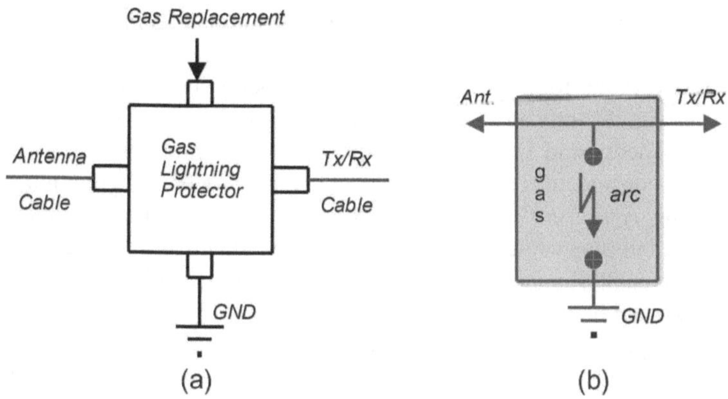

Fig. 16 Lightning grounding with a lightning gas protector. (**a**) External view and (**b**) inside view

The breakdown voltage V_b of a gas lightning protector with thin arcing electrodes mainly depends on the type of gas, the gas pressure p, and the separation s between arcing electrodes. Based on experimental studies [9], the breakdown voltage V_b is given by an empirical expression where k_1 and k_2 are constants that are dependent on the gas type as

$$V_b = \frac{k_1 (ps)}{k_2 + \ln(ps)} \tag{22}$$

In addition to gas protectors, simple air-spark gaps are used. For example, per experimental research, the minimum breakdown voltage V_b for air at standard atmospheric pressure is approximately 300 V when the arcing contacts are separated by approximately 7 μm.

5 Summary of Design Rules for Protection Against Lightning

When designing lightning protection for facilities and equipment, the following points are recommended:

1. An efficient low-impedance grounding system is essential for protection against *direct* and *indirect* lightning strikes, i.e., the quality of lightning grounding primarily depends on a low impedance path for lightning currents to the ground.
2. The low-impedance grounding system depends mainly on the grounding *cable inductance* and ground bonding resistance, as well as on an *underground electrode resistance* as a final load for the lightning energy.
3. A low conductor induction can be reached via multiple parallel thick conductors, keeping them as short as possible. Additionally, when possible, wide conductors such as metallic walls and other large grounded parts of a facility structure can be used.
4. Loops and round routings of the ground conductor should be avoided because this can significantly increase the conductor's inductance (see Chapter "EMI Filters—Theory and Design"). Additionally, large grounding loops can act as efficient loop antennas for electromagnetic fields produced by lightning.
5. Regarding *equipment's immunity* to lightning effects, first, the immunity of incoming building cables should be obtained by effective shielding and grounding, i.e., a cable should be placed in metallic conduits, and then equipment protection should be implemented.
6. For some specific situations, such as the isolation of devices and equipment sensitive to surges and transients, the higher inductance of conductors may have positive isolation effects.
7. Generally, as previously mentioned, lightning protection is most effective when implemented at the beginning of system design because more design options and opportunities are available.

In addition to the references [1–9], which influenced the writing of this chapter, we also refer the readers, who are interested in further reading lightning physics and protection against lightning, to references [3–5, 10–12].

References

1. V. Cooray, *An Introduction to Lightning* (Springer, 2015)
2. N. Cianos, E.T. Pierce, *A Ground Lightning Environment for Engineering Usage* (Stanford Research Institute, 1972)
3. V. Mazur, *Principles of Lightning Physics* (Institute of Physics Pub Inc., 2017)
4. V.A. Rakov, *Fundamentals of Lightning* (Cambridge University Press, 2016)
5. P. Parker, *An Introduction to the Lightning Process* (Hay House, 2012)
6. R.H. Golde, Protection of Structures against Lightning, Proceedings of the Institute of Electrical Engineers, 1968
7. U.S. MIL-STD-1757 (1980)
8. A. Plumer, Lightning, in *Handbook of EMC*, ed. by R. Perez, (1995)
9. R. Holm, *Electric Contacts* (Springer-Verlag, Theory and Applications, 1967)
10. A. McAdie, *Protection from Lightning* (Creative Media Partners, LLC, 2023)
11. E.M. Bazelyan, Y.P. Raizer, *Lightning Physics and Lightning Protection* (CRC Press, 2019)
12. M.A. Uman, *Lighting* (Publisher Dover, 2011)

Solar and Other Cosmic Radiations with Effects on HF Radio and Satellite Data Communications

1 Introduction

We discuss lightning phenomena in Chapter "Lightning Physics and Impact on Radio Systems on the Earth" as natural sources of EMI and EM disturbances. These phenomena originate exclusively on our planet (in Earth's atmosphere), but other natural or cosmic EMI phenomena originate outside our planet. The cosmic EM has several origins: solar, lunar, stellar, and galactic. Electromagnetic emissions from outer space were discovered in 1932 by physicist Karl Jansky. This discovery led to the establishment of a new scientific discipline called *radio astronomy*.

The level of EMI noise from the sun and cosmos is expressed in noise temperature, i.e., in Kelvins (K). The minimal cosmic noise radiation (as noise background) is approximately 2.7 K, which is always present. Most stars and galaxies radiate EMI noise in the broadband range from 10 MHz to 10 GHz. On the basis of many measurements of cosmic radiation, maps and charts of the spectral distributions of galactic EMI noise are made. These maps are useful for space research and research on aviation and satellite communications.

The Sun's radiation can be both constructive and destructive for HF radio ionospheric communications but may be destructive for satellite communications [1, 2]. The constrictive radiation supports the formation of the regular ionospheric layers needed for normal radio ionospheric communications. In contrast, the Sun's radiation can be a major contributor to EM disturbances and EMI noise, even up to $10^{11}\,K$ in some radio frequency bands and geographical locations. The primary source of solar radiation and disturbance occurs when the Sun releases solar flares. The sun radiates intense UV radiation and X-rays when solar flares occur, causing a temporary increase in the density and speed of charged particles (also called solar "wind") that travel toward the Earth's atmosphere. This phenomenon has dominant effects

M. Pajovic, *Fundamentals of EMI Phenomena in Radio Communication Systems*, https://doi.org/10.1007/978-3-031-88748-2_13

on HF radio system communications on the Earth and low-orbit (LO) and mid-orbit (MO) satellite communications, strongly influencing the Earth's magnetosphere and even affecting noncommunication systems and telemetry instruments.

Sun activity is highly dynamic and has two extreme periods: a quiet period and a highly disturbed period. During the quiet period, the solar UV and X-ray radiation and ambient solar "wind" are moderate. However, during highly disturbed periods, solar radiation and solar wind are intense enough to interfere with and even temporarily disrupt the operations of communication systems on Earth and satellites. Notably, the intensity of solar activity is usually associated with isolated solar flares that emit intense radiation and wind from charged particles to Earth for durations ranging from seconds to hours [2, 3]. For example, this intensity may exceed more than a thousand times the average solar radiation.

Therefore, intense flares can disrupt radio-communication operations on Earth, especially HF radio systems and satellite communications. For microwave radio systems and satellite links, the solar noise effects also depend on the RX antenna direction and its radiation pattern, as well as on the relative angular position of the Sun to Earth. The level of noise received from the Sun by an RX antenna can vary from ~3 K (when the antenna is directed to a quiet space zone overnight) to 5,000 K or higher when the RX antenna is directed toward the Sun with high activity.

Compared with EMI noise and solar wind from the sun, EMI noise from other cosmic objects, such as the moon and stars, poses a lower risk for the regular operation of radio communication systems on Earth. The effects of the Moon may be critical when an RX antenna is directed toward the full moon, which maximally reflects the sunlight transmitted toward Earth. Therefore, we focus more on the sun as a primary extraterrestrial source of EMI, which has a dominant influence on radio communication services on Earth. However, full-moon radiation and radiation from other cosmic objects cannot be ignored in satellite communications.

The writing of this chapter is motivated and influenced by the references [1–6]. We also recommend references [7–13] to readers interested in further studies of this topic.

2 Origins of Solar Radiation

Ionized and electrified regions (layers) are created in the Earth's upper atmosphere due to the Sun's activity, causing the refraction of HF waves to return toward Earth. Therefore, ionospheric regions act as waveguides for HF waves in conjunction with the Earth's surface. The regular ionosphere does not significantly influence the propagation of microwave waves that easily pass through regions of the regular ionosphere without returning to Earth.

The ionosphere is created primarily by the sun's ultraviolet (UV) radiation, X-rays, and less by other cosmic radiations. Because the Sun's activity varies, the ionosphere changes hourly, seasonally, and geographically, depending on the Earth's position to the Sun in time. Yearly, the conditions in the ionosphere change by approximately 11-year cycles on average. These changes are due to changes in the number and sizes of sunspots on the sun surface.

As previously noted, sunspots are turbulent areas on the Sun's surface that transmit mainly ultraviolet rays and X-rays followed by streams of charged particles toward the Earth. Because the intensity of the Sun's emissions depends on the number and size of sunspots, more and larger sunspots produce electrically robust ionospheric regions that sufficiently support HF radio communications for medium and long ranges. When the number of sunspots decreases, the HF channels become poorer or may even collapse. For example, this happens when the Sun's radiation stops moving toward the Earth during a Solar eclipse, i.e., when the Moon is positioned between the Sun and Earth, causing the so-called "Moon-blockage" effect. On the basis of many measurements from Earth stations and satellites during the Solar eclipse, the ionosphere almost disappears in regions affected by the solar eclipse, which provides evidence that the Sun is a principal contributor to the creation of the ionosphere.

2.1 Solar Effects

Solar effects are categorized into three groups:

(a) Direct effects of sunspots.
(b) Indirect effects of ionosphere region ionization, and
(c) Cosmic noise effects.

Electromagnetic waves from the outer cosmos are captured by the Sun's magnetic field, which results in the creation of sunspots. Their intensity can be approximately calculated by knowing the ionization intensity in a given ionosphere region or by measuring the variation in the magnetic field at the Earth's poles. The main properties of the sunspots are that they appear in clusters on the Sun's surface and that their appearance is periodic over short- and long-term periods. The short-term period is equal to one cycle of the Sun's rotation around its axis, which takes 27.5 days since the long-term period is between 7 and 16 years, with an average of 11 years.

The number of sunspots (known as the wolf number) is determined by the international sunspot number (ISN), which may be between 0 and 200. Sometimes, the solar radio flux (SF) is considered in a radio system design instead of the ISN number. The SF can be used as solar noise power because solar noise is categorized as colored noise whose power density is considerable at a frequency of 2.8 GHz (wavelength of 10.7 cm).

2.2 Geomagnetic Field Effects

The geomagnetic field has a considerable effect on the ionosphere, which is generally an anisotropic medium. When a plane wave penetrates an ionospheric region, it is divided into *ordinary* and *extraordinary* wave modes. Leaving the ionospheric region, these modes become a new plane wave with a different angle of polarization. This phenomenon is called Faraday's rotation, which causes signal loss at a receiver's antenna due to the polarization mismatch between the received wave and the antenna polarizations.

The incident planar wave is actually divided into right-hand and left-hand elliptical polarized waves in an ionosphere region that propagate inside the region with different propagation constants k_1 and k_2. The outgoing plane wave is a combination of the two wave modes $\hat{E}_x$ and $\hat{E}_y$ expressed as

$$\hat{E}_{out} = \left(\hat{E}_x - \hat{E}_y\right)e^{-ik_1 l} + \left(\hat{E}_x + \hat{E}_y\right)e^{-ik_2 l} \tag{1}$$

Using Euler's identity and trigonometric rearrangement, Eq. (1) becomes

$$\hat{E}_{out} = \hat{E}_x \cos\left[\left(k_2 - k_1\right)l/2\right] + \hat{E}_y \sin\left[\left(k_2 - k_1\right)l/2\right] \tag{1a}$$

where the new polarization angle ϕ is expressed as

$$\tan(\phi) = \frac{\sin\left[\left(k_2 - k_1\right)l/2\right]}{\cos\left[\left(k_2 - k_1\right)l/2\right]} = \tan\left[\left(k_2 - k_1\right)l/2\right] \rightarrow \\ \phi = a\tan\left[\left(k_2 - k_1\right)l/2\right] \tag{1b}$$

where l is the region thickness.

Note that the Faraday rotation is negligible at higher frequencies where the ionospheric region becomes isotropic and relatively homogeneous and where the propagation constants k_1 and k_2 are approximately equal.

Figure 1 shows (a) the incoming wave $\hat{E}_{in}$ before entering the region, (b) the two waves elliptically polarized (right-hand $\hat{E}_{rh}$ and left-hand $\hat{E}_{lh}$) developed inside the ionospheric region, and (c) the outgoing wave $\hat{E}_{out}$ with its orthogonal components $\hat{E}_x$ and $\hat{E}_y$ that left the region with different polarization angles by ϕ relative to the polarization of the incoming wave $\hat{E}_{in}$.

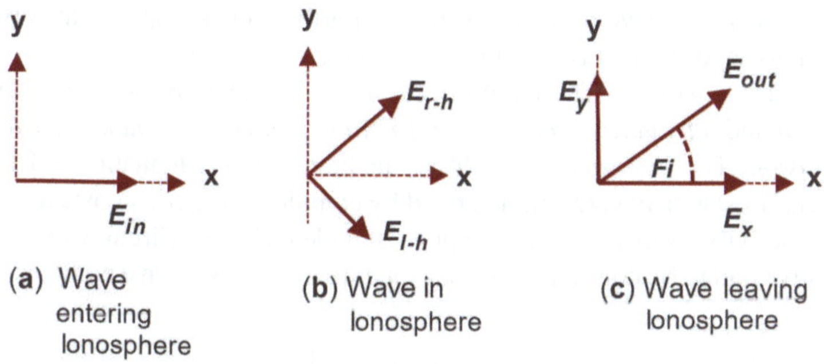

Fig. 1 (**a**) Incoming wave $\hat{E}_{in}$ in the ionosphere, (**b**) two elliptically polarized waves $\hat{E}_{rh}$ and $\hat{E}_{lh}$ inside the ionosphere, (**c**) resulting in an outgoing linearly polarized wave $\hat{E}_{out}$ with the changed polarization by the angle ϕ (Fi)

Note that the incoming ionospheric wave randomly changes its polarization angle while passing through an anisotropic and inhomogeneous ionospheric region whose properties randomly change.

3 Structure of Earth's Atmosphere

The atmosphere extends into the space above Earth's surface up to 150,000 km and consists of different gases, vapors, hydrometeors, and dust particles. Some of these matters permanently exist since others exist temporarily. Permanent matter exists in specific regions and is created mainly by the Sun's radiation. The Earth's main atmospheric regions are the troposphere, stratosphere, ionosphere, and magnetosphere. For more information about this topic, we refer readers to [2–5].

Figure 2 shows that the Earth's atmosphere is composed of these regions [4].

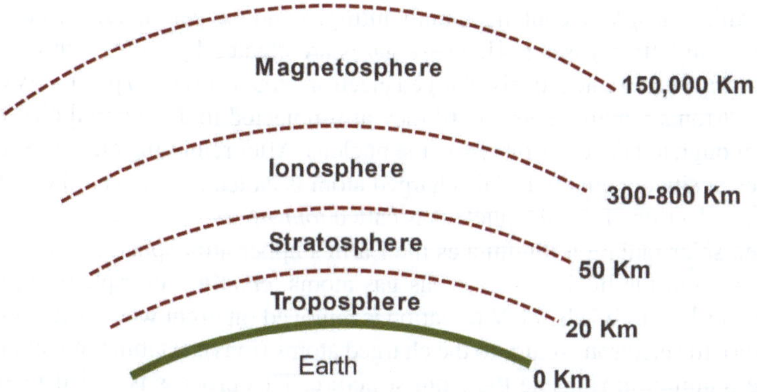

Fig. 2 Structure of the Earth's atmosphere (not to scale)

The *troposphere* layer spreads from the Earth's surface approximately up to 20 km. This layer is mainly a region of climatic events such as clouds, fog, rain, snow, wind, and storms. The main effects of the *troposphere* on radio waves are *absorption* and *attenuation, reflection, refraction, polarization changes, scattering*, and *diffusion*. The *troposphere* is a lossy medium whose attenuation is frequency dependent. Lossy media are characterized by complex dielectric constants and limited conductivity, as discussed in Chapter "Nonideal Passive Circuit Components". The relative dielectric constant of a lossy medium is expressed in a complex form as

$$\varepsilon_r = \varepsilon_r' - i\varepsilon''_r = \varepsilon' - i\left(\frac{\sigma}{\omega}\right) \tag{2}$$

where the real part ε'_r causes wave reflection and scattering since the imaginary part $\varepsilon''_r\left(\frac{\sigma}{\omega}\right)$ causes wave absorption and attenuation.

The *stratosphere* is a layer above the *troposphere* that extends from 20 km to approximately 50 km. The characteristics of this layer are a low variation in temperature per height and a large amount of atmospheric gases.

The *ionospheric* layers are located above the *stratosphere* and extend from 50 km to ~800 km (approximately). The *ionosphere* has positive and negative effects on wave propagation and the operation of HF radio and satellite communications. Owing to the importance of this layer, its creation and electrical characteristics will be discussed separately in the next section.

The *magnetosphere* extends beyond the *ionosphere,* from approximately 800 km to 150,000 km. This layer acts as a magnetic shield for the Earth against intense radiation.

3.1 Formation of the Ionospheric and Its Regions

The Earth's *ionosphere* contains mainly nitrogen and oxygen in addition to helium, hydrogen, and other gases [14]. These gases are created by atoms that are electrically neutral because negatively charged electrons rotate around a positively charged nucleus. Atoms remain neutral until they are subjected to an external electric field strong enough to take electrons from the nucleus. After removing electrons, an atom becomes positively charged. This charged atom is called an ion, and the process of removing electrons from the nucleus is called *ionization.*

When solar radiation illuminates the Earth's upper atmosphere, many electrons are taken from the nucleus of various gas atoms, creating ionosphere regions (or layers). Furthermore, when UV radiation is removed (at night when there is no solar radiation), the electrons return to the charged atoms (ions), establishing an electrical balance (equilibrium) inside the atom structure. This process is called *recombination* [2].

The general ionization process of an atom under external radiative forces is shown in Fig. 3.

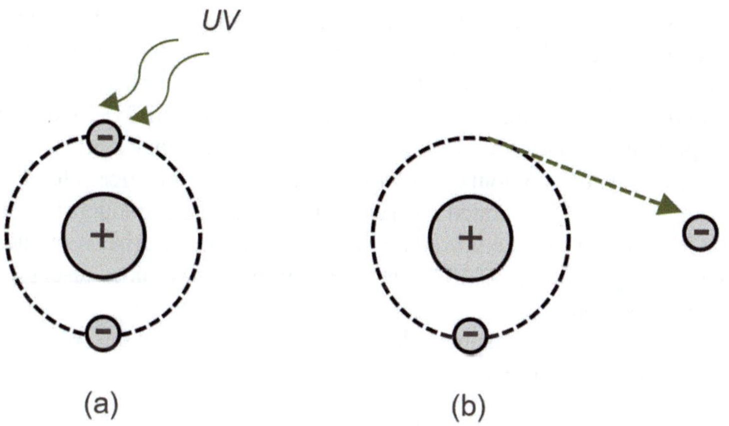

Fig. 3 Gas ionization by an external force or radiation

The source of atmospheric ionization is not only UV radiation from the Sun (although it is a principal source) but also other types of solar radiation, such as X-rays and cosmic rays, which are discovered as contributors to ionosphere creation. Additionally, meteors entering the Earth's atmosphere produce atmospheric ionization.

When solar radiation penetrates the Earth's atmosphere, it ionizes gases in the upper atmosphere with a lower gas concentration. As the radiation penetrates deeper into the atmosphere, it meets a greater density of gases, and consequently, their ionization increases. When the ionization reaches a maximum, the ionization starts to decay to a complete stop when an ionization layer is formed. Thus, the maximum ionization occurs at approximately the center of the ionospheric layer.

Figure 4 shows the relationships among the intensity of *UV* radiation, the molecular gas density *D*, and the intensity *I* of gas ionization. All the mentioned parameters are presented as functions of the height above Earth.

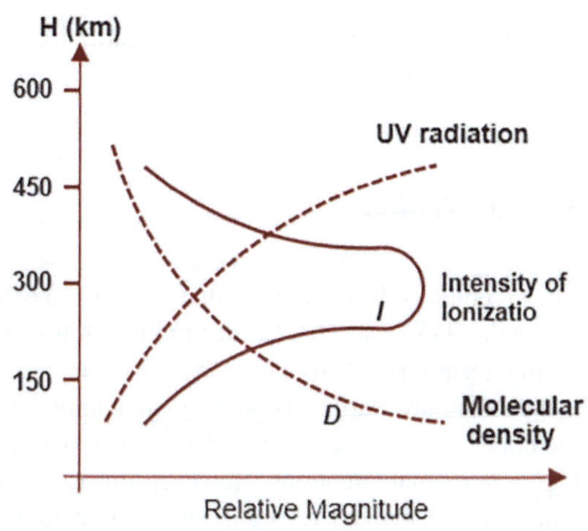

Fig. 4 UV radiation, molecular gas density (D), and intensity of ionization (I) versus height (H) above Earth

UV radiation has a relatively broad frequency spectrum, and because atmospheric gases respond differently to different UV and X-ray frequencies, the ionization layers are concentrated at different levels of the atmosphere, approximately between 50 km and 800 km. The ionized layers are designated D, E, and F when the F layer splits into two sublayers, F1 and F2, during the daytime.

The term "layers" for ionized regions is accepted in practice, although these regions are not entirely separated from each other. They are continuously distributed between their ionization maxima and minimums. The characteristics and heights of these layers vary from day to night and from season to season, as discussed in the following subsections.

Figure 5 shows the ionospheric layers during days and nights when the solar intensity substantially differs.

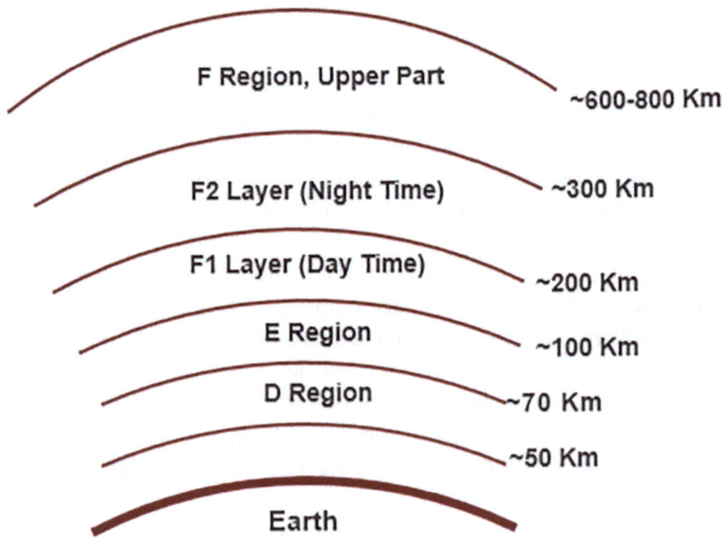

Fig. 5 Approximate positions of ionosphere regions

In the following subsections, ionization regions *D*, *E*, *F1*, *F2*, and *F* are briefly discussed [2, 4, 5].

3.2 D Region

The D region is the closest region to Earth and extends approximately 50 km to 70 km above the Earth's surface (shown in Fig. 5). It exists during daylight hours, with maximum ionization (10^8–10^9 electrons/m^3) occurring at approximately noon, when the Sun is highest above Earth. The average ionization level in this region is lower than that in the upper ionospheric regions. The mid-frequency and HF waves pass through the D region in regular situations. However, this region can be critical for wave propagation when one is ionized enough, resulting in high wave absorption. As noted in the preceding section, D-region absorption occurs locally during *ionospheric storms* (ISs) and can be heavy enough to prevent waves from reaching the upper ionospheric regions.

3.3 E Ionospheric Regions: Regular and Sporadic

The E region exists above the D region during the day's hours at heights from 70--100 km approximately above the Earth's surface (as shown in Fig. 5). The height of the E region varies from season to season and from location to location, but it remains largely constant during daylight hours. The ionization intensity in the E region is more significant than that in the D region and depends on the Sun's position in the sky during the day. For the D region, the maximum E ionization occurs at approximately noon, when the solar radiation perpendicularly penetrates the ionosphere. The recombination process in the ionosphere starts during sunset, and this region disappears during darkness.

3.3.1 Sporadic E Region

In addition to the discussed regular E region, another region occurs sporadically and irregularly. The height of this region is approximately 100 km. This region consists of sporadically ionized atmosphere fragments or patches. Since its height is close to that of the regular E region, this sporadic phenomenon is called the sporadic E region.

The sporadic E region mainly occurs in the daily hours of the summer months during intense ionospheric disturbances but may also occur at night during the winter. It appears as an intensely ionized region that is less than a kilometer thick, has a limited geographical spread, and drifts from location to location. This sporadic region or "ionization cloud" can be large, ranging from 80 km to 150 km in diameter, and may last a few hours. Radio communications above 50 MHz to several hundred MHz are enhanced during sporadic E occurrence. The sporadic-E region is often associated with auroras in the Northern Hemisphere, which occur in the winter months and are caused by *ionospheric disturbances.*

3.4 F Ionospheric Regions

The *F* regions are the most significant for long-distance HF radio communications. Two distinct subregions, *F1* and *F2,* exist during the daytime hours. The F1 subregion starts above the E region and extends to approximately 200 km (daytime), as shown in Fig. 5. This subregion behaves similarly to the E region: the maximum ionization occurs at approximately noon (~ $^{2\times1011}$ electrons/m^3) and it disappears at night.

The *F2* subregion is the highest and most ionized region, which changes its height seasonally from approximately 180–220 km during the winter months to 300 km during the summer months. The electronic concentration in this layer varies from 5×10^{11} electrons/m^3 at night to 20×10^{11} electrons/m^3 during the day. Above the F2 subregion, the upper part of the F region extends up to 600 km and beyond. This upper region has a significantly lower ionization level than the F1 and F2 regions.

The *F1* region practically disappears at night, after which the *F2* subregion remains active, with heights varying from 250–300 km above the Earth. The *F2* region is the most ionized region at night and is the most practical region for long-distance and low-frequency HF communications. Owing to its slow recombination, it exists all night regardless of the absence of solar radiation.

During regular ionosphere conditions (without excessive IS or SID), the F regions are fairly continuous, with regular cyclic variations in hourly, seasonal, and geographical terms. The regular changes in the ionosphere are summarized in the following subsection.

3.5 Summary of F Ionospheric Regions

The ionosphere is affected by solar radiation; its intensity depends on the position of the Earth relative to the Sun. Because of the Earth's rotation about its axis and revolution around the Sun, the pattern of solar radiation changes, and the ionosphere varies with time and ionospheric location. The regular changes in solar radiation can be more or less predictable and categorized into four groups of radiation: (a) hourly or day–night, (b) seasonal, (c) geographical, and (d) yearly cyclical.

(a) *Hour-to-hour changes* in different ionospheric regions are caused by the Earth's rotation around its axis, resulting in nights and days. Another cause of hour-to-hour changes is a corresponding change in the UV and X-ray radiation of the Sun penetrating the ionosphere at given points. Solar radiation is strongest during daylight hours when the ionosphere is highly ionized. However, during night hours, solar radiation does not exist, and the ionosphere is reduced to a single F layer. This layer has slow recombination rates and consequently remains during the night hours without the sun's light. Hence, the F regions are the most ionized layers with relatively high operable frequency for long-distance communications in the HF bands. The operable frequencies for long-distance communications vary from approximately several MHz at night to 30 MHz during the day.

(b) *Seasonal changes* caused by constant changes in the position of the Earth relative to the Sun during its 1-year-long rotation around the Sun. The ionization of the ionospheric regions E is mainly regular and depends mostly on the Sun's activity at a given time. During the summer days, the F1 layer behaves regularly, similar to the E layer. During the summer and winter nights, the F1 and F2 layers merge and can be separately identified only at the Earth's equator region.

The ionization of the F2 region is greater than average during the summer days, and its operable frequencies are high. However, on winter days, during shorter days and with less sunheating, the F2 region has a lower concentration of ionization than it does during the summer. In terms of summer and winter nights, F ionization is somewhat greater on winter nights than on winter days.

(c) *Geographical changes* in the ionosphere significantly depend on the Earth's latitude. The radiation is greater in the equatorial region where the Sun is more perpendicular to the Earth than in northern latitudes. Owing to this, operable frequencies for the E and F1 regions change with the Sun's activity at a time, taking a maximum in the equatorial belt and decreasing toward the poles. The situation is rather complex in the F region because other radiation sources, such as X-rays and cosmic radiation, cause ionization in this region. The operable frequency in the F region also follows a pattern of being higher in the equatorial region and lower in the polar regions. F2 region ionization varies along meridians of longitude, along which the local time is the same or similar due to the influence of the geomagnetic field. For example, F2 ionization variations are more common in Asia and Australia than in the Western Hemisphere and Europe [5].

(d) *Yearly cycles* occur on average every 11 years and are directly caused by solar activity on a year-to-year basis.

4 Solar Flares, SID, IS, and Polar Blackouts

As noted earlier, there are periods when the ionosphere has irregular variations that occur randomly, making radio communications difficult or even impossible. This phenomenon is called *ionospheric disturbance (ID),* which commonly affects HF radio communications, weakening signals to the level that they completely disappear, suddenly or gradually.

Ionospheric disturbances are classified into three groups: *sudden ionospheric disturbances* (SIDs), ionospheric *storms* (ISs), and *polar blackouts,* which occur between the IS and the SID. The cause of these disturbance phenomena is solar flares that start on the sun's surface and travel toward the Earth. The SID, IS, and polar blackout effects are similar in disrupting HF radio propagations, which may result in a drastic increase in ionospheric absorption. For more details of these topics, we refer readers to [2, 5].

4.1 Solar Flares

Hence, solar flares are the main causes of ionospheric disturbances (IDs) produced by sudden changes in the Sun's magnetic field within and close to large sunspots. As an example, Fig. 6 shows the National Science Foundation's photography of sunspots as dark spots on the Sun's surface.

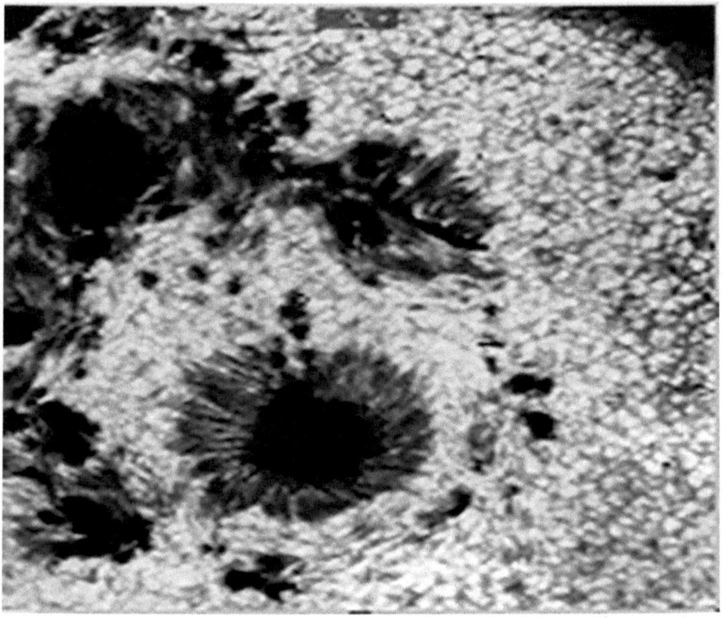

Fig. 6 Sunspots (dark ones) at the sun's surface as sources of solar flares

Solar flares start with explosions on larger sunspots when high-intensity ultraviolet and X-ray radiation starts to travel at the speed of light toward the Earth, as shown in Fig. 7a. Flares typically reach the Earth in approximately 8 min (the time needed for the light to travel from the Sun to the Earth).

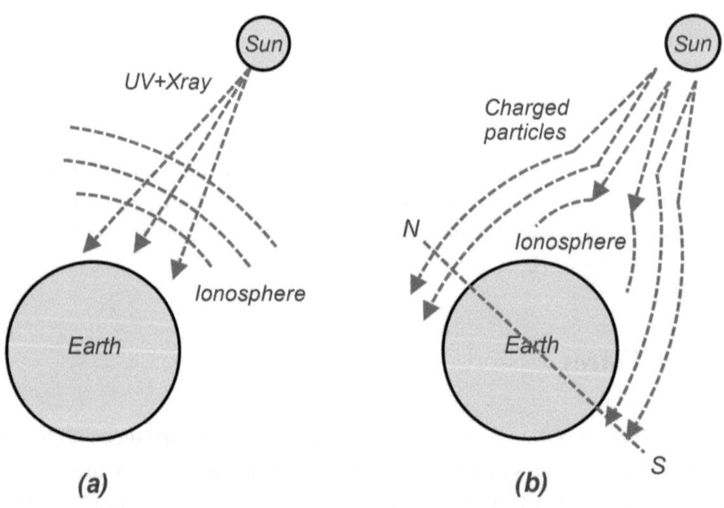

Fig. 7 (**a**) UV/X-ray radiation and (**b**) charged particles from solar flares

Raderated from solar flares also travels in the form of charged particles (as shown in Fig. 7b) with large amounts of energy. However, charged particles travel much slower than the speed of light, reaching the Earth in 18–36 h after the flare occurred on the sun's surface.

Solar flares produce radio noise that can be detected below 300 MHz. Moreover, solar flares may also produce different kinds of severe ionospheric disturbances in the Earth's atmosphere. Since solar flares depend primarily on the number and size of larger sunspots, they occur most frequently during years with high solar activity. Additionally, the Sun's radiation produce an absorbing "blanket" at lower levels of the ionosphere, causing heavy signal attenuation during SID and *Polar blackouts*. *Polar blackouts* occur mainly in the Earth's polar regions, as will be discussed in this section.

4.2 Sudden Ionospheric Disturbance (SID)

SID may cause radio communications to be completely nonfunctional (blackout) depending on the flare intensity and the relative position of the Earth toward the Sun. When a sunspot (a source of flare) is in line with Earth, the flare radiation penetrates perpendicularly into the Earth's ionosphere, causing a sharp increase in ionization even in the lowest ionospheric region (D). This causes increased absorption of signals in this region, resulting in the signals fading suddenly and gradually returning to the average level in several minutes to several hours.

Because SID is caused by solar radiation only, radio communications on the dark side of Earth are not impacted. Hence, the SID disturbance can be avoided by using the Earth's dark-side path for radio communications. In daily hours, SIDs are more intense in Earth's equatorial regions, where the flare radiation is more perpendicular to the ionosphere. Thus, SIDs affect communications more strongly over equatorial paths than at higher latitudes. During high-intensity SID, signal waves emitted from the Earth cannot reach higher ionospheric regions because the lowest D region blocks incoming waves. This phenomenon can temporarily block satellite communications, which will be discussed later in this chapter.

SID can also occur even when a solar flare is not visible on the sun's surface. This is a case of SID caused by high-temperature spots on the sun's surface that do not produce flares but high-energy electromagnetic radiation.

4.3 Ionospheric Storms (IS)

As mentioned before, solar flares are followed by delayed streams of charged particles that arrive at Earth in 16–36 h. This phenomenon results in the occurrence of an *ionospheric storm* (IS). The IS may also affect ionospheric propagation but differently than the SID. This will be discussed later in this chapter.

When a charged particle stream approaches the Earth, the Earth's magnetic field tends to push the particles toward the magnetic poles, as shown in Fig. 7b. The particle streams mainly disrupt the upper ionospheric F region rather than the lower ionospheric regions. This phenomenon may cause the F region to rapidly disappear or split into more subregions. The effect of this is the high loss of radio signals, which causes rapid signal fading and echoes due to multilayer propagations over the multilayered F region.

Extreme ionospheric storms (ISs) cause a weak ionosphere, and increased absorption results in a blackout of long-distance radio communications. This is mainly observed on radio paths crossing polar regions where the concentration of charged particles is greatest. Because of this, a signal path crossing through the polar regions should be avoided during ionospheric storms. In contrast, transequatorial paths can be used because the concentration of charged particles is significantly lower in the equatorial belt. Additionally, operations in lower radio bands can also be attempted during the IS to avoid this disturbance.

Note that using a dark-side path will not help much during the IS, as it does during the SID. The effects of *ionospheric storms* may last one to several days.

4.4 Polar Blackouts

The *polar blackouts* are produced by high-energy sun radiation diverted by the Earth's magnetic field toward polar regions (Fig. 7b). This radiation penetrates into the lower ionosphere layer, causing heavy absorption of HF waves in the polar regions. Polar blackouts are also associated with flares from large sunspots (such as the SID), which may last from several minutes to several hours, whereas *ionospheric storms* (ISs) may last longer, up to several days.

Radioabsorption caused by polar blackouts in the polar regions (called *Polar Cap Absorption*—PCA) affects only radio paths crossing the polar regions. When the IS intensifies, the polar blackouts are moved from the polar regions toward the south, where they may stay for several hours. Another type of strong polar blackout and absorption is associated with several consecutive ionospheric storms with *aurora* displays. This type of polar blackout is called an *aurora blackout.*

5 Ionospheric Parameters and Critical Region Frequencies

The two main ionospheric parameters are (a) the *12-month average number of sunspots* denoted by R_{12} and (b) the *12-month average value of solar radio noise at 2.8 GHz* (10.7 cm band), denoted by Φ_{12}. The main terms of solar variations are (a) a relationship with a *long-term period* of 7–17 years with an average value of 11 years, (b) a *quasiperiodic relationship* with a time duration somewhat *less* than a year, and (c) a *fluctuation period* equal to one cycle of the Sun's rotation about its axis of 27.5 days. For more information about the topics from this section, we refer interested readers to [2, 4].

5.1 Sunspots Number and Solar Radio Flux

The 12-month average number of sunspots, as a main component of the solar cycle, can be calculated via the following equation proposed by ITU-R:

$$R_{12}(n)=\frac{1}{12}\left[\sum_{n-5}^{n+5}R_k+\frac{1}{2}\left(R_{n+6}+R_{n-6}\right)\right] \tag{3a}$$

where R_k is the daily average value over month k of the Wolf number.

The *solar radio noise flux* is another helpful parameter of the ionosphere condition affected by solar activities. This flux is related to solar radiation at a wavelength of 10.7 cm (~ 2800 MHz), and its relationship with the sunspot number R_{12} is given by ITU-R as

$$\Phi_{12}=63.7+0.73\,R_{12}+8.9\cdot 10^{-4}\,R_{12}{}^{2} \tag{3b}$$

5.2 Ionosphere Critical Layer Frequencies

Owing to the random and stochastic nature of the ionosphere, the best way to investigate the ionosphere is by probing and performing measurements. There are different probing methods in which vertical radio probing is often used. This method investigates *critical* frequencies for specific layers. These frequencies are the highest frequencies at which vertically directed waves can reach an ionized region and be reflected to Earth. The essential critical frequencies of layers *E, F1*, and *F2* are f_0 *E*, f_0 *F1*, and f_0 *F2*, respectively.

The prediction of these frequencies is possible on the basis of ionospheric factors related to solar activity, such as the sessional factor, geographical factor (the factor of latitude), and daily or hourly factor. All these factors are statistical variables for analytical prediction, proposed by ITU-R, P.1239, or input data for computer prediction programs.

The ITU provided two programs called WOMAP and HRMNTH. The first program provides solar activity for various geographical locations at universal times and in a reference month. HRMNTH is a supplementary program that graphically obtains the ionosphere characteristics for different locations and years to the reference month, universal time zone, and associated solar period. Additionally, the ITU provides helpful data for the years with higher and lower solar activities for a given year as follows:

- The numerical coefficients for geographical, daily, and monthly variations in the E and F layers are designated f_0E and f_0F, respectively.
- Numerical maps of the virtual reflection height *h'* for vertical reflection from the F region at night, designated *h'F.*

- Numerical maps of the virtual reflection height vs. time for vertical reflection from the *F2* layer during the day, designated *h'F2*.
- Probability of dividing the F region into subregions F1 and F2 by mapping methods.

5.3 Geomagnetic Indices

Solar-charged particle emission (electrons and protons), caused by solar flares from sunspots, affects the geomagnetic field and normal propagation conditions of the ionosphere. The arrival of these particles results in ionospheric and magnetic storms that produce signal fluctuations, fading, and static noise. Several magnetic indices, such as α, K, A, and A_p, quantitatively indicate geomagnetic effects.

5.3.1 α and K Indices

The index a is measured by a given observatory and used to determine magnetic effects from solar particle radiation. Higher values for this index indicate more significant particle radiations associated with greater signal weakening and fading. For example, an α index less than 10 indicates quiet solar and magnetic periods, whereas an α greater than 30 indicates periods of greater magnetic disturbance and solar activity. Magnetic disturbances are most intense in polar regions, as noted earlier. Measurements of the α index determine geomagnetic conditions at 3 h intervals.

Another more practical magnetic index is the K index, which is provided by a single digit daily, although it is also determined in 3 h magnetic periods. For example, the K index is broadcast by NOAA (National Oceanic and Atmospheric Agency) at 18 min after each hour. Table 1 shows the conversion of the α indices to the K indices.

Table 1 α indices vs. K indices

α	0	3	7	15	27	48	80	140	240	400
K	0	1	2	3	4	5	6	7	8	9

5.3.2 A and A_p Indices

A more suitable method for publishing data on geomagnetic activity is a daily figure called the A index, which is measured by eight K indices at an observation station. K was measured every 3 h for 24 h. Furthermore, each of these K indices is converted to α indices via Table 1, and the average of these eight α indices leads to the A index at that observatory. When α indices provided by a worldwide network of stations are averaged every 3 h, the *planetary 3 h index* α_p is obtained. Furthermore, the planetary index A_p is obtained by averaging the eight α_p indices over 24 h. The average A_p index is a fairly reliable parameter for predicting and planning ionospheric radio links.

6 Optimization of the Ionospheric Radio Link

To optimize HF radio communication between two locations on Earth, fairly accurate prediction (forecast) of the ionosphere state should be available at times of interest and locations where the communication path will cross. It would also be good to have adaptive radio systems on both sides of a communication channel capable of dynamically adapting themselves to actual propagation conditions. Adaptive systems include synchronous and automatic changes in operating frequencies due to EMI from other radio systems and signal fadings in a dispersive medium such as the ionosphere. One of the adaptive radio systems discussed in Chapter "Overview of Broadband Radio Systems with High EMI Immunity" is a software-defined radio (SDR) system, which is flexible and adaptive to nonregular conditions on the HF propagation channel.

An additional option to overcome adverse multipath propagation effects is space diversity at a receiving site, which uses two or more antennas separated by a couple of operating wavelengths, as discussed in Chapter "Overview of Broadband Radio Systems with High EMI Immunity". In the end, using spread spectrum (SS) radio systems, which can communicate below the level of EMI and channel noise, is another option for radio ionospheric communications. This method is also discussed in Chapter "Overview of Broadband Radio Systems with High EMI Immunity".

In addition to solid propagation prediction (forecast) and adaptive radio systems, a type of signal modulation and a robust error-correction algorithm are vital for ionospheric communications, as discussed in Chapter "Introduction to Digital Communication Systems".

6.1 Prediction of Maximum Usable Frequency (MUF)

The most important parameter of the ionosphere is the *critical frequency* f_0, which is the maximal frequency of the echo of the radio wave emitted vertically from the Earth's surface toward the ionosphere. Each ionospheric region is characterized by its *critical* or *maximum usable frequency (MUF)*. For reliable radio contact, a practical MUF should be 10% to 15% lower than the critical layer frequency due to fluctuations associated with more or less unstable ionospheric conditions.

The practical use of the ionosphere for the communication link between a transmitting (TX) and a receiving (RX) location is shown in Fig. 8.

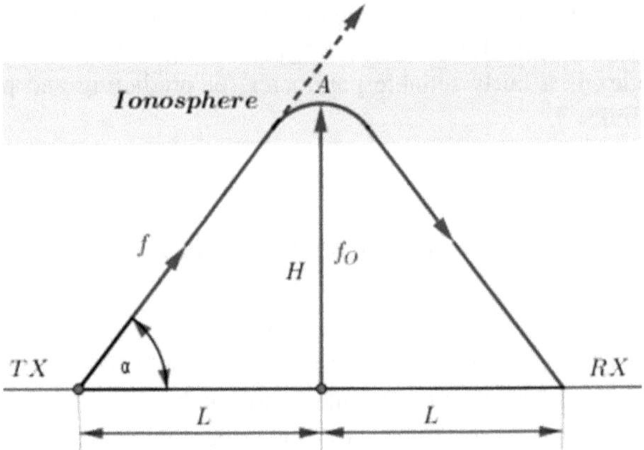

Fig. 8 Ionospheric model for MUF calculations

The oblique frequency f, which is used for communication over an oblique path between locations TX and RX, is relayed to the critical frequency via the following expression:

$$f = \frac{0.9 f_0}{\sin(\alpha)} \tag{4}$$

To obtain an ionospheric long-distance contact between two radio locations, the TX antenna must radiate waves at angle α toward the ionosphere. The proper radiation angle and the optimum frequency for the path of interest depend on the height H of the region and the distance 2 L between the TX and RX locations. Using the geometrical relationship from Fig. 8, Eq. (4) is modified to Eq. (5):

$$f = 0.9 f_0 \sqrt{1 + \left(\frac{L}{H}\right)^2} \tag{5}$$

where L is the half distance between the radio locations TX and RX and H is the height at which a wave refracts from the ionosphere. L and H should be given in the same units, for instance, in km or miles.

Equations (4) and (5) are numerically presented in the following example.

Example 1 This example illustrates the use of Eqs. (4) and (5) for the calculation of the MUF with data for the critical frequency f_0 of 6 MHz for a particular layer with height $H = 300$ km.

Therefore, we find the MUF for this ionospheric link and the optimal radiation angle toward the ionosphere when the distance 2 L between the TX and RX locations is 2000 km ($L = 1000$ km).

Solution From Eq. (5), there is

$$f_{muf} = 0.9 f_0 \sqrt{\left(\frac{L}{H}\right)^2 + 1} = 5\,MHz\sqrt{\left(\frac{1000}{300}\right)^2 + 1} \approx 19\,MHz$$

Per Eq. (4), the optimal angle of radio waves toward the layer of interest is as follows:

$$f_{muf} = \frac{0.9 f_0}{\sin(\alpha)} \rightarrow \sin(\alpha) = \frac{0.9 f_0}{f_{muf}} \rightarrow \alpha = \arcsin\left(\frac{0.9 f_0}{f_{muf}}\right) \approx 13.5°$$

Notably, this calculation is associated with the regularly changed ionosphere, but in the presence of ionospheric irregularities such as ISs or SIDs, the wave propagation in Example 1 will most likely be disrupted because of the increase in absorption in lower ionospheric layers and the disruption of F region. All these irregularities can be predicted by ionospheric broadcast services.

Hence, we can find the optimal (or maximum usable) frequency to obtain stable ionospheric radio communication between two radio locations during regular ionospheric conditions when we know the critical frequency for the region of interest and its height above the Earth, as well as the distance between two radio locations. The operating frequency should be somewhat below the MUF of the given region for radio-ionospheric contact. If the signal frequency f exceeds the MUF, the ionization at the point of refraction may not be strong enough to refract the wave back to Earth, which causes the wave to partially or completely pass through the ionosphere into space.

Since the MUF is an essential parameter of radio ionospheric communications, relatively straightforward graphical methods or computer calculators have been established on the websites of some professional organizations and institutions for estimating the MUF. For example, *Ionospheric Predictions* are published by the Institute of Telecommunications of Sciences of the U.S. Department of Commerce [14]. Similar data can also be found in the C.C.I.R. *Atlas of Ionospheric Characteristics*, published by the International Telecommunications Union [15]. Furthermore, a helpful source for these data is the National Oceanic and Atmospheric Administration (NOAA) website [16].

In addition to data for the MUF, the above organizations gather other valuable data relevant to predicting ionospheric conditions in further periods (e.g., short, medium, and long). As noted earlier, these data include the number of sunspots, magnetic indices, ionospheric absorption, lowest usable frequency (LUF), which we discuss in next subsection. Most of these data are available and open to the public and scientific communities.

6.2 Lowest Usable Frequency (LUF) and Related Ionospheric Absorption

Ionospheric absorption is the main cause of the reduction in radio wave strength while it passes through the ionosphere. The absorption increases with decreasing operation frequency, i.e., more absorption is observed at lower frequencies. In particular, the amount of ionospheric absorption changes inversely with the square of the frequency. The low-frequency ionospheric parameter is defined as the *lowest usable frequency* (LUF), which refers to the *lowest frequency* that can be used for satisfactory wave propagation over a specified ionospheric path at a specified time.

The minimum signal intensity level needed for satisfactory reception depends on the noise and EMI level at a receiving location, among other radio channel factors noted in Chapter "Performance of Radio Digital Systems in Presence of EMI". Atmospheric or static noise is the most dominant noise close to and below the LUF. Any reception is not possible at frequencies below the LUF because the prevailing noise covers the received signal. As the operating frequency increases above the LUF, the signal-to-noise ratio (SNR) improves. The optimal communication conditions are below and near the MUF, where the SNR and propagation quality can be very good.

Unlike the MUF, which depends entirely on present ionospheric conditions and characteristics, the LUF can be controlled to some level by adjusting the adequate radiated power or selecting different modulations of the transmitted signal. Since the absorption in the ionosphere increases when solar activity increases, the LUF is expected to be somewhat greater during peak solar activity.

6.3 HF Signal Fading

HF signal fading is signal fluctuation due to multipath wave propagation, reflection, and absorption in the unstable, inhomogeneous, and anisotropic ionosphere. Figure 9 shows multipath ionospheric propagation.

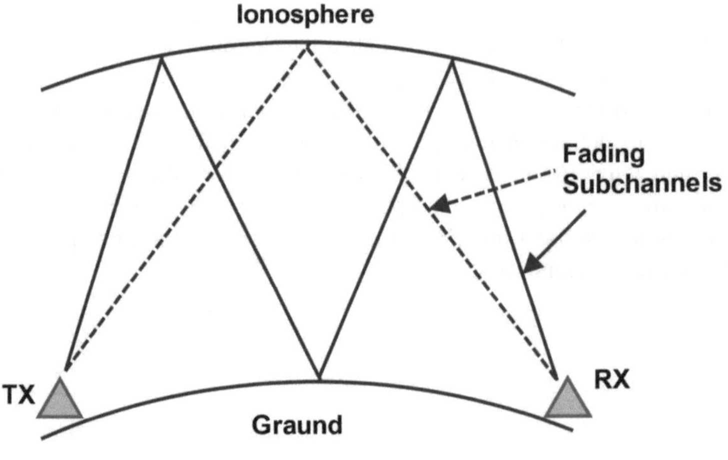

Fig. 9 Multipath propagation in the ground-ionosphere link

As mentioned in Chapter "Performance of Radio Digital Systems in Presence of EMI", there are two types of fading rates: fast (rapid) and *slow* fading. *Rapid* fading changes the signal amplitude in a few seconds, whereas *slow* fading lasts for a few minutes or longer. Both fadings, *slow* and *rapid*, can occur at the same time.

Slow fading is caused mainly by variations in absorption in the D region. Additionally, it is associated with random motion of charges within the ionosphere, mainly in the vertical plane.

Rapid fading is more critical than slow fading. It is caused by irregularities and instabilities in the ionosphere in equatorial regions, polar aurora zones, periods of ionospheric storms (ISs), and sudden ionospheric disturbances (SIDs). Some ionospheric irregularities cause a wave to be separated into several simultaneous propagation paths, arriving at the receiver as a group of independent signals with random phase relationships. This phase difference is caused by multipath propagation, which causes variations in signal strength. When the signaling components are in phase, a strong signal results. When the signaling components are out of phase, they cancel each other, i.e., the signal disappears entirely for a moment. One of the efficient methods to minimize the fading effects is the use of frequency or space diversity, as discussed in Chapter "Overview of Broadband Radio Systems with High EMI Immunity".

The writing of this section is influenced by reference [4], which we recommend to interested readers for more detailed study.

7 Ionosphere Effects on Satellite Communications

Ionospheric effects on satellite communications are significant for low-earth-orbit (LEO) and mid-earth-orbit (MEO) satellite frequencies below 3 GHz, such as mobile satellite services, climate monitoring satellites, and maritime communication networks (such as Inmarsat) [5, 6].

During the planning and design of satellite links, the following ionospheric effects should be considered:

- Excessive group delay and dispersion of the group velocity.
- Variation in the angle of signal arrival.
- Scintillation affects the amplitude, phase, and angle of wave arrivals.
- Doppler effect.
- Polarization (Faraday's) rotation.

Because it is not easy to analyze the effects of ionospheric effects on radio communication analytically because of their random nature, the best and most accurate way to study and model these phenomena is an empirical approach. Additionally, owing to their random behavior, ionospheric effects can be described only stochastically.

7.1 Ionosphere Ionization and Total Electron Content (TEC)

As noted before, solar radiation forms several ionized regions (layers), such as D, E, and F regions in the atmosphere, whose ionizations are nonhomogeneous and time-varying. These regions are characterized by the *total electron content* (TEC), which mainly depends on the Sun's activity in conjunction with the Earth's magnetic activity. The TEC refers to the total amount of electrons in the given surface and is expressed by an integral as

$$TEC = \int n_e(\mathrm{s})ds \quad \left[\mathrm{el/m^2}\right] \tag{6}$$

where $n_e(s)$ is the electron density in el/m^2 and s is the length of the propagation path in meters.

TEC modeling is typically performed in the solar zenith direction, i.e., along a vertical ionosphere cylinder with a unit area, where its changes are in the range of 10^{16} to 10^{19} electrons/m^2. The maximum TEC occurs during the daytime around noon. There are two common approaches for calculating and predicting TEC with acceptable accuracy: the International Reference Ionosphere (IRI) and the Ne Quick methods.

Irregular variations in the TEC in the ionosphere cause the ionosphere to become a dispersive (frequency-dependent) medium for radio waves.

7.2 Propagation Delay and Dispersion of Waves

The velocity of radio waves decreases due to the presence of charged particles in the ionosphere, which causes an additional delay Δt to the wave delay in free space. Δt is directly proportional to TEC and expressed as [IRU-R, P 531]

$$\Delta t \cong 1.345 \cdot 10^{-7} \cdot TEC \cdot f^{-2} \quad (s) \tag{7}$$

where f is the frequency in Hz.

The group delay varies with frequency inside the bandwidth of broadband satellite signals, causing unwanted *dispersion* in a received signal. For example, for the L band at approximately 1.6 GHz, the signal group delay varies from approximately 0.5 to 500 ns, whereas the TEC varies from 10^{16} to 10^{19} electrons/m^2.

7.3 Scintillation

Scintillation is the negative effect of the ionosphere on satellite links with frequencies below 3 GHz. The ionization of ionosphere regions can be irregular and variable in nature, causing *scintillation* that may seriously affect the amplitude, phase, and angle of arrival of satellite signals. In general, local irregularities in the ionosphere act as lenses that focus and defocus satellite wave signals. The higher *scintillation* level affects satellite communications depending on the utilized signal modulation. This phenomenon depends on solar activity at the geographical location of a satellite link at a time.

The estimation of *scintillation* effects can be obtained from statistical data provided by the reference of ITU-R, P-531 associated with the maximum peak-to-peak scintillation fluctuation in different months and Earth's locations, such as in Earth's eastern hemisphere, with an elevation angle of approximately 20 °C, and in the Earth's western hemisphere, with an elevation angle of approximately 30 °C. More details about this topic can be found in related ITU-R publications and websites.

7.4 Wave Polarization and Faraday's Rotation

A linearly polarized satellite wave changes gradually in the polarization plane, passing through the ionosphere due to the geomagnetic field effect and anisotropy of the ionosphere. The Faraday rotation depends on the intensity of geomagnetic field B, the factor TEC, and the frequency f of the wave signal that propagates through the ionosphere. Faraday's rotation is characterized by the change in signal polarization by an angle ϕ (in radians), whose maximal value is expressed per [IRU-R, P 531] as

$$\phi_{\max} \cong 2.36 \cdot 10^4 \cdot B_{av} \cdot TEC \cdot f^{-2} \tag{8}$$

where B_{av} is an average of magnetic induction in Wb/m^2 or T and f in Hz.

Since the polarization angle ϕ is predictable by knowing the above stochastic variables in a regular ionosphere, the polarization change can be compensated via the antenna setting. However, during intense geomagnetic storms and ionospheric disturbances (IS and SID), when the amplitude of satellite signals abruptly changes, the polarization angle is difficult to predict, especially in tropical regions.

The cross-polarization (XPD) of antennas caused by polarization rotation can be calculated via the following expression:

$$\mathrm{XPD} = -20\log\left|\tan\left(\phi_{\max}\right)\right| \; (dB) \tag{9}$$

Example 2 Per Eqs. (6), (7), (8), and (9) for TEC = 10^{18} el/m^2 and $B_{av} = 5\ 10^{-3}$ T, the time delay Δt, the change in the polarization angle ϕ_{max}, and the XPD in the L-band at 1.6 GHz are as follows:

$$\Delta t \approx 0.5\,ns; \quad \phi_{\max} \approx 52\,rad; \quad XPD \approx -3\,dB.$$

7.5 Frequency Dependency of Ionosphere Effects

The effects of different ionospheric factors and ionospheric dependencies on frequency are summarized by ITU-R, P531 in Table 2 for a frequency of 1 GHz, an elevation angle of approximately 30 °C, and TEC = 10^{18} *el/m*^{2}.

Table 2 The effects of different ionospheric parameters dependent on frequency

Effects	Magnitude	Freq. dependency
Faraday rotation	108°	f^{-2}
Propagation delay	0.25 us	f^{-2}
Refraction	< 0.17 mrad	f^{-2}
Variation of angle of arrival	0.2 min of arc	~ f^{-2}
Polar cap absorption	0.04 dB	f^{-3}
Dispersion	0–4 ns/MHz	f^{-2}

7.6 Effect of Ionospheric Refractive Index n(r) on a Satellite-Earth Path

The refractive index n(r) is not constant in the ionosphere but varies with distance from Earth, as shown in Fig. 10. This variation affects the Earth-satellite path, causing it to deviate from a straight line within the ionosphere.

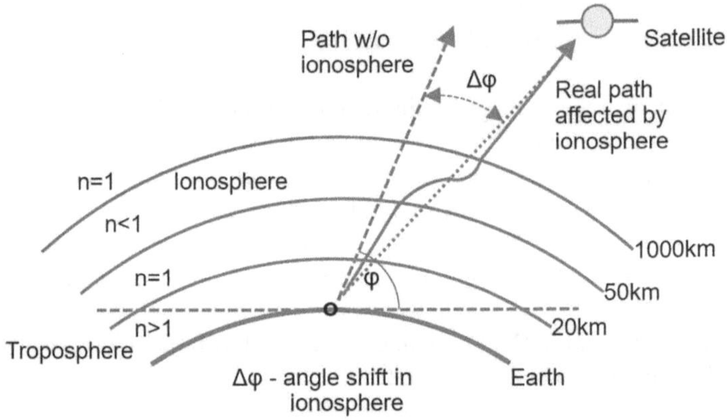

Fig. 10 A satellite signal path affected by ionosphere

$\Delta\varphi$ is an angle shift of satellite-Earth path due to the ionospheric effect. Because the ionosphere may change unpredictably, this angle shift becomes a random variable that affects satellite signals, especially during SID and IS discussed in Section 13.4 of this chapter.

References

1. R.K. Mishra, *The Effect of Solar Radiation on Communications* (Lambert Academic Publishing, 2019)
2. G. Jacobs, W3ASK, T. J. Cohen, N4XX, *The New Shortwave Propagation Handbook* (CQ Publishing, Inc., NY, 1995)
3. M. Massimo, B. Forte, A.J. Coster, S. Skone, *The Dynamical Ionosphere*, 1st edn. (Elsevier, 2019)
4. A. Ghasemi, A. Abedi, F. Ghasemi, *Propagation Engineering in Wireless Communications*, 2nd edn. (Springer, 2016)
5. H. Sizun, *Radio Wave Propagation for Telecommunication Applications* (Springer, 2005)
6. N. Blaunstain, C.G. Christodoulou, *Radio Propagation and Adaptive Antennas for Wireless Communication Networks—Terrestrial, Atmospheric, and Ionospheric*, 2nd edn. (Wiley, 2014)
7. C.J. Coleman, *Analysis and Modeling of Radio Wave Propagation* (Cambridge University Press, 2017)
8. K. Merriman, *Radio Wave Propagation in Ionosphere* (Wilford Press, 2016)
9. A.G. Kanatas, A.D. Panagopoulos (eds.), *Radio Wave Propagation and Channel Modeling for Earth-Space Systems* (CRC Press, 2016)
10. M.Y. Markov, V.D. Kuznetsov, Solar flares and impact on earth, in *Handbook of Cosmic Hazards and Planetary Defense*, (Springer, 2014)
11. B. Wolesi, L.R. Cander, *Ionospheric Prediction and Forecasting* (Springer, 2014)
12. J.A. Richards, *Radio Wave Propagation* (Springer, 2008)
13. K.G. Budden, *Radio Waves in the Ionosphere* (Cambridge University Press, 1988)
14. *Institute of Telecommunications of Sciences of the U.S. Department of Commerce*
15. C.C.I.R. *Atlas of Ionospheric Characteristics, The International Telecommunications Union (ITU)*
16. *National Ocean and Atmospheric Administration (NOAA)*

Appendix A. Origins of Electric and Magnetic Fields Governed by Maxwell's Equations

1 Introduction

The physical nature of electromagnetic interference (EMI) can be effectively explained by the fundamental principles of *electromagnetics* and by the origins and concepts of *electric* and *magnetic* fields. Because of this, we consider the fundamentals of electromagnetics and fields in this Appendix. There are many valuable textbooks on electromagnetics where these topics are well presented but sometimes from different aspects and purposes. Hence, in this book, we use fundamental principles of electromagnetics and fields to analyze and explain the background of EMI phenomena that affect radio system performance.

Because *Maxwell's equations* govern fundamental macroscopic electromagnetism laws and principles (first published in 1873), we start this chapter appendix by discussing these equations and their physical meaning. In addition to macroscopic electromagnetics, quantum electromagnetics considers fundamental physical laws at the subatomic level. Quantum electromagnetics is outside the scope of this book. Interested readers for this topic are referred to [1]. Additionally, the mathematical foundations used in this appendix chapter are covered in detail in textbooks [2, 3].

2 Maxwell's Equations

Hence, *Maxwell's equations* govern electromagnetic principles and laws. Specifically, they govern five constitutive electromagnetic laws: *Faraday's* law, *Ampere's* law, Gauss's laws (for *electric* and *magnetic* fields), and Ohm's law. There are two forms of Maxwell's equations: *differential* and *integral*. The *differential* form is more intuitive for understanding field phenomena, whereas the *integral*

M. Pajovic, *Fundamentals of EMI Phenomena in Radio Communication Systems*, https://doi.org/10.1007/978-3-031-88748-2

form is more understandable with respect to concrete and practical electromagnetic cases.

The following set of Maxwell *differential* equations, applied for any medium or material, is given in *SI* units as [4–9]

$$\vec{\nabla}\times\vec{E} = -\frac{\partial \vec{B}}{\partial t} \quad (Faradey's\,law) \tag{A.1a}$$

$$\vec{\nabla}\times\vec{H} = \left(\vec{J} + \frac{\partial \vec{D}}{\partial t}\right) \quad (Ampere's\,law) \tag{A.1b}$$

$$\vec{\nabla}\cdot\vec{D} = \rho_v \quad (Electric\ Gauss's\,law) \tag{A.1c}$$

$$\vec{\nabla}\cdot\vec{B} = 0 \quad (Magnetic\ Gauss's\,law) \tag{A.1d}$$

$$\vec{J} = \sigma\vec{E} = \frac{\vec{D}}{\varepsilon} \quad (Ohm's\,law) \tag{A.1e}$$

where $\vec{E}$ = Electric field intensity (in V/m) $\vec{D}$ = Electric flux intensity (in C/m^2)ρ_v = Volume charge density (in C/m^3) $\vec{H}$ = Magnetic field intensity (in A/m) $\vec{B}$ = Magnetic flux intensity (in Wb/m^2) $\vec{J}$ = Current density (in A/m^2)σ = Conductivity of a medium (in S)

The vectors $\vec{D}$, $\vec{E}$, $\vec{B}$, $\vec{H}$, and $\vec{J}$ represent the constitutive electromagnetic parameters of a medium, where the relationships between the fields $\vec{D}$ and $\vec{E}$ and between $\vec{B}$ and $\vec{H}$ are given as

$$\vec{D} = \varepsilon\vec{E} \tag{A.2a}$$

$$\vec{B} = \mu\vec{H}, \tag{A.2b}$$

where the proportional parameter ε is the dielectric permittivity (in F/m) and where μ is the magnetic permeability (in H/m).

Importantly, the sources of *electric* and *magnetic* fields are *moving charges* that are quantized by point, linear, or surface charge density ρ, as well as current density $\vec{J}$. The parameters ε and μ define medium electromagnetic properties, such as the *propagation speed* of electromagnetic waves and the intrinsic *impedance* of a medium. These media parameters will be considered in more detail later in this appendix.

In the following sections, we discuss the fundamentals of five constitutive laws of electromagnetics: *Faraday's* law, *Ampere's* law, *electric* and *magnetic Gauss's* laws, and Ohm's law, which are all covered and governed by *Maxwell's equations.*

3 Faraday's Law and the Induced Circular Electric Field

The main idea of Faraday's law is that changing magnetic flux through any surface surrounded by a conductive loop produces a circulating electric field along the loop that drives the electric current flow in that loop. This idea is presented by Maxwell's equation in *differential* and *integral* forms, as given in the following text.

3.0.1 Differential Form of Faraday's Law

Faraday's law is presented in *differential* form via Maxwell's equation (A.1a), given in the previous section as

$$\vec{\nabla}\times\vec{E} = -\frac{\partial\vec{B}}{\partial t} \quad (\textit{Faraday's law}). \tag{A.3}$$

Specifically, the meaning of this equation is that *a time-changing magnetic* field produces a *circular electric* field. The term $-\frac{\partial\vec{B}}{\partial t}$ in Eq. (A.3) is the *time rate* of *the changing magnetic* field, whereas the left term ($\vec{\nabla}\times\vec{E}$) is the analytical presentation of the *curl* of the *electric* field $\vec{E}$, which refers to the tendency of $\vec{E}$ lines to circulate around a field point. The minus sign on the right side of Eq. (A.3) is per *Lenz's law*: the circulating electric field and current induced by changing magnetic flux are always directed to oppose the magnetic flux change.

The *curl* of the field $\vec{E}$ at a specific point is considered the *electric field circulation* around an *infinitesimal* loop that surrounds that point. The mathematical definition of the *curl* $\vec{E}$ and Eq. (A.3) is

$$\vec{\nabla}\times\vec{E} = \lim_{\Delta S\to 0}\frac{1}{\Delta S}\oint_{0}^{2\pi}\vec{E}d\vec{l} = -\lim_{\Delta S\to 0}\int\frac{\partial}{\partial t}\vec{B}d\vec{S}, \tag{A.4}$$

where the circular integral is related to a closed line (a loop) around the point of interest and where ΔS is the surface area enclosed by that loop. The direction of the *curl* is normal to the surface where the field has maximum circulation. When the surface ΔS approaches zero, as shown in Eq. (A.4), it becomes the infinitesimal surface dS, within which there is an infinitesimal source of the *circulating electric* field, presented by the derivation $\frac{\partial}{\partial t}\vec{B}$. Accordingly, Eq. (A.4) becomes Eq. (A.3):

$$\vec{\nabla}\times\vec{E} = -\frac{\partial\vec{B}}{\partial t}. \tag{A.4a}$$

Figure A.1 shows several electric field structures with points where the curls are zero or nonzero.

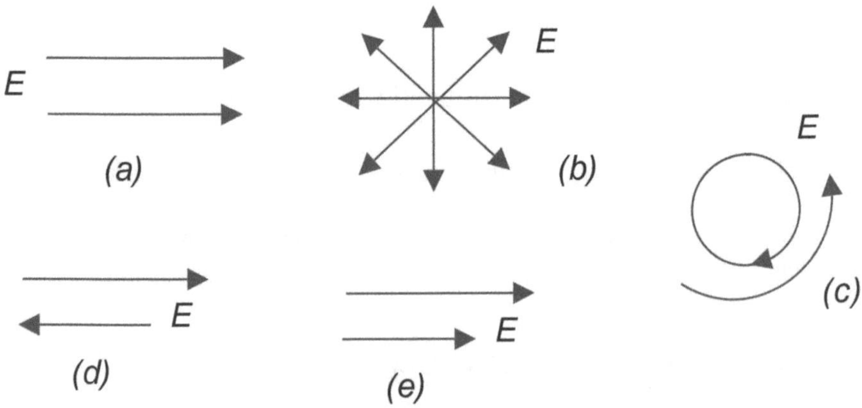

Fig. A.1 Origins of field curls: (**a**, **b**) zero curl; (**c**–**e**) nonzero curl

The zero curls in the present fields are at points where adjacent field vectors have the same magnitudes and directions. For example, the curl in field structure (a) is zero because the adjacent field lines (vectors) have the same magnitude and direction. Additionally, the curl is zero when the field vectors diverge from each other, as shown in field structure (b). For these structures, the fields are uniform (or stationary). Next, inside field structures (c), (d), and (e), the field curls differ from zero because the magnitudes and directions of adjacent field vectors differ from each other.

The vector of the electric field $\vec{E}$ can be expressed in the *Cartesian* coordinate system as

$$\vec{E} = E_x\vec{i} + E_y\vec{j} + E_z\vec{k}, \tag{A.5a}$$

where $\vec{i}$, $\vec{j}$, and $\vec{k}$ are the unit vectors of the field components E_x, E_y, and E_z. The differential operator $\vec{\nabla}$ is defined as

$$\vec{\nabla} \equiv \vec{i}\frac{\partial}{\partial x} + \vec{j}\frac{\partial}{\partial y} + \vec{k}\frac{\partial}{\partial z} \tag{A.5b}$$

and then the curl $\left(\vec{\nabla}\times\vec{E}\right)$ is expressed in *Cartesian* space as

$$\begin{aligned} \vec{\nabla}\times\vec{E} &= \left(\vec{i}\frac{\partial}{\partial x} + \vec{j}\frac{\partial}{\partial y} + \vec{k}\frac{\partial}{\partial z}\right)\times\left(E_x\vec{i} + E_y\vec{j} + E_z\vec{k}\right) \rightarrow \\ \vec{\nabla}\times\vec{E} &= \vec{i}\left(\frac{\partial E_z}{\partial y} - \frac{\partial E_y}{\partial z}\right) + \vec{j}\left(\frac{\partial E_x}{\partial z} - \frac{\partial E_z}{\partial x}\right) + \vec{k}\left(\frac{\partial E_y}{\partial x} - \frac{\partial E_x}{\partial y}\right). \end{aligned} \tag{A.5c}$$

The curl may also be presented in other coordinate systems when they are more suitable for the given field structure, such as *spherical* and *cylindrical* fields, which are usually considered in spherical and cylindrical coordinate systems, respectively.

3.0.2 Integral Form of Faraday's Law

Maxwell also expressed Faraday's law in the integral form as

$$EMF = \oint_{0}^{2\pi} \vec{E}d\vec{l} = -\int \frac{\partial}{\partial t} \vec{B}d\vec{S} = -\frac{d\phi}{dt} \text{ (V)}, \tag{A.6}$$

where EMF is the *electromotive force* or the *voltage* that drives current in a conductive loop, and the integral $\int \frac{\partial}{\partial t} \vec{B}d\vec{S}$ is the flux of the time-*changing magnetic* field through the surface S. Additionally, the last term $-d\phi/dt$ in Eq. (A.6) is the flux rate $d\phi$ that crosses the infinitesimal surface dS. In particular, $d\phi$ is the increment of the magnetic flux equal to $\vec{B}d\vec{S}$.

Thus, Eq. (A.6) represents Faraday's law in an integral form, where the time-changing magnetic flux $\phi(t)$ crossing a closed conductive loop produces an electromotive force or voltage V along the loop. The minus sign in this equation is according to Lenz's law: the currents induced by changing magnetic flux always flow in a direction to oppose the flux change. This fact is shown in the following subsection (Fig. A.2).

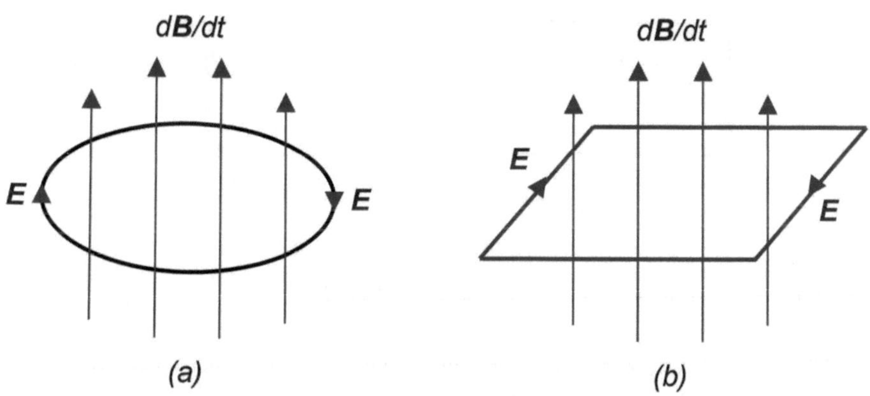

Fig. A.2 Induced electric field by a time-changing magnetic field in conductive (**a**) circular and (**b**) square loops

3.0.3 Changing Magnetic Flux and Applications of Faraday's Law

For example, Fig. A.2 visualizes *Faraday's* law for two differently shaped conductive loops through which the time-changing magnetic flux increases. When the magnetic flux decreases, the field $\vec{E}$ changes direction.

Additionally, the *magnetic flux* ϕ passing through a conductive loop is directly proportional to the current I in the loop:

$$\phi = L \cdot i(t), \tag{A.7}$$

where the proportional constant L represents the *inductance* of the loop.

By substituting Eq. (A.7) into Eq. (A.6), it becomes

$$V = -\frac{d\phi}{dt} = -L\frac{di(t)}{dt} \quad (\mathrm{V}). \tag{A.8}$$

Figure A.3 shows another example of the application of Faraday's law in a cylindrical solenoid with a current I and multiple turns on a ferromagnetic or ferrite core with permeability μ.

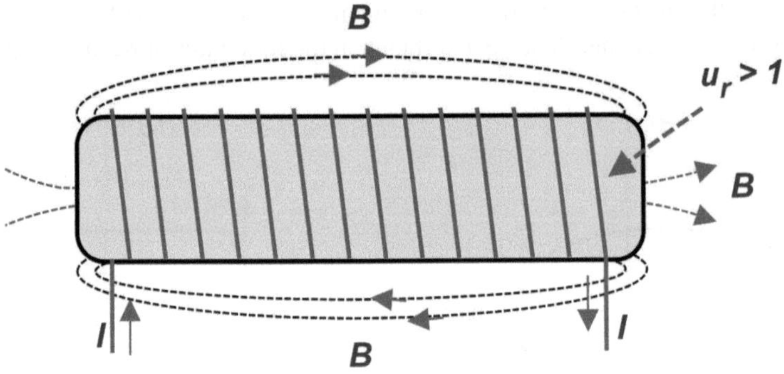

Fig. A.3 Ferromagnetic solenoid with increased inductance L

When the solenoid has N wire turns and the magnetic core has a permeability of $\mu = \mu_r\mu_0$, its inductance is expressed as

$$L = \frac{\phi}{I} = \mu\frac{N^2}{l}S = \mu_r\mu_0\frac{N^2}{l}S \quad (\mathrm{H}), \tag{A.9}$$

where μ_r is the relative permeability and where μ_0 is the permeability of air or free space (equal to $4\pi\ 10^{-7}$ H/m). Additionally, S is the cross-sectional area of the solenoid core, and l is the core length. When the solenoid has an air core, its inductance L_0 is expressed as

$$L_0 = \mu_0 \frac{N^2}{l} S \quad (\mathrm{H}), \tag{A.9a}$$

where $\mu_r = 1$ in the air. Furthermore, combining Eqs. (A.9) and (A.9a), the relation between the *inductance L* and L_0 is

$$L = \mu_r L_0. \tag{A.9b}$$

Because the relative permeability μ_r for ferromagnetic or ferrite materials is much greater than *1* ($\mu_r \gg 1$), the inductance of the solenoid with an air core can be significantly increased just by using ferromagnetic or ferrite materials instead of air. This method for increasing the inductance can be applied to any inductor structure, such as a toroid, E-profile, or other core structure.

The next practical examples show several sources of the *changing magnetic* field, which can induce *circular electric* fields. Figure A.4a, b shows (a) a rotating *conducting loop* in the *static magnetic* field and (b) a *moving* magnet through a conducting loop. In both cases, the conducting loops are effectively in the *changing* magnetic field, resulting in the current induction in present loops.

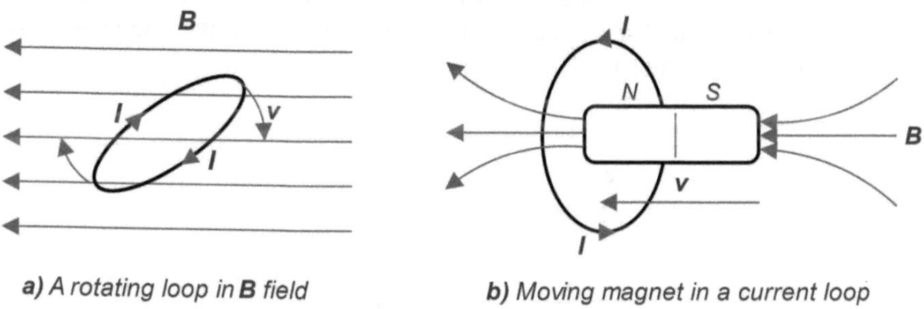

Fig. A.4 Some sources of induced electric currents in conductive loops

4 Ampere's Law and Induced Circular Magnetic Field

The main idea of Ampere's law is that a *steady* current flow in a conductor produces a *circulating magnetic* field around the conductor. Later, Maxwell concluded that the time-changing electric flux $\vec{D}$ may also generate a circulating magnetic field. This law is often called Ampere–Maxwell's law.

4.0.1 Differential Form of Ampere–Maxwell's Law

According to the above statement, Maxwell expressed Ampere's law in *differential* form as

$$\vec{\nabla} \times \vec{H} = \vec{J} + \frac{\partial \vec{D}}{\partial t} \quad (Ampere's\ law), \tag{A.10a}$$

where the left term $\vec{\nabla} \times \vec{H}$ refers to the *curl* of the *circulating magnetic* field, whose lines tend to circulate at a specific field point. The two right terms in Eq. (A.10a) are associated with the *steady current density* $\vec{J}$ and the *time-changing electric* flux $\frac{\partial \vec{D}}{dt}$. The mathematical definition of the curl $\vec{\nabla} \times \vec{H}$ in the Cartesian space is given as

$$\vec{\nabla} \times \vec{H} = \vec{i}\left(\frac{\partial H_z}{\partial y} - \frac{\partial H_y}{\partial z}\right) + \vec{j}\left(\frac{\partial H_x}{\partial z} - \frac{\partial H_z}{\partial x}\right) + \vec{k}\left(\frac{\partial H_y}{\partial x} - \frac{\partial H_x}{\partial y}\right). \tag{A.10b}$$

When more suitable, the curl can be presented in other coordinate spaces, such as *spherical* and *cylindrical* spaces.

Magnetic field structures with zero-curl and nonzero-curl electric fields are similar to structures with zero-curl and nonzero curl electric fields, as shown in Fig. A.1.

Figure A.5 visualizes Ampere–Maxwell's law in circuits with capacitor *C*. A *steady conductive current* I_c induces a *circular magnetic* field $\vec{H}$ around the circuit conductor, whereas the displacement current I_d between the C plates induces a *circular magnetic* field inside C.

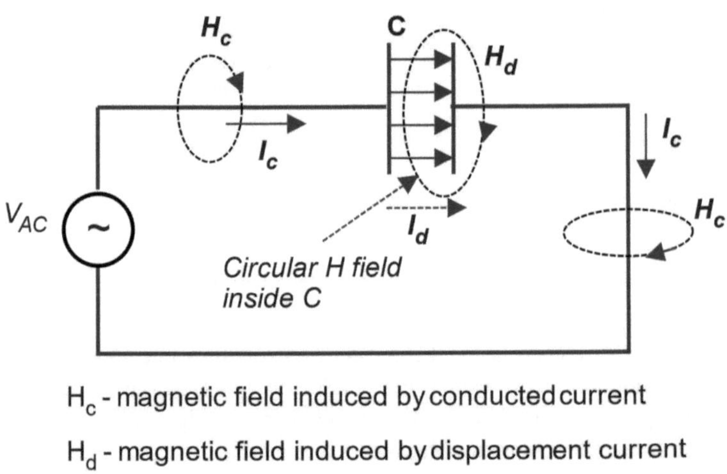

Fig. A.5 Sources of induced circular magnetic field in C

Here, the *time-changing magnetic* field $\vec{H}$ is produced by an AC voltage source.

4.0.2 Integral Form of Ampere–Maxwell's Law

Thus, two types of electric currents may produce the *circulating magnetic* field: the *steady conducted* current (Ampere based) and the *displacement* current (Maxwell based), both of which are associated with a change in *electric flux*. In this context, Ampere–*Maxwell's law* can also be expressed in *integral* form as

$$\begin{gathered}\oint \vec{H}d\vec{l} = I_c + \frac{d}{dt}\int \vec{D}d\vec{S} \rightarrow \\ \oint_0^{2\pi} \overrightarrow{H}\,d\vec{l} = I_c + \varepsilon\frac{d}{dt}\vec{D}d\vec{S} \rightarrow \int \vec{E}d\vec{S} = I_c + I_d,\end{gathered} \tag{A.11}$$

where the left term $\oint \vec{H}d\vec{l}$ represents the *circulating magnetic* field along a closed loop (real or imaginary loop), while the right terms represent the sum of the *steady conducted* current I_c and the *time-changing electric* flux, $\frac{d}{dt}\int \vec{D}d\vec{S}$, i.e., the *displacement* current I_d.

The simplest case of the *circular magnetic field* produced around a long linear wire that carries the *steady current I* is visualized in Fig. A.6.

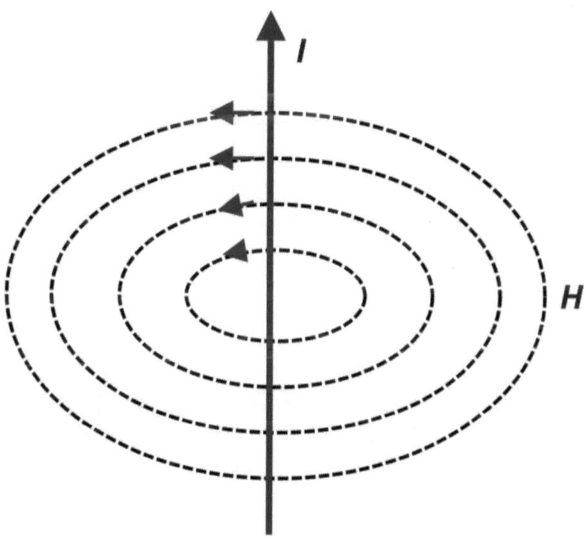

Fig. A.6 Induced circular magnetic field by current *I*

This figure shows that the $\vec{H}$ field and the current *I*, which produces the magnetic field, follow the *right-hand rule* where the fingers follow the direction of the $\vec{H}$ field rotation since the thumb is aimed at the *I* flow direction.

Applying *Ampere's law* for this simple case, shown in Fig. A.6, results in the following expression

$$\oint \vec{H}d\vec{l} = H.\ dl = H \cdot (2\pi r) = I \rightarrow$$
$$H = \frac{I}{2\pi r} \quad (\mathrm{A/m}), \tag{A.12}$$

where H is the intensity of the *circular magnetic* field at an observation point at distance r from the conductor. Therefore, a circular *magnetic* field line has a radius r and a length, or circumference, $2\pi r$.

5 Electrostatic Gauss's Law

5.0.1 Nature of the Electrostatic Field

Between two or more *electric* charges, *electrostatic forces* exist that push aside charges with the same polarity or drag toward each other with opposite polarity. *Faraday* and *Maxwell* named the space where *electrostatic* forces exist the *electrostatic* field. Thus, the *electrostatic field* results from the *electrostatic forces* between *electric charges*. The following expression states that the *electrostatic* field $\vec{E}$ is the *electrostatic force* $\vec{F}_e$ per punctual charge q:

$$\vec{E} = \frac{\vec{F}_e}{q} \quad \left(\mathrm{V/m}\right). \tag{A.13}$$

Figure A.7 shows two separated punctual charges, positive and negative (*a* and *b*), with the *electrostatic field* $\vec{E}$ they produce.

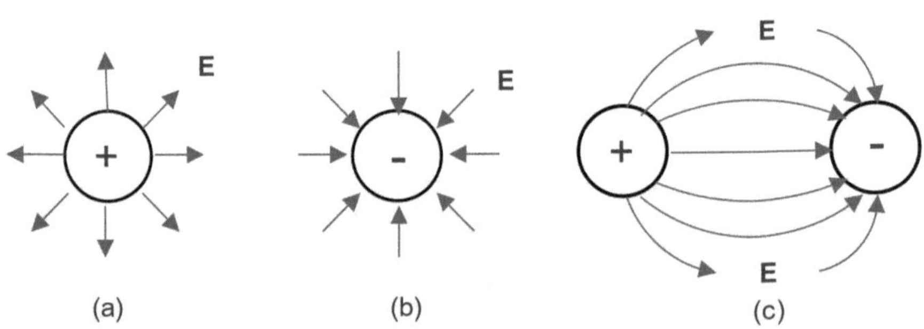

Fig. A.7 Structures of electrostatic fields are from (**a**) positive charge, (**b**) negative charge, and (**c**) electric dipole

The *electrostatic* model *c* shows an *electric dipole* created by a pair of closely separated charges with opposite polarities. Therefore, regarding convention, *electric field* lines always originate on a *positive* charge and *converge* to a *negative* charge.

5.0.2 Differential Form of Electric Gauss's Law

Gauss's law in differential form states that an electric field flux $\vec{D}$, which leaves a small closed surface dS, is produced by a single electric charge q within that surface, as shown in Fig. A.8.

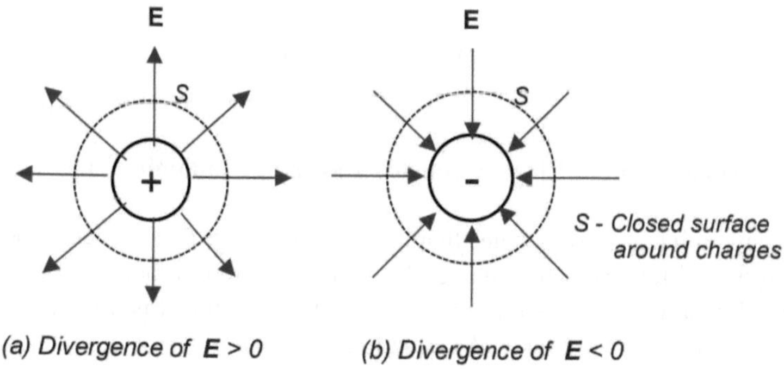

Fig. A.8 Divergences of electric fields

With respect to this statement, *Maxwell's differential* form of *Gauss's law* is given as

$$\vec{\nabla}\cdot\vec{D} = \vec{\nabla}\cdot\varepsilon\vec{E} = \rho_v \quad \left(\textit{Gauss's law for electric field}\right), \tag{A.14a}$$

where the left term $\vec{\nabla}\cdot\vec{D}$ is the *divergence* of the *electric flux density* $\vec{D}$ through a surface dS and the right term ρ_v is the *volume density* of charge. The formal mathematical definition of the *divergence* is

$$\vec{\nabla}\cdot\vec{D} \equiv \lim_{\Delta V\to 0}\frac{1}{\Delta V}\oint_0^{4\pi}\overline{D}\,d\vec{s} = \rho_v. \tag{A.14b}$$

This means that when the closed volume ΔV approaches zero, its surface approaches the infinitesimal surface dS that surrounds a single charge, resulting in the flux divergence $\vec{D}$ being equal to the volume charge density ρ_v.

Equation (A.14a) can be expressed in rectangular Cartesian (x, y, z) space as

$$\vec{\nabla}\cdot\vec{D} = \frac{\partial D_x}{\partial x} + \frac{\partial D_y}{\partial y} + \frac{\partial D_z}{\partial z} = \rho_v, \tag{A.14c}$$

where the operator $\vec{\nabla}$ and vector $\vec{D}$ are defined as

$$\vec{\nabla} \equiv \vec{i}\frac{\partial}{\partial x} + \vec{j}\frac{\partial}{\partial y} + \vec{k}\frac{\partial}{\partial z}; \vec{D} = \vec{i}D_x + \vec{j}D_y + \vec{k}D_z. \tag{A.14d}$$

Therefore, the dot product of the vectors $\vec{\nabla}$ and $\vec{D}$ results in Eq. (A.14c). If suitable, the divergence can also be presented in other coordinate spaces, such as *spherical* and *cylindrical* spaces.

Therefore, per Gauss's law, the divergence of the electric field is zero in a charge-free space:

$$\vec{\nabla}\cdot\vec{D} = 0 \rightarrow \vec{\nabla}\cdot\vec{E} = 0. \tag{A.14e}$$

5.0.3 Integral Form of Electric Gauss's Law and Coulomb's Law

Gauss's law in integral form states that the electric charge that produces the flux of the electrostatic field $\vec{D}$, which passes through any closed surface S, is proportional to the total charge located within that surface. This statement is presented analytically in integral form via Maxwell's equation

$$\oint \vec{D}d\vec{S} = \oint \left(\varepsilon\,\vec{E}\right)d\vec{S} = \int \rho_v dV = Q \tag{A.15a}$$

or

$$\oint \vec{E}d\vec{S} = \frac{Q}{\varepsilon}, \tag{A.15b}$$

where $\vec{D} = \varepsilon\,\vec{E}$ and where Q is the total charge within the closed surface of interest.

By applying Eq. (A.15b) for Gauss's law to the model in Fig. A.8, where the positive charge is placed in the center of an imaginary sphere with radius r and surface $S = 4\pi r^2$, the expression of the *electric* field produced by a single charge q_1 is

$$\oint EdS = E\oint dS = E\left(4\pi r^2\right) = \frac{q_1(t)}{\varepsilon}. \tag{A.16a}$$

From this expression, the intensity of the *electric* field $\vec{E}$ is

$$E = \frac{1}{4\pi r^2}\frac{q_1}{\varepsilon}. \tag{A.16b}$$

Furthermore, on the basis of Eq. (A.13), the force between two charges q_1 and q_2 is

$$F_{12} = q_2\vec{E}, \tag{A.17}$$

and substituting Eq. (A.16b) into Eq. (A.17), the expression of the *intensity* of the *mutual electrostatic force* F_{12} between the two charges q_1 and q_2 represents Coulomb's law:

$$F_{12} = \frac{1}{4\pi\varepsilon}\frac{q_1 q_2}{r^2}\ (\mathrm{N}), \tag{A.18}$$

where q_1 and q_2 represent the number of charges in coulombs (C) and r represents the distance (in meters) between the two charges.

In summary, the key difference between the *differential* and *integral* forms of Gauss's law for the *electrostatic* field is that the *differential* form is associated with the *divergence* of the *electric flux* $\vec{D}$ and charges q at an individual point in space, whereas the *integral* form is associated with the *electric* flux through a closed surface that contains the total amount of charges Q.

5.0.4 Electrostatic Field in Isotropic Dielectric Materials

Electrostatic fields in free space (air or vacuum) and a dielectric have different structures. A dielectric contains electric dipoles that are *randomly directed* when the dielectric is not exposed to an *external electric* field. When the dielectric is placed inside an *electric* field, for example, in the case of a dielectric within a polarized parallel-plate capacitor, all dipoles in the dielectric will be polarized (in one direction) by the effect of the *electrostatic forces* caused by the *external* field.

In particular, the electrostatic forces in the dielectric force positively charge in the direction of the external field, whereas negative charges are forced in the opposite direction. This process is called *dielectric polarization*. Figure A.9a shows a polarized dielectric in a capacitor with air by the electric field $\vec{E}_0$ applied between the capacitor's plates, where ρ_s is the *surface charge density* at the C plates.

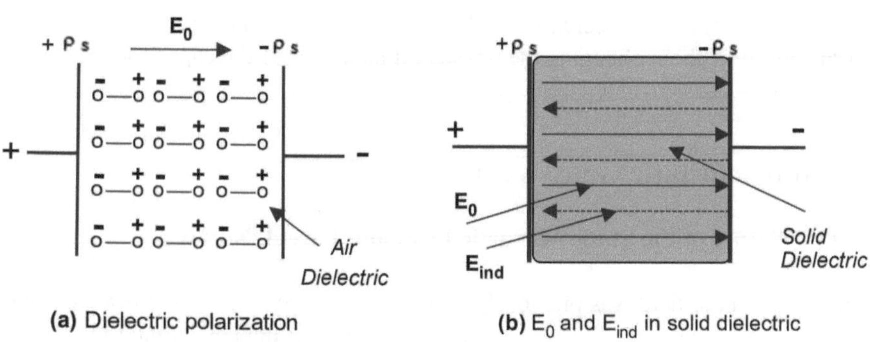

Fig. A.9 Polarization of a dielectric by an external field E_0

As a result of the *dielectric polarization,* a new *electric* field (dashed lines) is induced in the dielectric in the opposite direction to the external field (solid lines). Figure A.9b shows the electric field $\vec{E}_0$ with the *induced electric* field $\vec{E}_{ind}$ in the opposite direction.

The capacitor is characterized by the electric capacitance C as electrical energy storage. In general, the capacitance C of any capacitor is defined as the ratio of the *total charge* Q at its surface of the capacitor's electrode and the *voltage* V applied between the electrodes. In Fig. A.9a, the capacitance of a parallel plate capacitor with air is approximately given as

$$C_0 \equiv \frac{Q}{V} \cong \varepsilon_0 \frac{S}{d}. \tag{A.19}$$

In Eq. (A.19), ε_0 is the *permittivity* of free space (or air) equal to $\approx 8.854 \times 10^{-12}$ F/m, S is the surface of the capacitor's plate, and d is the spacing between the plates.

For the capacitor with a dielectric (case *b* in Fig. A.9b) with the same dimensions S and d as those in case *a*, the capacitance is

$$C = \varepsilon_d \frac{S}{d} = \varepsilon_r \varepsilon_0 \frac{S}{d}, \tag{A.20}$$

where ε_d is the dielectric permittivity equal to $\varepsilon_r\ \varepsilon_0$ and where ε_r is the relative dielectric permittivity $\varepsilon_r \geq 1$.

Comparing the capacitances C_0 and C, expressed by Eqs. (A.19) and (A.20), we can see that the capacitance C is ε_r-times greater than the capacitance C_0 even when both capacitors have the same physical dimensions:

$$C = \varepsilon_r \left(\varepsilon_0 \frac{S}{d} \right) = \varepsilon_r C_0. \tag{A.21}$$

Thus, a capacitance increase can be obtained just by increasing the dielectric permeability without changing the physical dimensions of the capacitor.

6 Magnetostatic Gauss's Law

6.0.1 Nature of the Static Magnetic Field and Lorentz's Law

The *electrostatic* field was previously defined as a space where *electrostatic* forces act on *electric charges*. Similarly, the *static magnetic* field is defined as a space where the magnetic *forces* act on *moving charged particles*. Thus, the *charged particles* experience *magnetic forces* only when they are in *motion*. With respect to this phenomenon, *Lorentz's law* gives the relation between the *magnetic force* $\vec{F}_m$ and *moving charged particle* q as

$$\vec{F}_m = q\left(\vec{v} \times \vec{B}\right), \tag{A.22}$$

where $\vec{v}$ is the *velocity of charged particles* and where $\vec{B}$ is the *magnetic induction* in the space where the particles move. *The magnitude* of the *magnetic force*, expressed by Eq. (A.22), is given as

$$\left|\vec{F}_m\right| = q\left|\vec{v}\right|\left|\vec{B}\right|\sin(\theta), \tag{A.23}$$

where $\left|\vec{v}\right|$ and $\left|\vec{B}\right|$ are the magnitudes of the vectors $\vec{v}$ and $\vec{B}$ and where θ is the angle between these vectors. For $\theta = 90°$, i.e., when the velocity vector $\vec{v}$ is perpendicular to the magnetic induction vector $\vec{B}$, the maximum magnetic force is produced. In contrast, when the charged particles move parallel to the vector $\vec{B}$, the *magnetic force* does not act on the particles at all.

Notably, the fundamental difference between the origins of *electrostatic* and *magnetostatic* fields is that *electric charges are produced an electrostatic field*, whereas the *static magnetic field* is produced by *moving charged particles or steady electric currents*.

Figure A.10 shows examples of *static magnetic* field structures induced by *steady electric currents* in a conductive loop.

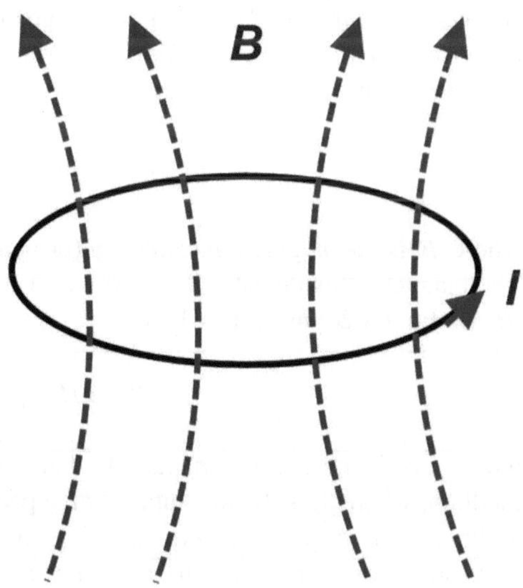

Fig. A.10 Origins of the magnetostatic field from the steady current I in a conductive loop

6.0.2 Differential Form of Gauss's Law for the Magnetostatic Field

In general, the *static magnetic* field and its origin are covered by magnetic Gauss's law, such as the electrostatic field. This law states that the total magnetic flux leaving any closed surface around the magnetic source is always zero. This statement is presented in the differential form of Maxwell's equation:

$$\vec{\nabla}\cdot\vec{B} = 0. \tag{A.24}$$

Equation (A.24) specifies that the *magnetic flux divergence* is zero through an infinitesimal surface *dS* surrounding a point magnetic source. Thus, the idea of magnetic Gauss's law for the *differential* form is that *the magnetic flux divergence is zero because the flux never diverges away from a magnetic source.*

Like in the *electrostatic* field case, the *divergence* of the *magnetic flux* $\vec{B}$ can be expressed in the rectangular Cartesian space as

$$\vec{\nabla}\cdot\vec{B} = \frac{\partial B_x}{\partial x} + \frac{\partial B_y}{\partial y} + \frac{\partial B_z}{\partial z} = 0. \tag{A.25}$$

Additionally, depending on the field configuration, it is suitable to present the field in other coordinate spaces such as *spherical* or *cylindrical* spaces.

6.0.3 Integral Form of Gauss's Law for the Magnetostatic Field

Because the *magnetic flux* does not diverge away from its source (such as a magnet or current loop), Maxwell's *integral* form of Gauss's law for the *magnetic flux* through a closed surface *S* is zero:

$$\oint_0^{4\pi} \vec{H}d\vec{S} = 0 \text{ or } \oint_0^{4\pi} \vec{B}d\vec{S} = 0, \tag{A.26}$$

where $\vec{B}$ is *the magnetic flux density* (or *magnetic inductance*) given in T (Tesla). The magnetic flux density $\vec{B}$ is given as a function of the magnetic field intensity $\vec{H}$ via Eq. (A.2c) in Sect. A.1 as

$$\vec{B} = \mu\vec{H} = \mu_r\mu_0\vec{H}, \tag{A.27}$$

where μ is the magnetic permeability that characterizes the property of a magnetic medium, where μ_r is the medium *relative* permeability, and where μ_0 is the permeability μ_0 in free space equal to $\mu_0 = 4\pi\cdot 10^{-7}\,(\mathrm{H/m})$.

In summary, Gauss's law for a *static magnetic* field states that the *total magnetic flux leaving any closed surface that contains a magnetic source is always zero.* For example, Fig. A.11 shows the zero-*magnetic* flux through a closed surface *S*.

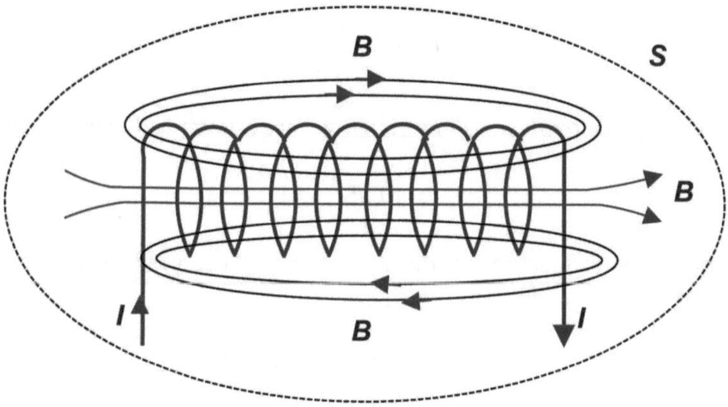

Fig. A.11 Source of static magnetic field and Gauss's law: $\vec{\nabla}\cdot\vec{B} = 0$

Here, the lines of *magnetic flux* $\vec{B}$ do not leave the closed surface *S*, resulting in the flux $\vec{B}$ being zero through the surface *S* as *Maxwell's* equation (A.26) states.

7 Steady Electric Currents and Ohm's Law

Here, we consider charges in *steady motion* that represent *electric currents*. There are several types of electric current: (1) currents of electrons and ions in a vacuum, (2) *electrolytic* currents of *positive* and *negative* ions in *electrolytic* media, and (3) *conduction* currents in conductors and semiconductors caused by the motion of *electrons* and *ions*. In this section, we consider the *conduction* currents.

The number (or amount) *of charges* ΔQ that cross through the surface $\Delta\vec{S}$ in time Δt with velocity $\vec{v}$ can be expressed as

$$\Delta Q = \rho_v \vec{v} \Delta\vec{S} \Delta t \quad (\mathrm{C}), \tag{A.28}$$

where ρ_v is the volume charge density. Next, the *electric* current ΔI is defined as the *time rate* of the charge change ΔQ, and takes into account Eq. (A.28),

$$\Delta I = \frac{\Delta Q}{\Delta t} = \rho_v \vec{v} \Delta\vec{S} \quad (\mathrm{A}). \tag{A.29}$$

From Eq. (A.29), the *surface current density* $\vec{J}$ is defined as the current per surface $\Delta\vec{S}$:

$$\vec{J} = \rho_v \vec{v} \quad \left(\frac{\mathrm{A}}{\mathrm{m}^2}\right). \tag{A.30}$$

Additionally, the *total current* I through surface S is defined in integral form as

$$I = \int \vec{J}\, d\vec{S} \quad (A). \tag{A.31}$$

For most good conductors, the *average velocity* of charges is proportional to the intensity of the *electric* field:

$$|\vec{v}| = -k_e |\vec{E}| \quad \left(\frac{m}{s}\right), \tag{A.32}$$

where the constant k_e is the *charge mobility*. Substituting Eq. (A.32) into Eq. (A.30) results in

$$\vec{J} = -\rho_v k_e \vec{E}, \tag{A.33}$$

and further substituting $\sigma = -\rho_v k_e$ into Eq. (A.33) yields Ohm's law:

$$\vec{J} = \sigma \vec{E} = \frac{\vec{E}}{R'} \quad \left(\frac{A}{m^2}\right), \tag{A.34}$$

where σ and R' are the *specific conductivity* and *specific resistivity* of a material, respectively. The relationship between σ and R' is $\sigma = 1/R'$.

Additionally, Eq. (A.34) is the origin of *Ohm's law*, which governs the electrical process in conducting and isotropic media in addition to other electromagnetic laws. For instance, the well-known *Ohm's law* for lumped electric circuits originates from expression (A.34), which shows that the current I through resistor R is given as

$$I = \frac{\Delta V}{R} \quad (A). \tag{A.35}$$

As an example, Fig. A.12 shows an RLC circuit and the voltage droop ΔV at R.

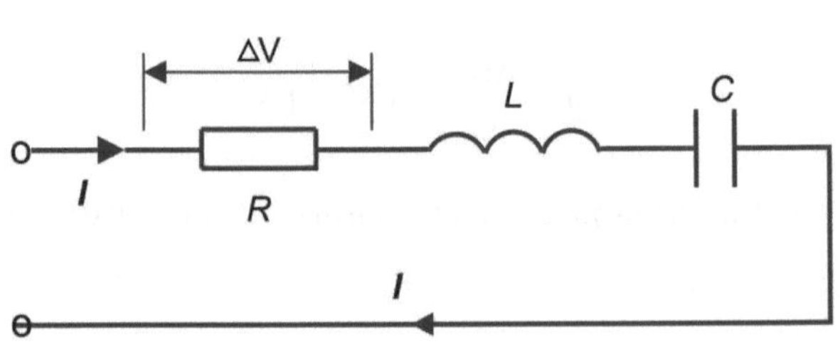

Fig. A.12 A circuit application of a circuit Ohm's law

8 Time-Harmonic Fields

The intensity of stationary *electric* and *magnetic* fields is expressed in complex exponential form or phasor form as

$$E(x,y,z,t) = |\vec{E}| \cdot e^{i\omega t} \tag{A.36a}$$

$$H(x,y,z,t) = |\vec{H}| \cdot e^{i\omega t}, \tag{A.36b}$$

where $|\vec{E}|$ and $|\vec{H}|$ are the magnitudes of the *electric* and *magnetic* fields, respectively, since ωt is the field phase. Furthermore, the time-harmonic form for these fields can be developed from real parts of the complex forms of the fields $\vec{E}$ and $\vec{H}$ as

$$\vec{E}(t) = \mathrm{Re}\{\vec{E}(\ldots)e^{i\omega t}\}; \vec{H}(t) = \mathrm{Re}\{\vec{H}(\ldots)e^{i\omega t}\}, \tag{A.37a, b}$$

and using Euler's identity

$$e^{i(\omega t)} = \cos(\omega t) + i \sin(\omega t), \tag{A.38}$$

Equation (A.37a) has the *time-harmonic* form

$$\begin{aligned} E &= \mathrm{Re}\{E_0 \cos(\omega t) + iE_0 \sin(\omega t)\} \rightarrow \\ E &= E_0 \cos(\omega t). \end{aligned} \tag{A.39}$$

In the same way, we can obtain the *time-harmonic* form for *the magnetic* field from Eq. (A.37b):

$$H = H_0 \cos(\omega t), \tag{A.40}$$

where E_0 and H_0 are the amplitudes of the harmonic vectors $\vec{E}$ and $\vec{H}$, respectively. Additionally, ω is the *angular* frequency equal to $\omega = 2\pi f$ where f is the frequency of a propagating signal.

Accordingly, a *steady-state voltage V* and *steady-state current I* can be presented in the same harmonic form:

$$V = V_0 \cos(\omega t) \text{and} I = I_0 \cos(\omega t). \tag{A.41a, b}$$

8.0.1 Maxwell's Equations for Time-Harmonic Fields

The time derivations of the field eqs. (A.36a), (A.36b) are as follows:

$$\vec{E}(t) = \vec{E}e^{i\omega t} \text{ and } \vec{H}(t) = \vec{H}e^{i\omega t} \tag{A.42a}$$

are given as

$$\frac{\partial}{\partial t}\vec{E} \equiv i\omega\vec{E} \text{ and } \frac{\partial}{\partial t}\vec{H} \equiv i\omega\vec{H} \tag{A.42b}$$

and from Eq. (A.42b), the operator identity is as follows:

$$\frac{\partial}{\partial t} \equiv i\omega. \tag{A.43}$$

On the basis of this operator identity, *Maxwell's* equations (Faraday and Ampere–Maxwell-based) in *differential* form in the *frequency domain* become

$$\vec{\nabla} \times \vec{E} = -\frac{\partial \vec{B}}{\partial t} \equiv -i\omega\vec{B} \quad (\text{Faraday}) \tag{A.44a}$$

$$\vec{\nabla} \times \vec{H} = \vec{J} + \frac{\partial \vec{D}}{\partial t} \equiv \vec{J} + i\omega\vec{D} \quad (\text{Ampere} - \text{Maxwell}). \tag{A.44b}$$

The vector $\vec{J}$ in the Ampere–Maxwell equation is the *surface current density* at a conductive surface $\vec{S}$, which is expressed as

$$\vec{J} = \frac{dI}{d\vec{S}}. \tag{A.45a}$$

Additionally, the electric current I is given in the *integral* form as

$$I = \oint \vec{J} \cdot d\vec{S} = -\frac{d}{dt}\int \rho_v dV \equiv -i\omega \int \rho_v dV \tag{A.45b}$$

and in the *differential* form as

$$I(t) = \frac{dq(t)}{dt} \equiv i\omega q(t). \tag{A.45c}$$

8.0.2 Electromagnetic Power Flow and Pointing Vector

The *electromagnetic power flow* through a surface S is given by the cross-product of the *electric* and *magnetic* fields $\vec{E}$ and $\vec{H}$ as

$$\vec{S} = \vec{E} \times \vec{H}. \tag{A.46}$$

The vector $\vec{S}$ is called the *Poynting* vector. It can be considered in the time and complex domains. Additionally, the *time-average* power flow is defined as half of the real part of the cross-product of the field $\vec{E}$ and the conjugation of the field $\vec{H}$:

$$S_{ave} = \frac{1}{2}\mathrm{Re}\left(\vec{E} \times \vec{H}^*\right) \quad \left(\mathrm{W/m^2}\right). \tag{A.47}$$

For *steady time-harmonic fields*, the *power* flow $\vec{S}_z$ in the +z direction is expressed as

$$\vec{E} = \vec{E}_x \cos(\omega t) \text{and} \left|\vec{H}_y\right| = \frac{1}{\eta}\left|\vec{E}_x\right|\cos(\omega t)$$

and $\vec{S}_z$ is given as

$$\begin{aligned} \left|\vec{S}_z\right| &= \left|\vec{E}_x \times \vec{H}_y\right| = \left[\left|\vec{E}_x\right|\cos(\omega t)\right] \cdot \left[\frac{1}{\eta}\left|\vec{E}_x\right|\cos(\omega t)\right] \rightarrow \\ \left|\vec{S}_z\right| &= \frac{1}{\eta}\left[E^+ \cos(\omega t)\right]^2, \end{aligned} \tag{A.48}$$

where η is the *intrinsic* medium impedance, which is defined as the ratio of the *electric* and *magnetic* field magnitudes:

$$\eta = \left|\frac{\vec{E}}{\vec{H}}\right|. \tag{A.49}$$

Over a complete cycle of the cosine field, the term $[\cos(\omega t)]^2$ in Eq. (A.48) is equal to 1/2, and the *time-average* value of the *Poynting* vector $\vec{S}_z$ is

$$\left|\vec{S}_z\right| = \frac{1}{2\eta}\left[E^+\right]^2. \tag{A.50a}$$

Additionally, in terms of the magnetic field $\vec{H}$, the *time-average* value of the *Poynting* vector magnitude $|\vec{S}_z|$ is

$$|\vec{S}_z| = \frac{\eta}{2}\left[H^+\right]^2. \tag{A.50b}$$

Figure A.13 shows the *electric* field $\vec{E}$ directed in the +x direction since the *magnetic* field vector $\vec{H}$ is directed in the +y direction. Both produce *transverse electromagnetic* (TEM) waves, where the *power flow* or *Poynting vector* $\vec{S}$ is directed in the +z direction. Thus, for the TEM fields, all vectors $\vec{E}$, $\vec{H}$, and $\vec{S}$ are orthogonal to each other.

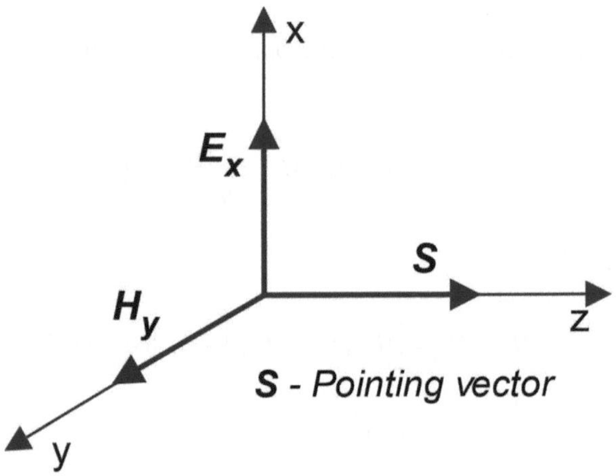

Fig. A.13 Power Poynting vector

9 Wave Propagation in a Source-Free Medium

9.0.1 Wave Equations

Electromagnetic wave propagation in a source-free and *lossless* space (where $\rho = 0$, $J = 0$) is governed by the following *wave equations* (A.51a)–(A.51d), which are developed on the basis of the modified Maxwell's equations (A.49) in source-free space and in the *time* domain [10–13]:

$$\nabla^2 \vec{E} - \mu\varepsilon \frac{\partial^2 \vec{E}}{\partial t^2} = \mu\sigma \frac{\partial \vec{E}}{\partial t} = 0 \tag{A.51a}$$

$$\nabla^2 \vec{H} - \mu\varepsilon \frac{\partial^2 \vec{H}}{\partial t^2} = \mu\sigma \frac{\partial \vec{H}}{\partial t} = 0 \tag{A.51b}$$

or in the *frequency* domain:

$$\nabla^2 \vec{E} + \mu\varepsilon\omega^2 \vec{E} = \nabla^2 \vec{E} + \gamma^2 \vec{E} = i\omega\mu\sigma \vec{E} \tag{A.51c}$$

$$\nabla^2 \vec{H} + \mu\varepsilon\omega^2 \vec{H} = \nabla^2 \vec{H} + \gamma^2 \vec{H} = i\omega\mu\sigma \vec{H}, \tag{A.51d}$$

where γ is the complex propagation constant in a lossy medium equal to $\alpha + i\beta$. α is the attenuation in the medium, and β is the propagation medium constant.

The general solutions of the wave equations for TEM waves in *lossy* media are as follows:

$$\tilde{E}(z,t) = E_0^{\;+} e^{-\gamma z} e^{i\omega t} + E_0^{\;-} e^{\gamma z} e^{i\omega t} \tag{A.51e}$$

$$\tilde{H}(z,t) = H_0^{\;+} e^{-\gamma z} e^{i\omega t} + H_0^{\;-} e^{\gamma z} e^{i\omega t}, \tag{A.51f}$$

where the field constants E_0^+, E_0^- , H_0^+, and H_0^- can be determined by the boundary conditions for the problem of interest. Equation (A.51e), (A.51f) represents the propagation and time-wiring fields $\vec{E}(z,t)$ and $\vec{H}(z,t)$ in the +z and −z directions, respectively. The solutions of the *wave equations* in a *lossless* medium (where $\alpha = 0$) in only the +z directions are given in complex exponential forms as

$$\tilde{E}(z,t) = E^{+} e^{-i\beta z} e^{i\omega t} \tag{A.52a}$$

$$\tilde{H}(z,t) = H^{+} e^{-i\beta z} e^{i\omega t}. \tag{A.52b}$$

Additionally, the same waves can be expressed in real-time harmonic form as

$$E(z,t) = E_0 \cos(\omega t - \beta z) \tag{A.52c}$$

$$H(z,t) = H_0 \cos(\omega t - \beta z) \tag{A.52d}$$

Figure A.14 shows the propagation of the time-harmonic *electric* wave characterized by the *propagation speed* ($\upsilon_p = \Delta z/\Delta t$) and the wavelength λ, both of which are shown in the present wave plot.

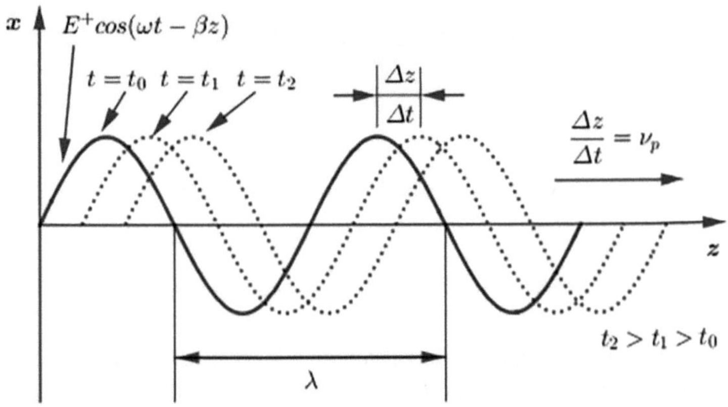

Fig. A.14 Time-harmonic wave

When the wave travels a small distance Δz and for a small duration of time Δt, the wave phase does not change significantly when it is looked at from an observing point located on the same wave. Therefore, we can approximate the wave phase in Eq. (A.52c), (A.52d) as close to zero (i.e., $\omega\Delta t - \beta\Delta z = \Delta\phi \cong 0$). The wave velocity, ν_p is defined as

$$v_p = \frac{\Delta z}{\Delta t} = \frac{\omega}{\beta}\left(\frac{\mathrm{m}}{\mathrm{s}}\right), \tag{A.53}$$

where ω is the *angular* frequency.

Additionally, the *wave velocity* ν_p in a medium other than free space is expressed in terms of the medium parameters μ and ε as

$$v_p = \frac{1}{\sqrt{\mu\varepsilon}} = \frac{1}{\sqrt{\mu_r\mu_0\varepsilon_r\varepsilon_0}} = \frac{c}{\sqrt{\mu_r\varepsilon_r}} \qquad (\mathrm{m/s}), \tag{A.54}$$

where $\mu_r > 1$ and $\varepsilon_r > 1$ are the medium *relative permeability* and *relative permittivity,* respectively, and μ_0 and ε_0 are the parameters of the free space medium.

The following important parameter of harmonic waves (or harmonic signals) is the frequency that is defined via the wave velocity ν_p and the corresponding wavelength λ:

$$f - \frac{v_p}{\lambda} \quad (\mathrm{Hz}). \tag{A 55}$$

Considering $\omega = 2\pi f$ and $\nu_p = c$ in free space and combining Eqs. (A.53) and (A.55), the expression for the *propagation* constant β becomes

$$c = \frac{\omega}{\beta} = \frac{2\pi f}{\beta} = \frac{2\pi(c/\lambda)}{\beta} \to \beta = \frac{2\pi}{\lambda} \quad \left(\mathrm{rad}/\mathrm{m}\right). \tag{A.56}$$

Additionally, the speed c in free space (i.e., speed of light) is defined in terms of the free-space electromagnetic parameters μ_0 and ε_0 as

$$c = \frac{1}{\sqrt{\mu_0 \varepsilon_0}} \qquad (\mathrm{m/s}), \tag{A.57}$$

where $\mu_0 = 4\pi \cdot 10^{-7}$ H/m and $\varepsilon_0 = 8.854 \cdot 10^{-12}$ F/m. Introducing these quantities into Eq. (A.57) results in a speed of light of $c \approx 3 \cdot 10^8$ m/s.

The next important medium parameter is the *intrinsic impedance* η, which is defined as

$$\eta = \frac{E}{H} = \sqrt{\frac{\mu_r \mu_0}{\varepsilon_r \varepsilon_0}} = \eta_0 \sqrt{\frac{\mu_r}{\varepsilon_r}}, \tag{A.58}$$

where η_0 is the *intrinsic* impedance in free space equal to $\sqrt{\mu_0 / \varepsilon_0} = 120\pi$.

Wave propagation in free-space media and closed media is discussed in, Appendices B and C: "Electromagnetic Radiators and Waves in Free and Open Space" and "Electromagnetic Waves guided by Transmission lines and Waveguides", respectively.

10 Media Properties and Boundary Conditions

10.0.1 Types of Media

The set of the following expressions is related to *linear*, *homogeneous*, and *isotropic* media, where the parameters ε, μ, and σ are constants:

$$\vec{D} = \varepsilon \vec{E}; \vec{B} = \mu \vec{H}; \vec{J} = \sigma \vec{E}. \tag{A.59}$$

For *nonlinear* media, the parameters ε, μ, and σ are not constants but are generally dependent on the intensity of fields E and H:

$$\varepsilon = \varepsilon(E), \mu = \mu(H), \sigma = \sigma(E). \tag{A.60}$$

An example of *nonlinear* media is *ferromagnetic* materials, where the magnetic flux $\vec{B}$ does not have a linear relationship with the magnetic field intensity $\vec{H}$ because ferromagnetic materials exhibit hysteresis.

There are also *anisotropic* media, where the vectors $\vec{D}$, $\vec{B}$, and $\vec{J}_c$ are not parallel to the corresponding vectors $\vec{E}$ and $\vec{H}$. Examples of *anisotropic* media are *ferrite* materials.

In the end, *inhomogeneous* media have parameters ε, μ, and σ that vary with location. These media parameters can be presented in general form by the following functions:

$$\mu = \mu(x,y,z), \varepsilon = \varepsilon(x,y,z),\ \sigma = \sigma(x,y,z). \tag{A.61}$$

10.0.2 Boundary Conditions Between Air and Perfect Electric Conductor (PEC)

The *boundary conditions* are associated with electromagnetic properties at the boundary between two different media (for example, air and a perfect conductor (PEC)). Figure A.15 shows field vectors $\vec{D}$, $\vec{E}$, $\vec{B}$, $\vec{H}$ at the boundary between medium #1 (air) with the parameters ε_1, μ_1, and σ_1 and medium #2 (PEC) with the parameters ε_2, μ_2, and σ_2, whose *electric conductivity is* $\sigma_2 = \infty$.

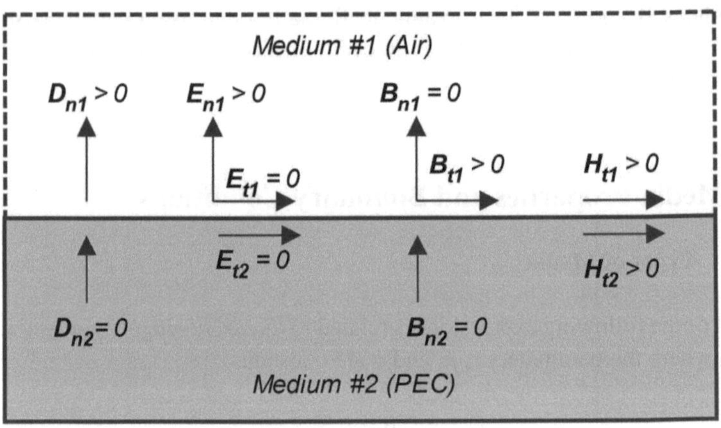

Fig. A.15 E and H fields at the boundary between air and PEC

Because the *conductivity* of the PEC, medium #2, is infinity, all electric field components ($\vec{D},\vec{E}$) inside the conductor are zero, but in medium #1 (in air), the *normal* components of the *electric field* on the PEC are not zero ($D_{n1} \neq 0$ and $E_{n1} \neq 0$) since both *tangential electric* components are zero ($D_{t1} = 0$ and $E_{t1} = 0$).

The situation for the *magnetic field* in media #1 (air) is quite the opposite: the *normal components* to the conductor surface, outside and inside the conductor, are zero ($B_{n1,2} = 0$ and $H_{n1,2} = 0$), whereas the *tangential components* in both media are nonzero, $B_{t1,2} \neq 0$ and $H_{t1,2} \neq 0$.

In general, when both media #1 and #2 are not perfect conductors nor air, with arbitrary parameters (ε_1, μ_1, σ_1) and (ε_2, μ_2, σ_2), the tangential and normal field components exist at the boundary between these two media (i.e., $D_{n1} \neq 0$, $E_{n1} \neq 0$,

$B_{t1} \neq 0$, $H_{t1} \neq 0$). In this case, the magnitudes and directions of the *D*, *E*, *B*, and *H* fields depend on the media parameters $(\varepsilon_1, \mu_1, \sigma_1)$ and $(\varepsilon_2, \mu_2, \sigma_2)$.

Readers interested in further studies of electromagnetic fields in linear, homogeneous, and isotropic media are referred to references [14–16].

References

1. W. Greiner, J. Reinhart, *Quantum Dynamics*, 4th edn. (Springer, 2008)
2. E. Kreyszig, *Advanced Engineering Mathematics*, 8th edn. (John Wiley & Sons, New York, 1999)
3. D.G. Dudley, Mathematical Foundations for Electromagnetic Theory (IEEE Press, 1994)
4. D. Fleischer, A Student's Guide to Maxwell's Equations (Cambridge University Press, 2008)
5. H.J. Eom, Primary Theory of Electromagnetics (Springer, 2013)
6. W.H. Hayt, JR, J.H. Buck, Engineering Electromagnetics, 9th edn. (McGraw-Hill Education, 2018)
7. E.J. Rothwell, M.J. Cloud, Electromagnetics (CRS Press, 2018)
8. J. Franklin, Classical Electromagnetism (Pearson, 2005)
9. J.D. Jackson, *Classical Elcctrodynamics*, 3rd edn. (Wiley, 1999)
10. D. Fleisch, L. Kinnaman, *A Student's Guide to Waves* (Cambridge University Press, 2015)
11. S. Ramo, J.R. Whinnery, T.V. Duzer, Fields and Waves in Communications Electronics, 3rd edn. (J. Wiley & Sons, 1994)
12. D.K. Cheng, Field and Wave Electromagnetics, 2nd edn. (Pearson, 1989)
13. P. Lorrain, D.R. Corson, Electromagnetic Fields and Waves (W.H. Freeman Company, 1988)
14. C.A. Balanis, Advanced Engineering Electromagnetics (John Wiley & Sons, New York, 1989)
15. R.K. Wangsness, Electromagnetic Fields, 2nd edn. (Wiley, 2007)
16. H.A. Hermann, J.R. Melcher, Electromagnetic Fields and Energy (Prentice-Hall, 1989)

Appendix B. Electromagnetic Radiators and Waves in Free and Open Space

1 Introduction

As noted in *Appendix A*, the essential electromagnetic radiation moves *electrical* charges along conductors in the form of electrical currents when conductors act as electromagnetic (EM) radiators. The elemental EM radiator is a small *electric* linear radiator and a small *magnetic* loop (or magnetic dipole). "Small" refers to the length of a *linear* radiator or the *circumference* of a magnetic loop, which is much smaller than the wavelength of the radiating signal. The small radiators are integral parts of larger EM radiators, linear and loop. Both radiators (small and large ones) and their radiations will be the subject of this appendix.

Many useful studies have influenced the writing of this appendix, such as references [1–12], which we refer to readers interested in further reading of electromagnetic wave phenomena.

2 Elemental Electric Radiator

The EM radiation source involves moving charges or currents that flow along a small linear conductor called the *electric* dipole. The small electric radiator has a length $l \leq 0.1\ \lambda_0$, where λ_0 refers to the signal wavelength in free space. Figure B.1 shows the *electric* and *magnetic* field components of a small *electric* radiator as vectors in a spherical coordinate system [1–5].

M. Pajovic, *Fundamentals of EMI Phenomena in Radio Communication Systems*, https://doi.org/10.1007/978-3-031-88748-2

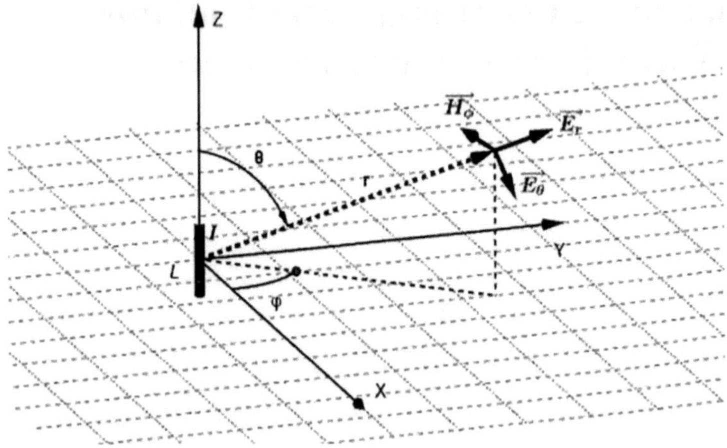

Fig. B.1 Small electric radiator in free space

To determine the field components shown in Fig. B.1, a first step is to choose the magnetic field component H_ϕ as a solution of the equation:

$$\vec{B} = \mu\vec{H} = \vec{\nabla}\times\vec{A} \rightarrow \vec{H} = \frac{1}{\mu}\vec{\nabla}\times\vec{A}, \tag{B.1}$$

where $\vec{A}$ is the *magnetic vector potential* associated with a linear current source long *dl* that carries a current *I*. The components of the $\vec{A}$ vector are given in spherical coordinates (*z*, *r*, θ, ϕ) as

$$\vec{A}_z = \vec{a}_z\frac{\mu Idl}{4\pi}\left(\frac{e^{-i|\beta|r}}{r}\right); \vec{A}_r = \vec{A}_z\cos(\theta); \vec{A}_\theta = \vec{A}_z\sin(\theta); \vec{A}_\phi = 0, \tag{B.2a}$$

where β is the propagation constant of the propagating medium, i.e., in air.

Hence, substituting Eq. (B.2a) into Eq. (B.1), the solution of Eq. (B.1) for the field component H_ϕ is

$$\vec{H}_\phi = -\vec{a}_\phi\frac{I_{av}dl\beta^2}{4\pi}\sin(\theta)\left(\frac{1}{i\beta r}+\frac{1}{(i\beta r)^2}\right)e^{-i\beta r}. \tag{B.2b}$$

The next step is to determine the E field components $\vec{E}_r$ and $\vec{E}_\theta$ on the basis of Ampere's law:

$$\vec{E} = \frac{1}{i\omega\varepsilon}\vec{\nabla}\times\vec{H}. \tag{B.3}$$

Next, substituting Eq. (B.2b) into Eq. (B.3) leads to the following solutions for $\vec{E}_r$ and $\vec{E}_\theta$:

$$\vec{E}_r = \vec{a}_r \frac{I_{av} dl \beta^2}{4\pi} \eta_0 \cos(\theta) \left(\frac{1}{(i\beta r)^2} + \frac{1}{(i\beta r)^3} \right) e^{-i\beta r} \tag{B.4a}$$

$$\dot{E}_\theta = \vec{a}_\theta \frac{I_{av} dl \beta^2}{4\pi} \eta_0 \sin(\theta) \left(\frac{1}{i\beta r} + \frac{1}{(i\beta r)^2} + \frac{1}{(i\beta r)^3} \right) e^{-i\beta r}, \tag{B.4b}$$

where $\vec{E}_\phi = 0$.

The parameters of the present equations are the following:

$\eta_0 = 120\pi\ \Omega$, the intrinsic impedance of free space

I_{av}= *Average current* at the radiator surface in amperes

l= Length of the radiator in meters

r=Radial distance (in meters) from the radiator's surface

β= Propagation constant in air, equal to → $\beta = 2\pi/\lambda_0$

λ_0= Signal wavelength in air, equal to →$\lambda_0 = c/f$

c= Speed of the light, equal to →$c = 3\ 10^8$ m/s

f= Signal frequency

θ= *Elevation* angle of the radiation

ϕ= *Azimuth* angle of the radiation

To estimate the intensity of EM emissions created by a small *electric* radiator, the above equations can be helpful. For example, a short and isolated wire in free space can act as a small *electric* radiator (dipole). The current distribution I at the small radiator has an approximately triangular form, i.e., the maximum I is at the radiator center, whereas the zero current is at the end of the radiator. Thus, the average current along a linear radiator is $I_{av} = \frac{I_{max}}{2}$, where I_{max} is the maximum current at the radiator center.

Notice from Eqs. (B.2b) and (B.4b) that the fields $\vec{H}_\phi$ and $\vec{E}_\theta$ have a maximum intensity at the elevation angle $\theta = \pi/2$ (*broadside* to the radiator), which is a worst-case scenario regarding EMI emissions. Additionally, according to Eq. (B.2b), the *electric* field $\vec{E}_r$ vanishes rapidly with increasing radial distance r from the radiator. Hence, to keep minimal EMI emissions, the conclusion from Eqs. (B.2b), (B.4a), and (B.4b) is to keep the radiator length l and its current I as small as possible.

3 Elemental Magnetic Radiator

Another elemental cause of *electromagnetic* radiation is the current flow in a small conducting loop. This case, shown in Fig. B.2, is based on a small *magnetic* radiator with a circumference C, which is smaller than the signal wavelength, i.e., $C = 2\pi a \leq 0.3\ \lambda_0$, where a is the loop radius.

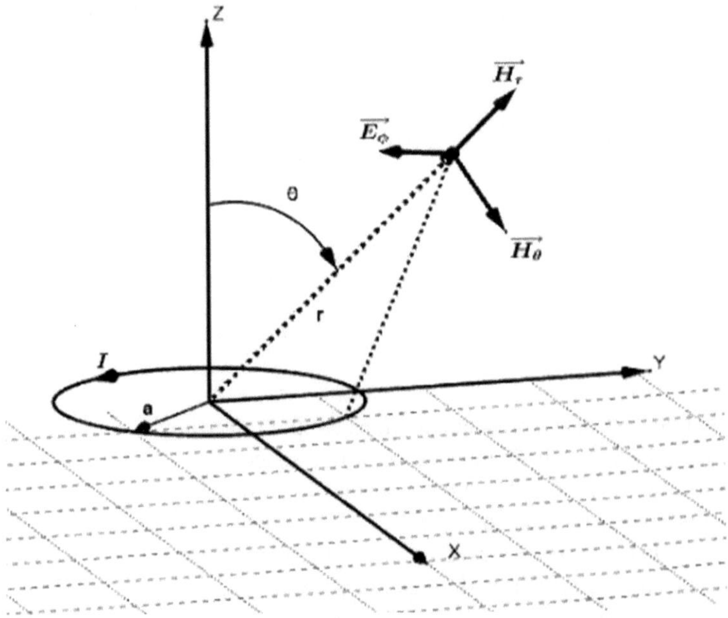

Fig. B.2 Small magnetic radiator in free space

The *magnetic vector potential* for this small loop is given in spherical coordinates as [1–5]:

$$\vec{A} \cong \vec{a}_\phi \frac{\mu IS}{4\pi r^2} \sin\theta\left(1+i\beta r\right)e^{-i\beta r}. \tag{B.5}$$

The intensity of the magnetic fields $\vec{H}_\theta$ and $\vec{H}_r$ can be determined by substituting Eq. (B.5) into Eq. (B.1), i.e., $\vec{H} = \frac{1}{\mu}\vec{\nabla}\times\vec{A}$, whereas the intensity of the electric field $\vec{E}_\phi$ can be determined by incorporating the results of H into Eq. (B.3), i.e., $\vec{E} = \frac{1}{i\omega\varepsilon}\vec{\nabla}\times\vec{H}$ (Ampere's law). The final result is a set of the following equations:

$$\vec{E}_\phi = \vec{a}_\phi \frac{i\omega\mu_0 IS\beta^2}{4\pi}\sin\left(\theta\right)\left(\frac{1}{i\beta r}+\frac{1}{\left(i\beta r\right)^2}\right)e^{-i\beta r} \tag{B.6a}$$

$$\vec{H}_\theta = -\vec{a}_\theta \frac{i\omega\mu_0 IS\beta^2}{4\pi\eta_0}\sin\left(\theta\right)\left(\frac{1}{i\beta r}+\frac{1}{\left(i\beta r\right)^2}+\frac{1}{\left(i\beta r\right)^3}\right)e^{-i\beta r} \tag{B.6b}$$

$$\vec{H}_r = -\vec{a}_r \frac{i\omega\mu_0 IS\beta^2}{4\pi\eta_0} 2\cos(\theta)\left(\frac{1}{(i\beta r)^2} + \frac{1}{(i\beta r)^3}\right)e^{-i\beta r}. \tag{B.6c}$$

Here, S is the loop area in square meters (S $=a^2\pi$), where a is the loop radius, whereas the other parameters from Eqs. (B.6a), (B.6b) are the same as those for Eqs. (B.2b) and (B.4a), (B.4b). Additionally, it was assumed that the current distribution I along the small loop is uniform. The above equations can be helpful for the estimation of the intensity of EMI emissions from a small *magnetic* loop.

Note that the field components E_ϕ and H_θ have a maximum intensity at an elevation angle $\theta = \pi/2$ (normal to the loop surface). Additionally, as seen from Eq. (B.6c), the *magnetic* field component $\vec{H}_r$ vanishes rapidly with the radial distance r from the loop. Note that the field of a small loop does not depend on a loop shape but only on the loop area S. Thus, per Eq. (B.3), we need to keep the loop area S and current I as small as possible to obtain zero or minimal EMI emissions from the current loop.

4 Near-Field Versus Far-Field Regions

According to the electromagnetic boundary conditions at the surface of a perfect conductor (PEC), as discussed in *Appendix A*, the *tangential electric* field is zero, $E_t \approx 0$, whereas the *normal* electric component is not zero, $E_n \neq 0$. The opposite is true for the *magnetic* field: its *normal* component is zero, $H_n = 0$, whereas the *tangential* component is not zero, $H_t \neq 0$. Thus, only E_n and H_t components exist in air at the surface of a perfect conductor.

Furthermore, when electromagnetic waves leave the surface of the EM radiator, they travel through three *regions* in space: the *reactive near-field* region, the *radiating* (Fresnel) *near-field* region, and the *far-field* (Fraunhofer) region. According to *Maxwell's* criterion, the theoretical boundary between the *near-field* and *far-field* regions lies at the radial distance $\lambda_0/2\pi$ from the radiator's surface. For elemental *electric* radiation to distance $\lambda_0/2\pi$, the higher exponents $\frac{1}{(r\beta_0)^2}$ and $\frac{1}{(r\beta_0)^3}$ dominate in the *reactive near-field* region, and Eqs. (B.2b) and (B.4b) obtain simpler forms for $\theta = \pi/2$:

Near field of a small *electric* radiator

$$\left|E_{\theta e}^{near}\right| = \eta_0 \frac{I_{av}\, l\lambda_0}{8\pi^2 r^3} \cong \frac{4.8 I_{av}\, l\lambda_0}{r^3} \quad \left(\frac{V}{m}\right) \tag{B.7}$$

$$\left|H_{\phi e}^{near}\right| = \frac{I_{av}\, l}{4\pi r^2} \quad \left(\frac{A}{m}\right), \tag{B.8}$$

where $\eta_0 = 377\ \Omega$, I_{av} is the average current in amperes, l is the length of the *electric* radiator in meters, λ_0 is the signal wavelength in free space in meters, and r is the radial distance of an observing point in space in meters.

Associated with the small *magnetic* loop, the higher exponents $\frac{1}{(r\beta_0)^2}$ in Eq. (B.6a) and $\frac{1}{(r\beta_0)^3}$ in Eq. (B.6b) dominate in the *reactive* near-field region, and the present equations obtain simpler forms:

Near field of a small current loop (*magnetic*) radiator

$$\left|E_{\phi m}^{near}\right| = \eta_0 \frac{SI}{2\lambda_0 r^2} \left(\frac{V}{m}\right) \tag{B.9}$$

$$\left|H_{\theta m}^{near}\right| = \frac{Sl}{4\pi r^3} \left(\frac{A}{m}\right), \tag{B.10}$$

where S is the loop area in m^2, I is the current in the loop in amperes, and r is the radial distance in space from the loop center in meters.

Furthermore, the *wave impedances* Z_{we} and Z_{wm} associated with the *electric* and *magnetic* fields in the *reactive near-field* region can be expressed as

$$Z_{we} = \left|\frac{E_{\theta e}^{near}}{H_{\phi e}^{near}}\right| \tag{B.11}$$

$$Z_{wm} = \left|\frac{E_{\phi m}^{near}}{H_{\theta m}^{near}}\right|. \tag{B.12}$$

By substituting Eqs. (B.7)–(B.8) into Eq. (B.11) and Eqs. (B.9)–(B.10) into Eq. (B.12), the *wave impedances* for small *electric* (Z_{we}) and *magnetic* radiators (Z_{wm}) in the *near-field* region are

$$Z_{we} = \eta_0 \frac{\lambda_0}{2\pi r} \tag{B.13}$$

$$Z_{wm} = \eta_0 \frac{2\pi r}{\lambda_0}. \tag{B.14}$$

Furthermore, for the radial distance $r > \lambda/2\pi$ (*far-field* region), the greater exponents in the field eqs. (B.2b), (B.4b), and Eqs. (B.6a), (B.6b) can be ignored except for the exponent $\sim \frac{1}{r}$, which leads to simplified expressions for the *electric* and *magnetic* fields for $\theta = \pi/2$:

Far-field region of a small *electric* radiator

$$\left|E_{\theta e}^{far}\right| = \eta_0 \frac{I_{av} l}{2\lambda_0 r} \left(\frac{V}{m}\right) \tag{B.15a}$$

$$\left|H_{\phi e}^{far}\right| = \frac{I_{av} l}{2\lambda_0 r} \left(\frac{A}{m}\right) \tag{B.15b}$$

Far-field region of a small loop radiator

$$\left|E_{\phi m}^{far}\right| = \eta_0 \frac{\pi SI}{\lambda_0^2 r} \left(\frac{V}{m}\right) \tag{B.16a}$$

$$\left|H_{\phi m}^{far}\right| = \frac{\pi SI}{\lambda_0^2 r} \left(\frac{A}{m}\right) \tag{B.16b}$$

Furthermore, substituting Eqs. (B.15a), (B.15b)–(B.16a), (B.16b) into Eqs. (B.11) and (B.12), the *wave impedance* for the *far-field* region becomes

$$Z_{we} = Z_{wm} = \eta_0 \approx 377\,\Omega. \tag{B.17}$$

Let's see now how the *wave impedance* behaves through all the radiating regions. On the basis of Eqs. (B.13), (B.14), and (B.17), plots for *wave impedance* Z_{we}, Z_{wm}, and η_0 are created, as shown in Fig. B.3.

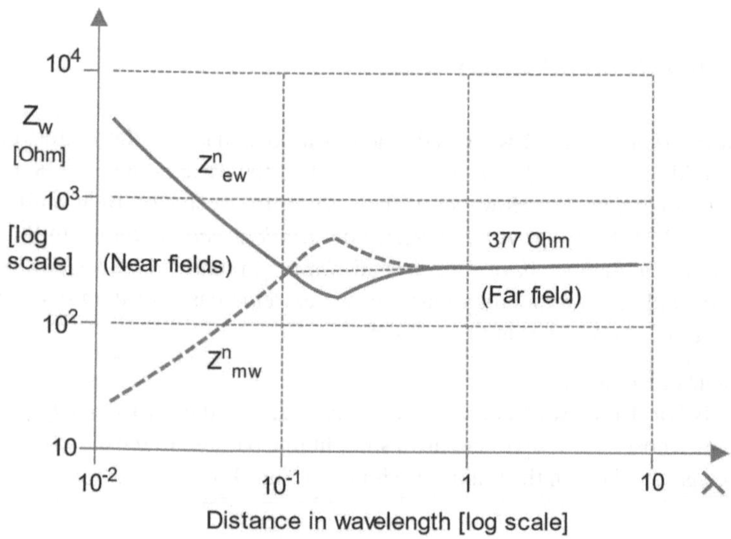

Fig. B.3 Electric and magnetic wave impedances

The present plots show that the *electric* radiator is associated with a *high wave impedance* in the *near-field* region (for $r < 0.1\ \lambda$), where Z_{we} decreases as r approaches the *transition* region (approximately $\lambda/6 < r < 1\ \lambda$). In contrast, a *magnetic* radiator is associated with a *low wave impedance* in the near-field region, where the impedance Z_{wm} increases as r approaches the transition region. Thus, *electric* and *magnetic* fields can be considered separately in the *near- field* region.

Figure B.3 represents the three field regions: the *near-field* region, transition region, and the *radiation far-field* region. Interestingly, the *magnetic* field is associated with a *higher wave impedance* than the *electric* field in the *transition* region.

Hence, in the *far-field* region (for $r > \lambda_0$), both radiators (*magnetic* and *electric*) produce a uniform plane field structure or *transverse electromagnetic* (TEM) waves in air with a unique wave impedance of $120\ \pi \approx 377$ ohms, as shown in Fig. B.3. The structure of the TEM waves is shown in Fig. B.4, where the field vectors $\vec{E}$ and $\vec{H}$ are in phase and orthogonal to each other.

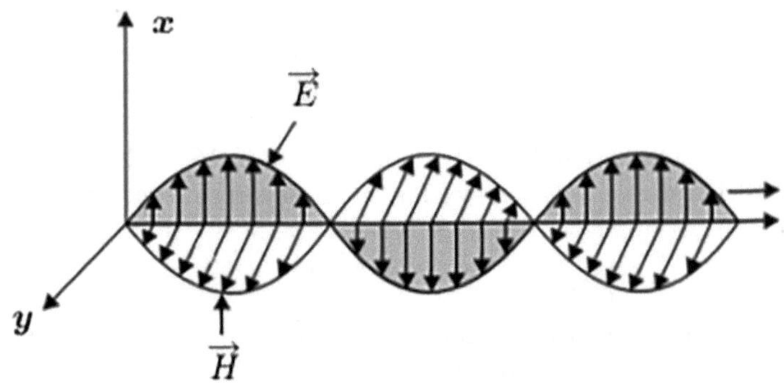

Fig. B.4 TEM waves (fields E and H in phase and orthogonal to each other)

5 Polarization of Plane Waves

The polarization of a TEM wave is defined by a trace made by the vector tip of the *electric* field in the wave propagation direction. If the trace is linear, it is a case of *linearly polarized* waves. However, if the trace of the vector tip rotates around the propagation line, forming *circular-helical* or *elliptical-helical* curves in the propagation direction, then we have *circular* and *elliptical* polarized waves, respectively. The *linear* and *circular* polarizations are two extreme cases of *elliptical* polarization, as will be shown later in this section.

(a) Linear Polarization

Figure B.5(a, b) shows field vectors $\vec{E}_x$ and $\vec{E}_y$ of the two *linearly polarized* waves whose vector tip traces are a straight line parallel to the direction of wave propagation, i.e., on the z-axis, as shown in Fig. B.5.

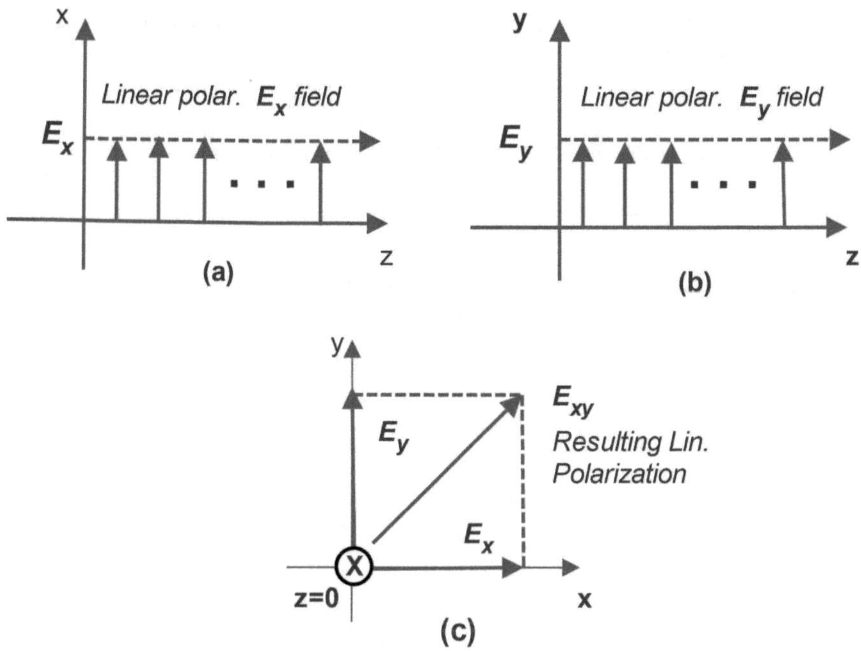

Fig. B.5 (**a**) and (**b**) Linearly polarized E_x and E_y fields and (**c**) resulting linear polarization

Additionally, Fig. B.5c shows the $\vec{E}_{xy}$ field vector of a *linearly* polarized wave at the start point of wave propagation (at $z = 0$) as a superposition of the two linearly polarized fields, $\vec{E}_x$ and $\vec{E}_y$. Therefore, on the basis of vector theory, the field $\vec{E}$ can be considered a superposition of the two orthogonal fields $\vec{E}_x$ and $\vec{E}_y$, which are in the *in-space quadrature* and in-time *phases*. The expression of the instantaneous field $\vec{E}_{xy}$ is given as

$$\vec{E}_{xy}(z,t) = \vec{a}_x E_x \cos(\omega t - \beta z) + \vec{a}_y E_y \cos(\omega t - \beta z), \tag{B.18}$$

where β is the propagation constant in the z-direction.

The fact that two waves orthogonal to each other may independently propagate in the same direction is effectively exploited in *polarization diversity* to minimize the undesired effects of signal fading and electromagnetic interference (EMI) in multipath propagation radio-wireless channels. This technique is discussed in chapter "Overview of Broadband Radio Systems with High EMI Immunity".

(b) Circular and Elliptical Polarizations

As mentioned above, when the tip of the electric field vector rotates circularly around the propagation line, forming a circular-helical trace in space, this is the case for *circular* polarization. The circular polarization originates from a combination of the two *in-space quadrature* fields shown in Fig. B.5c. Additionally, the fields are out of phase by 90°. Accordingly, the instantaneous *circularly polarized* field $\vec{E}_c$ can be expressed as

$$\begin{aligned}\vec{E}_c(z,t) &= \vec{a}_x E_x \cos(\omega t - \beta z) + \vec{a}_y E_y \cos\left(\omega t - \beta z - \frac{\pi}{2}\right) \\ &= \vec{a}_x E_x \cos(\omega t - \beta z) + \vec{a}_y E_y \sin(\omega t - \beta z)\end{aligned} \tag{B.19}$$

when the field amplitudes E_x and E_y are equal ($E_x = E_y$). However, if $E_x \neq E_y$, it is then the case of *elliptical* polarization.

Figure B.6a shows the circular polarization, where $E_x = E_y$ at $z = 0$.

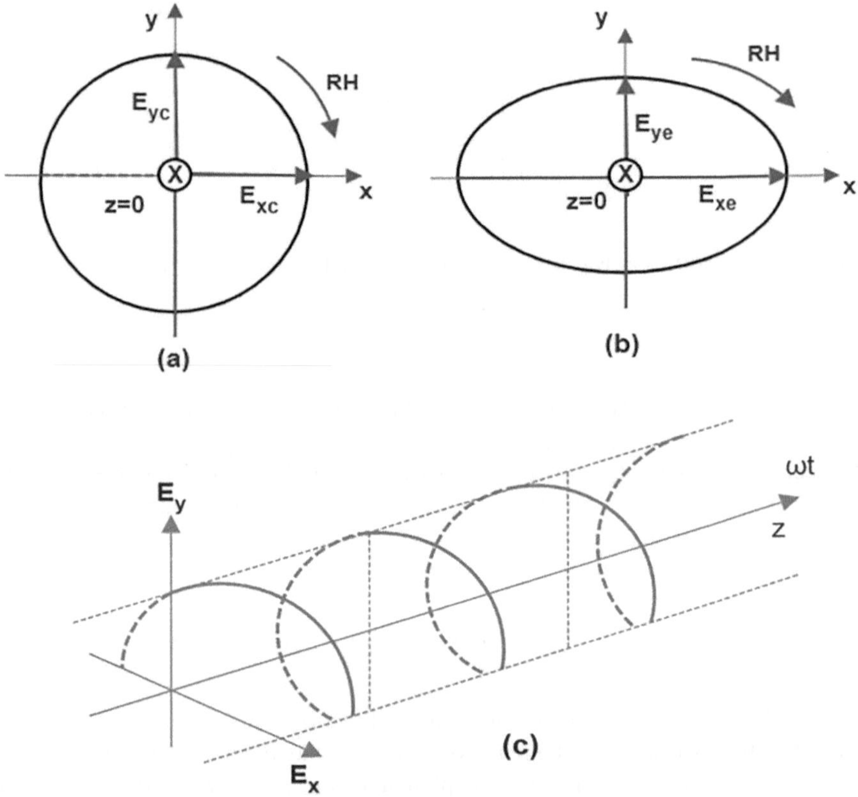

Fig. B.6 (**a**) Circular polarization when $E_{xc} = E_{yc}$, (**b**) elliptical polarization when the magnitude $E_{xe} \neq E_{ye}$ and (**c**) Trajectory of the tip of the vector E_{xy} in free space

For this case, the instantaneous angle α, in which the field $\vec{E}_{xy}(0,t)$ makes the z-axis, is $\alpha = \omega t$. Here, the $\vec{E}_{xy}$ field rotates with a uniform angular velocity ω in a *clockwise* direction. The *clockwise* rotation of the field $\vec{E}_{xy}$ vector is associated with the *right-hand* (*RH*) rule: the fingers of the right hand follow the rotation of the vector $\vec{E}_{xy}$, and the thumb points in the direction of wave propagation.

Additionally, there is *left-hand* (LH) circular polarization that is the opposite of the RH polarization when the $\vec{E}_{xy}$ field vector rotates in a *counterclockwise* direction. This happens when the amplitudes of the field components $\vec{E}_x$ and $\vec{E}_y$ have opposite polarities. Therefore, the expression for an RH-polarized wave is

$$\vec{E}_{RH}(z,t) = \vec{a}_x E_x \cos(\omega t - \beta z) + \vec{a}_y E_y \sin(\omega t - \beta z), \quad \text{(B.20a)}$$

whereas the expression for an *LH*-polarized wave is

$$\vec{E}_{LH}(z,t) = \vec{a}_x \cos(\omega t - \beta z) - \vec{a}_y E_y \sin(\omega t - \beta z). \quad \text{(B.20b)}$$

Typically, when the fields $\vec{E}_x$ and $\vec{E}_y$ have different amplitudes that are out of phase, this results in *elliptical polarization.* Figure B.6b shows this example.

Figure B.6a, b shows that when one of the field components is equal to zero, $E_x = 0$ or $E_y = 0$, the circular or elliptical polarization becomes linear. Additionally, linear polarization can be achieved when both the x- and y-field components are in phase or in the contraphase (when magnitude $E_x \neq E_y$), as previously noted. For the maximum reception of a linearly polarized wave, the receiving antenna should have a polarization parallel to the polarization of a transmitting antenna. With respect to the circular polarization and maximal reception, the circular polarizations of the receiving antenna and transmitting antenna should have the same direction.

From the EMI perspective, the above requirements are quite the opposite: the polarizations of the EMI source and a potential EMI receptor should differ as much as possible. For example, the minimum reception of an EMI plane wave, when one is linearly polarized, is achieved when the source and receptor polarizations are orthogonal to each other. With respect to circular polarization, a source and a receptor should have polarizations in opposite directions, i.e., the minimum reception occurs when the EMI source is RH-polarized since the EMI receptor is LH-polarized or vice versa.

6 Long Linear (Wire) Radiators

Wire radiators with a diameter much smaller than the wavelength of interest and lengths comparable to $\lambda_0/2$ or larger are classified as *large* radiators. In particular, the radiator length refers to the length of a linear radiator or the circumference of a radiating loop, as well as the dimensions of 2D or 3D radiators. Note that a large loop radiator with circumference $\geq\lambda_0/4$ is an *electric* radiator, not a *magnetic* radiator. Thus, as discussed previously in this appendix, only a small current loop behaves like a *magnetic* radiator or *magnetic* dipole. Additionally, the *large* radiators are considered *distributed* circuits, i.e., *transmission lines*, as discussed in Appendix C. Like a *transmission line*, a large radiator can be analyzed as a composition of small (elemental) radiators, which are discussed above in this section. More details about linear radiators (long wire antennas) can be found in [2, 6, 7].

The half-wave dipole with the *sine-current* distribution *I(x)* and far-field horizontal-plane radiation pattern shown in Fig. B.7a, b are commonly used as element antennas for more complex larger antenna systems (arrays).

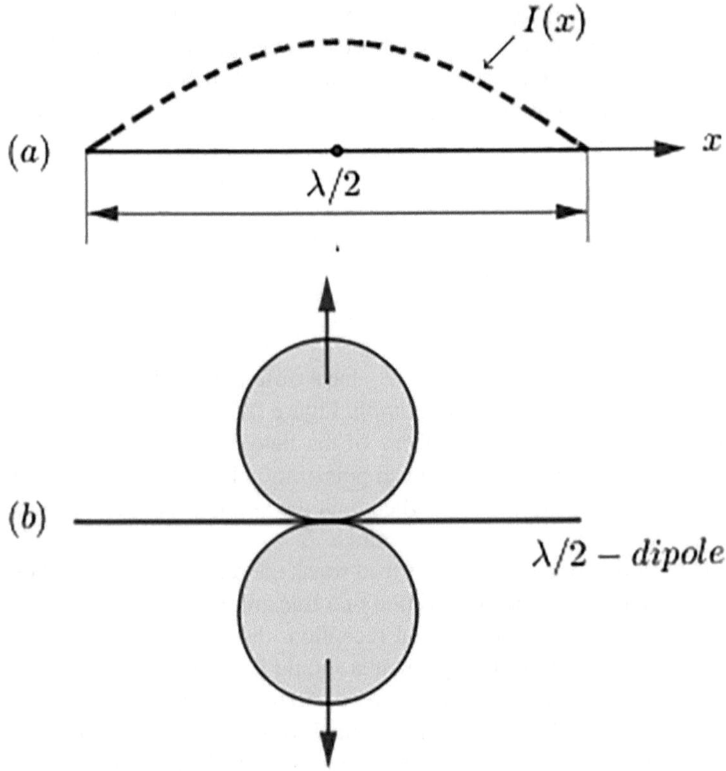

Fig. B.7 Half-wave dipole: (**a**) Current distribution and (**b**) horizontal-plane radiation pattern

Additionally, as an example of a longer wire radiator, Fig. B.8 shows the center-fed full-wave dipole (long λ_0) as a superposition of ten *small electric* radiators, each $0.1\ \lambda_0$ long.

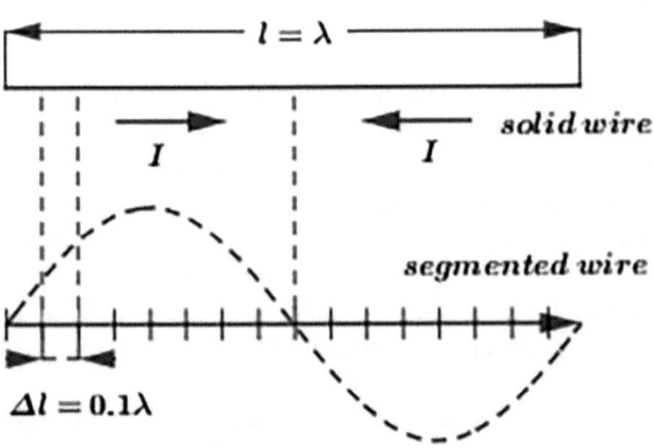

Fig. B.8 Segmentation of a full-wave linear radiator

The distribution of current I along each segment Δl is assumed to be uniform, whereas the total current distribution along the wire has a discrete (step) sine form that can approximate the continuous sine form approximately as

$$I(x) = I\sin\left[\beta(l - x)\right] \text{ for } 0 < x < l. \tag{B.21}$$

Furthermore, from Eq. B.21, the current distribution for the present full-wave dipole ($l = \lambda_0$) is

$$I(x) = I_m \sin\left[\beta(\lambda_0 - x)\right] \text{ for } 0 < x < \lambda, \tag{B.22}$$

where β is the propagation constant defined in *Appendix A*.

As a result of the sine current distribution, the horizontal-plane radiation pattern for the λ-dipole in the far-field region is shown in Fig. B.9.

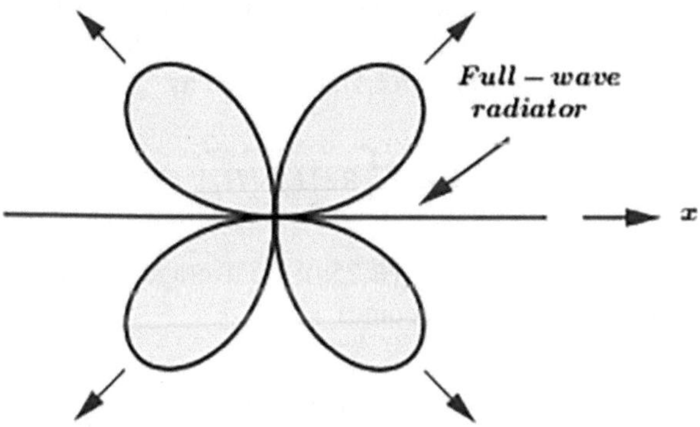

Fig. B.9 Horizontal radiation pattern of a full-wave dipole from Fig. B.8

Furthermore, Fig. B.10 shows the current sine distributions on a 4λ-long wire radiator as a composition of λ/2, λ, 2λ, 3λ, and 4λ dipoles.

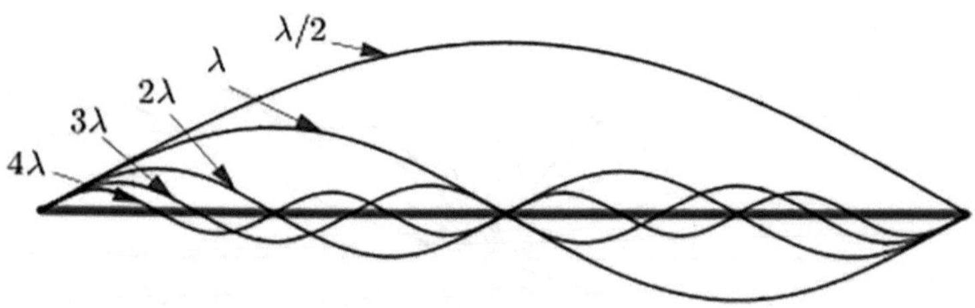

Fig. B.10 Current distribution along a 4λ-long wire

The field components E_θ and H_ϕ in the far-field region from a long wire l and with the sine current distribution are as follows:

$$I(x) = I\sin\left[\beta(l-x)\right] \tag{B.23}$$

can be derived and approximately expressed as [2]

$$\left|\tilde{E}_\theta(\theta)\right| \cong \eta_0 \frac{I}{2\pi r}\left[\frac{\cos(\beta l\cos\theta)-\cos(\beta l)}{\sin\theta}\right] \tag{B.24a}$$

$$\left|\tilde{H}_\phi(\theta)\right| = \frac{\left|\tilde{E}_\theta\right|}{\eta_0} \cong \frac{I}{2\pi r}\left[\frac{\cos(\beta l\cos\theta)-\cos(\beta l)}{\sin\theta}\right], \tag{B.24b}$$

where $\eta_0 = 120\pi$ Ohms, r is the radial distance of an observing point in the far-field region, and β is the propagation constant equal to $2\pi/\lambda_0$. The average power flow, or Poynting vector (see *Appendix A*), is defined as

$$\vec{S}_{av}(\theta) = \frac{1}{2}\mathrm{Re}\left\{\vec{E}_\theta \times \vec{H}_\phi\right\}. \tag{B.25a}$$

Substituting Eq. (B.24a) into Eq. (B.25a), the average power flow from a long-wire radiator with length $l = n\,\lambda$, becomes

$$\vec{S}_{av}(\theta) = \frac{1}{2}\eta_0\left[\frac{I}{2\pi r}\cdot\frac{\cos(\beta n\lambda\cos\theta)-\cos(\beta n\lambda)}{\sin\theta}\right]^2, \tag{B.25b}$$

where $n = 1, 2, 3, \ldots$.

On the basis of Eq. B.25b, the *power* radiation pattern of a 4λ wire ($n = 4$), whose current distribution is shown in Fig. B.10, is given in Fig. B.11.

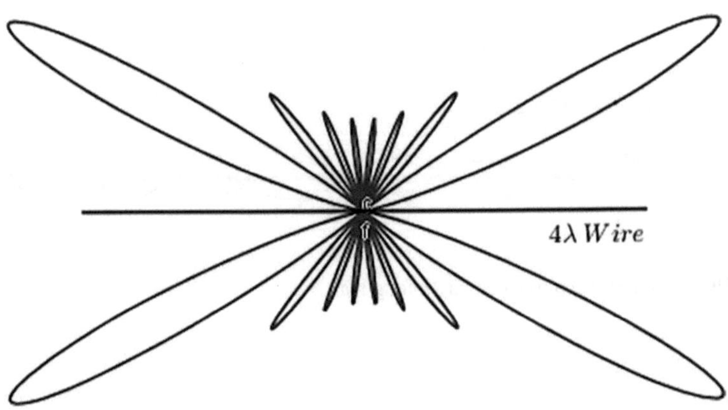

Fig. B.11 Power radiation pattern of a 4λ wire

A comparison of the radiation patterns of the λ/2 wire, λ wire, and 4λ wire shown in the previous figures reveals that the radiation pattern splits into more and more lobes when the radiator length increases. Additionally, when the wire length increases, the primary radiation lobs approach the wire, and its radiation intensity increases.

The regular antenna systems and arrays are large *electric* radiating systems whose elements are usually $\lambda_0/2$-dipole or $\lambda_0/4$-monopole antennas. Because a long wire antenna is most efficient at its resonance frequencies, we must avoid (when possible) the RF or other signal cables from being resonant at frequencies where EMI radiation may occur. Different parameters of electromagnetic radiators that are critical from an EMI standpoint can be indicated in the field equations described in the previous sections.

Larger electromagnetic radiators may have arbitrary shapes, such as thin and long linear conductors (discussed above), arbitrarily closed loops with arbitrary lengths and shapes, and a piece of an electrically large conductive sheet. Additionally, arbitrarily shaped large apertures or long slots in a conductive sheet can be efficient EM radiators. Some potential EMI radiators with different shapes related to signal cabling and EMI shielding are discussed in Chaps. 8 and 10: "Performance of RF Cables and IT Signals Lines in Radio Systems with EMI" and "EMI Shielding for Radio Equipment".

7 Rayleigh's Criteria for Radiation Regions for Large EM Radiators

In addition to the above-discussed Maxwell's criterion, *Rayleigh's* criterion also exists for electromagnetic regions in space around electrically large EM radiators [2, 6]. On the basis of this criterion, the space is also separated into the three regions shown in Fig. B.12.

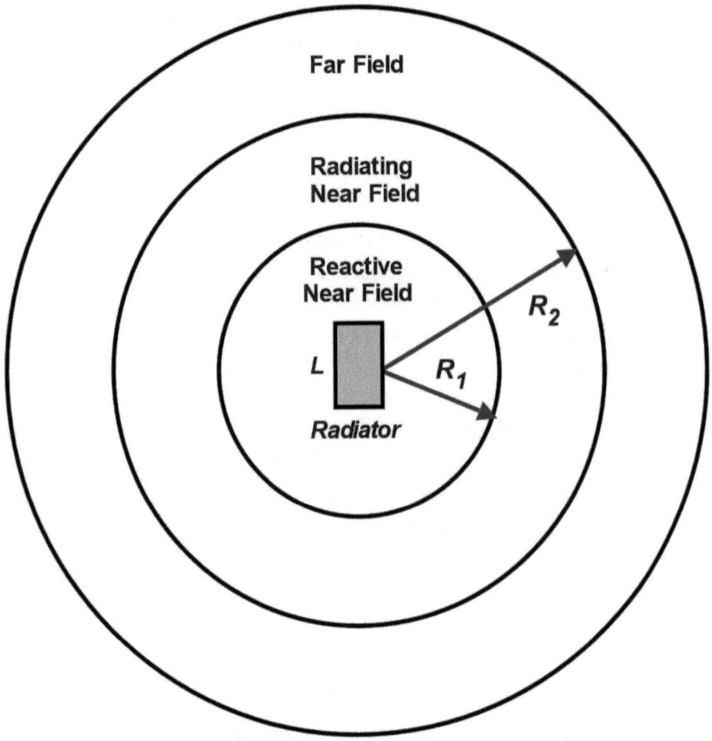

Fig. B.12 Rayleigh boundary regions

Here, the boundaries between the regions determined for different field structures are as follows:

1. The *reactive, near-field region* that exists up to the radial distance R_1 from the radiator is expressed as

$$R_1 = 0.62\sqrt{\frac{L^3}{\lambda_0}}, \tag{B.26}$$

where L is the largest dimension of the radiator in meters where the field is dominantly *reactive*.

2. *The radiating far-field (Fraunhofer)* region starts beyond the radial distance R_2 from the radiator surface:

$$R_2 = \frac{2L^2}{\lambda_0}. \tag{B.27}$$

In this region, the electromagnetic field has a *uniform* and *planar* structure with two orthogonal to each other field components (E and H) and with a unique wave impedance of 377 Ohms. Hence, this is the case of *TEM waves*, which are discussed in *Appendix A*, *Appendix C*, and previously in this appendix.

3. *The radiating near-field (Fresnel) region* is a *transition* region between the *near-field* and *far-field* (*Fraunhofer*) regions. This region exists between radial distances R_1 and R_2.

As noted earlier, we can consider a large radiator as a composition of electrically small segments between the field phases that differ (see Fig. B.8). Let us denote the phase difference $\Delta\phi$ between phases of two segments at two radiator ends separated by the maximal radiator dimension L. Here, Fresnel's criterion is defined by the maximal wave phase difference between the radiator ends, separated by the maximal radiator dimension L, as $\Delta\phi \approx \pi/8$ ($\approx 22°$) [2].

The *far-field* region is of the most practical interest for regular radio communication antennas. In contrast, in terms of EMI emissions, the *near-field* region is more interesting and critical, especially in the case of *unintentional mutual coupling* or *cross talk* between neighboring cables and equipment in radio systems.

8 Linear Radiators Above a Conductive Plane

In the previous sections, we considered electromagnetic radiators in free space. In reality, many radiators are close to large conductive planes or ground. Here, we consider two common cases in which *electric* linear radiators are placed parallel and orthogonal to a conductive plane or ground, as shown in Fig. B.13a, b. The figure also shows the radiator's electromagnetic images underneath the conductive plane, which may replace the conductive plane in the model according to the "image" theory. These cases are discussed in more detail in chapter "General Concept and Types of EMI Emissions".

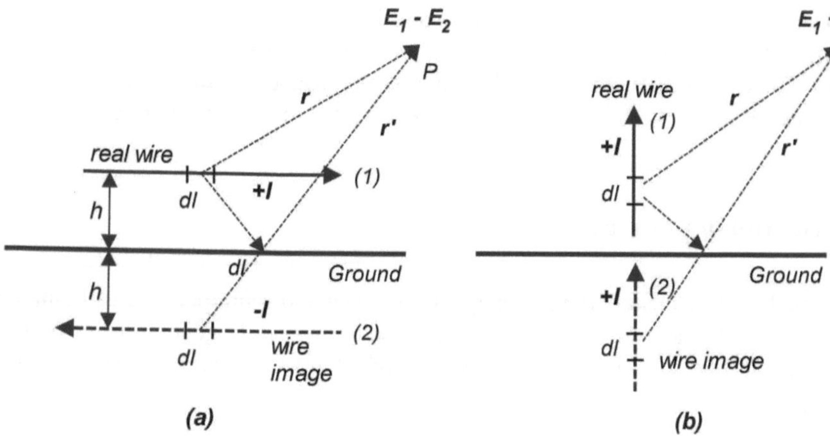

Fig. B.13 (**a**) Horizontal and (**b**) vertical radiators with images

The resulting field at observation point P above the conducting plane is produced by the incident (direct) wave of the horizontal real radiator and the reflected wave from the conducting plane or from the horizontal radiator's image as shown in Fig. B.13a, b.

With respect to the *horizontal* radiator, where the currents of the real radiator and its image have opposite directions, the resulting fields $\vec{E}$ and $\vec{H}$ at point P above the plane are expressed as

$$\vec{E} = \vec{E}_d - \vec{E}_r; \vec{H} = \vec{H}_d - \vec{H}_r. \tag{B.28}$$

Notably, the resulting intensities of the fields $\vec{E}$ and $\vec{H}$ are significantly reduced if the height h of the radiator above the conductive plane is much smaller than the wavelength of the radiating signal. This case is discussed in more detail in chapter "General Concept and Types of EMI Emissions".

Additionally, Fig. B.13b shows a *vertical* radiator above the conducting plane. The effect of the conducting plane can be replaced by the vertical radiator's image below the plane; however, the currents of the real radiator and its image have the same directions. Consequently, the strength of the resulting field above the plane is greater than the strength of the field of the actual radiator, as shown:

$$\vec{E} = \vec{E}_d + \vec{E}_r; \vec{H} = \vec{H}_d + \vec{H}_r. \tag{B.29}$$

Generally, the overall radiation from the real radiator above the conducting plane can have positive and negative effects, depending on the application. For example, for vertical radio communication antennas, the effect of the conducting plane or ground is an advantage because the antenna's gain is increased and almost doubled. However, when we need to lower EMI emissions, the horizontal conductor near the conducting plane is advantageous. Ideally, owing to the closeness and opposite directions of the currents of the actual radiator and its image, the resulting EMI emission can be practically canceled or maximally reduced.

The case of a perfectly conducting plane is considered in this section, but in practice, the real ground is not a perfect conductor, as we assumed in the above cases. The cases of nonperfect conducting planes (ground) are discussed in chapter "Grounding for Radio Systems and Facilities" with respect to grounding systems.

9 Scattering Waves

Previously, we discussed wave propagation in open and semiopen spaces bounded by a planar interface such as a conductive plane or ground. In general, wave propagation should be considered in the presence of arbitrary conductive and dielectric structures (called scatterers), especially when they are close to the source or the receptor of waves. Scatters may have geometries such as planar, cylindrical, spherical, wedges, or sharp edges. One of the common cases of scattering waves exists in cities because of wave reflections, diffractions, and refractions from building structures and street traffic. These cases are discussed in chapters "Performance of Radio Digital Systems in Presence of EMI" and "Overview of 4G-LTE and 5G-NR Mobile Radio Systems from EMI Prospect" regarding multipath wave propagation in urban wireless networks.

Hence, *scattering waves* can be created by reflections, refractions, and diffractions of incident waves from scatterers with arbitrary shapes and volumes. For example, one case of the scattering wave phenomenon is discussed in the previous section regarding horizontal and vertical radiators above the conductive plane, where incident and scattered (reflected) waves converge. However, in many cases, the incident and scattering waves do not converge at all or converge partially in space. This mainly depends on the incident wave's arrival angle to a scattering object, which can be treated as a new radiator with different types of radiation in terms of wave amplitude, direction, and polarization.

Therefore, generally, the total field ($\vec{E}_{tot}$ or $\vec{H}_{tot}$) created by a radiator at a point in space is the vector sum of the *incident* field E_i and *scattered fields* E_s from that object:

$$\vec{E}_{tot} = \vec{E}_i + \vec{E}_s; \vec{H}_{tot} = \vec{H}_i + \vec{H}_s. \tag{B.30}$$

The scattering radiation depends also on the shape and size of a scattering object. In addition to the conducting plane (or ground), scattering objects can be cylindrical, spherical, wedges, convex surfaces, sharp edges, etc. For example, Figs. B.14a, b and B.15a, b show scatter structures with different shapes.

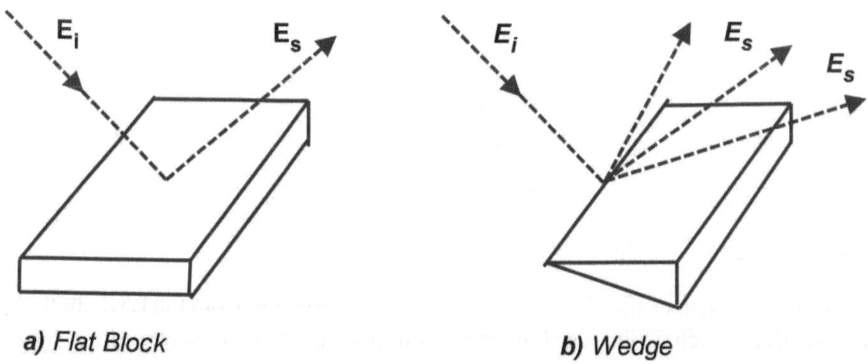

Fig. B.14 Scattering: (**a**) reflection from a flat block (or plane), (**b**) diffraction from a wedge

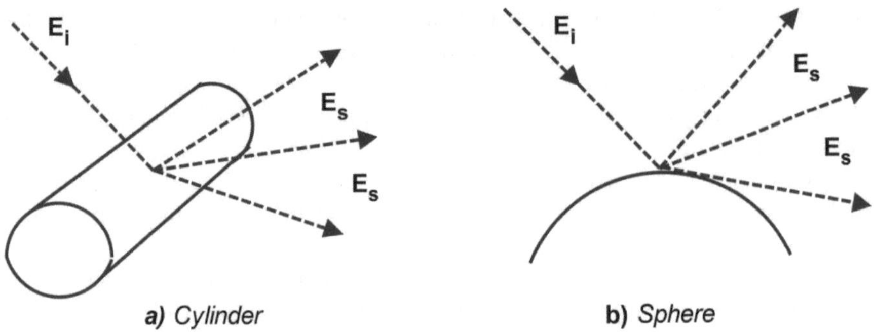

Fig. B.15 Scattering: (**a**) diffraction from a cylinder, (**b**) diffraction from a sphere

As mentioned above in this section, per *Huygens'* principle, any slot or aperture in a conducting (metal) sheet on the way of wave propagation can be treated as a new source of radiation, i.e., a source of scattered waves. Figure B.16 shows this phenomenon.

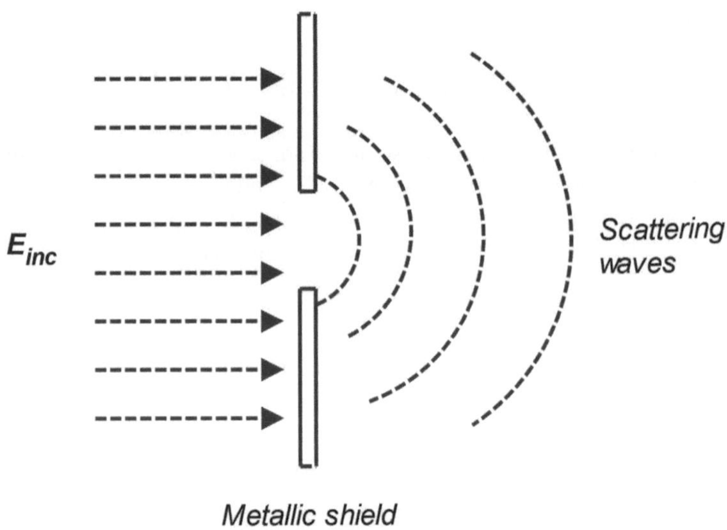

Fig. B.16 Huygens' principle of scattering radiation

The slots and apertures in the metal sheets are interesting from the EMI shielding perspective, which is discussed in more detail in chapter "EMI Shielding for Radio Equipment".

The readers interested in further reading about wave scattering and wave propagation are referred to references [8–12].

References

1. D.K. Cheng, *Field and Wave Electromagnetics*, 2nd edn. (Pearson, 1989)
2. C.A. Balanis, Antenna Theory, 3rd edn. (John Wiley & Sons, New York, 2005)
3. O. Heaviside, Electromagnetic Waves (Cambridge University Press, 2011)
4. M.S. Kao, C.F. Chang, Understanding Electromagnetic Waves (Springer, 2020)
5. U.S. Inan, A.S. Inan, R.K. Said, Engineering Electromagnetics and Waves, 2nd edn. (Pearson, 2015)
6. W.L. Stutzman, G.A. Thiele, Antenna Theory and Design, 3rd edn. (John Wiley & Sons, New York, 2013)
7. J.L. Volakis, Antenna Engineering Handbook, 4th edn. (McGraw Hill, 2007)
8. A. Ishimaru, Electromagnetic Wave Propagation, Radiation, and Scattering (IEEE Press, 2017)
9. A. Ghasemi, A. Abedi, F. Ghasemi, Propagation Engineering in Wireless Communications, 2nd edn. (Springer, 2016)
10. C.A. Balanis, *Advanced Engineering Electromagnetics* (John Wiley & Sons, New York, 1989)
11. J.A. Kong, *Electromagnetic Wave Theory*, 2nd edn. (John Wiley & Sons, 1990)
12. P. Lorrain, D.R. Corson, *Electromagnetic Fields and Waves* (W.H. Freeman Comp., New York, 1988)

Appendix C. Electromagnetic Waves Guided by Transmission Lines and Waveguides

1 Introduction

In Appendix B, *unguided electromagnetic waves* that may propagate in space in all directions are discussed. In this appendix, *electromagnetic waves* guided by conductive structures such as *transmission lines* (*TLs*) and *waveguides* are discussed. Knowing the theory of transmission lines and waveguides is also helpful from an EMI perspective. Some conductive structures may accidentally act as *TLs* or waveguides, guiding and enhancing undesired EMI emissions. Additionally, *TL* theory is important from an EMI perspective to recognize when a regular TL can be a source and/or victim of undesired EMI emissions.

Generally, a *transmission line* (*TL*) is made by two *electrically* long conductors separated by an approximately homogeneous dielectric with an approximately uniform thickness, whereas a *waveguide* is commonly made by one hollow conductor (a tube) with a dielectric or air inside. Therefore, *transmission lines* and *waveguide* structures are intentionally designed to transfer *desired* signals but can sometimes be *unintentionally* formed and used as paths for *EM* interference. Hence, knowing the fundamental properties of transmission structures can be useful for avoiding and solving various problems associated with waveguide structures (TLs and waveguides).

Figure C.1 shows four configurations of TLs.

M. Pajovic, *Fundamentals of EMI Phenomena in Radio Communication Systems*, https://doi.org/10.1007/978-3-031-88748-2

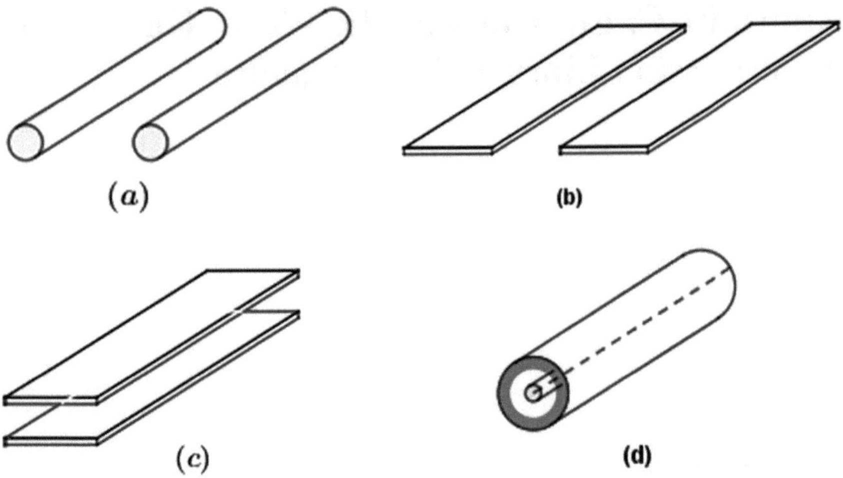

Fig. C.1 (**a**–**d**) TL configurations

Additionally, Fig. C.2 shows the two conductor *TLs* formed by single linear conductors with circular and plane shapes above an *electrically large conductive* plane or *ground*.

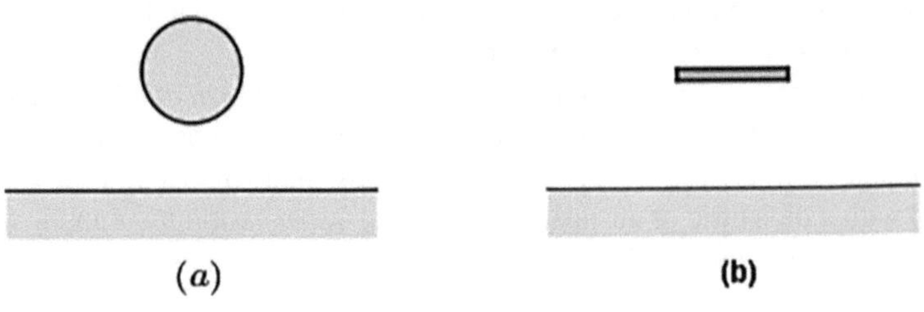

Fig. C.2 (**a**, **b**) Conductors above conductive planes form TLs

Any two-conductor line electrically longer than $\lambda/10$ acts as a *transmission line*, where λ is the shortest wavelength of a signal propagating along the *TL*. Additionally, when the spacing s between *TL* conductors is much smaller than the operating wavelength λ ($s \ll \lambda$), the *TL* supports *transverse electromagnetic* (*TEM*) waves consisting of *transverse* E and H field components that are orthogonal to each other (see Fig. C3a), such as the field components of planar waves in the *far-field* region, as discussed in Appendix B.

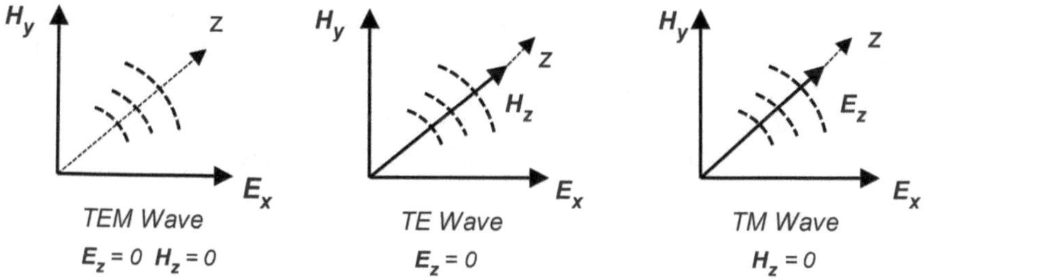

Fig. C.3 (**a**) TEM, (**b**) TE, (**c**) TM waves

TLs have a *cutoff frequency* below which only *TEM* waves may propagate. Above the *cutoff* frequency, higher modes propagate along TLs as *transverse electric* (TE) or *transverse magnetic* (*TM*) waves whose vector structures are shown in Fig. C.3b, c.

With respect to *waveguides*, these structures are *electrically long* guided systems and are commonly made of a *single conductor* (a tube) with rectangular or *circular* cross sections, as shown in Fig. C.4a, b.

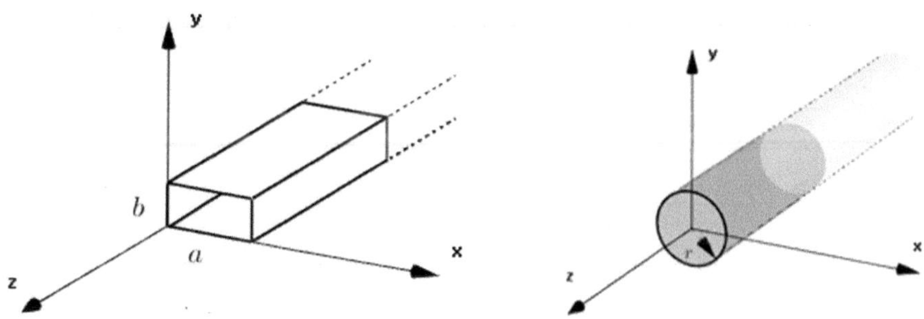

Fig. C.4 (**a**) Rectangular and (**b**) circular waveguides

Therefore, *waveguides* are transmission structures where at least one of their cross-sectional dimensions is comparable with $\lambda/2$, where λ is the shortest operating signal wavelength. This transmission structure transfers higher wave modes (*TE* and *TM*) with frequencies beyond its cutoff frequency. As a result, the waveguide acts as a high-pass filter because it cannot support any wave modes below its cutoff frequency. This is the opposite of TEM TLs, which act as low-pass filters.

The difference between the *TE* and *TM* modes is shown in Fig. C.3b, c, where the *TE* mode contains a *longitudinal magnetic* field component $\vec{H}_z$ and transverse components $\vec{E}_x$ and $\vec{H}_y$. On the other hand, the *TM* mode contains longitudinal *electric* field components $\vec{E}_z$ and transverse *magnetic* field components $\vec{E}_x$ and $\vec{H}_y$.

Notably, if electric circuits are connected by a short signal line (shorter than 0.1λ), then this line does not significantly affect the signals between the circuits. However, when the connecting line is longer than 0.1λ, it may act as a TL-distributed structure, as shown in Figs. C.1 and C.2. In this case, the line may significantly affect signals between circuits if some conditions are not satisfied, which will be discussed later in this chapter.

The classical theories of transmission lines and waveguides are presented well in literature [1–11], which motivated and influenced the writing of this chapter.

2 Transmission Lines

To analyze and design *TLs*, the following TL electrical parameters are essential: the *characteristic impedance, wave propagation velocity,* and related to the TL impedance mismatch, *reflection coefficient Γ*, *reflection loss RL,* and *voltage standing wave ratio* (*VSWR*). Additionally, the *attenuation (A)* and *insertion loss* (*IL*) of lossy TLs are considered.

2.0.1 TL Equations

A *TL* structure can be considered a composition of *electrically short* segments where each segment is represented by the equivalent *lumped* circuit shown in Fig. C.5.

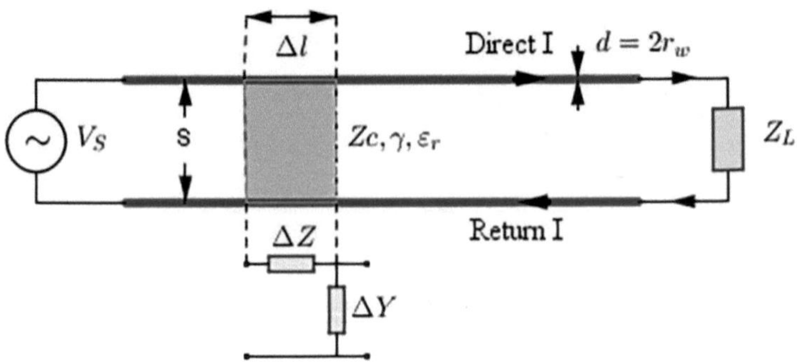

Fig. C.5 Two-conductor TL with an electrically small segment

Here, ΔZ and ΔY are the *impedance* and *admittance* of a TL segment, respectively, expressed as $\Delta Z = R'\Delta l + i\omega L'\Delta l$ and $\Delta Y = G'\Delta l + i\omega C'\Delta l$, where R', L', G', and C' are *TL*-distributed circuit parameters that include the per-unit-length *resistance, inductance, conductance,* and *capacitance,* respectively, as shown in the equivalent lumped circuit in Fig. C.6, where Δl is the length of a TL segment.

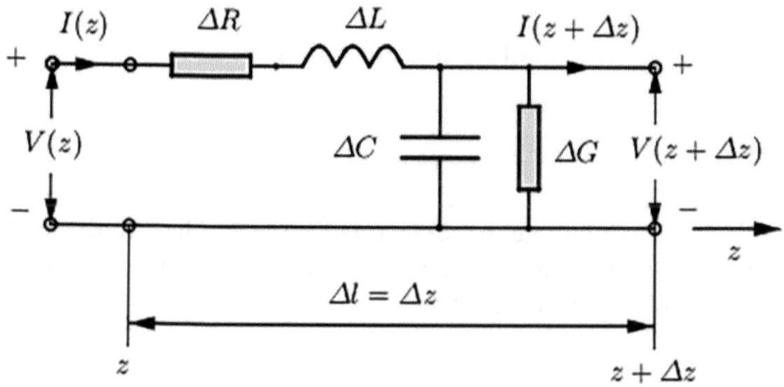

Fig. C.6 Lumped circuit for a small lossy TL segment

On the basis of circuit theory and *Kirchhoff's voltage* and *current laws* applied for the present lumped circuit, the *TL equations* are derived as

$$\frac{\partial V(z)}{\partial z} = -(R' + i\omega L')I(z) \quad \text{(C.1a)}$$

$$\frac{\partial I(z)}{\partial z} = -(G' + i\omega C')V(z). \quad \text{(C.1b)}$$

Furthermore, deriving Eq. (C.1a), (C.1b) with z leads to the *TL* voltage and current *wave equations* for lossy TLs:

$$\frac{\partial^2 V(z)}{\partial z^2} = \gamma^2 V(z) \quad \text{(C.2a)}$$

$$\frac{\partial^2 I(z)}{\partial z^2} = \gamma^2 I(z), \quad \text{(C.2b)}$$

where γ is the complex *propagation constant* expressed as

$$\gamma = \sqrt{Z' Y'} = \sqrt{(R' + i\omega L')(G' + i\omega C')} = \alpha + i\beta \quad (1/m). \quad \text{(C.3)}$$

Here, we can see the analogy between the *TL wave equations* (C.2a), (C.2b) and the *wave equations* (A.50a, b) in an open lossy medium that are discussed in *Appendix A*. Additionally, the real part α in Eq. (C.3) is the *attenuation constant* in (*Np/m*) since the imaginary part β is the *phase constant* in (*rad/m*).

Finally, the solutions of the TL *wave* equations, Eq. (C.2a), (C.2b), lead to the expressions for the voltage $V(z)$ and current $I(z)$ distributions along the TL:

$$V(z) = V^{+}_{0}\, e^{-\gamma z} + V^{-}_{0}\, e^{\gamma z} \tag{C.4a}$$

$$I(z) = I^{+}_{0} e^{-\gamma z} + I^{-}_{0}\, e^{\gamma z}, \tag{C.4b}$$

where "+" and "–" indicate the *direct* (*incident*) wave that propagates from the *TL* source to the *TL* load and the *reflected* wave from the TL load, respectively, which propagates back to the *TL* source.

2.0.2 TL Characteristic Impedance

The important *TL* parameter is the *characteristic impedance* determined by the ratio of the incident (or reflected) *voltage* $V(z)$ and the incident (or reflected) *current* $I(z)$ at any point on the *TL:*

$$Z_c = \frac{V(z)}{I(z)} = \sqrt{\frac{Z'}{Y'}} = \sqrt{\frac{(R' + i\omega L')}{(G' + i\omega C')}} \quad (\Omega), \tag{C.5}$$

or for lossless *TLs,* where $R' = 0$ and $G' = 0$, Eq. (C.5) becomes

$$Z_c = \sqrt{\frac{L'}{C'}} \quad (\Omega). \tag{C.6}$$

The equivalent lumped circuit for a segment of a lossless *TL* is shown in Fig. C.7.

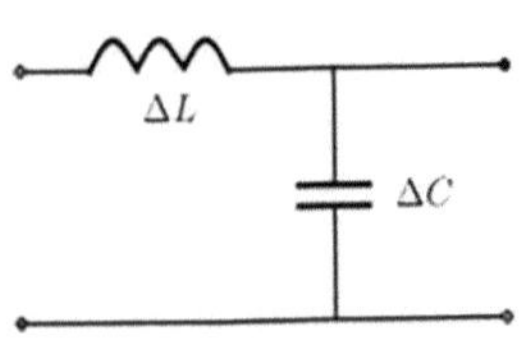

Fig. C.7 Lumped circuit for a lossless small TL segment, where $\Delta L = L'\Delta l$ and $\Delta C = C'\Delta l$. Notably, the above definition of the *characteristic impedance* is valid for any *TL*, finite or infinite

2.0.3 Propagation Velocity in the TL

The *propagation velocity* ν_0 in free space is the speed of light, $\nu_0 = c$. However, in a *TL* with a dielectric with a *relative* constant greater than *1* ($\varepsilon_r > 1$), the *propagation velocity* is lower than that in free space. As discussed in *Appendix A*, the *wave velocity* ν_p or the *phase velocity* is defined as the ratio of the *angular frequency* ω to the *phase constant* β:

$$v_p = \frac{\omega}{\beta}, \tag{C.7a}$$

where β is equal to

$$\beta = \omega\sqrt{L'C'} \tag{C.7b}$$

for lossless TLs.

Additionally, because the speed of the light in free space is $c = \frac{1}{\sqrt{\mu_0 \varepsilon_0}}$, ν_p in a homogeneous medium other than free space becomes

$$v_p = \frac{1}{\sqrt{L'C'}} = \frac{1}{\sqrt{\mu\varepsilon}} = \frac{1}{\sqrt{\mu_r \mu_0 \varepsilon_r \varepsilon_0}} = \frac{c}{\sqrt{\varepsilon_r}}. \tag{C.8}$$

Equation (C.8) shows that waves propagate more slowly in a dielectric medium where $\varepsilon_r > 1$ than in free space or air. Therefore, the signal *wavelength* λ_d in a dielectric space differs from the wavelength λ_0 in free space. The relationship between the wavelength λ_d in a dielectric and the wavelength λ_0 in free space is

$$\lambda_d = \frac{\lambda_0}{\sqrt{\varepsilon_r}}. \tag{C.9}$$

Equation (C.9a) becomes due to the identities $\mu_r = 1$, $\varepsilon_r \geq 1$, $\lambda_d = {}^{v_p}\!/_{f}$, and $\lambda_0 = {}^{c}\!/_{f}$, where $c = 1/\sqrt{\mu_0 \varepsilon_0}$.

Additionally, the following expression shows that the *electrical* length l_e of a conductor buried in a dielectric, with a relative constant $\varepsilon_r > 1$, is electrically shorter than its physical length l in free space by the factor k:

$$l_e = \frac{l}{\sqrt{\varepsilon_r}} = kl, \tag{C.10}$$

where k is known as the *velocity factor* ($k = 1/\sqrt{\varepsilon_r}$), which may have values between *0* and *1*.

Additionally, based on Eqs. (C.7a), (C.7b)

$$\beta = \frac{\omega}{v_p} \text{ and } \beta = \omega\sqrt{L'C'} \text{ where } v_p = \frac{1}{\sqrt{\mu\varepsilon}}$$

the important identity follows:

$$L'C' = \mu\,\varepsilon \rightarrow L' = \frac{\mu\,\varepsilon}{C'} \rightarrow C' = \frac{\mu\,\varepsilon}{L'}. \tag{C.11}$$

Equation (C.7b) can be expressed as

$$\beta = \omega\sqrt{L'C'} = \omega\sqrt{\mu\,\varepsilon}. \tag{C.12}$$

Equation (C.11) is useful because it may obtain a simple relationship between the per-unit-length TL parameters, *inductance* L' and *capacitance* C', in terms of media parameters μ and ε. Thus, when we know μ and ε and L', we can find C' via Eq. (C.11) or vice versa. We apply this identity later in the next section.

2.0.4 TL Input Impedance

Figure C.8 shows the finite *TL* with length l, terminated by the load with an impedance Z_L, which is different from the *TL characteristic* impedance Z_c.

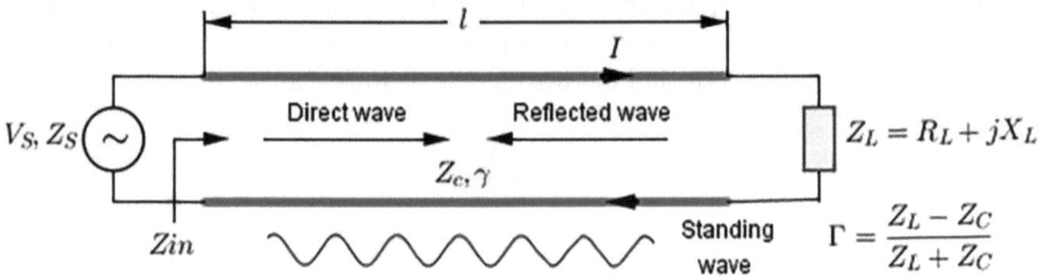

Fig. C.8 Finite TL terminated by load Z_L

Here, note that the two propagating waves (*direct* and *reflected*) exist on the *TL*, and from the solutions of the wave (Helmholtz) equations, Eqs. (C.2a), (C.2b) becomes

$$V(z) = V^+_0\, e^{-\gamma z} + V^-_0\, e^{\gamma z} \tag{C.13a}$$

$$I(z) = I^+_0\, e^{-\gamma z} + I^-_0\, e^{\gamma z} \tag{C.13b}$$

and the input impedance at point $z = 0$ at a TL with length l is expressed as

$$Z(z) = Z_{in} = \frac{V(l-z)}{I(l-z)} = \frac{V(l)}{I(l)} \quad \text{for } z = 0. \tag{C.14}$$

On the basis of Eqs. (C.13a), (C.13b) and (C.14), the input impedance of a lossy TL is derived as

$$Z_{in} = Z_c \frac{Z_L + Z_c \tanh(\gamma l)}{Z_c + Z_L \tanh(\gamma l)}, \tag{C.15a}$$

where Z_c is the TL *characteristic* impedance and where Z_L is the *TL* load impedance. For the special case when the TL is lossless, the TL input impedance becomes

$$Z_{in} = Z_c \frac{Z_L + iZ_c \tan(\beta l)}{Z_c + iZ_L \tan(\beta l)}, \tag{C.15b}$$

where the propagation constant γ for a lossless *TL* has the imaginary part only:

$$\gamma = \alpha + i\beta = 0 + i\beta = i\beta$$

According to Eq. (C.15a), (C.15b), a *TL* acts as an *impedance transformer*, i.e., the *TL* transforms the load impedance Z_L to the input impedance Z_{in}.

2.0.5 TL Impedance Matching

The special case of TL terminations (called *perfect impedance matching*) is when a *TL* is terminated at the source and the load ends with the impedance equal to its *characteristic* impedance Z_c ($Z_s = Z_L = Z_c$). Figure C.9 shows a finite *TL* with length l that is *perfectly* impedance-matched.

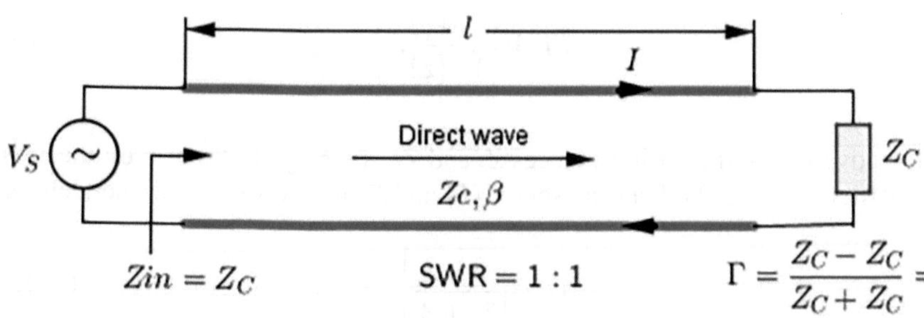

Fig. C.9 A TL with perfect impedance match

In this ideal case, only the *direct* wave propagates from the source to the load, and the input *TL* impedance Z_{in} is equal to the *TL characteristic* impedance Z_c ($Z_{in} = Z_c$). One of the significant contributors to *TL EMI* emission is a TL *impedance mismatch.* The best scenario to minimize the *EMI* emission from a TL is a *good impedance match* between the *TL* and its source and load.

Figure C.10 shows the scenario in which the *TL* is terminated by an arbitrary impedance Z_L and where the two waves propagate in opposite directions along the TL.

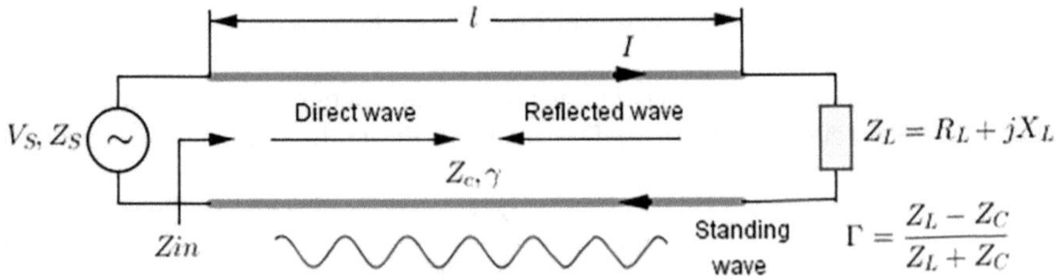

Fig. C.10 TL termination by an arbitrary load

The *direct* wave *propagates* from the source to the load, whereas the *reflected* wave propagates back from the load to the source. These waves are analytically presented by the *TL equations* (C.4a), (C.4b).

Four important TL parameters are established to quantify the effects of the TL *impedance mismatch* and TL loss, such as the *reflection coefficient, voltage standing wave ratio* (VSWR), *return* loss, and *insertion* loss. These parameters are discussed in the following subsections.

(a) Reflection coefficient

The *reflection coefficient* $\Gamma(z)$ at a point distant by z from the *TL* input is defined by the ratio of the voltage of the *reflected* wave V^- and the voltage of the *direct* wave V^+,

$$\tilde{\Gamma}(z) = \frac{\tilde{V}^-(z)}{\tilde{V}^+(z)}, \tag{C.16}$$

Alternatively, in terms of impedances Z_c and Z_L, the magnitudes of the reflection coefficients at the TL load and source (Γ_L and Γ_s) are expressed, respectively, as

$$\Gamma_L = \left|\frac{Z_L - Z_c}{Z_L + Z_c}\right| \tag{C.17a}$$

$$\Gamma_s = \left|\frac{Z_c - Z_s}{Z_c + Z_s}\right|. \tag{C.17b}$$

In general, when the *TL* is not perfectly terminated at its ends (i.e., when $Z_s \neq Z_c$ and/or $Z_L \neq Z_c$), *standing waves* exist along the *TL*, as shown in Fig. C.10. The term "*standing waves*" means that energy is not fully transferred to the TL load, but it stands and converts to heat at *lossy* TL conductors and dielectrics, and dissipates at a TL source, and partially *radiates* outside the TL.

Additionally, when a TL *impedance mismatch* at both TL ends exists (at the source and load), *multiple reflections* occur in the *TL*.

(b) *Voltage standing wave ratio (VSWR)*

The *VSWR* is defined as the ratio of the *maximal* and *minimal* voltages of *standing waves* on a *TL*:

$$VSWR = \frac{V_{\max}}{V_{\min}}. \tag{C.18}$$

For example, Fig. C.11 shows a graphical presentation of the *VSWR*.

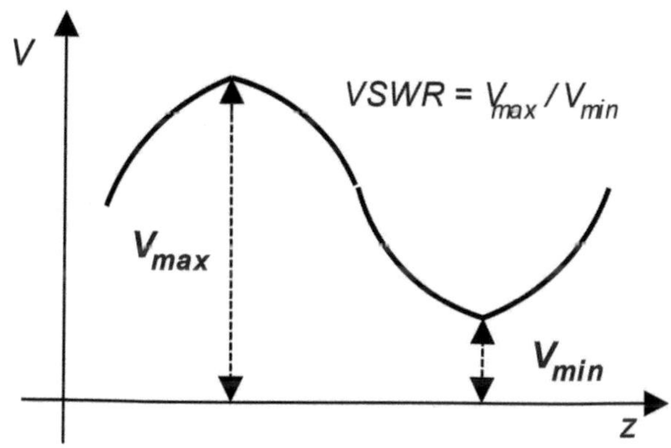

Fig. C.11 Voltage standing wave ratio (VSWR)

Additionally, in terms of the *reflection coefficient* at *load*, *VSWR* can be expressed as

$$VSWR = \frac{1+|\Gamma_L|}{1-|\Gamma_L|}. \tag{C.19}$$

Thus, a *perfect TL* impedance match exists when $\Gamma_L = 0$ and VSWR = 1.

(c) *Return loss and insertion loss*

Related to the *TL impedance mismatch*, the *return loss (RL)* and the *insertion* loss (*IL*) are defined in dB as

$$RL = -20\log\left(\left|\Gamma_L\right|\right) \tag{C.20a}$$

$$IL = -20\log\left(\left|T\right|\right), \tag{C.20b}$$

where the parameter T is the *transmission coefficient* expressed as $|T| = 1 - |\Gamma_L|$ when the TL is lossless.

2.0.6 Transients at Transmission Lines

Figure C.12 shows a transient traveling from a TL source and along a lossless finite Tl, and reflecting from the TL's load, whose impedance is different from the TL characteristic impedance. The change in the transient amplitude relative to the amplitude V_0 at the TL input can be observed.

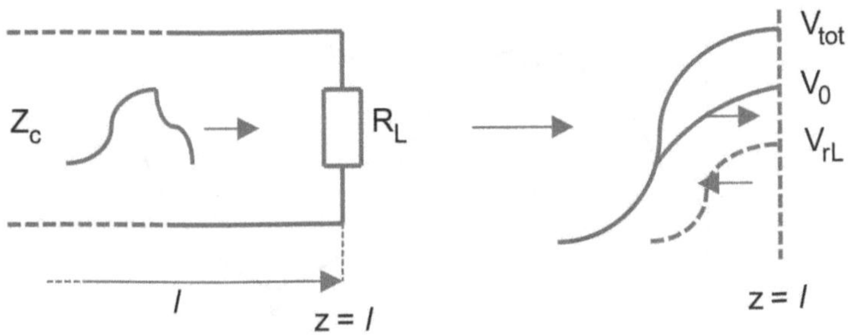

Fig. C.12 Reflection of a pulse transient from a TL load

Here, the reflected voltage from the load is $V_{rL} = \Gamma_L V_0$, and the total voltage is $V_{tot,L} = V_0 + \Gamma_L V_0 = V_0(1 + \Gamma_L)$ at the load, where Γ_L is the reflection coefficient at the load R_L expressed by Eq. (C.17a). For example, when $\Gamma_L = 0.3$, $V_{tot,L} = 1.3\ V_0$.

Additionally, Fig. C.13 shows the reflected transient wave from the TL load (V_rL) that arrives back to the TL source whose impedance is different from the TL characteristic impedance. Here, the amplitude of the reflected transients (V_{rL}) from the load and from the source (V_{rs}), relative to the amplitude (V_0) of the TL input transient, can be seen.

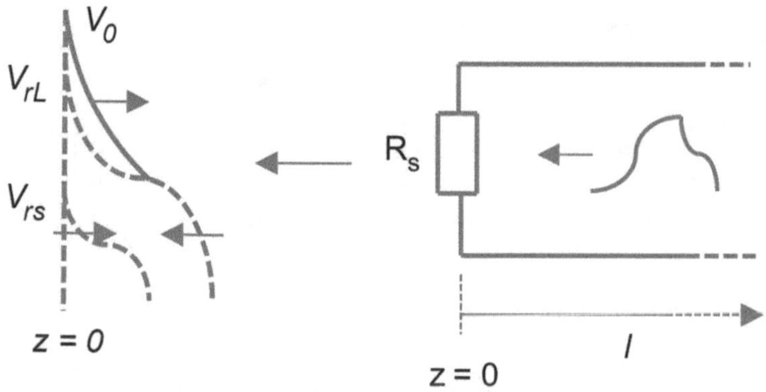

Fig. C.13 Reflection of a transient pulse from a TL source

Here, the reflected voltage from the source is expressed as $V_{rs} = \Gamma_s V_{rL} = \Gamma_s \Gamma_L V_0$, and the total voltage at the source is expressed as $V_{tot,\,s} = V_0 + V_{rL} + V_{rs} = V_0 + \Gamma_L V_0 + \Gamma_s \Gamma_L V_0 = V_0(1 + \Gamma_L + \Gamma_s \Gamma_L)$, where Γ_s is the reflection coefficient at the source impedance Rs, expressed via Eq. (C.17b). For example, when $\Gamma_L = \Gamma_s = 0.3$, $V_{tot,\,s} = 1.39\ V_0$.

Theoretically, the number of reflections from the load and source is infinite at a lossless TL, but in reality, all TLs are more or less lossy, and the number of reflections decays rapidly. This especially depends on the TL physical length and the conductivity of the TL conductors and dielectrics.

2.0.7 TL as Impedance Transformer

(a) *Perfect impedance-matched, open-ended, and short-ended TLs*

In terms of *TL* terminations, three special cases exist: (a) *TL* terminated by an impedance equal to its *characteristic* impedance (discussed in the previous section), (b) *TL* with no load (*open-ended TL*), and (c) *TL* with a *short* load end (*short-ended TL*). These cases are shown in Fig. C.14a–c, respectively, with the related reflection coefficients Γ.

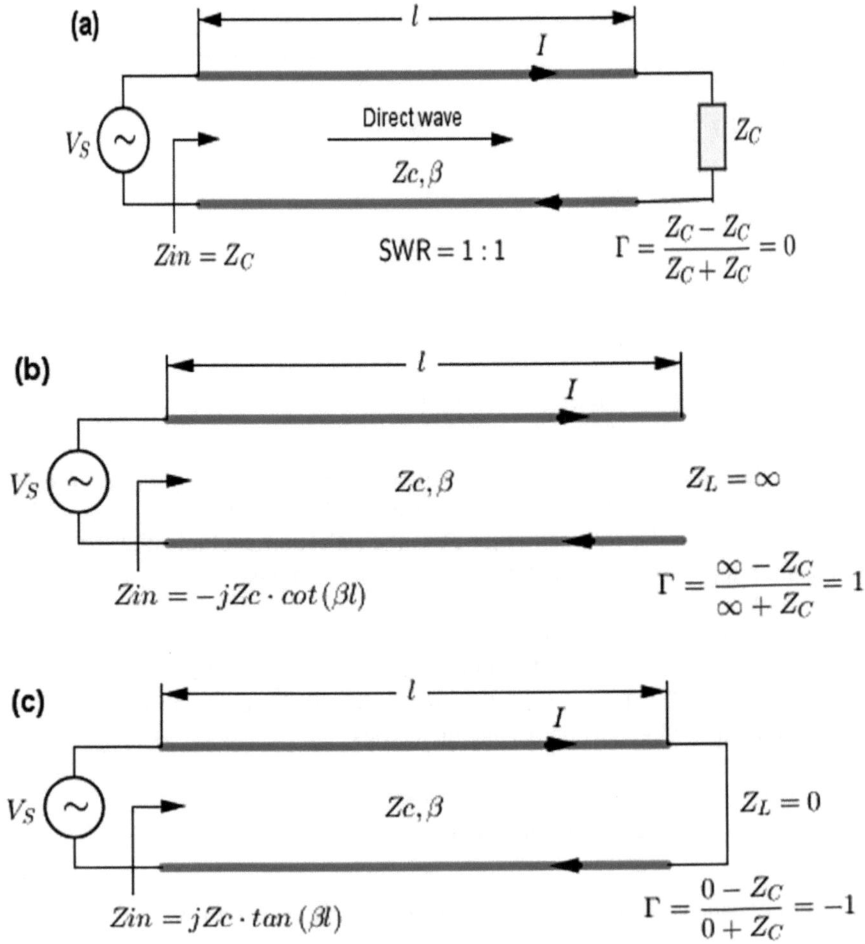

Fig. C.14 (**a**) Impedance-matched TL, (**b**) open-ended TL, (**c**) short-ended TL

As discussed before, when an arbitrarily long and lossless *TL* is terminated by the *load* equal to its *characteristic* impedance Z_c, $Z_{in} = Z_c$.

Additionally, on the basis of Eq. (C.15a), (C.15b), Z_{in} of *lossy* and *lossless open-ended TLs* ($Z_L = \infty$) can be expressed as C.21a and C21b:

$$Z_{in} = Z_c \coth\left[(\alpha + j\beta)l\right] \tag{C.21a}$$

$$Z_{in} = -i\, Z_c \cot(\beta l) \tag{C.21b}$$

since for a *short-ended TL* ($Z_L = 0$),

$$Z_{in} = Z_c \tanh\left[(\alpha + j\beta)l\right] \tag{C.22a}$$

$$Z_{in} = iZ_c \tan(\beta l). \tag{C.22b}$$

(b) *λ/2 TL as a (1:1) impedance transformer*

On the basis of Eq. (C.15b), a *lossless TL* long multiplied by λ/2 acts as a 1:1 *impedance transformer* for any load Z_L at frequency $f = v/\lambda$, where v and λ are the *wave velocity* and *wavelength* in the *TL dielectric*, respectively. Additionally, n is an integer with values of $n = 1, 2, 3$ .... This *TL* is shown in Fig. C.15.

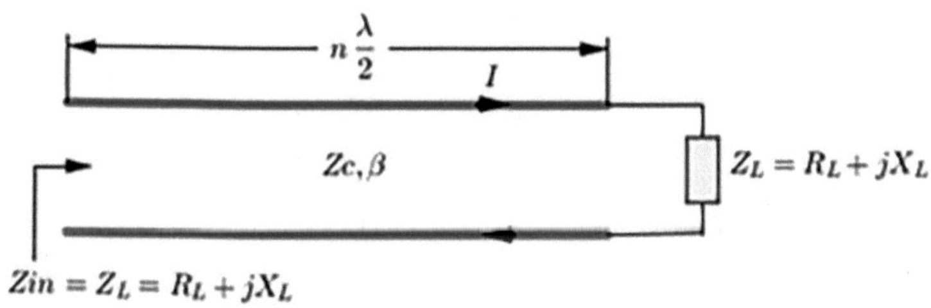

Fig. C.15 (1:1) λ/2 TL impedance transformer

Hence, a *TL* with length λ/2 and loaded by an arbitrary impedance Z_L acts as a 1:1 impedance transformer because

$$Z_{in} = Z_L. \tag{C.23}$$

Additionally, on the basis of Eq. (C.15b), the input impedance of a λ/2 *open-ended TL* is $Z_{in} = \infty$, whereas for a λ/2 *short-ended TL*, $Z_{in} = 0$, as shown in Fig. C.16a, b.

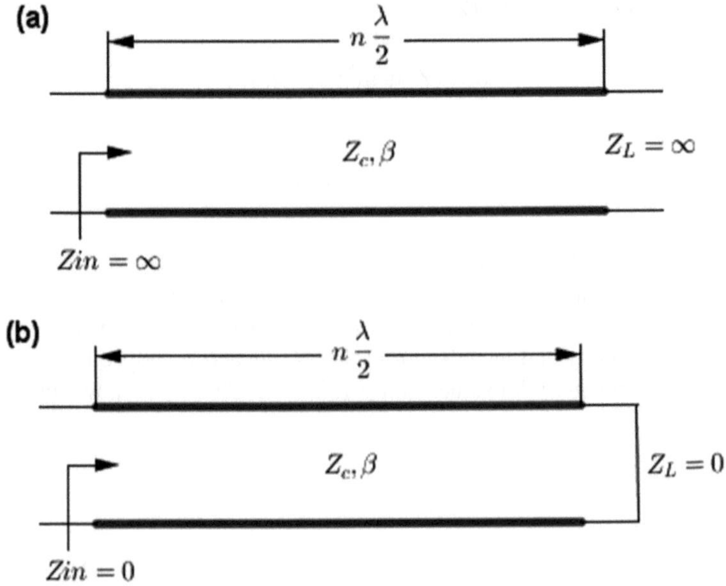

Fig. C.16 (**a**) Open-ended TL and (**b**) short-ended λ/2 TL

(c) *λ/4 TL impedance transformer*

The Z_{in} of a *lossless λ/4 open-ended TL*, long *odd* multiples of λ/4, is equal to *zero*, whereas for the same *TL* but *short-ended*, Z_{in} is *infinite*. Both *TLs* are shown in Fig. C.17a, b, respectively. The integer *n* has values *n* = 0, 1, 2, 3 ….

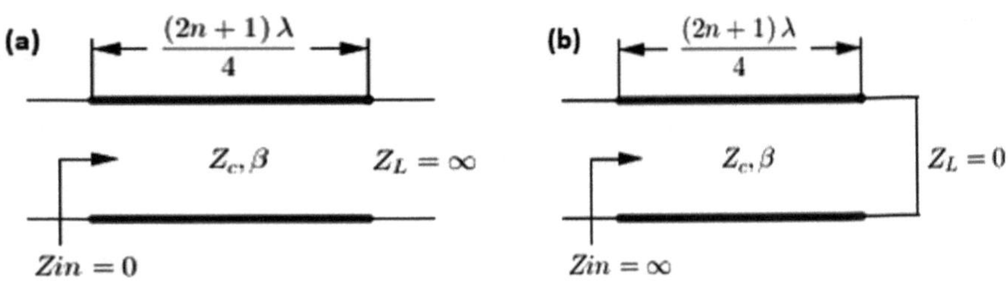

Fig. C.17 (**a**) Open-circuit and (**b**) short-circuit λ/4 TL

For example, per Eq. (C.15b),

$$Z_{in} = Z_c \frac{Z_L + iZ_c \tan(\beta l)}{Z_c + iZ_L \tan(\beta l)}, \tag{C.24a}$$

the input impedance Z_{in} for a *lossless, quarter-wave TL* terminated by the load Z_L is

$$Z_{in} = \frac{Z_c^2}{Z_L} \quad \text{Quater – wave transformer,} \tag{C.24b}$$

where $\beta = 2\pi/\lambda$ and for $l = \frac{\lambda}{4}$, $\tan(\beta l) \to \infty$.

2.0.8 Resonant Transmission Lines

TLs can resonate, and this fact is exploited to design high-frequency resonators and filters where discrete *LC* circuits cannot be efficiently used. In this section, we show that *open-ended* and *short-ended TLs* can act as efficient high-frequency resonators and antiresonators.

(a) *TL resonators*

Figure C.18a, b shows two cases of special terminated TLs that may act as efficient resonators. They are *λ/2 short-ended TLs* and *λ/4 open-ended TLs.*

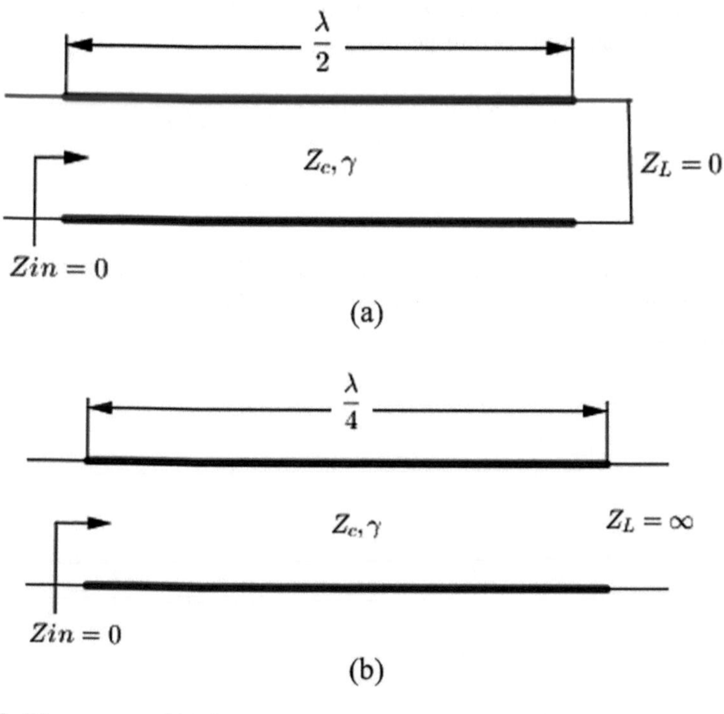

Fig. C.18 TL resonators: (**a**) λ/2 short-ended and (**b**) λ/4 open-ended

Figure C.19a shows the equivalent *series* circuit for both TL resonators, whereas Fig. C.19b shows the plot of the circuit input impedance Z_{in}, where Z_{in} has a minimum at the *resonance* frequency (f_r) equal to R_s or $Z_{in} \rightarrow 0$ for a lossless TL resonator.

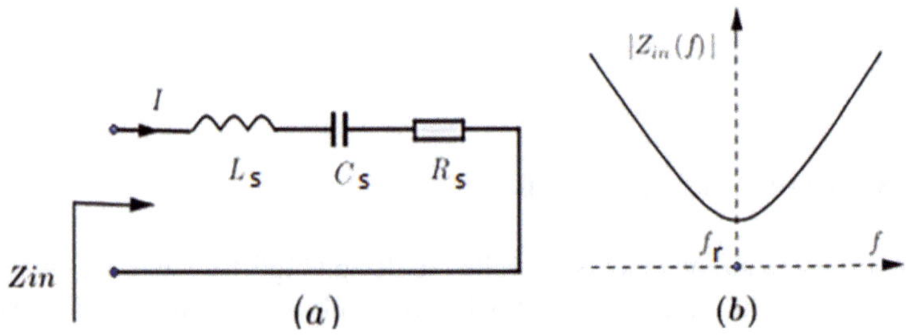

Fig. C.19 (**a**) Series-circuit resonator and (**b**) its $Z_{in}(f)$ plot

In circuit theory, the *resonance* frequency of this circuit is expressed as

$$f_r = 1/2\pi\sqrt{L_s C_s} \tag{C.25}$$

since the Q_s *factor* and *bandwidth BW* of the circuit are parameters related to lossy TL resonators, expressed respectively as

$$Q_s = \frac{2\pi f_r L_s}{R_s} = \frac{1}{2\pi f_r C_s R_s}; BW = 1/Q_s. \tag{C.26}$$

(b) *Antiresonant TLs*

With respect to *TL antiresonators* (or *parallel RLC* circuits), the two particularly terminated TLs can be considered efficient *anti-resonators*, as shown in Fig. C.20a, b.

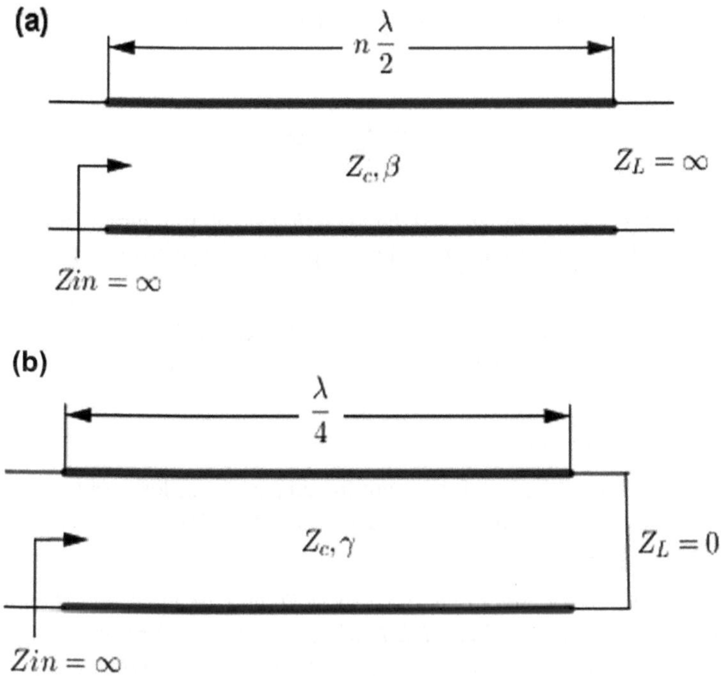

Fig. C.20 TL antiresonators

Figure C.21a shows the equivalent parallel circuit for the TL antiresonator from Fig. C.20a, while Fig. C.21b shows the plot of the circuit input impedance Z_{in}, where Z_{in} has a maximum ($Z_{in} = R_p$) at the *antiresonance* frequency f_{ar} or for a lossless *TL*, $Z_{in} \rightarrow \infty$. $Z_{in} \rightarrow \infty$

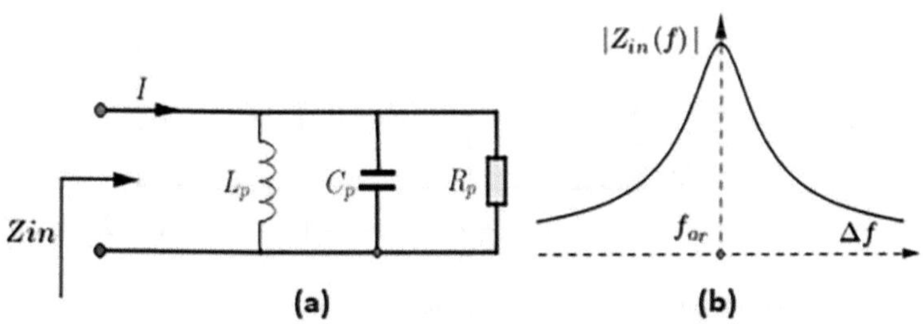

Fig. C.21 (**a**) Parallel circuit resonator and (**b**) $Z_{in}(f)$ plot

In circuit theory, the *antiresonance* frequency of this type of resonator is expressed as

$$f_{ar} = \frac{1}{2\pi\sqrt{L_p C_p}}. \quad \text{(C.27)}$$

Additionally, the Q_p *factor* and *bandwidth BW* of this circuit are important parameters of lossy TL resonators and are expressed, respectively, as

$$Q_p = \frac{R_p}{2\pi f_{ar} L_p} = 2\pi f_{ar} C_p R_p; BW = \frac{1}{Q_p}. \quad \text{(C.28)}$$

(c) *Multiple TL resonances and antiresonance*

Figure C.22 shows the plot of the input impedance Z_{in} of lossless arbitrarily *long open-ended TLs*, from Figs. (C.16a) and (C.17a), on the basis of Eq. (C.15b), $|\tilde{Z}_{in}| = Z_c \cot(\beta l)$

Here, *TL resonances* (where $Z_{in} = 0$) occur at *TL* lengths equal to *odd* multiples of a *quarter wavelength, (2n + 1)λ/4*, whereas the *TL antiresonance* ($Z_{in} \rightarrow \infty$) occurs at a *TL* length equal to multiples of a *half wavelength, n λ/2*.

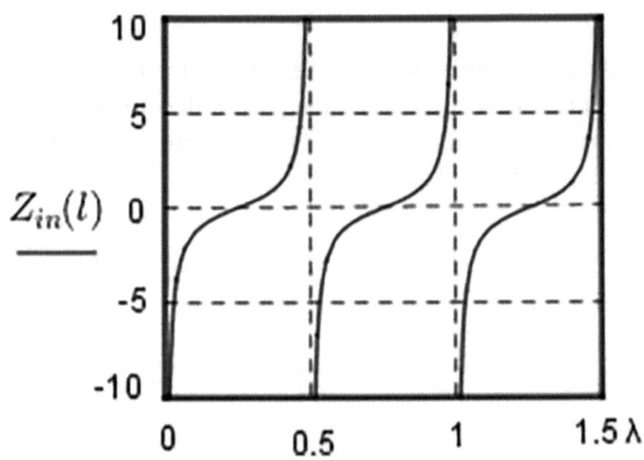

Fig. C.22 $|Z_{in}|$ of an open-ended TL ($Z_{in} = Z_c \cot(\beta l)$)

Furthermore, Fig. C.23 shows the plot of the input impedance Z_{in} of a *lossless short-ended TL* ($Z_L \rightarrow 0$) from Figs. (C.16a) and (C.16b), which is based on Eq. (C.16b), $|\tilde{Z}_{in}| = Z_c \tan(\beta l)$.

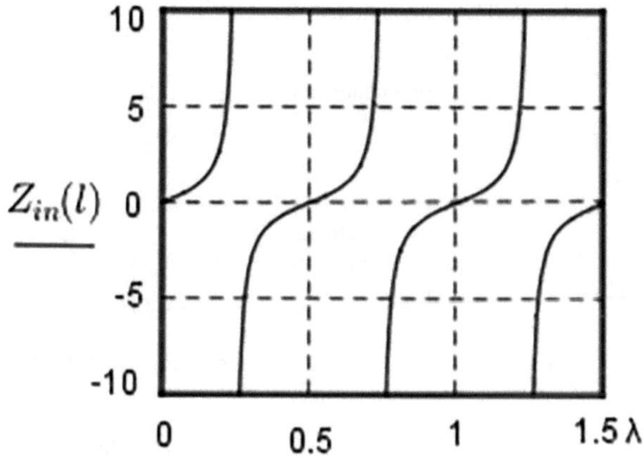

Fig. C.23 $|Z_{in}|$ of a short-ended TL ($Z_{in} = Z_c \, tan\,(\beta l)$)

Here, in contrast to the previous case, *TL antiresonances* (where $Z_{in} \rightarrow \infty$) occur at *TL* lengths equal to *odd* multiples of a *quarter wavelength, (2n + 1)*λ/4, whereas *TL resonances* ($Z_{in} = 0$) occur at *TL* lengths equal to multiples of a *half wavelength, n* λ/2.

In the end, *TLs* may unintentionally act as resonators and accidentally emit *EMI* or may act as useful high-frequency harmonic filters for suppressing discrete *EMI* at certain frequencies. These are the cases of *unintentional and intentional* resonators. Some conductive structures can accidentally act as *unintentional TL* resonators causing *EMI* problems. Therefore, understanding the basic *TL* resonator theory can be helpful for avoiding *EMI* problems resulting from the resonance and antiresonance effects of TLs.

2.0.9 Overview of Common TL Configurations

An overview of common *TL* configurations is given in Table C.1. They can be created intentionally and sometimes unintentionally in radio systems.

Table C.1 Common practical TL configurations

	TL	Z_c
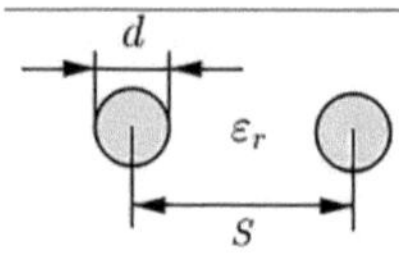	A two-conductor TL	$Z_c = \frac{120}{\sqrt{\varepsilon_r}} \cosh^{-1}\left(\frac{s}{d}\right)$
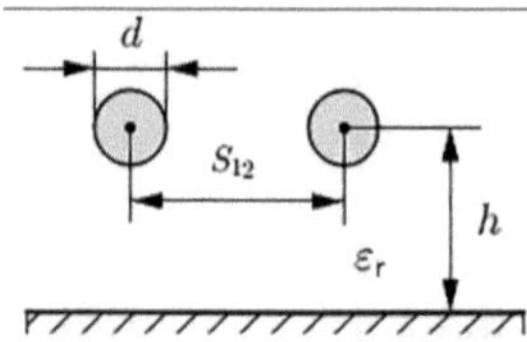	A two-conductor TL above conductive plane	$Z_c = \frac{60}{\sqrt{\varepsilon_r}} \begin{vmatrix} Z'_{11} & Z'_{12} \\ Z'_{21} & Z'_{22} \end{vmatrix}$ $Z'_{11} = Z'_{22} = \ln(4h/d)$ $Z'_{12} = Z'_{21} = \ln\left(S'_{12}/S_{12}\right)$ $s'_{12} = \sqrt{4h^2 + S_{12}^2}\,; d < (0.1h \wedge 0.1S_{12})$
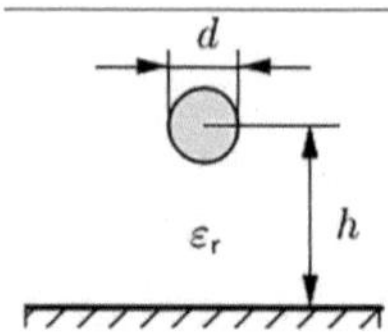	A single conductor above conductive plane	$Z_c = \frac{60}{\sqrt{\varepsilon_r}} \cosh^{-1}\left(\frac{2h}{d}\right)$
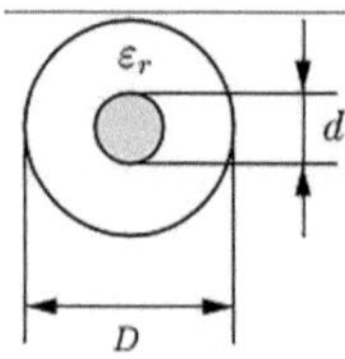	A coaxial TL	$Z_c = \frac{60}{\sqrt{\varepsilon_r}} \ln\left(\frac{D}{d}\right)$
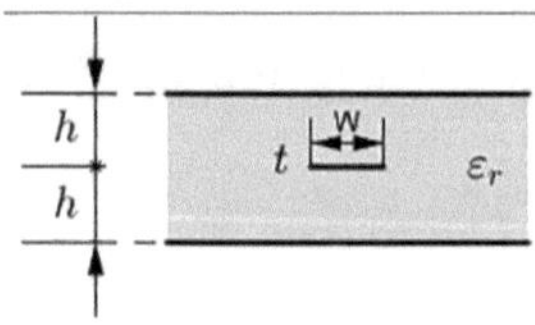	Stripline (t-strip thickness)	$Z_{in} = \frac{60}{\sqrt{\varepsilon_r}} \ln\left(\frac{4h}{0.67\pi(t+0.8w)}\right)$ for $\frac{w}{h} < 0.35; \frac{t}{h} < 0.25$
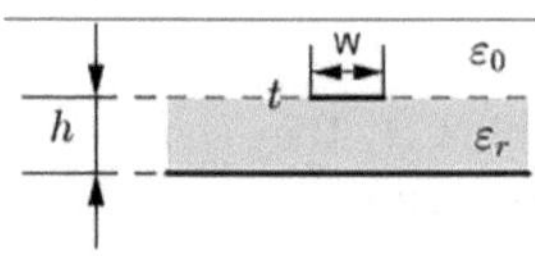	Microstrip (t-microstrip thickness)	$Z_{in} = \frac{87}{\sqrt{\varepsilon_r + 1.4}} \ln\left(\frac{6h}{(t+0.8w)}\right)$ for $0.1 < \frac{w}{h} < 2; 1 < \varepsilon_r < 15$
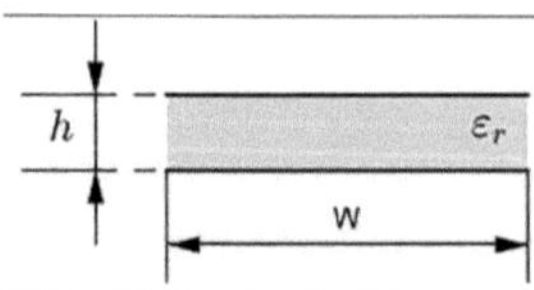	Long parallel plate TL	$Z_c = \frac{120\pi}{\sqrt{\varepsilon_r}} \ln\left(\frac{h}{w}\right)$ for $h < 0.1w$

3 Waveguides

A waveguide is primarily intended for microwave applications. As mentioned at the beginning of this chapter, it acts as a high-pass filter, allowing a signal to propagate only above the waveguide *cutoff* frequency and blocking signal frequencies below this frequency. This property of waveguides can be applied for suppressing EMI frequencies below their cutoff frequency. For example, this concept is used for blocking EMI emissions through equipment ventilation holes and cable openings, which are the subject of this section. Depending on the shape of the cross-sectional area, common waveguides may be rectangular and circular.

3.0.1 Rectangular Waveguides

An electrically long waveguide with a uniform *rectangular* cross section is shown in Fig. C.24, where a and b are the cross-sectional dimensions in the x-direction and y-direction, respectively.

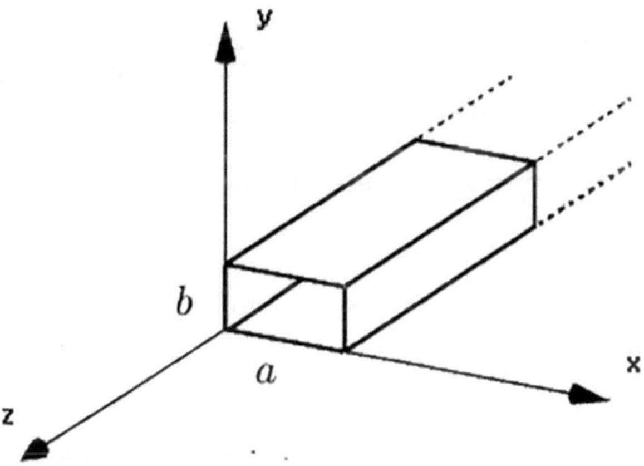

Fig. C.24 Rectangular waveguide

As discussed earlier, the *rectangular* waveguides support the propagation of *TM* and *TE* modes only. Multiples of these modes may exist simultaneously in a waveguide and are designated TE_{mn} and TM_{mn} modes, where m and n have integer values of 0, 1, 2, ...

As noted before, transverse electric (*TE*) waves do not have a longitudinal electric field in the direction of propagation, whereas transverse magnetic (*TM*) waves do not have a longitudinal magnetic field in the direction of propagation.

Assuming that a waveguide is lossless, the *TE* waves can be expressed in rectangular (Cartesian) space as

$$\vec{E}_z = 0; \vec{H}_z(x,y,z) = \vec{H}_z(x,y)e^{-i\beta z} \tag{C.29a}$$

since the *TM* waves are expressed as

$$\vec{H}_z = 0; \vec{E}_z(x,y,z) = \vec{E}_z(x,y)e^{-i\beta z}, \tag{C.29b}$$

where $\vec{E}_z(x, y, z)$ and $\vec{H}_z(x,y,z)$ satisfy the wave equations from *Appendix A*, where the waveguide medium conductance σ = 0,

$$\nabla^2\vec{E} + \omega^2\mu\varepsilon\vec{E} = 0; \nabla^2\vec{H} + \omega^2\mu\varepsilon\,\vec{H} = 0, \tag{C.29c}$$

or these equations in the Cartesian space as

$$\nabla^2{}_{xy}\vec{E} + \left(\omega^2\mu\varepsilon - \beta^2\right)\vec{E}_z = \nabla^2{}_{xy}\vec{E} + k^2\vec{E}_z = 0 \tag{C.30a}$$

$$\nabla^2{}_{xy}\vec{H} + \left(\omega^2\mu\varepsilon - \beta^2\right)\vec{H}_z = \nabla^2{}_{xy}\vec{H} + k^2\vec{H}_z = 0, \tag{C.30b}$$

where the parameter k is the *wavenumber*. Furthermore, applying the boundary conditions for the electric field on the waveguide walls results in the following solutions for Eq. (C.30a), (C.30b):

$$E_z(x,y) = E_0 \sin(k_x x)\sin(k_y y) \quad (TM) \tag{C.30c}$$

$$H_z(x,y) = H_0 \cos(k_x x)\cos(k_y y) \quad (TE) \tag{C.30d}$$

or

$$E_z(x,y) = E_0 \sin\left(\frac{m\pi}{a}x\right)\sin\left(\frac{n\pi}{a}x\right) \quad (TM) \tag{C.31a}$$

$$H_z(x,y) = H_0 \cos\left(\frac{m\pi}{a}x\right)\cos\left(\frac{n\pi}{a}x\right) \quad (TE), \tag{C.31b}$$

where $k_x = m\pi/a$ and $k_y = n\pi/b$ for $m, n = 1, 2, 3, \ldots$ and where m and n are the numbers of the electric field maximus in the x and y directions, respectively. Additionally, from Eq. (C.30a), (C.30b),

$$k^2 = \left(\omega^2\mu_0\,\varepsilon_0 - \beta^2\right) = \left(\frac{\omega^2}{c^2} - \beta^2\right), \tag{C.32a}$$

where c is the speed of light equal to $1/\sqrt{\mu_0 \varepsilon_0}$.

The *cutoff wavenumber* k_c for a waveguide is determined from Eq. (C.32a) when the propagation constant $\beta = 0$:

$$k_c^2 = \omega_c^2 \mu\varepsilon \rightarrow k_c = \omega_c \sqrt{\mu\varepsilon} \rightarrow \omega_c = \frac{k_c}{\sqrt{\mu\varepsilon}}, \tag{C.32b}$$

where ω_c is the wave angular *cutoff* frequency.

Additionally, the cutoff frequency f_c, below which the waves cannot propagate, is expressed as

$$f_c = \frac{1}{2\pi\sqrt{\mu\varepsilon}} k_{cmn}, \tag{C.33a}$$

where the cutoff wavenumber k_{cmn} is expressed as

$$k_{cmn} = \sqrt{k_x^2 + k_y^2} = \sqrt{\left(\frac{m\pi}{a}\right)^2 + \left(\frac{n\pi}{b}\right)^2}, \tag{C.33b}$$

based on $k_x = m\pi/a$ and $k_y = n\pi/b$ for $m, n = 1, 2, 3, \ldots$.

Additionally, by substituting Eq. (C.33b) into Eq. (C.33a), the waveguide *cutoff frequency* f_{cmn} can be expressed as

$$f_{cmn} = \frac{1}{2\pi\sqrt{\mu\varepsilon}} \sqrt{\left(\frac{m\pi}{a}\right)^2 + \left(\frac{n\pi}{b}\right)^2}. \tag{C.33c}$$

3.0.2 Absolute Dominant Mode in Rectangular Waveguide

3.0.2.1 Dominant Mode in the Rectangular Waveguide

The lowest possible propagating mode in the *rectangular* waveguide is the TE_{10} mode ($m = 1$, $n = 0$), called the *absolute dominant* mode, which has the lowest *cutoff* frequency corresponding to the *largest* waveguide's cross-sectional dimension. For example, per Eq. (C.33c), the *cutoff* frequency f_{c10} of the TE_{10} mode is expressed as

$$f_{c10} = \frac{1}{2\sqrt{\mu\varepsilon}} \sqrt{\left(\frac{1}{a}\right)^2 + \left(\frac{0}{b}\right)^2} = \frac{1}{2a\sqrt{\mu\varepsilon}} \quad \text{for} \quad a > b. \tag{C.34a}$$

Additionally, the TM_{11} mode ($m = 1$ and $n = 1$) has the lowest cutoff frequency of all TM modes, expressed on the basis of Eq. (C.33c) as

$$f_{c11} = \frac{1}{2\sqrt{\mu\varepsilon}}\sqrt{\left(\frac{1}{a}\right)^2 + \left(\frac{1}{b}\right)^2}. \tag{C.34b}$$

The *cutoff* frequency of the TM_{11} mode is higher than the *cutoff* frequency of the TE_{10} mode, which confirms that the TE_{10} mode is the absolute *dominant mode* in rectangular waveguides for $a > b$.

Figure C.25 shows the cross-sectional distribution of the dominant mode TE_{10} in a rectangular waveguide.

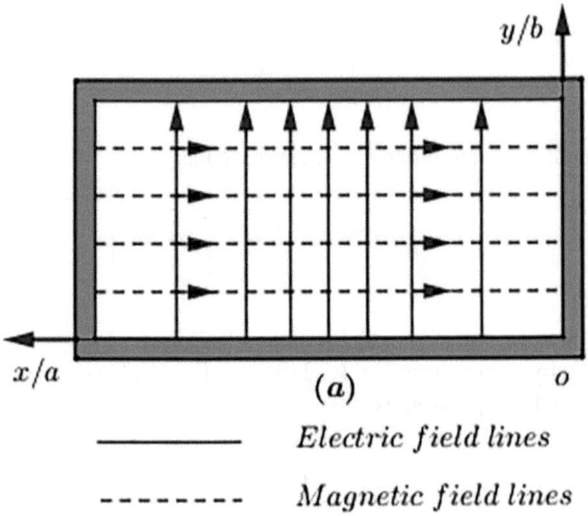

Fig. C.25 x–y field distributions of the TE_{10} dominant mode in a rectangular waveguide

3.0.2.2 Attenuation in Rectangular Waveguides

Per Eq. (C.32a), the propagation constant β is expressed as

$$\beta = \sqrt{\omega^2\mu\varepsilon - k_c^{\,2}}, \tag{C.35a}$$

and for a waveguide with air, where the wave propagates by the speed of the light, $c = 1/\sqrt{\mu_0\,\varepsilon_0}$, Eq. (C.35a) can be written as

$$\beta = \sqrt{\left(\frac{\omega}{c}\right)^2 - k_c^{\,2}}. \tag{C.35b}$$

Thus, because the field $\vec{E}_z(x,y,z) = \vec{E}_z(x,y)e^{-i\beta z}$ propagates along a rectangular waveguide with frequencies higher than the *cutoff* frequency $\left[\text{or}\left(\frac{\omega}{c}\right) > k_c\right]$, the *propagation constant* β is *real*. However, for frequencies below the cutoff frequency, where propagation does not exist, the propagation constant β is *imaginary*, i.e., $\beta = i\beta_e$. On the basis of Eq. (C.35b),

$$\beta = -\sqrt{-\left[k_c^{\,2} - \left(\frac{\omega}{c}\right)^2\right]} = -i\beta_e, \tag{C.36a}$$

where β_e is the propagation constant of nonpropagating waves expressed as

$$\beta_e = \sqrt{k_c^{\,2} - \left(\frac{\omega}{c}\right)^2} = \sqrt{\left(\frac{\omega_c}{c}\right)^2 - \left(\frac{\omega}{c}\right)^2} = \left(\frac{\omega}{c}\right)\sqrt{\left(\frac{\omega_c}{\omega}\right)^2 - 1}, \tag{C.36b}$$

where $k_c = \frac{\omega_c}{c}$.

A nonpropagating wave below the waveguide cutoff frequency is also called an *evanescent* wave. When we substitute $\beta = -i\beta_e$ into $\vec{E}_z(x,y,z) = \vec{E}_z(x,y)e^{-i\beta z}$,

$$\vec{E}_z(x,y,z) = \vec{E}_z(x,y)e^{-i(-i\beta_e)z} = \vec{E}_z(x,y)e^{-\beta_e z}. \tag{C.37a}$$

Furthermore, the *attenuation* A below the cutoff frequency is defined as the ratio between the field at the waveguide input (at $z = 0$) and at the waveguide end (at $z = l$):

$$A = \frac{E_z(x,y)e^{-\beta_e(0)}}{E_z(x,y)e^{-\beta_e(l)}} = \frac{1}{e^{-\beta_e l}}, \tag{C.37b}$$

where l is the length of the waveguide. The attenuation A can also be expressed in dB as

$$A(dB) = 20\log\left(\frac{1}{e^{-\beta_e l}}\right) \approx 8.7 l \beta_e, \tag{C.38a}$$

and substituting Eq. (C.36b) into Eq. (C.38a), there is

$$A(dB) \cong 8.7 l\left(\frac{\omega}{c}\right)\sqrt{\left(\frac{\omega_c}{\omega}\right)^2 - 1}. \tag{C.38b}$$

For the frequencies ω much below the cutoff frequency ω_c, $\omega \ll \omega_c$, the attenuation of the *dominant* mode TE_{10} from Eq. (C.38b) becomes

$$A(dB) \cong 8.7l\left(\frac{\omega_c}{c}\right) \cong 8.7l\left(\frac{\pi}{a}\right) \rightarrow$$
$$A(dB) \cong 27\left(\frac{l}{a}\right) \quad (TE_{10}), \tag{C.38c}$$

where $\frac{\omega_c}{c} = \frac{\pi}{a}$ for the TE_{10} mode.

3.0.3 Circular Waveguides

3.0.3.1 Absolute Dominant Mode in the Circular Waveguide

A *circular* waveguide is actually a conductive circular tube with a uniform cross section, as shown in Fig. C.26, where r is the waveguide radius.

Owing to the field distribution, it is more convenient to consider a cylindrical waveguide in cylindrical coordinates (r, ϕ, z).

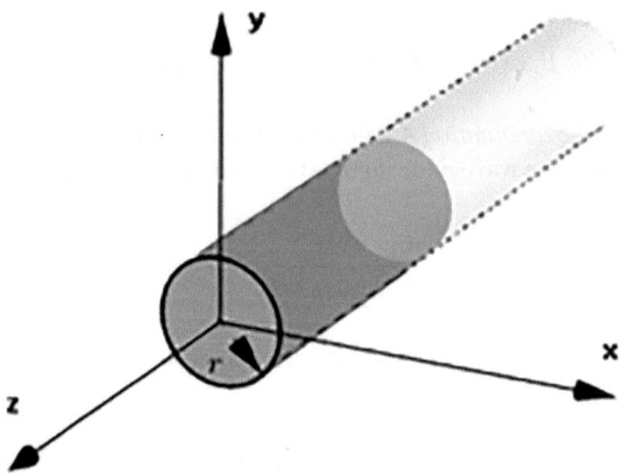

Fig. C.26 Circular waveguide

(a) *TE modes in the circular waveguide*

Hence, for *TE* waves in cylindrical coordinates, the *longitudinal* field components are

$$\vec{E}_z = 0 \text{ and } H_z(r,\phi,z) = H_z(r,\phi)e^{-i\beta z} \quad (TE), \tag{C.39a}$$

where these equations satisfy the *wave equations* given in cylindrical coordinates as

$$\nabla^2_{r\phi}\vec{H}_z + k^2\vec{H}_z = 0, \tag{C.39b}$$

where k is the *wavenumber* expressed via Eq. (C.32a) as

$$k^2 = \left(\frac{\omega^2}{c^2} - \beta^2\right). \tag{C.39c}$$

The complete solution of this *wave equation* in *cylindrical* coordinates is given in terms of the derivation of *Bessel's* function $J'_n(rk)$ of the nth kind as

$$H_z(r,\phi) = K'_n J'_n(rk)\cos(n\phi), \tag{C.39d}$$

where K'_n is an arbitrary constant dependent on the intensity of the field H_z.

When $J'_n(rk) = 0$ in Eq. (C.39d), there is no wave propagation in the waveguide. Hence, from the expression

$$J'_n(rk) = 0, \tag{C.40a}$$

we can determine the *cutoff* frequency for TE_{mn} modes in the circular waveguide with air via Eq. (C.33c) as

$$f_{cmn} = \frac{c}{2\pi}k_{cmn}, \tag{C.41}$$

where k_{cnm} is the *wavenumber* at the *cutoff* frequency f_{cmn}. The parameters k_{cnm} and f_{cmn} can be determined via the following expressions:

$$k_{cmn} = \frac{p'_{mn}}{r};\, f_{cmn} = \frac{c}{2\pi}\frac{p'_{mn}}{r}, \tag{C.42a, b}$$

where p'_{mn} are the roots of the derivation of the *Bessel's* function, $J'_n(x)$, and c is the speed of light. The zeros of the function $J'_n(rk_{cmn})$ are commonly listed in standard mathematical tables. Therefore, per the tabulated value $p'_{11} = 1.84$ and the cutoff frequency for the TE_{11} mode expressed by Eq. (C.42a, b),

$$f_{c11} = \frac{c}{2\pi}\frac{p'_{11}}{r} = \frac{c}{2\pi}\frac{1.84}{r} \quad (TE_{11}). \tag{C.43}$$

Here, the conclusion is that the TE_{11} mode is the absolute *dominant* mode in a *circular* waveguide, which has the *lowest* cutoff frequency.

Figure C.27 shows the cross-sectional field distribution of the *dominant mode* TE_{11} in a circular waveguide.

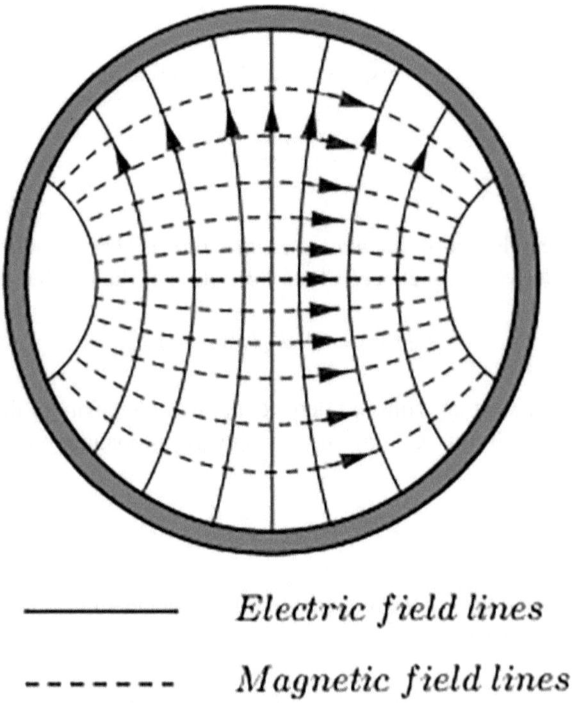

Fig. C.27 Cross-sectional distribution of the dominant mode TE_{11} in the circular waveguide

(b) *TM modes in a circular waveguide*

Regarding the TM_{mn} modes in the circular waveguide with air, the longitudinal field component is $\vec{H}_z = 0$ and where the equation

$$E_z(r,\phi,z) = E_z(r,\phi)e^{-\beta z}, \quad \text{(C.44a)}$$

satisfies the *wave equation* given in cylindrical coordinates as

$$\nabla^2_{r\phi}\vec{E}_z + k^2\vec{E}_z = 0. \quad \text{(C.44b)}$$

The solution of Eq. (C.44b) is

$$E_z(r,\phi) = K_n J_n(rk)\cos(n\phi), \quad \text{(C.44c)}$$

where $J_n(rk)$ is Bessel's function of the nth kind. When $J_n(rk) = 0$, the cutoff frequency for the *TM* mode is

$$f_{cnm} = \frac{c}{2\pi}\frac{p_{mn}}{r}, \tag{C.45a}$$

where p_{nm} are the roots of the Bessel's function $J_n(x)$. Many values of p_{nm} are also tabulated. Specifically, for the *TM dominant* mode TM_{01}, $p_{01} = 2.4$ and the cutoff frequency for this mode is

$$f_{c01} = \frac{c}{2\pi}\left(\frac{2.4}{r}\right) \qquad (TM_{01}). \tag{C.45b}$$

Equation (C.45b) shows that the cutoff frequency of the TM_{01} mode is higher than that of the TE_{11} mode; accordingly, the TE_{11} mode is *the absolute dominant mode for circular waveguides.* Figure C.27 shows the cross-sectional distribution of the dominant mode TE_{11} in a circular waveguide.

3.0.3.2 Attenuation in a Circular Waveguide

Owing to the lowest waveguide cutoff frequency being of the TE_{11} mode, we consider waveguide attenuation below this cutoff frequency.

Per Eq. (C.39a), the expression of the propagating TE_{mn} field in a circular waveguide is

$$H_z(r,\phi,z) = H_z(r,\phi)e^{-i\beta z}, \tag{C.46}$$

where the *propagation constant* β is based on Eq. (C.36b) as follows:

$$\beta = \sqrt{\left(\frac{\omega}{c}\right)^2 - (k_{cmn})^2} = \sqrt{\left(\frac{\omega}{c}\right)^2 - \left(\frac{p'_{mn}}{r}\right)^2}, \tag{C.47a}$$

where k_{cmn} is the *cutoff wavenumber* determined from Eq. (C.47a) for $\beta = 0$ as

$$k_{cmn} = \frac{\omega_c}{c} = \frac{p'_{mn}}{r}. \tag{C.47b}$$

p'_{mn} are the roots of derivation of the Bessel's function, $J'_n(rk)$.

Additionally, for the nonpropagating (*evanescent*) field, the propagation constant β is imaginary:

$$\tilde{\beta} = \sqrt{-\left[\left(\frac{p'_{mn}}{r}\right)^2 - \left(\frac{\omega}{c}\right)^2\right]} = i\beta_e, \tag{C.48a}$$

where β_e is the *real* propagation constant.

By substituting Eq. (C.47b), $\left(\frac{\omega_c}{c} = \frac{p_{mn}}{r}\right)$, into Eq. (C.48a), β_e can be written as

$$\beta_e = \left(\frac{\omega}{c}\right)\sqrt{\left(\frac{\omega_c}{\omega}\right)^2 - 1}. \tag{C.48b}$$

Furthermore, the *attenuation in a circular waveguide* long l is defined as the ratio between the field $H_z(r,\phi)e^{-\beta_e(z)}$ at the waveguide input ($z = 0$) and output ($z = l$):

$$A = \frac{H_z(r,\phi)e^{-\beta_e(0)}}{H_z(r,\phi)e^{-\beta_e l}} = \frac{1}{e^{-\beta_e l}}, \tag{C.49a}$$

where β_e is expressed via Eq. (C.48b). Therefore, after arranging, Eq. (C.49a) becomes in dB as

$$A(dB) = 20\log\left(\frac{1}{e^{-\beta_e l}}\right) \approx 8.7l\left(\frac{\omega}{c}\right)\sqrt{\left(\frac{\omega_c}{\omega}\right)^2 - 1}. \tag{C.49b}$$

For $\omega_c \gg \omega$, Eq. (C.49b) becomes

$$A(dB) \approx 8.7l\left(\frac{\omega_c}{c}\right). \tag{C.49c}$$

Furthermore, substituting $\omega_c = c\frac{p_{mn}}{r}$ into Eq. (C.49c) in

$$A(dB) \approx 8.7l\frac{p'_{mn}}{r} = 8.7l\frac{p'_{11}}{r} = 8.7l\frac{1.84}{r}, \tag{C.49d}$$

where $p'_{11} = 1.84$ for the dominant mode TE_{11}. Finally, the wave attenuation $A(dB)$ in a circular waveguide can be written as

$$A(dB) \approx 16\left(\frac{l}{r}\right) \quad \text{for } f \ll f_c \quad (TE_{11}). \tag{C.50}$$

3.0.4 Resonant Cavities

3.0.4.1 Empty Cavity Resonator

A resonant cavity is a closed waveguide similar to an electronic equipment box, as shown in Fig. C.28.

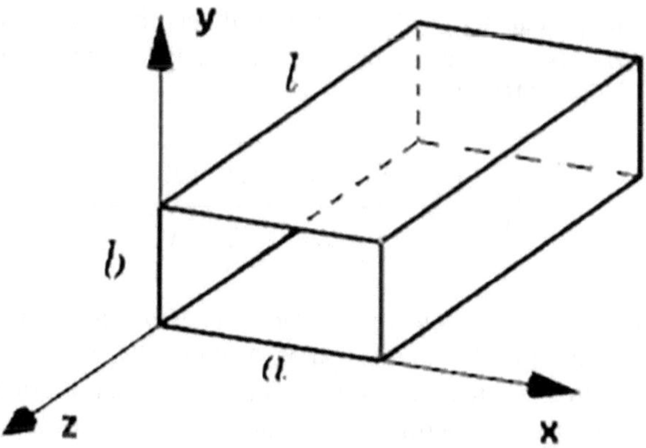

Fig. C.28 Resonant cavity

Standing electromagnetic waves can be created in a closed conductive cavity excited by an internal or external source or accidentally by outer or inner electromagnetic interference (*EMI*) through openings at the cavity. However, a wave inside the cavity reflects from one internal wall to another, resulting in *standing waves* in the cavity. Because a cavity is a 3D structure, we need to consider *standing waves* in three directions: x, y, and z.

In a cavity or equipment enclosure with air, multiple reflections inside the cavity can produce multiple resonant modes, such as TE_{mnp} and TM_{mnp} modes. For the rectangular cavity shown in Fig. C.28, the resonance frequencies are expressed as

$$f_{mnp} = \frac{c}{2\pi}\sqrt{\left(\frac{m\pi}{a}\right)^2 + \left(\frac{n\pi}{b}\right)^2 + \left(\frac{p\pi}{l}\right)^2}, \tag{C.51}$$

where *m*, *n*, and *p* integers (all may go 1, 2, 3 …) are designated for resonance. Additionally, *a*, *b*, and *l* are the cavity dimensions shown in Fig. C.28.

On the basis of Eq. (C.51), the cavity's *lowest resonance frequency* is associated with the *largest* cavity dimension d_{max}. In the case of a linear resonator with $d_{\max} = a = \frac{\lambda}{2}$, its resonance is

$$f_{100} = \frac{c}{2a}. \tag{C.52}$$

In particular, this is the case for the mode TE_{100} or TM_{100}.

3.0.5 Nonempty Cavity Resonator and Perturbation Theory

The previous analysis of a cavity is valid for the idealized case when the cavity is empty and where the average amount of *magnetic* and *electric* energy is equal at the *resonance* frequency. Even small objects (conductive or dielectric) added to the cavity can perturb the energy balance, and the mode resonant frequencies will be shifted to reestablish this balance. Therefore, according to cavity perturbation theory, any change in the energy balance of the resonator affects the resonance frequency f_r, changing it by Δf_r. The relationship between the change in the *energy balance* and the shift Δf_r is given as [10, 11]:

$$\frac{\Delta W_H - \Delta W_E}{W} = \frac{\int_0^{\Delta V}\left(\mu H^2 - \varepsilon E^2\right)dV}{\int_0^{V}\left(\mu H^2 - \varepsilon E^2\right)dV} = \frac{\Delta f_r}{f_r}, \tag{C.53}$$

where ΔW_H and ΔW_E are the *magnetic* and *electric* energy changes, respectively, and W is the *total energy stored* in the resonator. Additionally, V is the volume of the unperturbed resonator, whereas ΔV is the V reduction by an inserted object. As an example, Fig. C.29 shows a perturbed resonator where the lowest resonance frequency f_r is considered because the perturbation effects on higher resonance modes are similar.

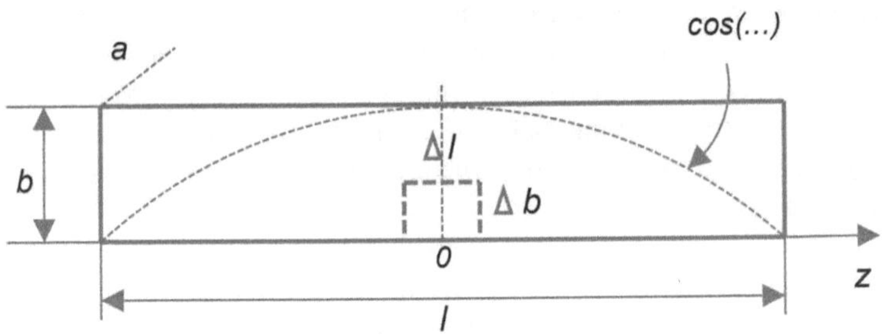

Fig. C.29 Field perturbation in a resonator

a, b, and l are the dimensions of the resonator, and a, Δb, and Δl are the dimensions of the inserted block. Accordingly, $V = abl$ and $\Delta V = a\Delta b\Delta l$.

Let us suppose that only the electric field exists in the center of the resonator and that the reduced electric energy by the inserted block is derived below as follows. The unperturbed electric field in Fig. C.29 is expressed as

$$E(z) = \cos\left(\frac{\pi z}{l}\right) \tag{C.54}$$

and the total energy stored in the resonator via Eqs. (C.53) and (C.54) is

$$W = 2\Delta W_E = 2ab\int_{-l/2}^{l/2} \frac{\varepsilon E^2}{4} \cos^2\left(\frac{\pi z}{l}\right) dz = abl\left(\frac{\varepsilon E^2}{4}\right). \tag{C.55}$$

Additionally, the changed field energy in the perturbed resonator is

$$\Delta W_E = a\Delta b\Delta l\left(\frac{\varepsilon E^2}{4}\right), \tag{C.56}$$

and for Eqs. (C.53), (C.55), and (C.56), there is

$$\frac{\Delta f_r}{f_r} = -\frac{\Delta b\Delta l}{bl}. \tag{C.57}$$

Therefore, the change in the resonance frequency is proportional to the change in the dimensions b and l of the cavity resonator by the dimensions of the inserted object, Δb and Δl.

Hence, in reality, boxes or enclosures of electronic equipment are never entirely empty, commonly containing subcavities, where waveguide–resonator theory can be applied and when the boundary conditions can be defined or fairly approximated. However, the most accurate method for predicting cavity resonance inside enclosures with electronic components and modules is computational modeling.

References

1. S.H. Hall, H.L. Heck, *Advanced Signal Integrity High-Speed Digital Designs* (John Wiley & Sons, NJ, 2009)
2. D.M. Pozar, Microwave Engineering, 3rd edn. (John Wiley & Sons, NJ, 2005)
3. H. Johnson, M. Graham, High-Speed Signal Propagation (Pearson, 2003)
4. R. Collins, Foundation for Microwave Engineering (McGraw-Hill, 1992)
5. F.A. Benson, T.M. Benson, Fields, Waves, and Transmission Lines (Springer, 1991)
6. B.C. Wadell, Transmission Line Design Handbook (Artech, 1991)
7. D.K. Cheng, Field and Wave Electromagnetics, 2nd edn. (Pearson, 1989)
8. E.F. Kuester, Theory of Waveguides and Transmission Lines (CRS Press, 2021)
9. U.A. Bakshi, Transmission Lines & Waveguides (Amazon Digital Service LLC KDP Print USA, 2020)
10. R.F. Harrington, Time-Harmonic Electromagnetic Fields (McGraw-Hill, NY, 1961)
11. S. Ramon, J.R. Whinnery, T.V. Duzer, Fields and Waves in Communication Electronic (John Wiley & Sons, 1994)

Index

M. Pajovic, *Fundamentals of EMI Phenomena in Radio Communication Systems*, https://doi.org/10.1007/978-3-031-88748-2

MIX
Papier aus verantwortungsvollen Quellen
Paper from responsible sources
FSC® C105338

If you have any concerns about our products,
you can contact us on
ProductSafety@springernature.com

In case Publisher is established outside the EU,
the EU authorized representative is:
Springer Nature Customer Service Center GmbH
Europaplatz 3, 69115 Heidelberg, Germany

Printed by Libri Plureos GmbH
in Hamburg, Germany